Design, Fabrication, Properties and Applications of Smart and Advanced Materials

Design, Fabrication, Properties and Applications of Smart and Advanced Materials

Editor

Xu Hou

Xiamen University, Xiamen, China
Collaborative Innovation Center, China
Harvard John A. Paulson School of Engineering and
Applied Sciences, Cambridge, MA, USA

CRC Press
Taylor & Francis Group
Boca Raton London New York

CRC Press is an imprint of the
Taylor & Francis Group, an **informa** business

A SCIENCE PUBLISHERS BOOK

CRC Press
Taylor & Francis Group
6000 Broken Sound Parkway NW, Suite 300
Boca Raton, FL 33487-2742

First issued in paperback 2020

© 2016 by Taylor & Francis Group, LLC
CRC Press is an imprint of Taylor & Francis Group, an Informa business

No claim to original U.S. Government works

ISBN-13: 978-1-4987-2248-3 (hbk)

ISBN-13: 978-0-367-78296-2 (pbk)

Library of Congress Cataloging-in-Publication Data

Names: Hou, Xu
Title: Design, fabrication, properties, and applications of smart and advanced materials / [edited by] Xu Hou.
Description: Boca Raton : Taylor & Francis, 2016. | "A CRC title." | Includes bibliographical references and index.
Identifiers: LCCN 2015048869 | ISBN 9781498722483 (hardcover : alk. paper)
Subjects: LCSH: Smart materials.
Classification: LCC TA418.9.S62 D46 2016 | DDC 620.1/1--dc23
LC record available at http://lccn.loc.gov/2015048869

Visit the Taylor & Francis Web site at
http://www.taylorandfrancis.com

and the CRC Press Web site at
http://www.crcpress.com

Preface

Advances in the research and technology of materials offers great promise. Materials science forms the foundation for those engaged in materials development because the intramolecular and intermolecular interactions, the interface properties, the mechanical properties, the chemical properties, the electrical properties, the thermal properties, the optical properties, the magnetic properties, the structures, and the components of materials that scientists study and engineers design with are all based on these material properties.

How to select the "BEST" material is usually a challenging task, requiring tradeoffs between properties. Designing smart materials is one of the important and effective ways to develop advanced functional materials that have one or more properties which can be significantly changed in a controlled fashion by external stimuli, such as light, stress, temperature, moisture, pH, electric or magnetic fields, etc. In this book, the authors introduce various smart and advanced materials, the strategies for design and preparation of novel materials from macro to micro/nano or from biological, inorganic, organic to composite materials. Meanwhile, the authors will make systematic introduction in each chapter about their latest research progress and the latest applications in the related fields.

The development of these smart and advanced materials and their potential applications is a burgeoning new area of research, and a number of exciting breakthroughs may be anticipated in the near future from the concepts and results reported in this book. The book can also be used as a textbook for undergraduate and graduate education.

January 2016
Xu Hou
Xiamen University

Contents

1

Introduction

Xu Hou

ABSTRACT

This book is organized into 14 chapters. The first chapter gives a brief introduction to smart and advanced materials, and then shows an example of the strategy for the design and fabrication of smart and advanced materials. Chapters 2–7 introduce smart materials for DNA, Protein, Controlled Drug Release, Cancer Diagnosis & Treatment, Fluorescence Sensing and Controlled Droplet Motion. Chapters 8 and 9 summarize a comprehensive overview of advanced materials for Thermoelectric and Thermo-Responsive Application. Chapter 10 presents an overview of Self-Healing Polymeric Materials, while Chapter 11 examines advanced materials for Soft Robotics. Finally Chapters 12–14 introduce a comprehensive overview of advanced materials for Biomedical Engineering, Solar Energy Harnessing & Conversion, and Reflective Display Applications. The authors in each chapter will systematically introduce their current research progress and report the latest applications in their related fields.

Materials science is an interdisciplinary field which deals with the discovery and design of novel materials. The material of choice in a given era is often a defining point, such as Stone Age, Bronze Age, Iron Age, Steel Age, Cement Age, and the Silicon Age (Naumann 2010). With further advances in the

---

Xiamen University, China, Collaborative Innovation Center of Chemistry for Energy Materials, China, Harvard John A. Paulson School of Engineering and Applied Sciences, USA.
Email: houx@xmu.edu.cn; houx@seas.harvard.edu

field, our society has entered into a new era in the 21st century. Now we cannot just use a single type material to represent this era, but various and different types of materials, such as nanomaterials, composite materials, and smart materials. This indicates that a completely new substance civilization of human society has started (Kumar 2010, Schwartz 2008).

A material is defined as a substance that is intended to be used for certain applications (Hummel 2004). Materials exhibit a myriad of properties, such as interface properties, mechanical properties, chemical properties, electrical properties, thermal properties, optical properties, and magnetic properties. These properties determine a material's usability and hence its real-world applications. According to the specific application requirements, materials can generally be divided into four classes: inorganic (ceramics, glass, etc.), organic (polymer, biomolecular, etc.), metal (elemental metals and alloys) and composite materials (Pedro and Sanchez 2006).

Selecting the "BEST" material is usually a challenging task, requiring tradeoffs between various material properties. Designing smart and advanced materials (also called "intelligent" or "responsive" materials) is one of the key ways to develop advanced functional materials, those that have one or more properties or functions that can be significantly tunable by external stimuli, such as light, stress, temperature, pH, electric, moisture or magnetic fields. In this book, we introduce various smart and advanced materials, as well as the strategies for design and preparation of novel materials, from macroscopic to nanoscopic, from biological, inorganic, organic or composite materials.

This book is organized into 14 chapters. The first chapter gives a brief introduction to smart and advanced materials, and then shows an example of the strategy for the design and fabrication of smart and advanced materials. Chapters 2–7 introduce smart materials for DNA, Protein, Controlled Drug Release, Cancer Diagnosis & Treatment, Fluorescence Sensing and Controlled Droplet Motion. Chapters 8 and 9 summarize a comprehensive overview of advanced materials for Thermoelectric and Thermo-Responsive Application. Chapter 10 presents an overview of Self-Healing Polymeric Materials, while Chapter 11 examines advanced materials for Soft Robotics. Chapters 12–14 introduce a comprehensive overview of advanced materials for Biomedical Engineering, Solar Energy Harnessing & Conversion, and Reflective Display Applications. The authors in each chapter will make systematically introduce their current research progress and report the latest applications in their related fields.

Various methods and strategies have been proposed to develop smart and advanced materials. This chapter is intended to utilize a specific research example to present one of the design strategies and fabrication processes of bio-inspired nanochannel materials. The strategies and processes may also be extended to other smart and advanced materials.

'Learning from Nature' provides various biological materials with an assortment of smart functions over millions of years of evolution. These serve as a major source of bio-inspiration for smart and advanced materials (Jiang and Feng 2010, Hou et al. 2015)

The bio-inspired study of the design and development of smart and advanced materials has been receiving a great deal of attention. For instance, ion pumps and ion channels are used by cells to transport ions across membranes. Various components of biological channels are not uniform in distribution, and their structures are also asymmetric (Kew and Davies 2010). Based on bio-inspired asymmetric design ideas, the following presents the design and fabrication of smart and advanced nanochannel materials.

Herein, I will suggest three routes for the design and fabrication of assorted smart and advanced nanochannel materials (Fig. 1). The first step is materials selection. According to the specific application requirements, various materials can be selected for building smart and advanced nanochannels, such as biological materials, inorganic materials, organic materials and composite materials (Hou 2013). After materials selection, the second step is to obtain different shapes and structures of nanochannel materials by utilizing various fabrication technologies, such as physical & chemical etching, laser cutting and photolithography. The final step

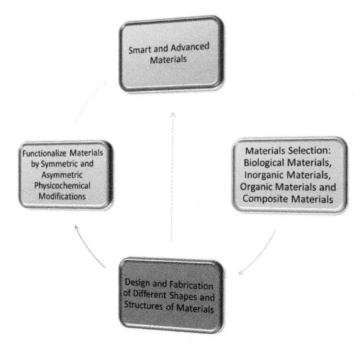

Figure 1. The design and fabrication of smart and advanced materials.

is to functionalize nanochannel materials by symmetric and asymmetric physicochemical modifications. It is worth mentioning that if the selection materials have certain functional properties, the nanochannels prepared by them will have the same characteristic.

By using symmetric & asymmetric shapes/structures and symmetric & asymmetric physicochemical modifications, various smart and advanced nanochannels have been developed (Hou et al. 2011, Hou et al. 2012). Following on from this there are two typical examples that show the key points of the symmetric and asymmetric design approach of smart and advanced nanochannels.

Before introducing these examples, there are two basic concepts: ionic gating and ionic rectification. Ionic gating is defined to evaluate the performance of ion passing through the nanochannel, which is observed the close state as the ionic current value approaches to zero under the same voltage (Fig. 2a) (Yameen et al. 2009). Ionic rectification of nanochannels is observed as an asymmetric current-voltage curve (Fig. 2b) (Yameen et al. 2009). The current recorded for one voltage polarity is higher than the current recorded under the same absolute value of voltage of opposite polarity. Therefore, this asymmetric curve indicates a certain extent of the preferential direction of ionic flow inside the nanochannels.

Azzaroni et al. developed pH gating ionic transport properties inside the single nanochannel by using the chemical modification of the inner surface

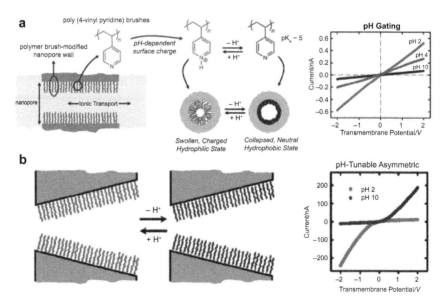

Figure 2. Current-voltage (*I–V*) curves of symmetric **(a)** and asymmetric **(b)** single nanochannels after the functionalization. Reprinted with permission from Ref. (Yameen et al. 2009). Copyright 2009 American Chemical Society.

of the symmetric cylindrical-shaped nanochannel (Fig. 2a). Meanwhile, they also developed pH-tunable asymmetric ionic transport properties inside the single nanochannel by utilizing the chemical modification of the inner surface of the asymmetric conical-shaped nanochannel (Fig. 2b). The above two workings show good examples of how to prepare simple-function pH controllable nanochannel systems.

Inspired by ion channels, in which the components are asymmetrically distributed between the cell membrane surfaces, the pH gating and pH-tunable asymmetric ionic transport properties can also be combined together in the only one system. Figure 3 shows a smart nanochannel, which displays the advanced feature of providing simultaneous control over the pH gating and pH-tunable asymmetric ionic transport properties (Hou et al. 2010). Its gating property inside the channel is caused by the pH response of the functional molecule, and the pH-tunable asymmetric ion transport property is caused by the symmetrically shaped nanochannel with the asymmetric chemical modification. Here the design idea is to prepare symmetrically shaped nanochannels for asymmetric chemical modification approaches to functionalize diverse specific local areas precisely with functional molecules. Based on this design, the nanochannels with a gradual structural transformation from asymmetric to symmetric shapes can obtain a gradual change in pH responsivity, which has the advantage of providing continuous ionic transport control, including asymmetric

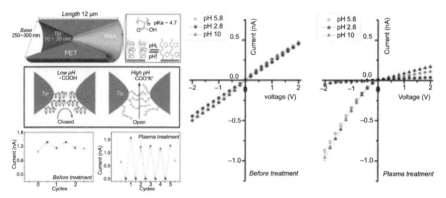

Figure 3. Scheme of the single hour-glass shaped nanochannel after plasma-induced graft polymerization, and hypothetical conformations of hydrogen bonding between the copolymers and water which reveal two kinds: the intramolecular hydrogen bond among the carboxylic acid groups in the polymer chains when the pH is below pKa, and the intermolecular hydrogen bonds between PAA chains and water molecules when the pH is above pKa. Explanation of pH-dependant water permeation through the single hourglass shaped nanochannel. Reversible variation of the ionic current transport of the single nanochannels before (left bottom) and after (right bottom) plasma treatment at 2 V. *I–V* properties of the single nanochannel under different pH conditions (pH 5.8, ■; pH 2.8, ●; pH 10, ▲). Reproduced from Ref. (Hou et al. 2010) by permission of John Wiley & Sons Ltd.

shape and asymmetric ionic transport; asymmetric shape but symmetric ionic transport; and symmetric shape and symmetric ionic transport under certain pH conditions (Fig. 4) (Zhang et al. 2015). Moreover, by using the asymmetric chemical modification approaches to achieve symmetric pH gating behaviors inside the asymmetric nanochannels, the foundation is laid to build diverse, stimuli-gated artificial asymmetric shaped nanochannels with symmetric ionic gating features (Fig. 5) (Zhang et al. 2015).

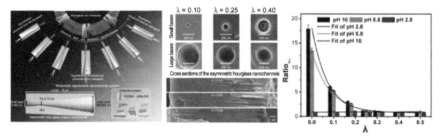

Figure 4. Bio-inspired single asymmetric hourglass nanochannel system demonstrating continuous shape and ionic transport transformation. Asymmetric hourglass biological ion channels give a hint to the process of creating the artificial nanochannel with evolving structure. Precise control of the tip position along the longitudinal axis of the channel allows us to transform the nanochannel structure from an asymmetric cone to a totally symmetric hourglass. (left) Geometry and pH responsive surface properties of a specific asymmetric hourglass nanochannel. (center) Ionic rectification degrees (ratio$_{+/-}$) of the sequential asymmetric nanochannels under pH 2.8, pH 5.8, and pH 10 conditions. Ratio$_{+/-}$ = I_{+2V}/ | I_{-2V} |. (right) Reproduced from Ref. (Zhang et al. 2015) by permission of John Wiley & Sons Ltd.

The inner surface chemical properties and shapes of the nanochannels are two key factors to control ionic transport properties inside the channels. There are two strategies for designing multiple responsive nanochannel materials (Hou 2013).

According to above two factors, I suggest two strategies for designing multiple responsive nanochannel materials. The first strategy focuses on the design and synthesis of functional molecules with multiple responsive properties on the inner surfaces of the nanochannel materials. For example, the light and pH cooperative nanochannel system was been developed by modifying the inner surface of the asymmetric nanochannel with light responsive molecules (Fig. 6) (Zhang et al. 2012). When light was on, the nanochannel could be either cation selective or anion selective according to the pH. However, when light was off, the channel was non-selective, due to the neutral molecules on the inner surface of the channel. At the same time, the asymmetric shape of the nanochannel allowed the system to exhibit ionic rectification as well. Therefore, this nanochannel material can be used as a nanofluidic diode that displays both light-gated and pH-tunable transport properties.

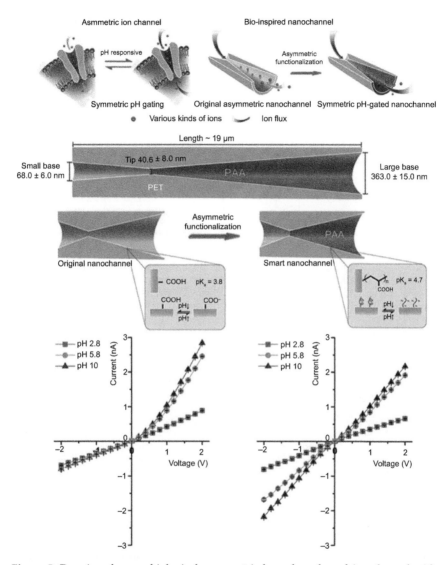

Figure 5. Drawing shows a biological asymmetric hourglass-shaped ion channel with symmetric pH-gating ionic transport property. The bio-inspired smart single asymmetric hourglass-shaped nanochannel illustrates symmetric pH responsive ionic transport property after asymmetric functionalization of the large base side of the channel. Schematic representation of the cross section of the asymmetric nanochannel asymmetrically modified with PAA. *I–V* properties of the asymmetric nanochannel before (bottom left) and after (bottom right) modification under different pH conditions (pH 2.8, ■; pH 5.8, ●; pH 10, ▲). Reproduced from Ref. (Zhang et al. 2015) by permission of John Wiley & Sons Ltd.

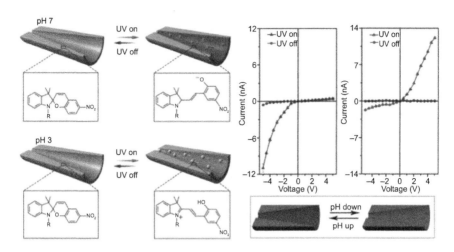

Figure 6. Scheme of the light-gated nanofluidic diode system at different pH values. The nanochannel is initially in the closed state due to the neutral and hydrophobic spiropyran; At pH 7, under UV light irradiation, the nanochannel turns into the open state due to the spiropyran being hydrophilic and negatively charged. Therefore, cations are the majority carriers and prefer to flow from the tip to the base. At pH 3, under UV light irradiation, the nanochannel is positively charged. Therefore, anions are the majority carriers and prefer to flow from the base to the tip. *I–V* curves of the spiropyran-modified nanochannels when UV light is off and under UV light irradiation. At pH 7, the nanochannel was in the closed state when UV light is off. Under UV light, the nanochannel was negatively charged and cations were the majority carries. The cations preferred to flow from the tip to the base to maintain the lower resistance, leading to current flowing in the same direction. At pH 3, the nanochannel was in the closed state when UV light is off. Under UV light, the nanochannel was positively charged and anions were the majority carriers. The anions preferred to flow from the tip to the base to maintain the lower resistance, leading to current flowing in the opposite direction. Explanation of the pH-tunable nanofluidic diode induced by the different polarities of the excessive surface charge. Reproduced from Ref. (Zhang et al. 2012) by permission of John Wiley & Sons Ltd.

The second strategy is to prepare various symmetric and asymmetric shaped nanochannels for different chemical modification approaches, to functionalize diverse specific local areas precisely with different functional molecules. Based on above two factors, once the channels are prepared, it is difficult to change them within a wide range of shapes. Therefore, the chemical modification of the inner surface of the channels with functional molecules is more flexible in the advancement of smart nanochannels, due to the fact that physicochemical modification of the channels can change both the sizes and physicochemical properties of the channels. For instance, an asymmetric responsive symmetric hourglass-shaped nanochannel system was developed, which displayed the advanced feature of simultaneous control over both pH and temperature tunable asymmetric ionic transport properties (Fig. 7) (Hou et al. 2010).

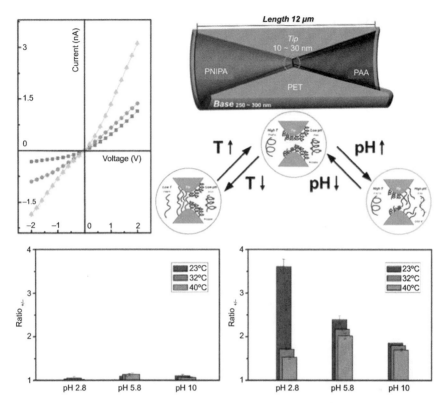

Figure 7. Scheme of a biomimetic asymmetric temperature/pH-responsive single nanochannel (right). *I–V* properties of the single nanochannel after asymmetric modification (left). Ionic current rectification of the single nanochannel before (bottom left) and after (bottom right) asymmetric chemical modification at 2 V. Reprinted with permission from Ref. (Hou et al. 2010). Copyright 2010 American Chemical Society.

Even if the shapes and structures of the nanochannels are all symmetrical or asymmetrical, these channels can still have very different design profiles, such as the symmetric hourglass-shaped channels (Hou et al. 2010) and the symmetric cigar-shaped channels (Zhang et al. 2013). The symmetric hourglass-shaped one has a very narrow center part and two larger ends. In contrast, the symmetric cigar-shaped one has a larger center part and two narrow ends. An artificial ion pump was developed based on a cooperative pH response double-gate nanochannel (Fig. 8) (Zhang et al. 2013). Here, two functional molecules, as two separate acid- and base-driven gates, immobilized on the small tip ends of a cigar-shaped nanochannel. The channel tip ends could be alternately/simultaneously opened and closed under symmetric and asymmetric pH stimuli. Such a system, as an example, demonstrates a development of the nanochannels from the simple-function, controllable channels systems to the complicated-

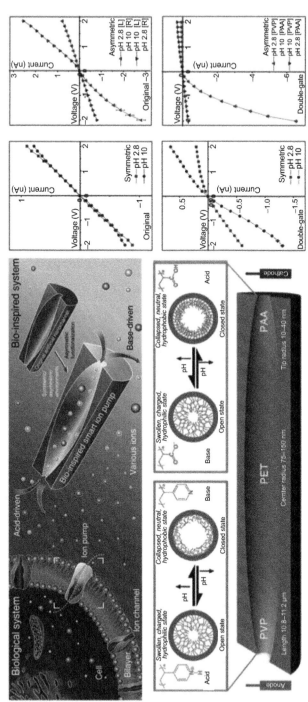

Figure 8. Bio-inspired artificial functional single ion pump. The drawing shows a biological system, in which ion pump as a membrane-spanning pore has two separate gates located separately on the both side of the pathway, and the bio-inspired smart single ion pump by integrating two acid- and base-driven artificial gates into a single cigar-shaped nanochannel. Schematic representation of the cross-section of the artificial cooperative response double-gate nanochannel in which acid activating PVP gate and base activating PAA gate were respectively immobilized on the inner surface of the left and right sides of the nanochannel. During the pH changes within a certain range on the PVP side, PVP underwent pH-responsive conformation changes from the open state (acid condition) to the closed state (base condition). At the same time, PAA underwent pH-responsive conformation transition between the closed state (acid condition) and the open state (base condition). I–V curves of the original and double-gate nanochannel under symmetric pH 2.8 (squares) and 10 (circles), and asymmetric pH 2.8/10 (up triangle) and 10/2.8 (down triangle) conditions illustrated functional ionic transport control inside the nanochannel by the double opposite pH-responsive gates. Reprinted with permission from Ref. (Zhang et al. 2013). Copyright 2013 American Chemical Society.

function, controllable channels systems, and may lead to a new generation of active transportation controlling nanofluidics, biosensors and energy conversion devices.

The above bio-inspired nanochannel materials are intended for general introduction for the design and fabrication of smart and advanced materials. The following chapters will be the systematic introduction and detailed discussion of design, fabrication, properties and applications of various smart and advanced materials.

Acknowledgements

X. Hou acknowledges the support of Recruitment Program for Young Professionals, China, and Research Institute for Biomimetics and Soft Matter, Fujian Provincial Key Laboratory for Soft Functional Materials Research, Xiamen University, supported by the 111 Project (B16029).

References

Hou, X., Y. Hu, A. Grinthal, M. Khan and J. Aizenberg. 2015. Liquid-based gating mechanism with tunable multiphase selectivity and antifouling behavior. Nature 519: 70–73.
Hou, X. 2013. Bio-inspired asymmetric design and building of biomimetic smart single nanochannels. Springer Science & Business Media.
Hou, X., W. Guo and L. Jiang. 2011. Biomimetic smart nanopores and nanochannels. Chem. Soc. Rev. 40(5): 2385–2401.
Hou, X., H. Zhang and L. Jiang. 2012. Building bio-inspired artificial functional nanochannels: from symmetric to asymmetric modification. Angew. Chem. Int. Ed. 51(22): 5296–5307.
Hou, X., Y. Liu, H. Dong, F. Yang, L. Li and L. Jiang. 2010. A pH-gating ionic transport nanodevice: asymmetric chemical modification of single nanochannels. Adv. Mater. 22(22): 2440–2443.
Hou, X., F. Yang, L. Li, Y. Song, L. Jiang and D. Zhu. 2010. A biomimetic asymmetric responsive single nanochannel. J. Am. Chem. Soc. 132(33): 11736–11742.
Hummel, R.E. 2004. Understanding materials science: history, properties, applications. Springer Science & Business Media.
Jiang, L. and L. Feng. 2010. Bioinspired intelligent nanostructured interfacial materials. World Scientific.
Kew, J.N. and C.H. Davies. 2010. Ion channels: from structure to function. Oxford University Press.
Kumar, C.S. 2010. Biomimetic and bioinspired nanomaterials (Vol. 3). John Wiley & Sons.
Naumann, R.J. 2011. Introduction to the physics and chemistry of materials. CRC Press.
Pedro, G. and C. Sanchez. 2006. Functional hybrid materials. John Wiley & Sons.
Schwartz, M. 2008. Smart materials. CRC Press.
Yameen, B., M. Ali, R. Neumann, W. Ensinger, W. Knoll and O. Azzaroni. 2009. Synthetic proton-gated ion channels via single solid-state nanochannels modified with responsive polymer brushes. Nano Lett. 9(7): 2788–2793.
Yameen, B., M. Ali, R. Neumann, W. Ensinger, W. Knoll and O. Azzaroni. 2009. Single conical nanopores displaying pH-tunable rectifying characteristics. Manipulating ionic transport with zwitterionic polymer brushes. J. Am. Chem. Soc. 131(6): 2070–2071.

Zhang, H., X. Hou, Z. Yang, D. Yan, L. Li, Y. Tian, H. Wang and L. Jiang. 2015. Bio-inspired smart single asymmetric hourglass nanochannels for continuous shape and ion transport control. Small 11(7): 786–791.

Zhang, H., X. Hou, J. Hou, L. Zeng, Y. Tian, L. Li and L. Jiang. 2015. Synthetic asymmetric-shaped nanodevices with symmetric pH-gating characteristics. Adv. Funct. Mater. 25(7): 1102–1110.

Zhang, M., X. Hou, J. Wang, Y. Tian, X. Fan, J. Zhai and L. Jiang. 2012. Light and pH cooperative nanofluidic diode using a spiropyran-functionalized single nanochannel. Adv. Mater. 24(18): 2424–2428.

Zhang, H., X. Hou, L. Zeng, F. Yang, L. Li, D. Yan, Y. Tian and L. Jiang. 2013. Bioinspired artificial single ion pump. J. Am. Chem. Soc. 135(43): 16102–16110.

2

Smart Materials for DNA-Based Nanoconstructions

Weina Fang, Jianbang Wang and *Huajie Liu**

ABSTRACT

Since its introduction by Nedrian Seeman 30 years ago, DNA has now been proved as a smart material for nanofabrication and nanoengineering. Based on the programmable, nano-addressable self-assembly ability and the nanosize of DNA molecules, they are of high utility in constructing elaborate information-encoded static nanostructures and dynamic nanomachines with designed size and geometry. Especially, after the invention of DNA origami by Rothemund in 2006, this field has witnessed an explosion of interest in the construction of functional DNA nanodevices. With DNA nanostructures as templates, plasmonic nanoclusters, metallic nanocircuits, as well as *in vitro* enzymatic cascades could be built with nano-addressability. Dynamic DNA nanomachines, on the other hand, could be regulated with external stimuli, leading to various applications in biosensor and logic operations. In this chapter, we will illustrate some smart DNA materials and focus on recent advances in DNA-based nanoconstructions. The design and fabrication of these smart materials will be summarized from a structural point of view, while their important applications with great potentials will be highlighted. Finally, the future perspectives will be discussed.

Shanghai Institute of Applied Physics, Chinese Academy of Sciences, No. 2019, Jialuo Road, Shanghai 201800, China.
  Emails: fangweina1989@163.com; 2011wangjianbang@sina.com
* Corresponding author: liuhuajie@sinap.ac.cn

1. Introduction

DNA has played an extraordinary, important role in life science since the establishment of Watson-Crick model. As the carrier of genetic information, DNA has long been considered as one of the most important biological substances. However, tracing back to its chemical essence, DNA is definitely a kind of macromolecules; it possesses many interesting properties, such as unique base-pairing, programmable sequence, various conformations and nanoscale-size (Seeman 2003). These extraordinary features precisely meet the requirements of the fast-growing science of materials. First, the nanosize-feature offers DNA inherent nano-addressability which can hardly be found from other materials. Second, the rigorous Watson-Crick base-pairing and programmable sequence-design abilities guarantee the robust and predictable self-assembly for the construction of nanostructures with expected geometries. Third, the ability of conformational exchange in response to external stimuli enables DNA as an ideal material for the building of prototype nano-mechanical devices. Fourth, the biological nature of DNA molecules bridges materials science with life science. Fifth, the well-established DNA synthesis technique makes DNA more and more convenient for materials scientists.

Therefore, pioneered by Nedrian Seeman from 1980s (Seeman 1982), more and more materials scientists have chosen DNA as a kind of smart material for nanoconstructions and nanoengineering (Jones et al. 2015). Static self-assembled DNA nanostructures, ranging from 1D nanotubes, 2D lattices, to 3D polyhedrons and other diverse objects, can now be easily fabricated via designed hybridization processes (Seeman 2010). These nano-addressable and information-encoded structures have been proven powerful in directing the precise assembly of functional nanomaterials, such as nanoparticles, quantum dots, enzymes, as well carbon nanotubes and polymers, for the construction of promising, integrated nanodevices. On the other hand, dynamic DNA nanomachines (Liu et al. 2011), that can perform mechanical movements at nanoscale in response to external stimuli, have also been invented and found various applications. Especially, the clearer structures, established synthesis and modification methods and clearer driven mechanisms of DNA nanomachines have proved interesting to material research as well as theoretical studies. Besides the above simple static and dynamic DNA self-assembled objects, Rothemund's revolutionary work on the invention of DNA origami in 2006 (Rothemund 2006) opened a new era in this field, since it solved the long-term problem in building finite-sized and sub-micron-scaled DNA structures with fully nanoscale-addressability and high complexity. This technique has been a constant focus of research and great efforts have been made to build new 2D and 3D DNA origami structures, improve assembly strategy, study inherent properties and develop new applications (Saccà and Niemeyer 2012). Undoubtedly, DNA has produced fascination in the material sciences and we have witnessed an explosion of interest in the last decade.

In this chapter, we will illustrate some smart DNA materials and focus on recent advances in DNA-based nanoconstructions. The chapter will start with DNA-nanoparticles conjugates, which are elaborate building blocks for constructing functional DNA nanodevices. Then, several DNA polyhedrons will be demonstrated and we will especially focus on DNA tetrahedrons (Goodman et al. 2004), since interesting applications of these materials have emerged recently. DNA origami will be highlighted in the fourth section with detailed descriptions on its invention, structural evolution and applications. In addition, recent progress on DNA nanomachines will be summarized. Finally, we will try to give some perspectives on the development of DNA-based materials for nanoconstructions in the near future. We note that DNA materials include but are not limited to these types. Early works on 1D nanotubes and 2D lattices assembled from DNA tiles and recently developed DNA hydrogels (Um et al. 2006, Park et al. 2009, Cheng et al. 2009, Jin et al. 2013, Lee et al. 2012c) and crystals (Zheng et al. 2009, Rusling et al. 2014) will not be discussed and we suggest some reviews for readers (Rangnekar and LaBean 2014, Roh et al. 2011).

2. DNA-Nanoparticles Conjugates

Nanoparticles (NPs) have unique physical and chemical properties. On the other hand, the development of DNA nanotechnology enables us to build up two- to three-dimensional, precisely addressable nanostructures and devices based on DNA sequence design and programmable assembly. Nowadays, advances in conjugating DNA with nanoparticles bridge these two technologies and provide scientists with a new tool to study material transportation and energy transfer at nanometer scale, while these smart DNA-nanoparticle conjugates also provide basic components for further directing assembly and constructing desired nanostructures. A variety of nanoparticles, ranging from inorganic nanoparticles, organic components, to biological molecules, have been successfully attached onto DNA via covalent or non-covalent bonds. The methods of preparation of DNA-nanoparticles will be the focus of this section. Especially, discrete DNA-nanoparticles (Zhang et al. 2011c), which have controlled position and density of DNA modifications on AuNPs surface, will be highlighted.

2.1 Conjugation of DNA with Inorganic Nanoparticles

Gold nanoparticle is one of the most important inorganic nanoparticles (Saha et al. 2012). DNA-AuNPs conjugates are promising building-blocks with programmable molecular recognition ability. The most straightforward way to conjugate DNA with AuNPs is the thiol chemistry which forms covalent Au-S bond (Mirkin et al. 1996, Loweth et al. 1999). Through increasing the number of Au-S bonds from mono-, di-, tri-, to tetra-thiol groups, studies have proved the improved stability of the conjugation (Zhang et al. 2011a, Zhang et al. 2011c). However, one critical challenge remained, namely to control the position and density of DNA modifications on an AuNPs surface, that is, the spatial control. The positional control aims to achieve the addressable surface modification on AuNPs, either symmetric/isotropic or asymmetric/anisotropic. The density control, on the other hand, aims to modify AuNPs with an exact number of DNA strands, and to control the surface coverage density on an AuNP with uniform DNA modifications. In principle, the spatial control could not only increase the precision of positioning assembly but also improve the hybridization efficiency.

Early efforts were based on the careful control of the stoichiometry with subsequent, proper post-synthesis separation (Zanchet et al. 2001). Discrete DNA-AuNPs with an exact number of DNA modifications could be obtained (Fig. 1a), although the yield was low. In order to overcome this shortage, new ideas have been developed through utilizing the steric hindrance. DNA duplex with two adjustable thiol groups as a geometrical template was reported for the hetero-bi-modification of a AuNP (Suzuki et al. 2009). Zhang et al. reported another DNA-templated approach to bimodify AuNPs with two different DNA strands on two opposite faces of AuNPs (Zhang et al. 2011b). It is noted that one DNA was attached to an AuNP through 5' end and the other one through 3' end, making the conjugates unidirectional. Besides DNA molecules, big templates could also be used as hindrance to block partial surface of an AuNP and force a DNA molecule to attach this AuNP on the opposite direction. Kim et al. reported the up to six fold symmetric DNA modifications on AuNPs, which was carried out on a silica gel template (Kim et al. 2011). Through combining both DNA and microspheres as dual steric hindrance, Tang and Liu et al. achieved the mono-DNA modification on AuNPs with greatly improved yield (Fig. 1b) (Li et al. 2011b). The high-yield modification can also simplify the post-synthesis separation. On the other hand, to precisely adjust the surface DNA density and at the same time let attached DNA strands adopt upright conformations is very difficult. In a recent work reported by Pei et al. (Pei et al. 2012a), they proposed a new, nonthiolated DNA

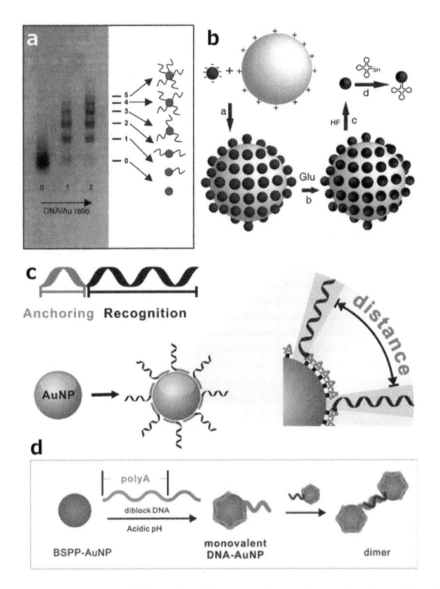

Figure 1. Conjugation of DNA with gold nanoparticles. **(a)** Conjugation through Au-S interaction with careful control of the stoichiometry. Reprinted with permission from ref. (Zanchet et al. 2001). Copyright 2001 American Chemical Society. **(b)** Preparation of mono-DNA modifications on AuNPs by combining both DNA and microspheres in a dual steric hindrance. Reprinted with permission from ref. (Li et al. 2011b). Copyright 2011 American Chemical Society. **(c)** Spatially isolated and highly hybridizable functionalization of gold nanoparticles with diblock oligonucleotides. Reprinted with permission from ref. (Pei et al. 2012a). Copyright 2012 American Chemical Society. **(d)** Preparation of mono-DNA modifications on AuNPs with diblock oligonucleotides. Reprinted with permission from ref. (Yao et al. 2015b). Copyright 2015 Nature Publishing Group.

modification strategy and suggested that a unique polyA block could specifically absorb on AuNPs surface while also acting as a hindrance to force the rest sequence to be upright (Fig. 1c). The upright conformations can avoid the nonspecific absorption of DNA on AuNPs and speed up the hybridization behavior of DNA-AuNPs. Very recently, this strategy has also been proved effective in preparing monovalent DNA-AuNPs conjugates (Fig. 1d) (Yao et al. 2015b).

Apart from AuNPs, conjugation of DNA with other noble metal nanoparticles and quantum dots can also be obtained. Deng et al. reported the discrete modifications of AgNPs (Zheng et al. 2012) and PtNPs (Saha et al. 2012) with DNA. Since these nanoparticles are not as stable as AuNPs, they optimized the capping strategy and made these nanoparticles suitable for DNA conjugations. For discrete modifications on quantum dots, Kelley et al. reported a one-pot synthesis approach and the DNA was connected with CdTe quantum dots through phosphorothioate linkages (Tikhomirov et al. 2011). The spectral properties of the quantum dots could be tuned by using specific sequences and the number of modifications could be adjusted by changing the reaction conditions. In addition, monovalent DNA-QDs conjugates were obtained through the steric exclusion strategy (Farlow et al. 2013). A technique towards separating DNA-QDs conjugates with magnetic bead was reported recently (Uddayasankar et al. 2014).

2.2 Conjugation of DNA with Organic Nanoparticles

The conjugations of DNA molecules with synthetic organic nanoparticles, such as polymers and dendrimers, often involve organic reactions. Generally, there are two main strategies: solid-phase synthesis and solution coupling. These two strategies are complementary to each other and, in case the DNA cannot be conjugated with organic nanoparticles in a straightforward manner, a combination of both methods can be utilized.

For the first strategy, the development of the conjugation of DNA with organic nanoparticles attributes to the DNA synthesis technology (Kedracki et al. 2013). The organic components are first modified with 2-cyanoethyl-N,N-diisopropylphosphoramidite (CEPA) group, and then reacted with detritylated 5' end DNA immobilized on controlled-pore glasses (CPG) (Wang et al. 2011, Kwak and Herrmann 2011). After cleavage from the CPG, DNA-organic hybrid materials are obtained. This method is applicable to not only water-soluble but also water-insoluble organic moieties. It is also noted that the stoichiometry could be well controlled.

However, it is inconvenience if there is no DNA synthesizer available in the lab. Other disadvantages include the solvent, which is typically acetonitrile, which is not mild for fragile materials and the cleavage, which needs ammonia, that hinders the conjugation of DNA with ester or base-labile compounds. Solution coupling is an alternative option; especially as it can be conducted in aqueous solution and is suitable for conjugating DNA with hydrophilic molecules. This method employs the amide bond formation by conjugating amino-DNA with NHS esters on functional groups, Michael addition via reaction of a sulfhydryl terminal group on DNA with maleimide, disulfide bond formation between the thiol groups of both components, and Huisgen-Meldal-Sharplessazide-alkyne "click" chemistry (Hermanson 2008). One challenge for the solution coupling is to control the stoichiometry, since on each organic nanoparticle, e.g., dendrimers, there may be many reactive groups. Sun et al. and Chen et al. reported a strategy to obtain 1:1 DNA-dendrimer conjugates by placing the only one reactive group at the core of the dendrimer molecule (Sun et al. 2010, Chen et al. 2010). Both solid-phase synthesis and solution coupling strategies require post-synthesis separation, such as polyacrylamide gel electrophoresis (PAGE) or high performance liquid chromatography (HPLC) to purify the compounds (Liu et al. 2010).

2.3 Conjugation of DNA with Protein Nanoparticles

The conjugation, especially the site-specific conjugation, of DNA to protein or peptide is another important issue (Niemeyer 2010, Sacca and Niemeyer 2011) and may be promising for bioorthogonality (Sletten and Bertozzi 2009). Basically, there are two major strategies: chemical conjugation and biological coupling. Chemical conjugation is traditionally and widely used for bioconjugations (Hermanson 2008). Typical bioconjugate chemistry includes amine reactions, thiol reactions, as well as "click" reactions. Protein or peptide could be covalently coupled to DNA strands directly or using homo- or hetero-multi-functional linkers. It is important to achieve site-specific coupling on the DNA strand, as well as on protein surface. One method is to incorporate the vinylsulfonamide or acrylamide modified cytosine in the process of prime extension and therefore a cysteine-containing peptide could be coupled to the DNA strand site-specifically on cytosines (Dadova et al. 2013). The advantage of this technique is that these groups are more stable than the commonly used maleimide. Using a similar strategy, the same group also reported the conjugation between aldehyde-DNA and lysine-peptide (Raindlova et al. 2012). Niemeyer et al. described the coupling of amino-DNA to genetically engineered fluorescent protein with cysteine residue through a bifunctional linker (Lapiene et al. 2010).

Chemical conjugation strategy utilizes strong covalent bonds to conjugate DNA and protein molecules. However, this strategy has difficulty meeting the increasing requirements of reversibility, simplicity and biocompatibility. Therefore, biological coupling methods, such as bioaffinity coupling or enzymatic ligation, are important alternatives.

Biotin-streptavidin pair, His-tag, apoenzyme-cofactor interaction, are all commonly used for bioaffinity coupling. Comparatively, enzymatic ligation (Sunbul and Yin 2009, Rashidian et al. 2013) is an emerging strategy for DNA-protein conjugation. Kobatake et al. used a specific enzyme Gene-A* protein, which can site-selectively cleave its recognition sequence and then couple to the 5′ end of this sequence, as the linker for DNA and protein (Mashimo et al. 2012). Kjems et al. reported the use of terminal deoxynucleotidyl transferase (TdT) for direct enzymatic ligation of native DNA (through 3′ end) to nucleotide triphosphates coupled to proteins and other large macromolecules (Sorensen et al. 2013). In an interesting work published by Tao and Zhu (Tao and Zhu 2006), they built an *in vitro* translation system in which the newly synthesized protein or peptide could be *in situ* coupled to a puromycin-modified ssDNA. This method has found applications for *in situ* protein chip fabrication (Tao and Zhu 2006) and *in vitro* display (Ishizawa et al. 2013).

3. DNA Polyhedrons

Early works on DNA-based nanoconstructions focused on 1D and 2D DNA lattices (Winfree et al. 1998, Rangnekar and LaBean 2014) assembled from DNA tiles (Seeman 2001). However, a crucial goal is to extend the success achieved in two dimensions to three dimensions. As early as 1982, Seeman envisioned three-dimensional molecular construction with DNA in his seminal proposal (Seeman 1982), but it was not until the 1990s that DNA 3D self-assembly has been seriously developed (Simmel 2008). The most important 3D DNA materials are DNA polyhedrons with finite size and defined shape that can essentially be regarded as nanosized cages. These materials are important since they are not only nano-addressable but also can act as containers and carriers for functional chemicals and nanomaterials. In this section, we will discuss the design principles and fabrication methods of DNA polyhedrons. Especially, DNA tetrahedrons will be highlighted since they have been proven as promising candidates for various applications including bio-sensors and drug delivery.

3.1 Design and Fabrication

Most DNA polyhedrons are built from DNA tiles. According to the final structure, there have two design principles: a face-centered approach and a vertex-centered approach. The face-centered approach means that each DNA strand runs round one face, thus, every edge of the polyhedron is double-helical DNA, and each vertex is a nicked multi-arm junction. In addition, some hinge bases between subsequences usually remain unmatched for flexibility. In 1991, Chen et al. used this strategy to build a topologically cube-like molecular complex. It is the first example of a closed polyhedral object constructed from DNA (Chen and Seeman 1991), though the real shape of this cube is uncertain. Turberfield's group then contributed a lot in constructing DNA tetrahedrons (Fig. 2a) (Goodman et al. 2004, Goodman et al. 2005, Erben et al. 2006, Goodman et al. 2008).

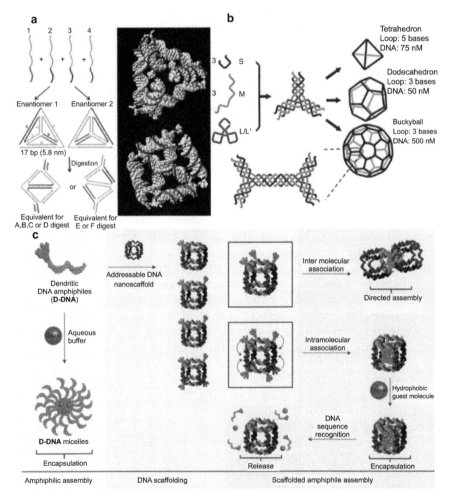

Figure 2. Conjugation of DNA polyhedrons. (a) DNA tetrahedrons formed through a face-centered approach. Reprinted with permission from ref. (Goodman et al. 2004). Copyright 2004 the Royal Society of Chemistry. (b) DNA polyhedrons formed through a vertex-centered approach. Reprinted with permission from ref. (He et al. 2008). Copyright 2008 Nature Publishing Group. (c) Hierarchical self-assembly of DNA cages by dendritic alkyl chains. Reprinted with permission from ref. (Edwardson et al. 2013). Copyright 2013 Nature Publishing Group.

Importantly, they measured the axial compressibility of double-helical DNA by compressing the upstanding edges of the DNA tetrahedron with AFM tips and proved the geometry and mechanical stability of their tetrahedrons (Goodman et al. 2005). They also introduced the toehold reaction to the tetrahedron system, thus achieving actively controllable 3D structures, whose shapes change precisely and reversibly in response to specific nucleic acid signals (Goodman et al. 2008). Other polyhedrons with more complexity and less symmetry such as a trigonal bipyramid have also been created (Erben et al. 2007). In addition, Andersen et al. designed and constructed a DNA octahedron (Andersen et al. 2008b), and Sleiman's group used a similar but slightly different scheme to access a prism family, whose height can be switched reversibly in response to external strands (Aldaye and Sleiman 2007). Recently they achieved the hierarchical self-assembly of DNA cages through site-specific modification of dendritic alkyl chains (Fig. 2c) (Edwardson et al. 2013).

The vertex-centered approach can be depicted as follows. A multi-way junction motif serves as each vertex, and its branches each associate with a branch from another motif to form the edges of the polyhedron. In principle, any tile with multiple arms can be used for vertex material. In 1994, a truncated octahedron was assembled from DNA 4-arm junctions on a solid support by Seeman's group (Zhang and Seeman 1994). The most systematic work has come from Mao's group. He et al. constructed three symmetric supramolecular polyhedrons from 3-point-star tiles (He et al. 2008). All three 3D structures, tetrahedrons, dodecahedrons and buckyballs, were well-formed and with relatively high yields (Fig. 2b). He also highlighted three factors which would promote polyhedron formation: abandoning corrugation design (Yan et al. 2003), whereby all tiles face in the same direction to increase curvature; lowering DNA concentrations; and elongating the loops in the central strand to increase tile flexibility (He and Mao 2006). By using the same principle, Zhang et al. assembled icosahedrons and nanocages from 5-point stars (Zhang et al. 2008), and cubes from 3-point stars (Zhang et al. 2009). The latter approach can restrict polyhedral faces so that they consist of only even numbers of vertices, by incorporating a two-tile system or two-face directions for adjacent tiles. They also succeeded in regulating the surface porosity of a DNA tetrahedron (Zhang et al. 2012a). Other work like the construction of a dodecahedron consisting of 20 trisoligonucleotides, with 15 bases per arm (Zimmermann et al. 2008), also belongs to the vertex-centered approach.

3.2 Applications in Biosensors

Another important application of DNA tetrahedrons is to improve the sensitivity of electrochemical DNA (E-DNA) sensors. It is critically important to precisely control the surface density of E-DNA sensors to obtain high reproducibility. 3D DNA nanostructures might provide a new route to the solution of these challenging problems. In 2010, Fan's group developed the first 3D DNA nanostructure-based electrochemical DNA sensor by using an exquisite, rigid DNA tetrahedron (Fig. 3a) (Pei et al. 2010). This tetrahedron-structured probe (TSP) was synthesized from four designed oligonucleotides with self-complementary sequences. The three vertices of the tetrahedron were modified with thiol to anchor the structure to the Au surface. The fourth vertex at the top of the bound tetrahedron was

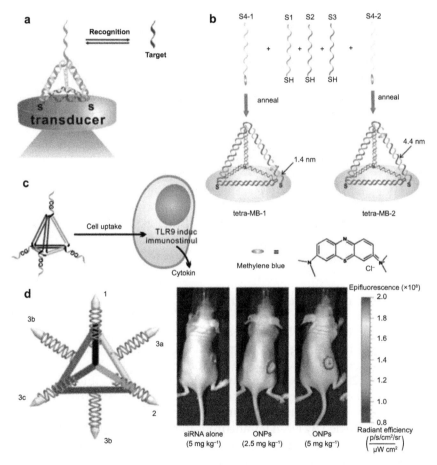

Figure 3. Applications of DNA tetrahedrons. **(a)** E-DNA sensor system. Reprinted with permission from ref. (Pei et al. 2010). Copyright 2010 John Wiley & Sons Ltd. **(b)** For charge-transfer mechanism studies. Reprinted with permission from ref. (Lu et al. 2012). Copyright 2012 American Chemical Society. **(c)** As CpG carriers. Reprinted with permission from ref. (Li et al. 2011a). Copyright 2011 American Chemical Society. **(d)** As siRNAs carriers. Reprinted with permission from ref. (Lee et al. 2012b). Copyright 2012 Nature Publishing Group.

appended with a pendant ssDNA probe. Owing to the presence of three thiol legs, TSP could be rapidly and firmly assembled onto the gold surface, which showed approximately 5000-times greater affinity compared with monothiolated DNA. More importantly, the high mechanical rigidity of TSP allowed it to stay on the Au surface with highly ordered upright orientation, even in the absence of the "helper" molecule MCH. AFM studies of Leitner et al. confirmed this directed surface attachment (Leitner et al. 2011). The surface density was measured to be ~$4.8*10^{12}$ TSP/cm^2, corresponding to an interstrand spacing of ~4 nm (Pei et al. 2010).

Based on this TSP platform, the unoptimized sensor has a picomolar sensitivity (Pei et al. 2010). This TSP-based DNA sensor possessed very high selectivity for single-base mismatches, exceeding that of ssDNA probe-based sensors by a 25–100-fold increase in the discrimination factor. They also developed an ultrasensitive electrochemical miRNA sensor (EMRS) for reliable quantitative detection of attomolar (< 1000 copies) miRNAs with high sequence specificity (Wen et al. 2012). Replacement of the DNA probe at the top vertex with an antithrombin aptamer sequence turns the E-DNA sensor into an immunological sensor for proteins (Pei et al. 2010), that is, thrombin, with a low detection limit of 100 pM that excels that of the conventional aptamer-based sensor by 3 orders of magnitude. Similarly, by incorporating a split aptamer of cocaine, they developed an electrochemical sensor for small molecules (Wen et al. 2011). Other examples include single nucleotide polymorphism (SNPs) genotyping (Ge et al. 2011), mercury ion detection (Bu et al. 2011), ATP detection (Bu et al. 2013), immunological sensing (Pei et al. 2011), and charge-transfer mechanism studies (Pei et al. 2010, Lu et al. 2012).

3.3 Applications in Drug Delivery

As aforementioned, DNA tetrahedrons are nanosized DNA cages that have several interesting advantages, such as high mechanical rigidity, structural stability for nuclease degradation and cell entry ability (Walsh et al. 2011). By exploiting these properties, Fan's group employed DNA tetrahedron nanostructures for the purpose of drug delivery. First, cytosine-phosphate-guanosine (CpG) oligonucleotides have been successfully carried by DNA tetrahedrons into macrophage-like RAW264.7 cells (Fig. 3c) (Li et al. 2011a). Since CpG sequence has high frequency in bacterial genomes but occasional frequency in mammalian genomes, it is considered to be a signal of pathogen invasion by the immune system. CpG can be recognized by Toll-like receptor 9 (TLR9) and induces immune responses (Hemmi et al. 2000). However, natural CpG oligonucleotides can be easily degraded by nucleases and are difficult to enter cells and reach the target sites. DNA tetrahedrons can act as effective CpG carriers because they not only protect CpG oligonucleotides from degradation but show no side effects. Second, apart from CpG and siRNAs, aptamers can also be delivered by DNA tetrahedrons (Pei et al. 2012b). The significant therapeutic effects of aptamers were due to their high specificity and affinity to specific targets. Like CpG, aptamers are artificially selected oligonucleotides which are much easier for DNA tetrahedrons to load. In their work, DNA tetrahedrons were used as anti-ATP aptamer carriers for monitoring the ATP level in cell. In addition, Charoenphol et al. incorporated AS1411 aptamers into DNA tetrahedrons to enhance the intracellular uptake, which showed high inhibition ability to the growth of cancer cells (Charoenphol and Bermudez 2014).

Besides the studies at the cell level, DNA tetrahedrons have also been used as drug carriers at the animal level, such as being used to carry small interfering RNAs (siRNAs) (Fig. 3d) (Lee et al. 2012b). siRNAs make it possible to suppress the expression of target genes for the therapy of diseases, while RNA molecules themselves can hardly resist enzymatic degradation and have much shorter half-life *in vivo*. In that work, with the help of folate acid, siRNAs-loaded DNA tetrahedrons were successful in transporting into tumor cells, resulting in improving the efficacy of RNAi and increasing the blood circulation time. Yan's group utilized DNA tetrahedrons to deliver a synthetic vaccine complex which contained a model antigen, streptavidin (STV) and a representative adjuvant, CpG oligo-deoxynucleotides (ODN) (mouse-specific ODN-1826), into mice (Liu et al. 2012). They tested their immunogenicity of the assembled tetrahedron-STV-CpG ODN with STV only and STV mixed with CpG, both *in vitro* and *in vivo*, and found that the assembled tetrahedron-STV-CpG ODN complexes induced the strongest and longest-lasting antibody responses against the antigen while DNA nanostructures alone did not stimulate any response. This study not only was an excellent example for DNA nanostructures as a generic platform for the rational design and construction of vaccines but also showed synergic effect arising from DNA nanostructure-based co-delivery.

4. DNA Origami

The term "DNA origami" was proposed by Paul Rothemund in 2006 to describe his invention of a new type of DNA nanomaterial (Rothemund 2006). In that revolutionary work, he showed the ability of controlled folding of a long, single-stranded scaffold DNA, with the help of hundreds of short staple strands, into exquisite nanopatterns. After his invention, this technique has been a constant focus in the field of DNA nanotechnology for the past few years. Great efforts have been made to build new 2D and 3D DNA origami structures, improve assembly strategy, study inherent properties, and develop new applications.

4.1 Invention and Structural Evolution

In 2006, the invention of DNA origami by Rothemund greatly increased the complexity and size of man-made DNA nanostructures, as well as largely simplified the design and preparation processes. Inspired from the same name Japanese paper-folding art, Rothemund used term "origami" to describe this new milestone strategy. In brief, DNA origami involves raster-filling the desired shape with a long single-stranded scaffold with the help of hundreds of short oligonucleotides, called staple strands, to hold the scaffold in place (Fig. 4a) (Sanderson 2010). Rothemund chose the genomic DNA from the virus M13 mp18 with more than 7000 bases as the scaffold. More than 200 staple strands were used to help folding, and different shapes were assembled from different sets of staple strands, such as square, triangle, star, disk, and so on (Fig. 4b). The resulting DNA structures all conform well to the design and have a diameter of roughly 100 nm and a spatial resolution of 6 nm. DNA origami is considered as a breakthrough in structural DNA nanotechnology, which has produced two main achievements. The first is the amazing nanoarchitecture it has made possible. The second achievement is experiment simplification. Since DNA origami has so many advantages, several studies have undertaken the construction of a number of 2D (Fig. 4c) (Qian et al. 2006, Andersen et al. 2008a) and 3D (Andersen et al. 2009, Kuzuya and Komiyama 2009) intricate and creative architectures.

The above DNA origami nanostructures were all constructed by folding planar sheets. This design principle is simple and straightforward but since planar DNA origami has intrinsic flexibility, it would be difficult to build rigid and various 3D nano-objects. The next breakthrough in this field was reported by Shih's group (Douglas et al. 2009). They used a different strategy to achieve the building of custom 3D shapes (Fig. 4d). The key in their design principle is that the 3D shapes are composed of honeycomb lattice. This design could be conceptualized as stacking corrugated sheets of antiparallel helices. The resulting structures resemble bundles of double helices constrained to a honeycomb lattice. The shape and size could be adjusted by changing the number, arrangement and lengths of the helices in the lattice. In addition, hierarchical assembly of structures can be achieved by programming staple strands to link separate scaffold strands. Based on this design, they also engineered complex 3D shapes with controlled twist and curvature at the nanoscale, by targeted insertions and deletions of base pairs (Dietz et al. 2009). Later on, Shih's group collaborated with Yan's group to achieve a more compact design which used square lattice instead of honeycomb lattice (Ke et al. 2009).

Following the effort to increase the complexity of DNA origami shapes, Yan's group reported the first topological DNA origami architecture—a

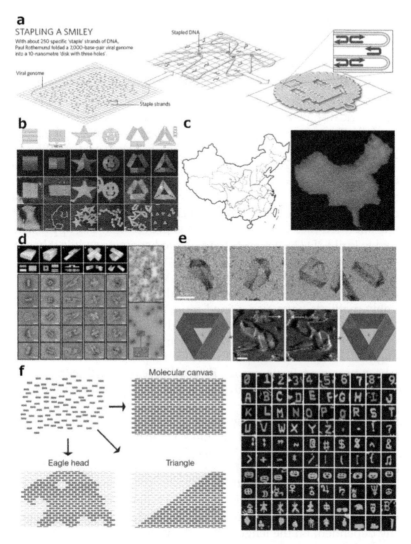

Figure 4. Structural evolution of DNA origami technique. **(a)** Folding principle of 2D DNA origami. Reprinted with permission from ref. (Sanderson 2010). Copyright 2010 Nature Publishing Group. **(b)** 2D DNA origami patterns created by Rothemund. Reprinted with permission from ref. (Rothemund 2006). Copyright 2006 Nature Publishing Group. **(c)** A 2D DNA origami mimicking the shape of China map. Reprinted with permission from ref. (Qian et al. 2006). Copyright 2006 Springer Science + Business Media. **(d)** 3D objects created by Shih's group. Reprinted with permission from ref. (Douglas et al. 2009). Copyright 2009 Nature Publishing Group. **(e)** A Möbius strip. Reprinted with permission from ref. (Han et al. 2010). Copyright 2010 Nature Publishing Group. **(f)** 3D objects assembled from single-stranded tiles. Reprinted with permission from ref. (Wei et al. 2012). Copyright 2012 Nature Publishing Group.

Möbius strip (Fig. 4e) (Han et al. 2010). It is a topological ribbon-like structure that has only one side. Recently, in an escape from the rigid lattice model used for conventional DNA origami nanostructures, they also reported a new strategy for the building of 3D DNA origami with complex curvatures (Han et al. 2011). Following this design principle, a series of 2D and 3D DNA nanostructures with high curvature, such as concentric rings, spherical shells, ellipsoidal shells, and a nanoflask were assembled successfully. To engineer wireframe architectures and scaffolds of increasing complexity, they presented a design strategy to create gridiron-like DNA structures (Han et al. 2013). A series of four-arm junctions were used as vertices within a network of double-helical DNA fragments. The new milestone in this field was created by Yin's group (Wei et al. 2012). They used a conceptually new technique which was based on the assembly of "single-stranded tiles" (SST) (Fig. 4f). It is proved that this approach can be used to build 3D DNA objects with defined geometry while it has more advantages (Wei et al. 2013, Ke et al. 2014, Wei et al. 2014).

While building new complex DNA origami structures is the prominent goal in the field, efforts towards optimizing and developing assembly methods have also gained more and more attentions. Distinct from the conventional annealing methods, isothermal assembly technique (Jungmann et al. 2008), room temperature assembly technique (Zhang et al. 2013), and magnesium-free technique (Martin and Dietz 2012) have been proposed. Furthermore, Fu et al. have proved that the assembly could be finished in as short as 10 min (Fu et al. 2013). Double-stranded DNA scaffolds have also been shown able to prepare two distinct DNA origami shapes in a one-pot reaction (Hogberg et al. 2009). New purification methods for DNA origami were reported by Shih's group based on a modified DNA electroelution (Bellot et al. 2011), while Dietz's group described a method based on poly (ethylene glycol)-induced depletion of species with high molecular weight (Stahl et al. 2014).

Although DNA origami technique shows superior ability in preparing arbitrary nanostructures with high complexity, the size of DNA origami is strictly dependent on the length of long scaffold strand. In most reported cases, ~7 kb M13 strand was used and therefore for 2D DNA origami, its size should be ~7000 nm^3 which is still too small for possible practical applications. One simple way to solve this problem is using longer scaffold strand. Fan's group prepared a 26 kilobase single strand DNA fragment, which was obtained from long-range PCR amplification and subsequent enzymatic digestion, for folding large DNA origami (Zhang et al. 2012b). The results showed that this strand could fold into a super-sized DNA with a theoretical size of 238×108 nm^2. Yan's group reported a more complex design by using a double-stranded scaffold to fabricate integrated DNA origami structures that incorporate both of the constituent ssDNA molecules

(Yang et al. 2012). In a recent publication by LaBean's group, they described two methods that overcome some of the major challenges for future progress of the DNA origami field (Marchi et al. 2014).

Different from the long scaffold strategy, a more efficient and practical strategy is the higher-order assembly of individual DNA origami units into large arrays. This idea was first proposed by Rothemund in his pioneer work, in which he designed extended staples on DNA triangle edges to induce the assembly of six triangles into a big hexagon (Rothemund 2006). This principle was then followed by other researchers. Endo and Sugiyama proposed a programmed-assembly system using DNA jigsaw pieces where each jigsaw piece contains sequence-programmed connection sites, a convex connector, and a corresponding concavity (Endo et al. 2010). Another strategy for scaling up was invented by Yan's group. In that design (Zhao et al. 2010), they suggested that instead of ssDNA staples used in DNA origami, an origami itself may also mimic the function of a staple if single-stranded overhangs are extended at the four corners. By introducing bridge strands, each origami-based "staple tile" could hybridize with the scaffold and form large structures that contain several small staple tiles. On another hand Simmel's group demonstrated that by electrostatically controlling the adhesion and mobility of DNA origami structures on mica surfaces, by the simple addition of monovalent cations, large ordered 2D arrays of origami tiles can be generated (Aghebat Rafat et al. 2014). Rothemund recently also reported the self-assembly of DNA origami rectangles into two-dimensional lattices based on the stepwise control of surface diffusion, implemented by changing the concentrations of cations on the surface (Woo and Rothemund 2014).

4.2 As Templates for Nanoparticles Self-Assembly

DNA nanostructures, especially DNA origami, receive close interest because of the programmable control over their shape and size, precise spatial addressability, easy and high-yield preparation, mechanical flexibility and biocompatibility. Therefore, DNA origami structures have been widely used as templates for precise geometrical control over the positioning of nanoscale objects.

Metal nanoparticles, including Au and Ag nanoparticles, have been successfully positioned on DNA origami by many groups. The first work is the assembly of lipoic acid-modified DNA-AuNPs conjugates onto a rectangular DNA origami template (Sharma et al. 2008). Later on, DNA origami has been demonstrated to organize different-sized Au nanoparticles to form a linear structure with well-controlled orientation and < 10 nm spacing (Ding et al. 2010). This structure could be used to generate extremely high field enhancement and thus work as a nanolens. Discrete and well-ordered AgNPs nanoarchitectures on DNA origami structures of triangular shape, by using AgNPs conjugated with chimeric phosphorothioated DNA (ps-po DNA) as building blocks, have been achieved (Pal et al. 2010). Discrete monomeric, dimeric, and trimeric AgNP structures and an AgNP–AuNP hybrid structure could be constructed reliably in high yield. Besides, Yan's group achieved programmable positioning of one-dimensional (1D) gold nanorods (AuNRs) by DNA directed self-assembly (Pal et al. 2011). AuNR dimer structures with various predetermined inter-rod angles and relative distances were constructed with high efficiency. Recently, Fan's group reported a jigsaw-puzzle-like assembly strategy mediated by gold nanoparticles (AuNPs) to break the size limitation of DNA origami (Yao et al. 2015a). They demonstrated that oligonucleotide-functionalized AuNPs function as universal joint units for the one-pot assembly of parent DNA origami of triangular shape to form sub-microscale super-origami nanostructures. AuNPs anchored at predefined positions of the super-origami exhibited strong interparticle plasmonic coupling.

Fluorescent semiconductor quantum dots are another kind of NPs. Bui et al. first reported assembly of streptavidin functionalized QDs on DNA origami tube at predetermined locations with full control over the number of QDs on each origami and the separating distance between them (Bui et al. 2010). Ko et al. investigated the binding kinetics of these QDs to DNA origami quantitatively and established some thumb rules that affect binding efficiency (Ko et al. 2012). Wind and co-workers organized disparate functional nanomaterials, namely, semiconducting QDs and metallic nanoparticles, on a single DNA origami template for the first time (Wang et al. 2012). Liedl's group used rigid DNA origami scaffolds to assemble metal nanoparticles, quantum dots and organic dyes into hierarchical nanoclusters that have a planet–satellite-type structure (Schreiber et al. 2014). The nanoclusters have a tunable stoichiometry, defined distances of 5–200 nm between components and controllable overall sizes of up to 500 nm.

Besides inorganic nanoparticles, biological particles and carbon nanotubes have also been positioned on DNA origami templates in recent years. Kuzyk et al. explored two general approaches to the utilization of DNA origami structures for the assembly of streptavidin-binding pattern (Kuzyk et al. 2009). Kuzuya et al. designed a punched DNA origami assembly that can selectively capture exactly one streptavidin tetramer each of any of its predetermined wells and stably accommodate them (Kuzuya et al. 2009). On the basis of biotin–streptavidin (STV) interactions, Niemeyer's group demonstrated that DNA nanostructures can be site-specifically decorated with several different proteins by using coupling systems orthogonal to the biotin–STV system (Saccá et al. 2010).

Another way to assemble biological materials on origami is DNA hybridization. Yan and co-workers used peptides immobilized on the surface of a DNA nanotube to template-direct the *in situ* nucleation and growth of gold nanoparticles from soluble chemical precursors. The prepared peptide–DNA conjugate was hybridized to complementary DNA capture probes present on the surface of preformed DNA nanotubes (Stearns et al. 2009). Besides, Francis and co-workers adopted the same method to immobilize virus capsids on DNA origami with nanoscale precision (Stephanopoulos et al. 2010). Other techniques, including nitrilotriacetic acid (NTA) and Histidine-tag metal linked interaction (Shen et al. 2009) and Zinc-Finger proteins for site-specific protein positioning on DNA origami structures (Nakata et al. 2012), were also reported.

Single-walled carbon nanotubes were assembled on DNA origami templates through streptavidin-biotin interaction. Furthermore, this method is a general method for arranging single-walled carbon nanotubes in two dimensions (Maune et al. 2010). Yan's group reported a convenient, versatile method to organize discrete length single-walled carbon nanotubes (SWNT) into complex geometries using 2D DNA origami structures (Zhao et al. 2013).

4.3 Applications in Plasmonics

The fast-growing field of plasmonics, which study the abilities of metal nanostructures to localize, guide, and manipulate electromagnetic waves beyond the diffraction limit, down to the nanometer-length scale, is an emerging research area (Gramotnev and Bozhevolnyi 2010). Morphological parameters for metal nanostructures such as size, shape, geometry, and interparticle distance could substantially influence the plasmonic coupling and the subsequent optical and electrical properties (Rycenga et al. 2011).

With the help of DNA origami techniques, it is possible to render more exquisite control for nanopatterning, since DNA origami techniques can meet the critical demands of both structural robustness and nanoscale addressability in 1D–3D. Yan's group, as mentioned above, contributed significantly in this direction by establishing some reliable metal nanoparticle immobilization techniques on DNA origami. They then reported the first DNA origami-based plasmonic nanostructure by arranging gold nanorods dimer with variable orientations and observed plasmonic coupling induced band shifts in UV-vis spectra (Pal et al. 2011). They also studied the distance dependent local electric field enhancement of both monomeric and dimeric AuNPs structures and their effects on the photophysics of the fluorophores in close proximity to the AuNPs (Pal et al. 2013).

Some other groups were focusing on the circular dichroism effect originated from chiral geometry of gold nanoparticles arrays. Liedl and co-workers showed that DNA origami-based gold nanoparticles' nano-helices exhibited defined circular dichroism (CD) and optical rotatory dispersion effects at visible wavelengths (Fig. 5a) (Kuzyk et al. 2012). Ding et al. (Shen et al. 2012, Shen et al. 2013) and Wang et al. (Lan et al. 2013) observed pronounced circular dichroism from chiral gold nanoparticles and nanorod arrays assembled on planar rectangular DNA origami templates, respectively. The Liedl group reported switchable CD responses by toggling the orientation of the same construct, as shown in their previous paper, with respect to incident light (Schreiber et al. 2013). Most recently two groups, in collaboration, fabricated a reconfigurable plasmonic nanostructure with AuNRs placed onto two, interlinked DNA origami bundles at specific angles (Kuzyk et al. 2014). This smartly designed dynamic system is capable of switching between two conformations to tune the angle between the two bundles by DNA strand displacement, which was reflected in distinctive

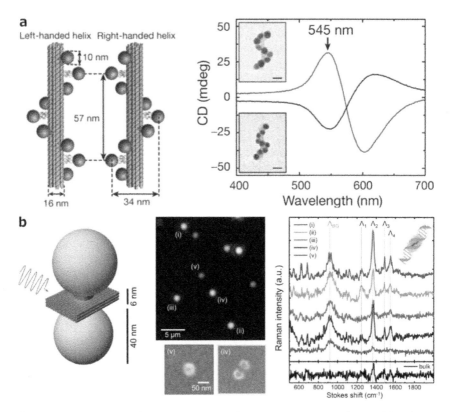

Figure 5. Applications of DNA origami in plasmonics. **(a)** Directed assembly of gold nanoparticle arrays with defined chiral geometry. Reprinted with permission from ref. (Kuzyk et al. 2012). Copyright 2012 Nature Publishing Group. **(b)** Characterization and SERS spectra of individual origami-AuNP dimmers. Reprinted with permission from ref. (Kühler et al. 2014). Copyright 2014 Nature Publishing Group.

alteration in the CD spectra. They envisioned this to be an *in situ* probe for monitoring dynamic biological process.

Another important progress in this direction is the regulation of fluorescence efficiency by plasmonic arrays on DNA origami. Tinnefeld et al. prepared nanoantennas by attaching one or two gold nanoparticles to a DNA origami pillar structure (Acuna et al. 2012), which also incorporated docking sites for a single fluorescent dye next to one nanoparticle, or in the gap between two nanoparticles. They studied the dependence of the fluorescence enhancement on nanoparticle size and number and obtained a maximum of 117-fold fluorescence enhancement. Liu et al. studied a similar fluorescence enhancement effect by using a simple DNA origami triangle as the template (Pal et al. 2013). Recently, Kuang et al. extended the application of this technique to waveguides based on linear gold nanoparticle arrays assembled on a multiscaffold DNA origami ribbon (Klein et al. 2013). This showed the great potential of this technique for building more functional devices.

Another major application of plasmonic NPs is in enhancing resonance Raman signal. It was discovered decades ago that a rough surface of silver or gold can induce significant enhancement of Raman signal, a phenomenon called surface enhanced Raman scattering (SERS). With the unique addressability of DNA origami, two or more large Au/AgNPs placed in close proximity behaved as a nanoantenna, creating a plasmonic hot spot with an intense local electric field at their junction. Three recently published papers used DNA templated assembly to place Raman molecules (RhodamineSYBR-Gold or 4-aminebenzenethiol) in between two or more AuNPs and showed a several magnitude fold enhancement of the SERS signal (Thacker et al. 2014, Pilo-Pais et al. 2014, Kühler et al. 2014). Feldmann's group characterized SERS spectra from individual origami-AuNP dimers (Fig. 5b).

4.4 Applications in Nano-Reactors

In living cells, metabolism is spatially regulated through the site-specific compartmentalization of multi-enzymatic cascades in subcellular organelles (Agapakis et al. 2012). The high level of spatial organization and integration make metabolism an optimized network of interconnected biological reactions, guiding the production, transportation and consumption of nutrients, and allowing the maintenance, growth and reproduction of life. Inspired by nature, synthetic biologists are aiming at building biomimetic nano-/micro-factories for positional immobilization of multi-enzymatic cascades with nanoscale precision (Lee et al. 2012a, Schoffelen and van Hest 2013, Chen and Silver 2012). These artificial systems are expected to regulate reaction pathways in high efficiency, and hold great promise in producing expensive medicines and materials that are difficult for common synthetic chemistry, as well as generating cheap renewable fuels.

Towards this goal, Gothelf's group carried out the first DNA origami-templated chemical reaction for the addressable coupling and cleavage of some chemical bonds (Voigt et al. 2010). Then, Liu et al. conducted the directed polymerization of macromolecular patterns (Liu et al. 2010). In this design, dendrimers were used as model molecules and assembled on a rectangular DNA origami template. Then, through covalently coupling between adjacent dendrimers, a polymerized macromolecular pattern can be obtained (Fig. 6a). DNA origami nanostructures have also been proved

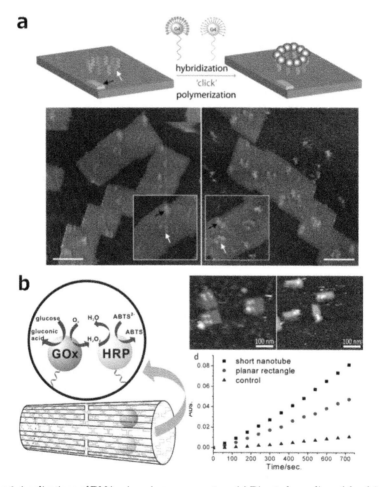

Figure 6. Applications of DNA origami as nano-reactors. **(a)** Directed coupling of dendrimers into polymerized patterns. Reprinted with permission from ref. (Liu et al. 2010). Copyright 2010 American Chemical Society. **(b)** Regulating GOx-HRP bienzyme cascade. Reprinted with permission from ref. (Fu et al. 2012). Copyright 2012 American Chemical Society.

as effective templates for protein immobilization and enzymatic cascades. Yan's group reported the enhanced GOx-HRP cascade on a rectangular DNA origami (Fu et al. 2012). It is noted that a protein bridge could be placed between GOx and HRP for promoting intermediate diffusion and increasing cascade efficiency. Furthermore, Fu et al. developed a rapid self-assembly approach for DNA origami and constructed a new 3D DNA origami nanotubes (Fu et al. 2013). The cascade efficiency has been compared between on planar DNA origami and in 3D nanotubes. The cascade in nanotubes showed higher activity (Fig. 6b), implying that the confined space of a DNA nanotube may be utilized as a nanoscale reactor.

Since DNA origami could be a platform for more exquisite controls on positions, distance, and even orientations, it is expected that this technique could be further explored for the regulation of cascade pathways both spatially and temporally.

4.5 Applications in Drug Delivery

DNA origami has also found applications in drug delivery. Since it is much larger than DNA tetrahedrons, in principle it can act as carrier for more drugs. Up to now, various kinds of model materials, such as CpGs, small molecule drugs, fluorescent dyes and peptides have been successfully delivered into cells by DNA origami carriers.

Liedl and co-workers utilized hollow 30-helix DNA origami nanotubes for the delivery of immune-activating CpGs into freshly isolated spleen cells that targeted the endosome (Schuller et al. 2011). As the DNA origami nanotubes were much larger than DNA tetrahedron, it can bear up to 62 CpG oligonucleotides. Recently, Fan's group assembled a series of DNA origami nanoribbons to bear CpG motifs into mammalian cells which showed enhanced immunostimulatory activities (Ouyang et al. 2013).

Compared to DNA tetrahedron, DNA origami nanostructures are more suitable to carry intercalation molecules since they contain more DNA duplex. Doxorubicin (Dox) is such a DNA helix intercalating molecule which is also a potent anti-cancer drug to inhibit macromolecular biosynthesis in cancer therapy (Bagalkot et al. 2006, Xiao et al. 2012). Recently, Ding and coworkers reported a DNA origami nanostructures based Dox delivery system (Jiang et al. 2012). They utilized two dimensional triangle DNA origami and three dimensional DNA tubes to load Dox, which had an increased loading efficiency by calculation and partly proven by electrophoresis gels and AFM images. Hogberg et al. reported that DNA origami nanostructures could deliver Dox to three different breast cancer cell lines (Zhao et al. 2012).

Another important progress in this direction is the success in constructing a DNA origami-based intelligent drug carrier that can release drug after receiving the orders. Church's group designed an edge flexible DNA hexagonal barrel nanostructure as an autonomous DNA nanorobot (Douglas et al. 2012). By modifying staples with three well characterized aptamer, the DNA hexagonal barrel nanostructure generated six robots with different locks. They chose fluorescently labeled antibody to human HLA-A/B/C Fab as payloads to probe 6 different cell lines. After corresponding "keys" were added, the nanorobot opened for the exposure of the antibody which bound to cell-surface receptors and inhibited the growth of the target cells. Although this study focused on logic gate functions of payloads to cellular surfaces, this was still an excellent prototype for intelligent nanocarriers in drug delivery.

5. DNA Nanomachines

Nanomachines, or molecular machines, can be regarded as nanosize-devices that convert energy into mechanical tasks. As the most famous example, and an essential part of life, ATP-powered motor protein is a class of natural nanomachine that can perform mechanical movements in nanoscale. Likely, the conformational switch-ability of DNA also enables the fabrication of nanoscale molecular machines. From the structural point of view, DNA nanomachines are made up of assembled DNA structures which contain both rigid and switchable parts. The switchable part is responsive to external stimuli, such as "fuel" DNA strands and environmental changes, and therefore can generate force and perform movements. In this section we will summarize present efforts on constructions of DNA nanomachines based on different driven mechanisms and further discuss their important applications and future development directions.

5.1 Prototypes

Up until now all established DNA nanomachines could be sorted into several catalogues by their original driven mechanisms. The most commonly used one can be considered as the "fuel-strands" strategy. It is well known that a short DNA strand could be replaced by a longer strand to form a more stable duplex, which is called "chain-exchange reaction" or "strand-exchange reaction". This reaction was firstly employed by Yurke et al. in 2000 (Yurke et al. 2000) to induce motions to DNA-based nanostructures. The device is assembled by three single-strands which can form two rigid duplex arms connected by a hinge section and two dangling ends linked to arms, resembling a pair of tweezers (Fig. 7a). Upon adding "fuel" and "anti-fuel" strands alternately, the tweezers can be switched between "open" and "closed" states. This fuel-strands

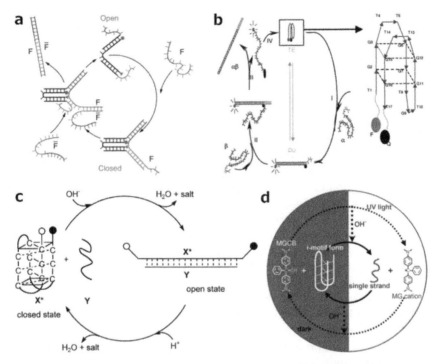

Figure 7. DNA nanomachines. **(a)** A DNA tweezers powered by fuel-strands. Reprinted with permission from ref. (Yurke et al. 2000). Copyright 2000 Nature Publishing Group. **(b)** A G-quadruplex-based DNA nanomachine. Reprinted with permission from ref. (Li and Tan 2002). Copyright 2002 American Chemical Society. **(c)** An i-motif-based DNA nanomachine which is responsive to pH change. Reprinted with permission from ref. (Liu and Balasubramanian 2003). Copyright 2003 John Wiley & Sons Ltd. **(d)** A light-driven i-motif-based DNA nanomachine. Reprinted with permission from ref. (Liu et al. 2007). Copyright 2007 John Wiley & Sons Ltd.

strategy has also been employed to drive other kinds of DNA assemblies to move, including PX-JX$_2$ complex (Yan et al. 2002), G-quadruplex (Fig. 7b) (Li and Tan 2002), and DNA origami-based nanomachines. In 2010, Seeman et al. demonstrated that a nanoscale assembly line can be realized by the judicious combination of three known DNA-based modules (Gu et al. 2010). In the meantime, Yan and co-workers developed molecular robots can walk along the prescriptive landscapes, which can also autonomously carry out sequences of actions such as 'start', 'follow', 'turn' and 'stop' (Lund et al. 2010). In 2011, Turberfield and co-workers assembled a 100-nm-long DNA track on a two-dimensional scaffold, and observed stepwise movement of a synthetic molecular transporter directly by AFM (Wickham et al. 2011). The next year, they achieved another DNA-based molecular motor that can navigate a network of tracks (Wickham et al. 2012).

In principle, the above fuel-strands strategy could be applied to all strand-exchange reaction-powered DNA nanomachines, since, as we have mentioned, hybridization is the common feature of DNA. However, the main disadvantage is these reactions will result in cumulated duplex wastes. These useless duplexes may compete with surrounding nanomachines and, from the point of entropy flow, the accumulation of waste DNA will increase the entropy of the system, making the machine dead in final.

To avoid duplex wastes, non-DNA stimuli should also be choices for controlling motions. In fact, this approach has already been proposed in the construction of the first DNA-based nanomechanical device (Mao et al. 1999), in which case ethidium ions were used as intercalators to induce branch point migration in a tetramobile branched junction structure. G-quadruplex, in some cases (Fahlman et al. 2003, Miyoshi et al. 2007, Monchaud et al. 2008), can also be responsive to environmental changes and can be switched by metal ions.

However, for almost every chemical or biological system, pH value is a very important factor. For some DNA structures, it is also true. Therefore, it will be very interesting to use pH change for driving DNA nanomachines. Towards this goal, Liu and Balasubramanian (Liu and Balasubramanian 2003) invented a C base-rich DNA nanomotor (Fig. 7c) which is based on the pH-responsive folding and unfolding of a quadruple structure called i-motif. The operation of this machine, which is based on the addition of acid or base to switch solution pH values, can be completed in less than 5s and the wastes are only salt and water from neutralization reaction. The advantages of this proton-driven nanomachine are obvious: it is clear, quick, reliable and efficient. In order to further simplify the experiment by freeing our hands from adding fuels manually, light- (Fig. 7d) (Liu et al. 2007) and electricity-controlled (Yang et al. 2010) i-motif nanomachines have also been developed. In addition, the pH-sensitive DNA triplex-duplex transitions (Chen et al. 2004), and photo-switchable G-quadruplex (Wang et al. 2010) have also been made to build DNA nanomachines.

5.2 Spatial Control of Nanoparticles by DNA Nanomachines

One important application of DNA nanomachines is to regulate the movement of nanoparticles. By controlling the change of DNA structures, the spatial arrangement of nanoparticles can be tuned accordingly (Sun et al. 2010, Chen et al. 2010). For example (Meng et al. 2009), the gold electrode is modified with i-motif DNA machine which can fold into a four-stranded structure under acidic condition and unfold under basic condition. The other end of DNA is connected with CdSe/ZnS core–shell quantum dots. At acidic and basic condition, the i-motif machine can change its configuration and brought quantum dots near/far to the gold electrode, respectively, providing a strategy to dynamically control the photoelectric conversion. Similarly, i-motif DNA can be anchored to AuNPs to tune the assembly of AuNPs. For instance, ssDNAs containing half stretch of i-motif DNA are modified onto AuNPs, and at high pH, showing a random coil structure. In contrast, at low pH, the formation of interparticle i-motifs leads to the assembly of AuNPs into aggregates. The two states can be switched by varying pH (Wang et al. 2007). Differently, when the full stretch of i-motif DNA are modified onto AuNPs, there is no interparticle interaction at low pH, although the i-motif DNA folds into compact structure, and AuNPs aggregate when pH is increased (Sharma et al. 2007). Both examples show that by modifying AuNPs with DNA machines, it is possible to reversibly switch the assembly and disassembly of AuNPs. There are also other means, such as using the DNA strand displacement to control the assembly of nanoparticles and their relative positions (Song and Liang 2012, Elbaz et al. 2013). As the process is dynamic, and it allows the control of the movement of nanoparticles on a designed DNA origami pattern towards a desired direction (Gu et al. 2010).

Besides controlling spatial arrangement of inorganic nanoparticles by DNA machine, the same strategy can be applied to organic nanoparticles and more. For example, two amphiphilic dendrons are covalently connected with an i-motif DNA machine and the amphiphilic dendron contains two parts. The yellow region represents the inner hydrophobic poly(arylether) and the blue regions represent the peripheral hydrophilic oligoethylene glycol (OEG) (Sun et al. 2010). At a basic condition, the rigid double helix formed by hybridization of DNA extend two dendrons apart for about 5.8 nm. As the pH decreases, the i-motif DNA machine folds into a compact quadruplex structure, bringing two dendrons close, so that they merge together. This process is reversible and the association of the two dendron parts increases the stability of i-motif DNA. Their increased melting temperature (T_m) is an indicator or used to probe the strength of the interaction between two nanoparticles. With the versatile modification of DNA, the dendron and protein can be bound to both sides of DNA and the conformation of the dendron-DNA-protein hybrid molecular system can be controlled by the i-motif DNA machine reversibly (Chen et al. 2010). The incorporation of DNA machines into synthetic molecules or biological components provides a new platform to achieve well-defined supramolecular assemblies, manipulate nanoparticles at single molecular level in a controllable manner and study the interaction between nanoparticles.

5.3 Regulation of Protein Binding Affinity

Biology usually involves polyvalent interactions, i.e., the simultaneous binding of multiple ligands on one biological entity to multiple receptors on another. Such interaction strongly relates to the spatial position of multiple ligands (Mammen et al. 1998). A DNA machine enables the study of complex interactions in three dimensions. In the early stage of this field, scientists employed ssDNA and dsDNA forms to control the spatial distance (Choi and Zocchi 2006, Röglin et al. 2007, Williams et al. 2009, Furman et al. 2010). The protein entities are covalently attached to DNA strands. As ssDNA is usually considered flexible and dsDNA is rigid with a persistence length of about 50 nm, the change of the rigidity from ssDNA to dsDNA can adjust the approximation of two proteins. It is noted that the more rigid DNA structures, such as double-crossover (DX), Holliday junction and origami scaffolds may provide more controllability and precision. For example, a tweezers-like DNA machine (Fig. 8a) can be opened (stem-loop structure) and closed (double helix structure) via adding fuel and antifuel strand, respectively (Zhou et al. 2012). Two ligands which can bind the target protein, thrombin, are introduced at both terminals of the DNA tweezers. In the closed state, the two ligands are at cooperative position and catch thrombin, however, as the DNA tweezers open, two ligands are apart and the thrombin is released. Such a target-responsive DNA machine can be repeatedly used and provides a promising strategy to study spatial dependent interactions of nanoparticles.

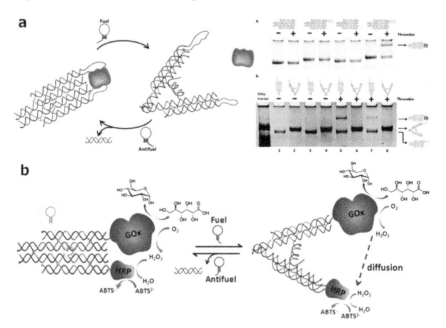

Figure 8. Regulation of biomolecules by DNA nanomachines. **(a)** Reversible regulation of target binding affinity. Reprinted with permission from ref. (Zhou et al. 2012). Copyright 2012 American Chemical Society. **(b)** Regulation of enzyme cascade. Reprinted with permission from ref. (Xin et al. 2013). Copyright 2013 John Wiley & Sons Ltd.

5.4 Regulation of Enzyme Cascade Reactions

In living systems, a series of enzymatic reactions in a cascade way is conducted to mediate biological functions. The efficiency and specificity of enzyme cascade reactions rely on the appropriate spatial arrangement. DNA nanotechnology provides predictable and programmable design of positioning nanoparticles in a precise manner. By taking advantage of this, previous studies have shown that it is feasible to locate multiple enzymes on static DNA nanostructures (Barrow et al. 2012). However, the distance between enzymes is determined and fixed by the underlying DNA structures therefore, in order to mimic the enzyme cascade reaction *in vitro* where the location of enzymes is not consistent, scientists turned to dynamic DNA nanostructures (Xin et al. 2013, Liu et al. 2013). Following the work of using DNA tweezers to regulate protein binding, the similar design is used to attach two enzymes of glucose oxidase (GOx) and horse-radish peroxidase (HRP) to both arm ends of DNA tweezers (Fig. 8b). As the reaction occurs, GOx first catalyzes oxidization of glucose to generate gluconic acid and H_2O_2, which is as the substrate of HRP and reduced into H_2O. The turnover rate of GOx is much slower than HRP, therefore, the diffusion distance of H_2O_2 determines the rate of this enzyme cascade reaction. By deliberately tuning the state of DNA tweezers in closed and open state, the distance of GOx and HRP can be adjusted from several nanometers to about eighteen nanometers. In the closed state, GOx and HRP are brought into proximity, leading to a much closer diffusion pathway for H_2O_2. In contrast, in the open state, both enzymes are spatially separated, thus lowering enzymatic efficiency. It is shown that several cycles of regulation of enzyme cascade reaction are successfully conducted. Furthermore, this proof-of-concept is also confirmed in another system to actuate the activity of an enzyme/cofactor pair. A dehydrogenase and NAD^+ cofactor are fixed on DNA tweezers, and the enzyme inhibition and activation is reversibly turned on and off upon tuning the open and closed state of DNA tweezers, respectively. This approach opens up the design of dynamically regulating other enzyme systems and mimicking enzyme cascade reaction outside of living organisms.

6. Perspectives

In this chapter, we have summarized the development of DNA-based smart materials for nanoconstructions in recent decades. The development of this field has demonstrated again that DNA is an important multipurpose material in nanotechnology other than its biological characters. Especially, after the invention of DNA origami in 2006, we have witnessed an explosive growth in research interest. Compared with Seeman's early works in the last century, people are now able to build complex 2D and 3D artificial DNA nanostructures with defined geometry, look into its inherent physical, mechanical, electrical and biological properties, assemble higher-order and larger patterns, and narrow the gap between top-down and bottom-up. Applications of DNA-based smart materials are also a fast-growing research field.

Looking towards the future, the current research on DNA-based materials is still in its early stages and more challenges remain to be solved.

First, for DNA-nanoparticles conjugates, though various methods of combining DNA and nanoparticles have been established, there is still a big challenge to develop new conjugation strategies in order to realize spatially controlled and anisotropic conjugation on nanoparticles. In addition, further improvement to the methods of conjugating DNA with biological components, which usually contain multiple functional groups on the surface, and how to connect DNA to specific site without changing its biological functions need to be addressed.

Second, for static DNA nanostructures including polyhedrons and origami, there are more questions that are waiting for answers. Fundamentally, we are still not clear about the formation mechanism of many DNA nanostructures, especially DNA origami, and need to improve the yield, quality and stability, simplify the preparation, and make it cheaper. From the structural point of view, it is constantly expected to build more complex, larger and stronger structures with controlled addressability and flexibility that could be manipulated by present scientific instruments. In terms of future applications, several aspects are foreseen for further investigation. One is artificial bio-nanoreactor: it is expected that DNA-biomolecule conjugates anchored on spatially addressable DNA template to form controlled cascades, which could mimic *in vivo* bioreactions in compartments. Addressable nanocircuits, based on self-assembled DNA-meal nanoparticle patterns, should also be studied to overcome the limits of optical lithography. Furthermore, nanophotonic devices, also DNA-metal nanoparticles assembled on DNA 2D/3D templates to form metal nanoparticle arrays with controlled plasmonic modes, are promising devices. More efforts should also be devoted to developing structures from nano to micro even macro scale, and use DNA-nanoparticle conjugates as artificial atoms for the construction of finite or infinite, periodic or aperiodic large structures.

Third, for DNA nanomachines, we here consider that the following issues might be the most important challenges for future development. Experimental and theoretical studies on the single DNA nanomachine should be carried out to understand its energy conversion mechanism and entropy exchange with environment. New power supplying methods, which could be easily incorporated into current silicon based nanodevices, should be developed. The reliability of DNA nanomachines and multi-component DNA nanomachines are also expected to be studied. In addition, the construction of dynamic structures for incorporation of multiple nanoparticles and study of the complex, multiple interactions is needed. Most importantly, to continuously and precisely control the spatial distance in nanoscale by DNA nanomachines should be highlighted, which will provide a new platform of mimicking polyvalent interactions *in vitro*.

Finally, in the long term, the cost and quality of DNA is an issue. So far, the DNA for constructing nanostructures and conjugating with nanoparticles is programmed and synthesized. Lowering the cost and synthesis on a large scale would promote the application of DNA-based materials. The quality of DNA structures, such as uniformity, stability and toxicity also need to be improved in future studies.

In summary, with the fast development of DNA-based smart materials for nanoconstructions, the awaited evolution process in this field will be fascinating and we believe DNA will play a more important role in materials science in the coming decade.

References

Acuna, G.P., F.M. Moller, P. Holzmeister, S. Beater, B. Lalkens and P. Tinnefeld. 2012. Fluorescence enhancement at docking sites of DNA-directed self-assembled nanoantennas. Science 338: 506–510.

Agapakis, C.M., P.M. Boyle and P.A. Silver. 2012. Natural strategies for the spatial optimization of metabolism in synthetic biology. Nat. Chem. Biol. 8: 527–535.

Aghebat Rafat, A., T. Pirzer, M.B. Scheible, A. Kostina and F.C. Simmel. 2014. Surface-assisted large-scale ordering of DNA origami tiles. Angew. Chem. Int. Ed. 53: 7665–7668.

Aldaye, F.A. and H.F. Sleiman. 2007. Modular access to structurally switchable 3D discrete DNA assemblies. J. Am. Chem. Soc. 129: 13376–13377.

Andersen, E.S., M. Dong, M.M. Nielsen, K. Jahn, A. Lind-Thomsen, W. Mamdouh, K.V. Gothelf, F. Besenbacher and J. Kjems. 2008a. DNA origami design of dolphin-shaped structures with flexible tails. ACS Nano 2: 1213–1218.

Andersen, E.S., M. Dong, M.M. Nielsen, K. Jahn, R. Subramani, W. Mamdouh, M.M. Golas, B. Sander, H. Stark, C.L. Oliveira, J.S. Pedersen, V. Birkedal, F. Besenbacher, K.V. Gothelf and J. Kjems. 2009. Self-assembly of a nanoscale DNA box with a controllable lid. Nature 459: 73–76.

Andersen, F.F., B. Knudsen, C.L. Oliveira, R.F. Frohlich, D. Kruger, J. Bungert, M. Agbandje-McKenna, R. McKenna, S. Juul, C. Veigaard, J. Koch, J.L. Rubinstein, B. Guldbrandtsen, M.S. Hede, G. Karlsson, A.H. Andersen, J.S. Pedersen and B.R. Knudsen. 2008b. Assembly and structural analysis of a covalently closed nano-scale DNA cage. Nucl. Acids Res. 36: 1113–1139.

Bagalkot, V., O.C. Farokhzad, R. Langer and S. Jon. 2006. An aptamer-doxorubicin physical conjugate as a novel targeted drug-delivery platform. Angew. Chem. Int. Ed. 45: 8149–8152.

Barrow, S.J., X. Wei, J.S. Baldauf, A.M. Funston and P. Mulvaney. 2012. The surface plasmon modes of self-assembled gold nanocrystals. Nat. Commun. 3: 1275.

Bellot, G., M.A. McClintock, C. Lin and W.M. Shih. 2011. Recovery of intact DNA nanostructures after agarose gel-based separation. Nat. Methods 8: 192–194.

Bu, N.N., A. Gao, X.W. He and X.B. Yin. 2013. Electrochemiluminescent biosensor of ATP using tetrahedron structured DNA and a functional oligonucleotide for Ru(phen)3(2+) intercalation and target identification. Biosens Bioelectron. 43: 200–204.

Bu, N.N., C.X. Tang, X.W. He and X.B. Yin. 2011. Tetrahedron-structured DNA and functional oligonucleotide for construction of an electrochemical DNA-based biosensor. Chem. Commun. 47: 7689–7691.

Bui, H., C. Onodera, C. Kidwell, Y. Tan, E. Graugnard, W. Kuang, J. Lee, W.B. Knowlton, B. Yurke and W.L. Hughes. 2010. Programmable periodicity of quantum dot arrays with DNA origami nanotubes. Nano Lett. 10: 3367–3372.

Charoenphol, P. and H. Bermudez. 2014. Aptamer-targeted DNA nanostructures for therapeutic delivery. Mol. Pharm. 11: 1721–1725.

Chen, A.H. and P.A. Silver. 2012. Designing biological compartmentalization. Trends Cell Biol. 22: 662–670.

Chen, J.H. and N.C. Seeman. 1991. Synthesis from DNA of a molecule with the connectivity of a cube. Nature 350: 631–633.

Chen, P., Y. Sun, H. Liu, L. Xu, Q. Fan and D. Liu. 2010. A pH responsive dendron-DNA-protein hybrid supramolecular system. Soft Matter 6: 2143–2145.

Chen, Y., S.-H. Lee and C. Mao. 2004. A DNA nanomachine based on a duplex–triplex transition. Angew. Chem. Int. Ed. 43: 5335–5338.

Cheng, E., Y. Xing, P. Chen, Y. Yang, Y. Sun, D. Zhou, L. Xu, Q. Fan and D. Liu. 2009. A pH-triggered, fast-responding DNA hydrogel. Angew. Chem. Int. Ed. 48: 7660–7663.

Choi, B. and G. Zocchi. 2006. Mimicking cAMP-dependent allosteric control of protein kinase a through mechanical tension. J. Am. Chem. Soc. 128: 8541–8548.

Dadova, J., P. Orsag, R. Pohl, M. Brazdova, M. Fojta and M. Hocek. 2013. Vinylsulfonamide and acrylamide modification of DNA for crosslinking with proteins. Angew. Chem. Int. Ed. 52: 10515–10518.

Dietz, H., S.M. Douglas and W.M. Shih. 2009. Folding DNA into twisted and curved nanoscale shapes. Science 325: 725–730.

Ding, B., Z. Deng, H. Yan, S. Cabrini, R.N. Zuckermann and J. Bokor. 2010. Gold nanoparticle self-similar chain structure organized by DNA origami. J. Am. Chem. Soc. 132: 3248–3249.

Douglas, S., H. Dietz, T. Liedl, B. Hogberg, F. Graf and W. Shih. 2009. Self-assembly of DNA into nanoscale three-dimensional shapes. Nature 459: 414–418.

Douglas, S.M., I. Bachelet and G.M. Church. 2012. A logic-gated nanorobot for targeted transport of molecular payloads. Science 335: 831–834.

Edwardson, T.G.W., K.M.M. Carneiro, C.K. McLaughlin, C.J. Serpell and H.F. Sleiman. 2013. Site-specific positioning of dendritic alkyl chains on DNA cages enables their geometry-dependent self-assembly. Nat. Chem. 5: 868–875.

Elbaz, J., A. Cecconello, Z. Fan, A.O. Govorov and I. Willner. 2013. Powering the programmed nanostructure and function of gold nanoparticles with catenated DNA machines. Nat. Commun. 4.

Endo, M., T. Sugita, Y. Katsuda, K. Hidaka and H. Sugiyama. 2010. Programmed-assembly system using DNA jigsaw pieces. Chem. Eur. J. 16: 5362–5368.

Erben, C.M., R.P. Goodman and A.J. Turberfield. 2006. Single-molecule protein encapsulation in a rigid DNA cage. Angew. Chem. Int. Ed. 45: 7414–7417.

Erben, C.M., R.P. Goodman and A.J. Turberfield. 2007. A self-assembled DNA bipyramid. J. Am. Chem. Soc. 129: 6992–6993.

Fahlman, R.P., M. Hsing, C.S. Sporer-Tuhten and D. Sen. 2003. Duplex pinching: a structural switch suitable for contractile DNA nanoconstructions. Nano Letters 3: 1073–1078.

Farlow, J., D. Seo, K.E. Broaders, M.J. Taylor, Z.J. Gartner and Y.-w. Jun. 2013. Formation of targeted monovalent quantum dots by steric exclusion. Nat. Methods 10: 1203–1205.

Fu, J., M. Liu, Y. Liu, N.W. Woodbury and H. Yan. 2012. Interenzyme substrate diffusion for an enzyme cascade organized on spatially addressable DNA nanostructures. J. Am. Chem. Soc. 134: 5516–5519.

Fu, Y., D. Zeng, J. Chao, Y. Jin, Z. Zhang, H. Liu, D. Li, H. Ma, Q. Huang, K.V. Gothelf and C. Fan. 2013. Single-step rapid assembly of DNA origami nanostructures for addressable nanoscale bioreactors. J. Am. Chem. Soc. 135: 696–702.

Furman, J.L., A.H. Badran, O. Ajulo, J.R. Porter, C.I. Stains, D.J. Segal and I. Ghosh. 2010. Toward a general approach for RNA-templated hierarchical assembly of split-proteins. J. Am. Chem. Soc. 132: 11692–11701.

Ge, Z., H. Pei, L. Wang, S. Song and C. Fan. 2011. Electrochemical single nucleotide polymorphisms genotyping on surface immobilized three-dimensional branched DNA nanostructure. Sci. Chin. Chem. 54: 1273–1276.

Goodman, R.P., R.M. Berry and A.J. Turberfield. 2004. The single-step synthesis of a DNA tetrahedron. Chem. Commun.: 1372–1373.

Goodman, R.P., M. Heilemann, S. Doose, C.M. Erben, A.N. Kapanidis and A.J. Turberfield. 2008. Reconfigurable, braced, three-dimensional DNA nanostructures. Nat. Nanotechnol. 3: 93–96.

Goodman, R.P., I.A. Schaap, C.F. Tardin, C.M. Erben, R.M. Berry, C.F. Schmidt and A.J. Turberfield. 2005. Rapid chiral assembly of rigid DNA building blocks for molecular nanofabrication. Science 310: 1661–1665.

Gramotnev, D.K. and S.I. Bozhevolnyi. 2010. Plasmonics beyond the diffraction limit. Nat. Photonics 4: 83–91.

Gu, H., J. Chao, S.-J. Xiao and N.C. Seeman. 2010. A proximity-based programmable DNA nanoscale assembly line. Nature 465: 202–205.

Han, D., S. Pal, Y. Liu and H. Yan. 2010. Folding and cutting DNA into reconfigurable topological nanostructures. Nat. Nanotechnol. 5: 712–717.

Han, D., S. Pal, J. Nangreave, Z. Deng, Y. Liu and H. Yan. 2011. DNA origami with complex curvatures in three-dimensional space. Science 332: 342–346.

Han, D.R., S. Pal, Y. Yang, S.X. Jiang, J. Nangreave, Y. Liu and H. Yan. 2013. DNA gridiron nanostructures based on four-arm junctions. Science 339: 1412–1415.

He, Y. and C. Mao. 2006. Balancing flexibility and stress in DNA nanostructures. Chem. Commun.: 968–969.

He, Y., T. Ye, M. Su, C. Zhang, A.E. Ribbe, W. Jiang and C. Mao. 2008. Hierarchical self-assembly of DNA into symmetric supramolecular polyhedra. Nature 452: 198–201.

Hemmi, H., O. Takeuchi, T. Kawai, T. Kaisho, S. Sato, H. Sanjo, M. Matsumoto, K. Hoshino, H. Wagner, K. Takeda and S. Akira. 2000. A Toll-like receptor recognizes bacterial DNA. Nature 408: 740–745.

Hermanson, G.T. 2008. Bioconjugate Techniques. 2nd ed. London: Academic Press.

Hogberg, B., T. Liedl and W.M. Shih. 2009. Folding DNA origami from a double-stranded source of scaffold. J. Am. Chem. Soc. 131: 9154–9155.

Ishizawa, T., T. Kawakami, P.C. Reid and H. Murakami. 2013. TRAP display: a high-speed selection method for the generation of functional polypeptides. J. Am. Chem. Soc. 135: 5433–5440.

Jiang, Q., C. Song, J. Nangreave, X. Liu, L. Lin, D. Qiu, Z.G. Wang, G. Zou, X. Liang, H. Yan and B. Ding. 2012. DNA origami as a carrier for circumvention of drug resistance. J. Am. Chem. Soc. 134: 13396–13403.

Jin, J., Y. Xing, Y. Xi, X. Liu, T. Zhou, X. Ma, Z. Yang, S. Wang and D. Liu. 2013. A triggered DNA hydrogel cover to envelop and release single cells. Adv. Mater. 25: 4714–4717.

Jones, M.R., N.C. Seeman and C.A. Mirkin. 2015. Nanomaterials. Programmable materials and the nature of the DNA bond. Science 347: 1260901.

Jungmann, R., T. Liedl, T.L. Sobey, W. Shih and F.C. Simmel. 2008. Isothermal assembly of DNA origami structures using denaturing agents. J. Am. Chem. Soc. 130: 10062–10063.

Kühler, P., E.-M. Roller, R. Schreiber, T. Liedl, T. Lohmüller and J. Feldmann. 2014. Plasmonic DNA-origami nanoantennas for surface-enhanced raman spectroscopy. Nano Letters 14: 2914–2919.

Ke, Y., S.M. Douglas, M. Liu, J. Sharma, A. Cheng, A. Leung, Y. Liu, W.M. Shih and H. Yan. 2009. Multilayer DNA origami packed on a square lattice. J. Am. Chem. Soc. 131: 15903–15908.

Ke, Y.G., L.L. Ong, W. Sun, J. Song, M.D. Dong, W.M. Shih and P. Yin. 2014. DNA brick crystals with prescribed depths. Nat. Chem. 6: 994–1002.

Kedracki, D., I. Safir, N. Gour, K.X. Ngo and C. Vebert-Nardin. 2013. DNA-polymer conjugates: from synthesis, through complex formation and self-assembly to applications. *In*: Bio-Synthetic Polymer Conjugates, edited by H. Schlaad.

Kim, J.-W., J.-H. Kim and R. Deaton. 2011. DNA-linked nanoparticle building blocks for programmable matter. Angew. Chem. Int. Ed. 50: 9185–9190.

Klein, W.P., C.N. Schmidt, B. Rapp, S. Takabayashi, W.B. Knowlton, J. Lee, B. Yurke, W.L. Hughes, E. Graugnard and W. Kuang. 2013. Multiscaffold DNA origami nanoparticle waveguides. Nano Lett. 13: 3850–3856.

Ko, S.H., G.M. Gallatin and J.A. Liddle. 2012. Nanomanufacturing with DNA origami: factors affecting the kinetics and yield of quantum dot binding. Adv. Funct. Mater. 22: 1015–1023.

Kuzuya, A., M. Kimura, K. Numajiri, N. Koshi, T. Ohnishi, F. Okada and M. Komiyama. 2009. Precisely programmed and robust 2D streptavidin nanoarrays by using periodical nanometer-scale wells embedded in DNA origami assembly. Chembiochem. 10: 1811–1815.

Kuzuya, A. and M. Komiyama. 2009. Design and construction of a box-shaped 3D-DNA origami. Chem. Commun.: 4182–4184.

Kuzyk, A., K.T. Laitinen and P. Torma. 2009. DNA origami as a nanoscale template for protein assembly. Nanotechnology 20: 235305.

Kuzyk, A., R. Schreiber, Z. Fan, G. Pardatscher, E.M. Roller, A. Hogele, F.C. Simmel, A.O. Govorov and T. Liedl. 2012. DNA-based self-assembly of chiral plasmonic nanostructures with tailored optical response. Nature 483: 311–314.

Kuzyk, A., R. Schreiber, H. Zhang, A.O. Govorov, T. Liedl and N. Liu. 2014. Reconfigurable 3D plasmonic metamolecules. Nat. Mater. 13: 862–866.

Kwak, M. and A. Herrmann. 2011. Nucleic acid amphiphiles: synthesis and self-assembled nanostructures. Chem. Soc. Rev. 40: 5745–5755.

Lan, X., Z. Chen, G. Dai, X. Lu, W. Ni and Q. Wang. 2013. Bifacial DNA origami-directed discrete, three-dimensional, anisotropic plasmonic nanoarchitectures with tailored optical chirality. J. Am. Chem. Soc. 135: 11441–11444.

Lapiene, V., F. Kukolka, K. Kiko, A. Arndt and C.M. Niemeyer. 2010. Conjugation of fluorescent proteins with DNA oligonucleotides. Bioconjugate Chem. 21: 921–927.

Lee, H., W.C. DeLoache and J.E. Dueber. 2012a. Spatial organization of enzymes for metabolic engineering. Metab. Eng. 14: 242–251.

Lee, H., A.K. Lytton-Jean, Y. Chen, K.T. Love, A.I. Park, E.D. Karagiannis, A. Sehgal, W. Querbes, C.S. Zurenko, M. Jayaraman, C.G. Peng, K. Charisse, A. Borodovsky, M. Manoharan, J.S. Donahoe, J. Truelove, M. Nahrendorf, R. Langer and D.G. Anderson. 2012b. Molecularly self-assembled nucleic acid nanoparticles for targeted *in vivo* siRNA delivery. Nat. Nanotechnol. 7: 389–393.

Lee, J.B., S. Peng, D. Yang, Y.H. Roh, H. Funabashi, N. Park, E.J. Rice, L. Chen, R. Long, M. Wu and D. Luo. 2012c. A mechanical metamaterial made from a DNA hydrogel. Nat. Nanotechnol. 7: 816–820.

Leitner, M., N. Mitchell, M. Kastner, R. Schlapak, H.J. Gruber, P. Hinterdorfer, S. Howorka and A. Ebner. 2011. Single-molecule AFM characterization of individual chemically tagged DNA tetrahedra. ACS Nano 5: 7048–7054.

Li, J., H. Pei, B. Zhu, L. Liang, M. Wei, Y. He, N. Chen, D. Li, Q. Huang and C. Fan. 2011a. Self-assembled multivalent DNA nanostructures for noninvasive intracellular delivery of immunostimulatory CpG oligonucleotides. ACS Nano 5: 8783–8789.

Li, J.J. and W. Tan. 2002. A single DNA molecule nanomotor. Nano Letters 2: 315–318.

Li, Z., E. Cheng, W. Huang, T. Zhang, Z. Yang, D. Liu and Z. Tang. 2011b. Improving the yield of Mono-DNA-Functionalized gold nanoparticles through dual steric hindrance. J. Am. Chem. Soc. 133: 15284–15287.

Liu, D. and S. Balasubramanian. 2003. A proton-fuelled DNA nanomachine. Angew. Chem. Int. Ed. 42: 5734–5736.

Liu, D., E. Cheng and Z. Yang. 2011. DNA-based switchable devices and materials. NPG Asia Mater. 3: 109–114.

Liu, H., T. Torring, M. Dong, C.B. Rosen, F. Besenbacher and K.V. Gothelf. 2010. DNA-templated covalent coupling of G4 PAMAM dendrimers. J. Am. Chem. Soc. 132: 18054–18056.

Liu, H., Y. Xu, F. Li, Y. Yang, W. Wang, Y. Song and D. Liu. 2007. Light-driven conformational switch of i-Motif DNA. Angew. Chem. Int. Ed. 46: 2515–2517.

Liu, M., J. Fu, C. Hejesen, Y. Yang, N.W. Woodbury, K. Gothelf, Y. Liu and H. Yan. 2013. A DNA tweezer-actuated enzyme nanoreactor. Nat. Commun. 4.

Liu, X., Y. Xu, T. Yu, C. Clifford, Y. Liu, H. Yan and Y. Chang. 2012. A DNA nanostructure platform for directed assembly of synthetic vaccines. Nano Lett. 12: 4254–4259.

Loweth, C.J., W.B. Caldwell, X. Peng, A.P. Alivisatos and P.G. Schultz. 1999. DNA-based assembly of gold nanocrystals. Angew. Chem. Int. Ed. 38: 1808–1812.

Lu, N., H. Pei, Z. Ge, C.R. Simmons, H. Yan and C. Fan. 2012. Charge transport within a three-dimensional DNA nanostructure framework. J. Am. Chem. Soc. 134: 13148–13151.

Lund, K., A.J. Manzo, N. Dabby, N. Michelotti, A. Johnson-Buck, J. Nangreave, S. Taylor, R. Pei, M.N. Stojanovic, N.G. Walter, E. Winfree and H. Yan. 2010. Molecular robots guided by prescriptive landscapes. Nature 465: 206–210.

Mammen, M., S.-K. Choi and G.M. Whitesides. 1998. Polyvalent interactions in biological systems: implications for design and use of multivalent ligands and inhibitors. Angew. Chem. Int. Ed. 37: 2754–2794.

Mao, C., W. Sun, Z. Shen and N.C. Seeman. 1999. A nanomechanical device based on the B-Z transition of DNA. Nature 397: 144–146.

Marchi, A.N., I. Saaem, B.N. Vogen, S. Brown and T.H. LaBean. 2014. Toward larger DNA origami. Nano Lett. 14: 5740–5747.

Martin, T.G. and H. Dietz. 2012. Magnesium-free self-assembly of multi-layer DNA objects. Nat. Commun. 3: 1103.

Mashimo, Y., H. Maeda, M. Mie and E. Kobatake. 2012. Construction of semisynthetic DNA-protein conjugates with Phi X174 Gene-A* protein. Bioconjugate Chem. 23: 1349–1355.

Maune, H.T., S.P. Han, R.D. Barish, M. Bockrath, W.A. Iii, P.W. Rothemund and E. Winfree. 2010. Self-assembly of carbon nanotubes into two-dimensional geometries using DNA origami templates. Nat. Nanotechnol. 5: 61–66.

Meng, H., Y. Yang, Y. Chen, Y. Zhou, Y. Liu, X.a. Chen, H. Ma, Z. Tang, D. Liu and L. Jiang. 2009. Photoelectric conversion switch based on quantum dots with i-motif DNA scaffolds. Chem. Commun.: 2293–2295.

Mirkin, C.A., R.L. Letsinger, R.C. Mucic and J.J. Storhoff. 1996. A DNA-based method for rationally assembling nanoparticles into macroscopic materials. Nature 382: 607–609.

Miyoshi, D., H. Karimata, Z.-M. Wang, K. Koumoto and N. Sugimoto. 2007. Artificial G-Wire Switch with 2,2'-Bipyridine units responsive to divalent metal ions. J. Am. Chem. Soc. 129: 5919–5925.

Monchaud, D., P. Yang, L. Lacroix, M.-P. Teulade-Fichou and J.-L. Mergny. 2008. A metal-mediated conformational switch controls G-quadruplex binding affinity. Angew. Chem. Int. Ed. 47: 4858–4861.

Nakata, E., F.F. Liew, C. Uwatoko, S. Kiyonaka, Y. Mori, Y. Katsuda, M. Endo, H. Sugiyama and T. Morii. 2012. Zinc-finger proteins for site-specific protein positioning on DNA-origami structures. Angew. Chem. Int. Ed. 51: 2421–2424.

Niemeyer, C.M. 2010. Semisynthetic DNA–Protein Conjugates for biosensing and nanofabrication. Angew. Chem. Int. Ed. 49: 1200–1216.

Ouyang, X., J. Li, H. Liu, B. Zhao, J. Yan, Y. Ma, S. Xiao, S. Song, Q. Huang, J. Chao and C. Fan. 2013. Rolling circle amplification-based DNA origami nanostructrures for intracellular delivery of immunostimulatory drugs. Small 9: 3082–3087.

Pal, S., Z. Deng, B. Ding, H. Yan and Y. Liu. 2010. DNA-origami-directed self-assembly of discrete silver-nanoparticle architectures. Angew. Chem. Int. Ed. 49: 2700–2704.

Pal, S., Z. Deng, H. Wang, S. Zou, Y. Liu and H. Yan. 2011. DNA directed self-assembly of anisotropic plasmonic nanostructures. J. Am. Chem. Soc. 133: 17606–17609.

Pal, S., P. Dutta, H. Wang, Z. Deng, S. Zou, H. Yan and Y. Liu. 2013. Quantum efficiency modification of organic fluorophores using gold nanoparticles on DNA origami scaffolds. J. Phys. Chem. C 117: 12735–12744.

Park, N., S.H. Um, H. Funabashi, J. Xu and D. Luo. 2009. A cell-free protein-producing gel. Nat. Mater. 8: 432–437.

Pei, H., F. Li, Y. Wan, M. Wei, H. Liu, Y. Su, N. Chen, Q. Huang and C. Fan. 2012a. Designed diblock oligonucleotide for the synthesis of spatially isolated and highly hybridizable functionalization of DNA-gold nanoparticle nanoconjugates. J. Am. Chem. Soc. 134: 11876–11879.

Pei, H., L. Liang, G. Yao, J. Li, Q. Huang and C. Fan. 2012b. Reconfigurable three-dimensional DNA nanostructures for the construction of intracellular logic sensors. Angew. Chem. Int. Ed. 51: 9020–9024.

Pei, H., N. Lu, Y. Wen, S. Song, Y. Liu, H. Yan and C. Fan. 2010. A DNA nanostructure-based biomolecular probe carrier platform for electrochemical biosensing. Adv. Mater. 22: 4754–4758.

Pei, H., Y. Wan, J. Li, H. Hu, Y. Su, Q. Huang and C. Fan. 2011. Regenerable electrochemical immunological sensing at DNA nanostructure-decorated gold surfaces. Chem. Commun. 47: 6254–6256.

Pilo-Pais, M., A. Watson, S. Demers, T.H. LaBean and G. Finkelstein. 2014. Surface-enhanced raman scattering plasmonic enhancement using DNA origami-based complex metallic nanostructures. Nano Letters 14: 2099–2104.

Qian, L., Y. Wang, Z. Zhang, J. Zhao, D. Pan, Y. Zhang, Q. Liu, C. Fan, J. Hu and L. He. 2006. Analogic China map constructed by DNA. Chin. Sci. Bull. 51: 2973–2976.

Röglin, L., M.R. Ahmadian and O. Seitz. 2007. DNA-controlled reversible switching of peptide conformation and bioactivity. Angew. Chem. Int. Ed. 46: 2704–2707.

Raindlova, V., R. Pohl and M. Hocek. 2012. Synthesis of aldehyde-linked nucleotides and DNA and their bioconjugations with lysine and peptides through reductive amination. Chem. Eur. J. 18: 4080–4087.

Rangnekar, A. and T.H. LaBean. 2014. Building DNA nanostructures for molecular computation, templated assembly, and biological applications. Acc. Chem. Res. 47: 1778–1788.

Rashidian, M., J.K. Dozier and M.D. Distefano. 2013. Enzymatic labeling of proteins: techniques and approaches. Bioconjugate Chem. 24: 1277–1294.

Roh, Y.H., R.C.H. Ruiz, S. Peng, J.B. Lee and D. Luo. 2011. Engineering DNA-based functional materials. Chem. Soc. Rev. 40: 5730–5744.

Rothemund, P.W.K. 2006. Folding DNA to create nanoscale shapes and patterns. Nature 440: 297–302.

Rusling, D.A., A.R. Chandrasekaran, Y.P. Ohayon, T. Brown, K.R. Fox, R. Sha, C. Mao and N.C. Seeman. 2014. Functionalizing designer DNA crystals with a triple-helical veneer. Angew. Chem. Int. Ed. 53: 3979–3982.

Rycenga, M., C.M. Cobley, J. Zeng, W. Li, C.H. Moran, Q. Zhang, D. Qin and Y. Xia. 2011. Controlling the synthesis and assembly of silver nanostructures for plasmonic applications. Chem. Rev. 111: 3669–3712.

Saccà, B. and C.M. Niemeyer. 2012. DNA origami: the art of folding DNA. Angew. Chem. Int. Ed. 51: 58–66.

Saccà, B., R. Meyer, M. Erkelenz, K. Kiko, A. Arndt, H. Schroeder, K.S. Rabe and C.M. Niemeyer. 2010. Orthogonal protein decoration of DNA origami. Angew. Chem. Int. Ed. 49: 9378–9383.

Sacca, B. and C.M. Niemeyer. 2011. Functionalization of DNA nanostructures with proteins. Chem. Soc. Rev. 40: 5910–5921.

Saha, K., S.S. Agasti, C. Kim, X. Li and V.M. Rotello. 2012. Gold nanoparticles in chemical and biological sensing. Chem. Rev. 112: 2739–2779.

Sanderson, K. 2010. What to make with DNA origami. Nature 464: 158–189.

Schoffelen, S. and J.C. van Hest. 2013. Chemical approaches for the construction of multi-enzyme reaction systems. Curr. Opin. Struct. Biol. 23: 613–21.

Schreiber, R., J. Do, E.M. Roller, T. Zhang, V.J. Schuller, P.C. Nickels, J. Feldmann and T. Liedl. 2014. Hierarchical assembly of metal nanoparticles, quantum dots and organic dyes using DNA origami scaffolds. Nat. Nanotechnol. 9: 74–78.

Schreiber, R., N. Luong, Z. Fan, A. Kuzyk, P.C. Nickels, T. Zhang, D.M. Smith, B. Yurke, W. Kuang, A.O. Govorov and T. Liedl. 2013. Chiral plasmonic DNA nanostructures with switchable circular dichroism. Nat. Commun. 4: 2948.

Schuller, V.J., S. Heidegger, N. Sandholzer, P.C. Nickels, N.A. Suhartha, S. Endres, C. Bourquin and T. Liedl. 2011. Cellular immunostimulation by CpG-sequence-coated DNA origami structures. ACS Nano 5: 9696–9702.

Seeman, N.C. 1982. Nucleic acid junctions and lattices. J. Theor. Biol. 99: 237–247.

Seeman, N.C. 2001. DNA nicks and nodes and nanotechnology. Nano Letters 1: 22–26.

Seeman, N.C. 2003. DNA in a material world. Nature 421: 427–431.

Seeman, N.C. 2010. Nanomaterials based on DNA. Annu. Rev. Biochem. 79: 65–87.

Sharma, J., R. Chhabra, C.S. Andersen, K.V. Gothelf, H. Yan and Y. Liu. 2008. Toward reliable gold nanoparticle patterning on self-assembled DNA nanoscaffold. J. Am. Chem. Soc. 130: 7820–7821.

Sharma, J., R. Chhabra, H. Yan and Y. Liu. 2007. pH-driven conformational switch of "i-motif" DNA for the reversible assembly of gold nanoparticles. Chem. Commun.: 477–479.

Shen, W., H. Zhong, D. Neff and M.L. Norton. 2009. NTA directed protein nanopatterning on DNA Origami nanoconstructs. J. Am. Chem. Soc. 131: 6660–6661.

Shen, X., A. Asenjo-Garcia, Q. Liu, Q. Jiang, F.J. Garcia de Abajo, N. Liu and B. Ding. 2013. Three-dimensional plasmonic chiral tetramers assembled by DNA origami. Nano Lett. 13: 2128–2133.

Shen, X., C. Song, J. Wang, D. Shi, Z. Wang, N. Liu and B. Ding. 2012. Rolling up gold nanoparticle-dressed DNA origami into three-dimensional plasmonic chiral nanostructures. J. Am. Chem. Soc. 134: 146–149.

Simmel, F.C. 2008. Three-dimensional nanoconstruction with DNA. Angew. Chem. Int. Ed. 47: 5884–5887.

Sletten, E.M. and C.R. Bertozzi. 2009. Bioorthogonal chemistry: fishing for selectivity in a sea of functionality. Angew. Chem. Int. Ed. 48: 6974–6998.

Song, T. and H. Liang. 2012. Synchronized assembly of gold nanoparticles driven by a dynamic DNA-fueled molecular machine. J. Am. Chem. Soc. 134: 10803–10806.

Sorensen, R.S., A.H. Okholm, D. Schaffert, A.L.B. Kodal, K.V. Gothelf and J. Kjems. 2013. Enzymatic ligation of large biomolecules to DNA. ACS Nano 7: 8098–8104.

Stahl, E., T.G. Martin, F. Praetorius and H. Dietz. 2014. Facile and scalable preparation of pure and dense DNA origami solutions. Angew. Chem. Int. Ed. 53: 12735–12740.

Stearns, L.A., R. Chhabra, J. Sharma, Y. Liu, W.T. Petuskey, H. Yan and J.C. Chaput. 2009. Template-directed nucleation and growth of inorganic nanoparticles on DNA scaffolds. Angew. Chem. Int. Ed. 48: 8494–8496.

Stephanopoulos, N., M. Liu, G.J. Tong, Z. Li, Y. Liu, H. Yan and M.B. Francis. 2010. Immobilization and one-dimensional arrangement of virus capsids with nanoscale precision using DNA origami. Nano Lett. 10: 2714–2720.

Sun, Y., H. Liu, L. Xu, L. Wang, Q.-H. Fan and D. Liu. 2010. DNA-molecular-motor-controlled dendron association. Langmuir 26: 12496–12499.

Sunbul, M. and J. Yin. 2009. Site specific protein labeling by enzymatic posttranslational modification. Org. Biomol. Chem. 7: 3361–3371.

Suzuki, K., K. Hosokawa and M. Maeda. 2009. Controlling the number and positions of oligonucleotides on gold nanoparticle surfaces. J. Am. Chem. Soc. 131: 7518–7519.

Tao, S.-C. and H. Zhu. 2006. Protein chip fabrication by capture of nascent polypeptides. Nat. Biotechnol. 24: 1253–1254.

Thacker, V.V., L.O. Herrmann, D.O. Sigle, T. Zhang, T. Liedl, J.J. Baumberg and U.F. Keyser. 2014. DNA origami based assembly of gold nanoparticle dimers for surface-enhanced Raman scattering. Nat. Commun. 5: 3448.

Tikhomirov, G., S. Hoogland, P.E. Lee, A. Fischer, E.H. Sargent and S.O. Kelley. 2011. DNA-based programming of quantum dot valency, self-assembly and luminescence. Nat. Nanotechnol. 6: 485–490.

Uddayasankar, U., Z. Zhang, R.T. Shergill, C.C. Gradinaru and U.J. Krull. 2014. Isolation of Monovalent Quantum Dot–Nucleic acid conjugates using magnetic beads. Bioconjugate Chem. 25: 1342–1350.

Um, S.H., J.B. Lee, N. Park, S.Y. Kwon, C.C. Umbach and D. Luo. 2006. Enzyme-catalysed assembly of DNA hydrogel. Nat. Mater. 5: 797–801.

Voigt, N.V., T. Torring, A. Rotaru, M.F. Jacobsen, J.B. Ravnsbaek, R. Subramani, W. Mamdouh, J. Kjems, A. Mokhir, F. Besenbacher and K.V. Gothelf. 2010. Single-molecule chemical reactions on DNA origami. Nat. Nanotechnol. 5: 200–203.

Walsh, A.S., H. Yin, C.M. Erben, M.J.A. Wood and A. Turberfield. 2011. DNA cage delivery to mammalian cells. ACS Nano 5: 5427–5432.

Wang, L., Y. Feng, Y. Sun, Z. Li, Z. Yang, Y.-M. He, Q.-H. Fan and D. Liu. 2011. Amphiphilic DNA-dendron hybrid: a new building block for functional assemblies. Soft Matter 7: 7187–7190.

Wang, R., C. Nuckolls and S.J. Wind. 2012. Assembly of heterogeneous functional nanomaterials on DNA origami scaffolds. Angew. Chem. Int. Ed. 51: 11325–11327.

Wang, W., H. Liu, D. Liu, Y. Xu, Y. Yang and D. Zhou. 2007. Use of the interparticle i-Motif for the controlled assembly of gold nanoparticles. Langmuir 23: 11956–11959.

Wang, X., J. Huang, Y. Zhou, S. Yan, X. Weng, X. Wu, M. Deng and X. Zhou. 2010. Conformational switching of G-Quadruplex DNA by photoregulation. Angew. Chem. Int. Ed. 49: 5305–5309.

Wei, B., M. Dai, C. Myhrvold, Y. Ke, R. Jungmann and P. Yin. 2013. Design space for complex DNA structures. J. Am. Chem. Soc. 135: 18080–18088.

Wei, B., M. Dai and P. Yin. 2012. Complex shapes self-assembled from single-stranded DNA tiles. Nature 485: 623–626.

Wei, B., L.L. Ong, J. Chen, A.S. Jaffe and P. Yin. 2014. Complex reconfiguration of DNA nanostructures. Angew. Chem. Int. Ed. 53: 7475–7479.

Wen, Y., H. Pei, Y. Shen, J. Xi, M. Lin, N. Lu, X. Shen, J. Li and C. Fan. 2012. DNA Nanostructure-based interfacial engineering for PCR-free ultrasensitive electrochemical analysis of microRNA. Sci. Rep. 2: 867.

Wen, Y., H. Pei, Y. Wan, Y. Su, Q. Huang, S. Song and C. Fan. 2011. DNA nanostructure-decorated surfaces for enhanced aptamer-target binding and electrochemical cocaine sensors. Anal. Chem. 83: 7418–7423.

Wickham, S.F.J., J. Bath, Y. Katsuda, M. Endo, K. Hidaka, H. Sugiyama and A.J. Turberfield. 2012. A DNA-based molecular motor that can navigate a network of tracks. Nat. Nanotechnol. 7: 169–173.

Wickham, S.F.J., M. Endo, Y. Katsuda, K. Hidaka, J. Bath, H. Sugiyama and A.J. Turberfield. 2011. Direct observation of stepwise movement of a synthetic molecular transporter. Nat. Nanotechnol. 6: 166–169.

Williams, B.A.R., C.W. Diehnelt, P. Belcher, M. Greving, N.W. Woodbury, S.A. Johnston and J.C. Chaput. 2009. Creating protein affinity reagents by combining peptide ligands on synthetic DNA scaffolds. J. Am. Chem. Soc. 131: 17233–17241.

Winfree, E., F. Liu, L.A. Wenzler and N.C. Seeman. 1998. Design and self-assembly of two-dimensional DNA crystals. Nature 394: 539–544.

Woo, S. and P.W. Rothemund. 2014. Self-assembly of two-dimensional DNA origami lattices using cation-controlled surface diffusion. Nat. Commun. 5: 4889.

Xiao, Z., C. Ji, J. Shi, E.M. Pridgen, J. Frieder, J. Wu and O.C. Farokhzad. 2012. DNA self-assembly of targeted near-infrared-responsive gold nanoparticles for cancer thermo-chemotherapy. Angew. Chem. Int. Ed. 51: 11853–11857.

Xin, L., C. Zhou, Z. Yang and D. Liu. 2013. Regulation of an enzyme cascade reaction by a DNA machine. Small 9: 3088–3091.

Yan, H., S.H. Park, G. Finkelstein, J.H. Reif and T.H. LaBean. 2003. DNA-templated self-assembly of protein arrays and highly conductive nanowires. Science 301: 1882–1884.

Yan, H., X. Zhang, Z. Shen and N.C. Seeman. 2002. A robust DNA mechanical device controlled by hybridization topology. Nature 415: 62–65.

Yang, Y., D. Han, J. Nangreave, Y. Liu and H. Yan. 2012. DNA origami with double-stranded DNA as a unified scaffold. ACS Nano 6: 8209–8215.

Yang, Y., G. Liu, H. Liu, D. Li, C. Fan and D. Liu. 2010. An electrochemically actuated reversible DNA switch. Nano Letters 10: 1393–1397.

Yao, G., J. Li, J. Chao, H. Pei, H. Liu, Y. Zhao, J. Shi, Q. Huang, L. Wang, W. Huang and C. Fan. 2015a. Gold-nanoparticle-mediated jigsaw-puzzle-like assembly of supersized plasmonic DNA origami. Angew. Chem. Int. Ed. 54: 2966–2969.

Yao, G.B., H. Pei, J. Li, Y. Zhao, D. Zhu, Y.N. Zhang, Y.F. Lin, Q. Huang and C.H. Fan. 2015b. Clicking DNA to gold nanoparticles: poly-adenine-mediated formation of monovalent DNA-gold nanoparticle conjugates with nearly quantitative yield. NPG Asia Mater. 7: e159.

Yurke, B., A.J. Turberfield, A.P. Mills, F.C. Simmel and J.L. Neumann. 2000. A DNA-fuelled molecular machine made of DNA. Nature 406: 605–608.

Zanchet, D., C.M. Micheel, W.J. Parak, D. Gerion and A.P. Alivisatos. 2001. Electrophoretic isolation of discrete Au nanocrystal/DNA conjugates. Nano Letters 1: 32–35.

Zhang, C., S.H. Ko, M. Su, Y. Leng, A.E. Ribbe, W. Jiang and C. Mao. 2009. Symmetry controls the face geometry of DNA polyhedra. J. Am. Chem. Soc. 131: 1413–1415.

Zhang, C., M. Su, Y. He, X. Zhao, P.A. Fang, A.E. Ribbe, W. Jiang and C. Mao. 2008. Conformational flexibility facilitates self-assembly of complex DNA nanostructures. Proc. Natl. Acad. Sci. USA 105: 10665–10669.

Zhang, C., C. Tian, X. Li, H. Qian, C. Hao, W. Jiang and C. Mao. 2012a. Reversibly switching the surface porosity of a DNA tetrahedron. J. Am. Chem. Soc. 134: 11998–12001.

Zhang, H., J. Chao, D. Pan, H. Liu, Q. Huang and C. Fan. 2012b. Folding super-sized DNA origami with scaffold strands from long-range PCR. Chem. Commun. 48: 6405–6407.

Zhang, T., P. Chen, Y. Sun, Y. Xing, Y. Yang, Y. Dong, L. Xu, Z. Yang and D. Liu. 2011a. A new strategy improves assembly efficiency of DNA mono-modified gold nanoparticles. Chem. Commun. 47: 5774–5776.

Zhang, T., Y. Dong, Y. Sun, P. Chen, Y. Yang, C. Zhou, L. Xu, Z. Yang and D. Liu. 2011b. DNA bimodified gold nanoparticles. Langmuir 28: 1966–1970.

Zhang, T., Z. Yang and D. Liu. 2011c. DNA discrete modified gold nanoparticles. Nanoscale 3: 4015–4021.

Zhang, Y.W. and N.C. Seeman. 1994. Construction of a DNA-truncated octahedron. J. Am. Chem. Soc. 116: 1661–1669.

Zhang, Z., J. Song, F. Besenbacher, M. Dong and K.V. Gothelf. 2013. Self-assembly of DNA origami and single-stranded tile structures at room temperature. Angew. Chem. Int. Ed. 52: 9219–9223.

Zhao, Y.X., A. Shaw, X. Zeng, E. Benson, A.M. Nystrom and B. Hogberg. 2012. DNA origami delivery system for cancer therapy with tunable release properties. ACS Nano 6: 8684–8691.

Zhao, Z., Y. Liu and H. Yan. 2013. DNA origami templated self-assembly of discrete length single wall carbon nanotubes. Org. Biomol. Chem. 11: 596–598.

Zhao, Z., H. Yan and Y. Liu. 2010. A route to scale up DNA origami using DNA tiles as folding staples. Angew. Chem. Int. Ed. 49: 1414–1417.

Zheng, J., J.J. Birktoft, Y. Chen, T. Wang, R. Sha, P.E. Constantinou, S.L. Ginell, C. Mao and N.C. Seeman. 2009. From molecular to macroscopic via the rational design of a self-assembled 3D DNA crystal. Nature 461: 74–77.

Zheng, Y., Y. Li and Z. Deng. 2012. Silver nanoparticle-DNA bionanoconjugates bearing a discrete number of DNA ligands. Chem. Commun. 48: 6160–6162.

Zhou, C., Z. Yang and D. Liu. 2012. Reversible regulation of protein binding affinity by a DNA machine. J. Am. Chem. Soc. 134: 1416–1418.

Zimmermann, J., M.P. Cebulla, S. Monninghoff and G. von Kiedrowski. 2008. Self-assembly of a DNA dodecahedron from 20 trisoligonucleotides with C(3h) linkers. Angew. Chem. Int. Ed. 47: 3626–3630.

3

Smart Materials for Protein-Based Nanoconstructions

Youdong Mao

ABSTRACT

The self-assembly of protein molecules into nanoscale supramolecular architectures or sophisticated nanomachines is a hallmark of biological systems. In living cells, proteins are the major components that execute intracellular functions and mediate intercellular communications. Proteins are essentially natural nanomaterials to maintain the living states of cells, tissues and organisms. The biochemical functions of these molecular nanomaterials have inspired methodological development and engineering of self-assembled protein nanostructures and nanodevices. Proteins represent an intriguing source of building blocks for smart material engineering on the nanoscale. Biomaterials made of these self-assembled protein nanoconstructions have a variety of applications in life science and biotechnology, such as bio-inspired nanodevice fabrications and applications, lab-on-a-chip, molecular diagnosis, biomolecular sensing, tissue engineering, regenerative medicine

Center for Quantitative Biology, School of Physics, Peking University, Beijing, China; Intel Parallel Computing Center for Structural Biology, Dana-Farber Cancer Institute, Boston, USA; and Department of Microbiology and Immunobiology, Harvard Medical School, Boston, USA. Email: youdong_mao@dfci.harvard.edu and immunotherapy, etc. Many protein complexes are natural nanodevices and nanomachines that exhibit amazing properties, such as the rotary protein motors that rotate with a speed of 100 cycles per second, the locomotive protein motors that walk bipedally along a microtubule track, and enzymatic proteins that make ratchet-like motions.

1. Introduction

The self-assembly of protein molecules into nanoscale supramolecular architectures or sophisticated nanomachines is a hallmark of biological systems (Alberts et al. 2007, Lewin et al. 2007). In living cells, proteins are the major components that execute intracellular functions and mediate intercellular communications (Lewin et al. 2007). Proteins are essentially natural nanomaterials to maintain the living states of cells, tissues and organisms. Proteins represent an intriguing source of building blocks for material engineering on the nanoscale. The biochemical functions of these molecular nanomaterials have inspired the methodological development and engineering of self-assembled protein nanostructures and nanodevices. Biomaterials made of these self-assembled protein nanoconstructions have a variety of applications in life science and biotechnology, such as bio-inspired nanodevice fabrications and applications, lab-on-a-chip, molecular diagnosis, biomolecular sensing, tissue engineering, regenerative medicine and immunotherapy, etc. Many protein complexes are natural nanodevices and nanomachines that exhibit amazing properties, such as the rotary protein motors that rotate with a speed of 100 cycles per second, the locomotive protein motors that walk bipedally along a microtubule track, and enzymatic proteins that make ratchet-like motions.

Proteins are large biological macromolecules consisting of one or more long chains of amino acid residues. In an aqueous solution, a linear chain of amino acid residues, called polypeptide, folds into a specific three-dimensional structure that determines its biological chemical activity. Through these structures, proteins execute a myriad of biological functions within living organisms, including catalyzing metabolic reactions, transforming energy, processing genetic information, responding to environmental stimuli and transporting small molecules or ions between cellular compartments. The execution of biological functions and activities of proteins are often accompanied by structural changes or conformational transitions in three-dimensional protein folds. The sequence of amino acid residues in a protein is defined by the sequence of its corresponding gene, which is encoded in the genetic code (Nelson 2008). The genetic code specifies 20 standard amino acids, including six non-polar hydrophobic residues, i.e., glycine (Gly, G), alanine (Ala, A), valine (Val, V), leucine (Leu, L), isoleucine (Ile, I), proline (Pro, P), three aromatic residues, i.e., phenylalanine (Phe, F), tyrosine (Tyr, Y), tryptophan (Trp, W), six polar residues, i.e., serine (Ser, S), threonine (Thr, T), cysteine (Cys, C), methionine (Met, M), asparagine (Asn, N), glutamine (Gln, Q), three positively charged residues, i.e., lysine (Lys, K), arginine (Arg, R), histidine (His, H) and two negatively charged residues, i.e., aspartate (Asp, D) and glutamate (Glu, E). The genetic code can include non-standard residues, such as selenocysteine in certain organisms and pyrrolysine in certain archaea. The genetic code is a set of three-nucleotide sets called codons and each three-nucleotide combination designates an amino acid; for example, AUG codes for methionine. As DNA consists of four types of nucleotides (A, T, C and G), the total number of possible codons is 64; there is some redundancy in the genetic code, with some amino acids specified by more than one codon. During protein synthesis, a ribosome produces a protein chain using messenger RNA (mRNA), transcribed from a gene, as a template. After synthesis, the amino acid residues may be chemically modified by posttranslational modification that extends the range of functions of the proteins. The posttranslational modification may covalently add other biochemical functional groups to specific amino acid residues, such as acetate, phosphate, lipids and carbohydrates, which changes the chemical nature of an amino acid, e.g., citrullination, as well as makes structural modifications such as formation of disulfide bridges. This alters the physical and chemical properties, folding, stability, activity, and ultimately, the function of the proteins. Some proteins are modified with non-peptide groups, which can be prosthetic groups or cofactors (Nelson 2008). Individual proteins can often work together to achieve a particular function; they often associate to form stable protein complexes or spontaneously assemble into large integrative particles, such as viral capsid particles (Johnson and Speir 2009).

The shape into which a protein naturally folds in living organisms represents its native conformation. Many proteins can fold themselves into native conformations in appropriate physiological conditions, driven by intramolecular hydrophobic interactions among the seven hydrophobic residues (alanine, isoleucine, leucine, phenylalanine, valine, proline, glycine); however, many others, such as certain multi-subunit protein complexes, require the assistance of molecular chaperones to fold into their native conformations. The protein structures are often analyzed at different levels. Primary structure is referred to the amino acid sequence, which may be measured through gene sequencing. Secondary structure is referred to regularly repeating local structures stabilized by hydrogen bonds. The most common secondary structures are the α-helix, β-sheets and turns. Tertiary structure is referred to the overall three-dimensional shape of a single protein chain, regarding how the secondary structures are organized in space. Tertiary structure is commonly stabilized by the formation of a hydrophobic core in a folded protein chain. Quaternary structure is referred to the architecture of supramolecular assembly formed by multiple protein chains through non-covalent interactions, such as hydrogen bonds, π-π interactions, and hydrophobic interactions. Such an assembly is called a complex, in which each protein chain is called a subunit. The tertiary and quaternary structures are often experimentally analyzed by X-ray crystallography, cryo-electron microscopy and nuclear magnetic resonance spectroscopy.

The amino acid sequence of a protein determines its native fold. This is known as Anfinsen's dogma (Anfinsen 1973, Anfinsen and Haber 1961). The hypothesis proposes that a native protein folds into a unique, stable structure on a kinetically accessible ground state of the free energy of the protein. Protein folding has been a major subject of investigation in the biophysical community. An important concept related to protein folding mechanism is called Levinthal's paradox (Levinthal 1968). The Levinthal's paradox states that the number of physically possible conformations for a given protein is astronomically large, so that even a small protein of only 100 residues would require more time than the universe has existed (10^{26} seconds) to explore all possible conformations and choose the unique ground state. As such, it would arguably make computational prediction of protein structures unfeasible if not impossible. The Levinthal's paradox raised general interests in the investigations of protein folding mechanism, which later suggested that the protein folding follows a funnel-like energy landscape and in fact explores only a limited number of folding pathways, allowing protein chains to fold efficiently into a unique, stable structure. Many globule-like protein foldings were found to follow Anfinsen's dogma. But there is certain violation. For example, prions can misfold into multiple stable conformations that differ from the native folding state; one of the prion's misfolded conformations can replicate, propagate and transmit itself to other prion proteins in tissues, which causes diseases. In Bovine spongiform encephalopathy, also known as mad cow disease, native proteins misfold into a stable filament structure causing fatal amyloid to build up. Eventually, protein structures are dynamic objects, in constant thermodynamic motion in solutions, although most known protein structures may sample only one or few conformations in its equilibrium state. Some such excursions are harmonic, such as stochastic fluctuations of chemical bonds and bond angles. Others are an harmonic, such as side chains that jump between separate discrete rotamers at local energy minima.

Proteins are essential parts of all organisms and play major roles in virtually every molecular process within cells. Many proteins are enzymes that catalyze biochemical reactions and are vital to metabolism. Some of these enzyme proteins are essentially molecular machines or motors, such as ATP synthase, myosin, kinesin and dynein. Some proteins serve as structural building blocks, such as actin that builds the strength of muscle and filament proteins in the cytoskeleton, which form a network of scaffolding that maintain cell shape and homeostasis. Protein structures that undergo functional work can exhibit dramatic conformational changes. The structural organization of multiple domains or subunits in proteins gives rise to considerable degree of freedom in protein domain dynamics. Domain motions can be appreciated by comparing different conformations of a protein under different functioning states. They can also be studied by conformational sampling in extensive molecular dynamics trajectories in computers. Domain motions are important for catalysis, regulatory activity, transport of metabolites, formation of protein assemblies, and cellular locomotion. Although small protein domain often assume unique single conformation following Anfinsen's dogma, large protein assembly or complex may achieve a less stable or metastable architecture that is required for their biological activity facilitated by protein dynamics (Mao et al. 2012). Some of these protein complexes may refold drastically from one conformation to another along their functional pathways (Mao et al. 2013). In such conformational transitions, β-sheets may refold into α-helix or vice versa; coiled coils may refold into a long α-helix.

From the standpoint of nanomaterials, natural proteins may be roughly classified into two general categories. One is called structural proteins, which can spontaneously assemble into various nanoscale architectures, such as helical filament, icosahedron, and elastic network. The other is called machinery proteins, which work as a nanoscale mechanical device like rotating rotor and walking motor. Some protein may possess the characteristics of both structural and machinery proteins. The universe of proteins is enormous so that they provide a large pool of molecular templates for rational nanomaterial engineering of both higher-order nanoarchitectures and sophisticated nanomachines. Our knowledge regarding how natural proteins assemble into nanostructures or function as nanomachines represents a tremendous source of inspiration for the engineering strategies for smart protein nanomaterials.

2. Self-Assembly of Proteins into Various Nanoscale Architecture

Proteins can spontaneously assemble into various forms of higher-order architectures on the nanometer length scale under appropriate biochemical and physiological conditions. In cells, many structural proteins self-assemble into filament, network or elastic mesh, which constitutes mechanical supports of cells or guides cellular activity including signal transduction and material transportation. Many of these structures were able to readily form *in vitro* given appropriate buffer conditions, making them suitable to various biomaterial applications (Galland et al. 2013, Hudalla et al. 2014, Reymann et al. 2010). The higher-order nanoarchitectures, such as helical assembly, icosahedral particle and elastic fiber, often exhibit unique properties that may not be readily achieved by other materials. They can be extremely resilient and versatile. Supramolecular assemblies of proteins are, therefore, highly suited for rational re-engineering and could be used as three-dimensional supports of smart biomaterials with desired properties.

2.1 Helical Assembly

The architecture of eukaryotic cells is supported mechanically by a hierarchical nanostructure called cytoskeleton, which is self-assembled from a collection of structural proteins (Fletcher and Mullins 2010). The cytoplasm of eukaryotic cells is in constant motion as cellular organelles are transported continuously from place to place along the tracks of the cytoskeleton. The cytoskeleton is the higher-order protein assemblies that form the rails of the cell's transportation system; many motor proteins move on the tracks of rails made from the different components of cytoskeleton. The cytoskeleton is composed mainly of three types of higher-order protein assemblies: microtubules, microfilament (actin) and intermediate filaments. Each type of protein assembly functions as a polymerized complex composed of many identical subunit proteins.

Microtubules provide mechanical supports for cells. They are the strongest component of the cytoskeletal polymers (Weisenberg 1972). Microtubules can resist strong compression. Microtubules are functionally required by cell migration, mitosis, gene regulation and morphogenesis. Microtubules also have a major structural role in eukaryotic cilia and flagella. Cells rely on the dynamic assembly and disassembly of microtubules to reorganize the cytoskeleton quickly in response to environmental stimuli. This allows cells to take advantage of both the adaptability and strength of microtubules. Different cells can have unique organizations of microtubules to suit specific needs. Microtubules are hollow rigid tubes whose outer diameter is about 24 nm while the inner diameter is about 12 nm (Fig. 1a). The basic building block of microtubules is a protein called tubulin heterodimers. Tubulin heterodimer is made up of two protein subunits, α- and β-tubulin, each of which has a molecular weight of approximately 50 kDa. The α/β-tubulin dimers polymerize end-to-end into linear protofilaments that associate laterally to form a single microtubule, which can then be extended by the addition of more α/β-tubulin dimers. Typically, microtubules are formed by a parallel association of thirteen protofilaments, although microtubules composed of fewer or more protofilaments have been

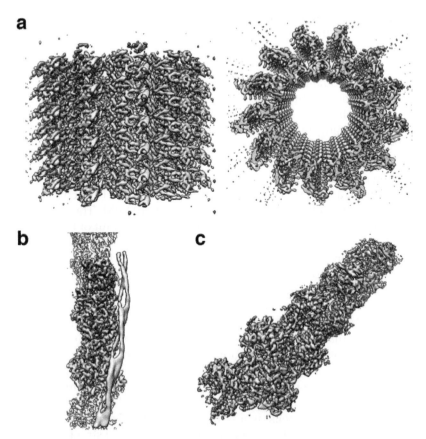

Figure 1. Helical assembly of protein nanomaterials. **(a)** 3D cryo-electron microscopic reconstruction of microtubule decorated by Ndc80 kinetochore complex at 8.6 Å resolution (EMDataBank entry: EMD-5223) from two orthogonal perspectives (Alushin et al. 2010). **(b)** 3D cryo-electron microscopic reconstruction of F-actin decorated with topomyosin at 3.7 Å resolution (EMDataBank entry: EMD-6124) (von der Ecken et al. 2015). **(c)** 3D cryo-electron microscopic reconstruction of dynactin at 3.5 Å resolution, which interacts with dynein and structurally resembles F-actin (EMDataBank entry: EMD-2857) (Urnavicius et al. 2015).

observed *in vitro*. Microtubules have a distinct polarity that is critical for their biological function. Tubulin polymerizes end to end, with the β-subunits of one tubulin dimer contacting the α-subunits of the next dimer. Therefore, in a protofilament, one end has only the α-subunits exposed while the other end has only the β-subunits exposed. These ends are designated the minus and plus ends, respectively. Although microtubule elongation can take place at both the plus and minus ends, it is more rapid at the plus end. The lateral association of the protofilaments creates a pseudo-helical structure. One turn of the helix contains 13 tubulin dimers, each from a different protofilament. Two distinct types of interactions take place between the subunits of lateral protofilaments within the microtubule, which are called the A-type and B-type lattices. In the A-type lattice, the lateral associations of protofilaments occur between adjacent α and β-tubulin subunits. Hence, an α-tubulin subunit from one protofilament can interact with a β-tubulin subunit from an adjacent protofilament. In the B-type lattice, the α and β-tubulin subunits from one protofilament interact with the α and β-tubulin subunits from an adjacent protofilament, respectively. The B-type lattice was suggested to be the primary arrangement within microtubules (Nogales 2000).

The second major component of the cytoskeleton is a globular multi-functional protein called actin that assembles into helical microfilament (Gunning et al. 2015). The microfilament measures approximately 7 nm in diameter with a helical pitch of 37 nm (Fig. 1b and 1c). The actin is found in all eukaryotic cells at concentrations of over 100 μM. The mass of an actin protein is about 42-kDa. It is the monomeric subunit of two types of filaments in cells: microfilaments, one of the three major components of the cytoskeleton, and thin filaments, part of the contractile apparatus in muscle cells. The actin proteins are observed as either a free monomer called G-actin or as part of a linear polymer microfilament called F-actin, both of which are essential for such important cellular functions as the mobility and contraction of cells during cell division. Actin participates in many important cellular processes, including muscle contraction, cell motility, mitosis, cytokinesis, vesicle and organelle movement, cell signaling, and the establishment and maintenance of cell junctions and cell shape. Many of these processes are mediated by extensive and intimate interactions of actin with cellular membranes. The actin filaments also contribute to the mechanical support of cells, whose dynamic properties are mediated via assembly and disassembly of actin filaments. The helical structure of actin filament are structurally polarized and tightly controlled within cells by a multitude of actin-binding proteins, such as profilin, Arp2/3 complex, myosin, flaming, villin, fimbrin, and cofilin. The actin-binding proteins regulate polymerization of new actin filaments, prevent polymerization of actin monomers, control filament length, and crosslink actin filaments. The interaction of signal transduction pathways with actin-binding proteins provides a mechanism for controlling the dynamics and structure of cytoskeleton. In vertebrates, three main groups of actin isoforms, α, β, and γ have been identified. The α-actins, found in muscle tissues, are a major constituent of the contractile apparatus. The β and γ-actins coexist in most cell types as components of the cytoskeleton, and as mediators of internal cell motility. The diverse range of structures formed by actin enables it to fulfill a large range of functions that are regulated through the binding of tropomyosin along the filaments (Fig. 1b). The actin filament has been used as a versatile scaffold to develop novel materials, such as three-dimensional electrical connections (Galland et al. 2013, Reymann et al. 2010).

Intermediate filaments are the third type of major components of the cytoskeleton and essential for maintenance of correct tissue structure and function (Fuchs and Cleveland 1998). They form robust networks of 10-nm-thick filaments, which in diameter is intermediate between 7-nm actin filaments and 25-nm microtubules. Most types of intermediate filament are in cytoplasm, but one type, called lamin, is in nucleus. The minimal building block of intermediate filament is a parallel dimer. The monomers are encoded by a large gene family and share a similar structure consisting of a central long α-helical domain divided into sections and flanked by a head and a tail domain, which are subject to posttranslational modifications and regulate assembly by their phosphorylation. The dimers form antiparallel, tetrameric assembly units, which *in vitro* rapidly associate laterally and longitudinally to form a 10-nm filament. This self-assembly mechanism establishes the nonpolarized nature of the intermediate filament, which is different from the polarized nature of microtubules and actin microfilaments (Hermann et al. 2007). Mature intermediate filament networks are highly strain resistant and exhibit strain hardening. Intermediate filaments are classified into six types based on similarities in amino acid sequence and protein structure. Different types of intermediate filaments are expressed in tissues of specific differentiation patterns. Most intermediate filament proteins are keratins, classified as either type I or type II, which are expressed in epithelia sheet tissues (Hanukoglu and Fuchs 1982). Simple keratins, K8/K18, are the least specialized and likely the oldest keratin evolutionarily, whereas structural trichocyte hair keratins of the most recently evolved. Type III and type IV groups, show some overlapping expression ranges and can be heteropolymeric in tissues; they are expressed in connective, nerve, muscle, and hematopoietic tissues. One of type III proteins, vimentin, represents minimal, nonepithelial intermediate filament expression and is characteristic of solitary cells. Type V proteins, the ubiquitous and ancient lamins, reinforce the nuclear envelope and interact with it via posttranslational modifications that produce membrane anchorage sites. Lamins, and probably all other intermediate filament proteins, are disassembled by phosphorylation during mitosis to allow chromosome separation and cytokinesis.

In multicellular organism, multiple intracellular signaling proteins can assemble into higher-order signaling machines that in some cases appear as helical architectures (Wu 2013). The formation and extension of the helical assemblies, also called signalosomes, transmits receptor activation information to cellular responses. During the process of the helical growth of signalosomes, infinite assemblies provides the potential of signal amplification by incorporating an over-stoichiometric number of signaling enzyme proteins into the complexes. In Drosophila, signalling mediator proteins, MyD88, IRAK4 and IRAK2 forms a specific signalosome called Myddosome complex, that assembled hierarchically into left-handed helical oligomer of ~10 nm diameter. Formation of these Myddosome complexes brings the kinase domains of IRAKs into proximity to trigger phosphorylation and activation in signal transduction (Lin et al. 2010). Helical symmetry may be especially suitable for regulating the pathway and the strength in signal transduction, because signalsomes with helical symmetries can evolve to accommodate a variable number of binding partners with specificity. Helical assemblies may also reduce biological noise in signal transduction both kinetically and thermodynamically, because the growth of signalosome under helical symmetry requires cooperativity in the protein subunit association, which could generate stable signals above the background noise of stochastic molecular interactions.

2.2 Icosahedral and Conical Assembly

Different from the higher-order self-assembly of proteins native in cell, which often exhibits helical filament structures, viral proteins often self-assemble into more closed architectures like icosahedral and conical particles. The shell of a virus, called capsid, is composed of a number of non-covalently associated protein subunits and often exhibits higher-order symmetry including helical and icosahedral assemblies. The icosahedral architecture has 20 equilateral triangular faces (Branden and Tooze 1991). The diameter of an icosahedral virus capsid is generally in the range of 50 to 150 nm. Each capsid face may consist of multiple protein subunits. On the foot-and-mouth disease virus capsid, each face consists of three proteins named VP1–3. The icosahedral symmetry is quite common among viral capsids of different sizes. A regular icosahedron has 60 rotational symmetries and a symmetry order of 120 including the transformations that combine a rotation with a reflection. Hence, given three asymmetric protein subunits on a triangular face of a regular icosahedron, 60 of such subunits are needed to assemble the icosahedron particle in an equivalent manner. Most viral capsid assemblies have more than 60 subunits and can be formed by more than one type of subunit. The variations in icosahedral capsid assemblies have been classified on the basis of the quasi-equivalence principle, in which it is postulated that the protein subunits do not interact equivalently with one another, but nearly equivalently in an icosahedron assembly (Caspar and Klug 1962). An example is the Norwalk virus capsid, where there are 180 identical subunits forming an icosahedral architecture in a quasi-equivalent manner. There are 22 subgroups in icosahedral symmetry so that icosahedron architecture may be achieved via different protein assembly topology. In general, an icosahedral structure may be constructed from 12 pentamers (Fig. 2a and 2b). A number of hexamers could appear between pentamers in an icosahedron. The number of pentamers is fixed but the number of hexamers can vary (Johnson and Speir 2009).

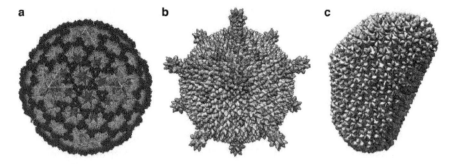

Figure 2. Icosahedral and conical assembly. **(a)** The icosahedron assembly of Aquareovirus (EMD-5160) (Zhang et al. 2010). **(b)** The capsid of sulfolobus turreted icosahedral virus, with spikes distributed on the capsid surface. (PDB ID: 3J31) (Veesler et al. 2013). **(c)** The full atomic model of a complete conical HIV capsid (PDB ID: 3J3Y) (Zhao et al. 2013).

These icosahedrons can be constructed from pentamers and hexamers by minimizing the triangulation number, T, of nonequivalent locations that subunits occupy. The T-number adopts the particular integer values 1, 3, 4, 7, 12, 13, and so on, according to the formula $T = h^2 + k^2 + hk$, where h and k are nonnegative integers.

Although most known virus capsid assemblies adopt the icosahedron symmetry, there are exceptions such as the retroviruses. On the retrovirus capsids, a "fullerence cone" architecture may be formed by insertion of around 12 pentamers into a curved hexagonal lattice from around 200 hexamers, which closes the ovoid (Fig. 2c). Recent studies combining cryo-electron microscopy and molecular dynamics simulation suggested that the interfaces between neighboring hexamers and between pentamer and hexamers exhibit quasi-equivalence in the capsid lattice (Zhao et al. 2013). These phenomena imply that protein-protein assembly following the quasi-equivalence principle could be an efficient, versatile approach for rational design of higher-order nanostructures.

2.3 Elastic Fiber Assembly

Elastic fibers represent intriguing biomaterials that are an integral component of the extracellular matrix in tissues (Muiznieks et al. 2010). They are highly elastic and repetitively stretchable. Such properties are characteristic of vertebrate tissues whose physiological role requires a high degree of resilience over a lifetime. Elastic fibers are found predominantly in connective and vascular tissue, lungs, and skin. 90% of the elastic fiber consists of a protein called elastin, which is a polymer of the monomeric precursor tropoelastin and is the dominant contributor to fiber elasticity. The self-assembly of elastin monomers into insoluble fibers involves two steps.

First, self-association of monomers occurs through hydrophobic domains in a process known as coacervation. Second, lysine residues of elastin are cross-linked by one or more members of the lysyl oxidase family. The non-elastin component of elastic fibers is collectively referred to as microfibrils and includes fibrillins, fibulins, and microfibril-associated glycoproteins. The peripheral microfibrils around elastin in elastic fibers may contribute to elastin fiber assembly as a guiding scaffold, as the microfibril appears prior to elastin deposition. The elastin can elastically extend and contract in repetitive motion when hydrated. It has been suggested that elastin functions as an entropic spring. Upon relaxation from a stretching, elastic recoil is driven by increasing entropy that complies with increased disorder of the polypeptide chain and of surrounding water molecules. This implies certain degree of structural heterogeneity and disorder within the elastin monomer, where the formation of extended secondary structures is restricted in favor of transient and fluctuating local motifs.

Water-swollen elastin fibers are about 5–8 mm in diameter. Negative staining electron microscopy of elastin fibers revealed distinct filaments aligned parallel to the long axis of the fiber (Gotte et al. 1974, Gotte et al. 1976). Similar filaments were found in α-elastin, coacervates of tropoelastin, synthetic elastin-like polypeptides, fibrous elastin, and tropoelastin (Muiznieks et al. 2010 and references therein). These filaments laterally associate into fibrils, which are bridged at regular intervals, and display a high propensity to aggregate into thick bundles (Pepe et al. 2007). Stretching of elastin fibers reveals a transition from a glassy, plastic, disordered network to well-ordered filaments aligned along the direction of the applied force. The solution structure of individual hydrophobic domains of tropoelastin and elastin-derived peptides is disordered and flexible in conformation. The fluctuating β-turns within hydrophobic elastin-like sequences are believed to be labile and may contribute to elasticity. The elastin monomer is flexible and highly dynamic. Structural disorder within monomer and aggregated elastin does not preclude the formation of preferred self-interactions or specific interactions with matrix and cell components, and thus the directional formation of fibers. A high degree of combined composition of proline and glycine residues prevents the collapse of hydrophobic residues into compact globular structures by limiting the formation of backbone self-associations. Thus, the elastin monomer remains more hydrated and flexible than globular proteins of similar size, where its conformational disorder plays an essential role in driving extensibility and elasticity.

3. Proteins Assemble into Dynamic Nanomachines

All living cells are full of molecular nanomachines that catalyze biological chemical reactions, transduce biological energy and maintain metabolic homeostasis. Most natural biological molecular nanomachines are enzyme proteins. Enzymes catalyze biochemical reactions in cells, converting substrates into products. Almost all metabolic processes in cells are accomplished by enzyme proteins. Binding substrate often induces conformational transition in an enzyme protein, which lowers the activation energy of the biochemical reaction. The substrate-bound conformation of an enzyme may form an electrostatic complementarity to the transition state of the substrate, or form a covalent intermediate to provide a lower energy transition state, or mechanically distort the substrate into their transition state. Enzymes often exhibit complex internal dynamics required by their functionality, a common property shared by all types of protein nanomachines.

All eukaryotic cells contain cyclic molecular motors, a special class of enzyme proteins that convert chemical energy into mechanical energy or vice versa. Different from structural proteins that assemble into relatively static high-order architecture, many protein nanomachines are highly dynamic and can generate forces. They may interact with high-order protein architecture to fulfill their cellular functions and perform mechanical work during enzymatic cycle. For instance, a golf-club shaped protein myosin uses actin filaments as tracks to crawl bipedally. Myosin, kinesin and polymerases are the remarkable examples of cyclic motors. They can take an infinite number of cyclic steps of conformational changes without damaging the integrity of their structures, given sufficient quantities of "fuel" molecules, such as adenosine triphosphate (ATP). Therefore, the natural protein nanomachines have been considered a primary source of nanoscale engineering of molecular machinery and studied extensively from the standpoint of nanomaterial engineering.

3.1 Nanochannels and Nanopumps

There are three principal classes of membrane transport proteins: channels, transporters, and pumps (Hille 2001). These proteins reside in the plasma membrane and in the membranes of intracellular organelles, lysosomes, and mitochondria. They are extremely important for a myriad of cellular functions, ranging from uptake of nutrients such as glucose to more complex physiological tasks such as the reabsorption of solutes by the kidney and the propagation of action potentials in neural cells. Ion channels catalyze the rapid, selective and passive transport of ions down their electrochemical gradients (Jentsch et al. 2004). They vary in their degree of selectivity: some are highly selective for K^+, Na^+, Ca^{2+}, Cl^- ions, or H_2O, whereas others are selective for certain anions or cations. Selectivity is enforced by a region in ion channels called the selectivity filter, which makes partial dehydration of the permeating ions energetically favorable over other ions of similar size. Ion channels undergo regulated opening and closing in a process called gating. They may be gated by ligand, voltage, mechanical stretch, or temperature change. The general architecture for several ion channel proteins have been characterized, and their atomic models have been built to account for their selectivity, transport energetics and gating mechanism.

Potassium (K^+) channels are the most widely distributed type of ion channels and found in nearly all living organisms. K^+ channels are found in most cell types and control a wide variety of cell functions. They form K^+-selective pores across cell membranes. Each K^+ channel is a homotetramer, with the four subunits together forming a central pore, through which ions pass (Doyle et al. 1998, Zhou et al. 2001). Each subunit has two transmembrane α-helices separated by a pore (P) loop. The P-loop contains a short helical element called the pore helix. The pore consists of two major components, the selectivity filter and the central cavity. The selectivity filter, which is the narrowest part of the pore and is 12 Å long and 3 Å wide, is near the extracellular opening and binds K^+ ions (Fig. 3a). Six ion-containing sites have been observed in the channel: four

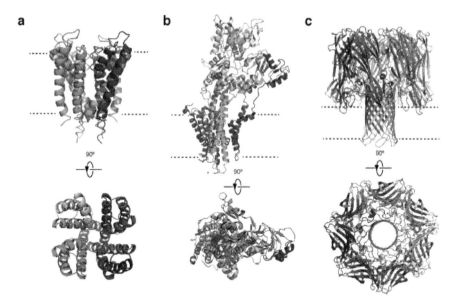

Figure 3. Nanochannels and nanopumps. **(a)** A crystal structure of the potassium channel from Streptomyces lividans, with sequence similarity to all known K^+ channels, particularly in the pore region (PDB ID: 1BL8) (Doyle et al. 1998). **(b)** A crystal structure of the Calcium ATPase pump from skeletal muscle sarcoplasmic reticulum (1IWO) (Toyoshima and Nomura 2002). **(c)** A crystal structure of α-hemolysin, a heptameric transmembrane pore (PDB ID: 7AHL) (Song et al. 1996, Mathé et al. 2005), which has been extensively used in nanopore devices for single-molecule analysis and biosensing (Clarke et al. 2009, Wang et al. 2011, Rodriguez-Larrea and Bayley 2013). In all panels, the dashed lines approximately mark the position of lipid bilayer membrane.

sites (P1–P4) within the selectivity filter, one on the extracellular side of the filter (P0) and one in the central cavity (P5). These sites represent the transport pathway of K^+ ions. K^+ ions in solution are hydrated. Dehydration of ions is energetically disfavored because it must remove the partial negative charges of water dipoles from positively charged ions. The partial negative charge of the oxygen atoms in the selectivity filter imitates water molecules, creating a hydrophilic environment that lowers the dehydration energy for the permeating potassium ions and mimics the weak negative charges of water dipoles forming the hydration shell (Berneche and Roux 2001).

K^+ channels open and close within milliseconds to maintain the negative resting membrane potential and prevents ion leak. Different subfamilies of K^+ channels are gated by different extracellular or intracellular signals. For some K^+ channels, the gate opens intrinsically with a sensing mechanism that links channel activity to the metabolic state of the cell. In another type of gating mechanism, ligand binding to the channel's intracellular domains causes a conformational change that opens the channel gate. Ca^{2+}, ATP, trimeric G proteins, and polyamines are examples of such ligands. In voltage-gated K^+ channels, changes in the membrane potential lead to conformational changes in the transmembrane segments that open the channel gate. An example of voltage gating is the voltage-sensing mechanism of voltage-dependent K^+ channels, which maintains the resting membrane potential at negative voltages and allows for the termination of action potentials in electrically excitable cells such as neurons and muscle cells (Jiang et al. 2003, Long et al. 2005a, Long et al. 2005b). Some K^+ channels are gated by both voltage change and ligand binding. These gating mechanisms of K^+ channels were also found similarly in other ion channels, such as Na^+ channels and Ca^{2+} channels, although others exhibited different structural change to close the channels (Stuhmer et al. 1989, Toyoshima et al. 2000, Toyoshima and Nomura 2002). For instance, voltage-dependent Na^+ channels are closed by specific hydrophobic residues that block the pore.

In contrast to ion channels that transport ions passively, transporters and pumps are solute-selective carrier proteins that use free energy to actively transport solutes against their electrochemical gradients. These transport proteins often alternate between two different conformations. In one conformation, the transport protein binds solute molecules on one side of the membrane; and in another conformation, it releases solutes on the other side. Uniporters moves solutes down their transmembrane concentration gradients, such as glucose transporters (Mueckler et al. 1985). Symporters and antiporters move a solute against its transmembrane concentration gradient; this movement is energetically coupled to the movement of a second solute down its transmembrane concentration gradient. Many transporters are part of the major facilitator superfamily (MFS) transport proteins that translocate sugars, sugar-phosphates, drugs, neurotransmitters, nucleosides, amino acids, peptides, and other solute across membranes. A typical MFS member, the bacterial lactose permease LacY functions as a monomeric oligosaccharide/H⁺ symporter, which uses the free energy released from the translocation of H⁺ down its electrochemical gradient to drive the accumulation of nutrients such as lactose against its concentration gradient.

Transporters and pumps use different sources of energy to accomplish transport. Transporters couple transport with the energy stored in electrochemical gradients across the membrane. In contrast, pumps use energy from ATP or environmental trigger such as light to drive transport. A typical molecular pump is Ca^{2+}-ATPase (Fig. 3b), which pumps Ca^{2+} into intracellular storage compartments using the free energy released by ATP hydrolysis (Toyoshima et al. 2000, Toyoshima and Nomura 2002). In each pumping cycle, the Ca^{2+}-ATPase changes its conformation between two states: the E_1 conformation, which binds Ca^{2+} on the cytosolic side with high affinity, and the E_2 conformation, which binds Ca^{2+} with a significantly reduced affinity and therefore releases Ca^{2+} from the Ca^{2+}-binding sites into the sarcoendoplasmic reticulum. Some molecular pumps are also rotary molecular motors. For example, H⁺-ATPase pumps protons out of cytosol across the membrane into the organelle lumen through rotational cycle driven by ATP hydrolysis (Wilkens et al. 1999).

In addition to these ion channels, transporters and pumps, a pore-forming β-barrel toxin family has been involved in bionanotechnology development. The first identified member of this toxin family, a heptameric transmembrane protein nanopore α-hemolysin (Fig. 3c) consists mostly of β-sheets with only about 10% α-helices contents (Song et al. 1996, Mathé et al. 2005). α-hemolysin has been used in lab-on-a-chip devices for single-molecule analysis and biosensing, translating sequences of nucleotides or peptides directly into electronic signatures (Clarke et al. 2009, Wang et al. 2011, Rodriguez-Larrea and Bayley 2013).

3.2 Rotary Motors

ATP synthase is a rotary molecular motor found on the thylakoid and inner mitochondrial membrane in eukaryotic cells, which couples the free energy of the electrochemical proton gradient across the membrane to ATP synthesis (Noji et al. 1997, Itoh et al. 2004, Stock et al. 1999, Rondelez et al. 2005). It is an extraordinary enzyme that provides energy source for the cells to consume through the synthesis of ATP. ATP is the most commonly used biological "fuel" in cells from most organisms. The ATP synthase join adenosine diphosphate (ADP) and phosphate (Pi) covalently to produce ATP. Energy is then released in the form of hydrogen ions or protons (H⁺), moving down an electrochemical gradient, such as from the lumen into the stroma of chloroplasts or from the inter-membrane space into the matrix in mitochondria.

The overall structure of ATP synthase is similar in all cells (Fig. 4a). ATP synthase consists of two structural regions: the F_0 domain within the membrane, which is involved in translocation of protons down their electrochemical gradient, and the F_1 domain above the membrane and inside the matrix of the mitochondria, which contains catalytic sites responsible for ATP synthesis (Allegretti et al. 2015). In bacteria, the F_0 and F_1 regions are composed of ab_2c_{10-14} and $\alpha_3\beta_3\gamma\sigma\epsilon$ subunits, respectively. The c-subunits of the F_0 domain form a ring that interacts with the a-subunit. The γ-subunit forms a central rotor stalk connecting to the c-ring at its base and penetrates

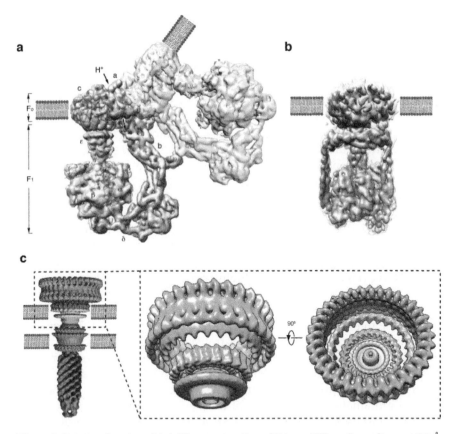

Figure 4. Rotational motors. **(a)** A 3D reconstruction of F-type ATP synthase dimer at 7.0 Å resolution by cryo-electron microscopy (EMDataBank entry: EMD-2852) (Allegretti et al. 2015). The F-type ATP synthase was found to form a dimeric complex on mitochondrial membranes. **(b)** 3D reconstruction of V-type ATP synthase at 9.7 Å by cryo-electron microscopy (EMDataBank entry: EMD-5335) (Lau and Rubinstein 2012). **(c)** 3D model of bacterial flagellar rotor with its basal architecture highlighted by two orthogonal views (EMDataBank entry: EMD-1887) (Thomas et al. 2006).

into the F_1 catalytic domain. The F_1 domain is assembled from three α- and three β-subunits arranged alternately in the form of a hexagonal cylinder around the γ-subunit. The peripheral stator stalk consists of $β_2$ σ-subunits, with the σ-subunit binding to the F_1 domain and $β_2$-subunit anchoring the F_0 domain in the mitochondrial membrane and interacting with the a-subunit. To use the energy of the transmembrane proton gradient to fuel ATP synthesis, the a- and c-subunits control proton transport in such a way that the c-ring rotates relative to the a-subunit, converting the energy of the electrochemical proton gradient into mechanical rotation of the c-subunits; the γ-subunit of the central stalk rotates with the c-ring and couples the transmembrane proton motive force over a distance of 10 nm to the F_1 domain. The mechanical energy of rotation is used to release ATP, whose synthesis is catalyzed by the β-subunits in the F_1 domain. Rotation of the c-ring and the central γ-subunit relative to the $α_3β_3$ subdomain is essential for coupling the proton motive force across the membrane to drive ATP formation and release. Since each c-subunit carries one proton, there are 10–14 protons transported per complete revolution of the c-ring, and roughly four protons are transported per ATP synthesized. The ATP synthase converts electrochemical energy to mechanical energy and back to chemical energy, with nearly 100% efficiency (Wang and Oster 1998). ATP synthesis can occur at a maximal rate on the order of 100 sec^{-1} and sustains millimolar ATP concentrations in cells.

In some bacteria, the ATP synthase works in the reverse direction, in which the energy released by ATP hydrolysis drives the translocation of protons out of the cell, generating a proton gradient across the cytoplasmic membrane (Noji et al. 1997). Another family of protein motors that are closely related to the F_0F_1-ATP synthases is the H$^+$-ATPases, which function in the direction opposite to ATP synthase in mitochondria, chloroplasts and bacteria (Fig. 4b) (Lau and Rubinstein 2012). H$^+$-ATPases transport protons from the cytosol to the vesicle lumen or to the extracellular space through the membrane integral V_0 domain. The cytoplasmic V_1 domain couples the free energy of ATP hydrolysis to proton transport. The energy of ATP hydrolysis by the V_1 domain is translated into rotation of the c-ring and proton transport. The a-subunit of the V_0 domain transports protons via the nine transmembrane segments located in its C-terminal region. A positively charged arginine residue in the a-subunit stabilizes a negatively charged glutamate residue in one of the c-subunits prior to protonation. The a-subunit uses two hemichannel structures to transport a proton and protonates the glutamate residue reversibly, which would release it from its interaction with the arginine residue in the c-subunit. The electrostatic attraction of the arginine residue with the unprotonated glutamate residue of the next c-subunit would allow rotation of the c-ring, bringing each c-subunit in contact with the a-subunit. This mechanism allows for ATP hydrolysis to drive unidirectional proton transport.

Another prominent rotatory motor is the bacterial flagellar motor, powered by proton motive force generated by the proton flux across the bacterial cell membrane (Block and Berg 1984, Thomas et al. 2006). The rotation is driven by an ensemble of torque-generating units containing the proteins MotA and MotB (Fig. 4c). The rotor is consecutively rotated in the process of proton transport across the membrane. The rotor can rotate at a speed of 6000–17000 rpm alone, but may reach only 200–1000 rpm with the flagellar filament attached. Like ATP synthase motor, the flagellar rotor uses free energy with high efficiency, generates power output of 1.5×10^5 pN nm s^{-1} and is the nanosclae engine driving the locomotion of bacteria in solutions (Ryu et al. 2000).

3.3 Locomotive Motors

There are two families of linear molecular motors, kinesins and dyneins, which move on microtubule rails, powered by ATP hydrolysis (Vale et al. 1996, Kon et al. 2012, Gee et al. 1997, Shingyoji et al. 1998). Kinesins usually move toward the plus ends of microtubules, and dyneins move toward the minus ends of microtubules. Both families of motor proteins share a common architecture: the motor head domain (Fig. 5a and 5b), which binds microtubules and generates force, and the tail domain, which binds cargo or membrane. The affinity that the head binds to the microtubule is regulated through binding of ATP and ADP to the head domain. For kinesin, binding to a microtubule is the strongest when ATP is bound. By changing the strength of kinesin's hold on a microtubule, ATP hydrolysis and nucleotide release regulate the attachment of the motor to the microtubules. ATP hydrolysis also causes a conformational change in the head domain, which is amplified to generate a larger movement of the whole motor molecule. Therefore, cycles of ATP hydrolysis and nucleotide release a couple of microtubule attachments with changes in the conformation of the motor's head domain. Through this mechanism the motor steps along the microtubule in a head-over-head fashion, taking one step of 8 nm for each ATP hydrolyzed (Schnitzer and Block 1997). The two heads of kinesin work in tandem, as it walks along a microtubule. At the beginning of a walking cycle, one head is tightly bound to the microtubule and has no nucleotide in its active site. The second head has ADP in its active site and is positioned behind the first head. Kinesin is then ready for its first step and the coordination between the two heads takes place. ATP binding at the forward head (head 1) causes its neck linker to swing forward toward the plus end of the microtubule. Head 2 then moves to the leading position over the next binding site in the microtubule. It binds weakly and releases its ADP. ATP hydrolysis at head 1, then strengthens the interaction between head 2 and the microtubule, resulting in an intermediate with both heads

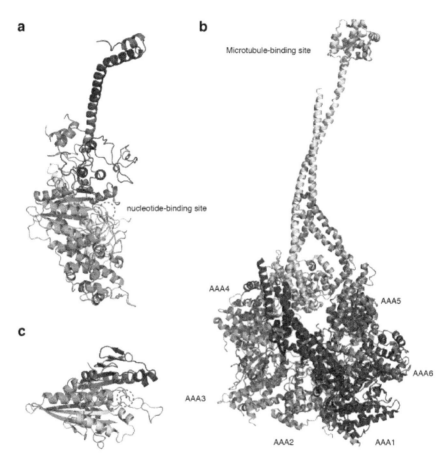

Figure 5. Linear motors. **(a)** The motor domain of kinesin (PDB ID: 1BG2) has a striking structural similarity to the core catalytic domain of the myosin, showing that it shares a similar force-generating strategy with that of myosin (Kull et al. 1996). **(b)** The motor domain of Dictyostelium discoideum cytoplasmic dynein (PDB ID: 3VKH), showing the ring-shaped AAA+ ATPase unit and a junction structure between the ring and microtubule-binding stalk, which are critical for the dynein mobility (Kon et al. 2012). **(c)** The myosin motor domain that contains both the actin and nucleotide binding sites (PDB ID: 2MYS) (Rayment et al. 1993).

strongly bound to the microtubule. Once head 2 is tightly bound, head 1 releases the phosphate group generated from hydrolyzed ATP. Release of phosphate causes head 1 to dissociate from the microtubule and results in a conformational change in head 2 that reopens its active state. This whole cycle of events returns kinesin to the starting condition, with the essential differences that head 2 is now in front and the kinesin molecule is 8 nm closer to the plus end of microtubule. A second cycle will begin when head 2 binds ATP, and the two heads will then alternate roles for hundreds or thousands of rounds, each round producing a step and moving the motor along the microtubule toward its plus end. Both kinesin and dynein walk along a microtubule in 8-nm steps, which is equal to the length of a single tubulin heterodimer. While kinesin walks regularly, stepping from one heterodimer to the next along a single protofilament, dynein walks stochastically, stepping randomly between protofilaments transversely toward the minus end of microtubule.

Myosins are motor proteins that also use the free energy released by ATP hydrolysis to generate force and power motility along actin filaments (Whittaker et al. 1995, Bershitsky et al. 1997). Myosins are best known for their role in muscle contraction. However, myosins are found in all eukaryotic cells and make up a large protein superfamily with at least eighteen classes. The structural and biochemical properties of the different myosin isoforms have evolved to carry out diverse roles that include powering muscles and cellular contractions, driving membrane and vesicle transport, regulating cell shape and polarity, and participating in signal transduction. Members of the myosin superfamily consist of three common domains, the head or motor domain, the regulatory domain, and the tail domain. The motor domain contains the ATP- and actin-binding sites and is responsible for converting the energy from ATP hydrolysis into mechanical work (Fig. 5c). The regulatory domain acts as a force transducing lever arm. The tail domain of myosin interacts with cargo protein or lipid and determines its biological function. The regulatory domain extends from the motor domain as a long α-helix, acting as a lever arm for force generation. Conformational changes in the nucleotide-binding site, which are determined by whether ATP, ADP-Pi or ADP is bound, are transmitted to the regulatory domain from the motor. These conformational changes cause rotation of the lever arm, resulting in a force-generating power stroke. Myosin's affinity for actin depends on whether ATP, ADP-Pi, or ADP is bound to the nucleotide-binding site of the motor domain. Myosins with bound ATP or ADP-Pi are in weak binding states, and rapidly associate with and dissociate from actin. ATP hydrolysis activates myosin and occurs when myosin is detached from actin. The force-generating power stroke accompanies phosphate release after myosin-ADP-Pi rebinds actin. Myosins with either bound ADP or with no nucleotide bound are in strong binding states. Myosins are able to convert ~50% of the total energy from ATP hydrolysis into useful work, generating ~40 pN-nm per ATP.

3.4 Membrane-Fusing Nanomachines

SNARE (soluble N-ethylmaleimide sensitive factor attachment protein receptor) proteins are a large protein superfamily consisting of more than 60 members in yeast and mammalian cells (Sutton et al. 1998).

Evolutionarily conserved SNARE proteins form a complex that mediate the fusion of vesicles with their target membrane-bound compartments such as a lysosome (Hu et al. 2003). The SNAREs are best known to mediate docking of synaptic vesicles with the presynaptic membrane in neurons. These SNAREs are the targets of the bacterial neurotoxins responsible for botulism and tetanus. SNAREs are often posttranslationally inserted into membranes via a C-terminal transmembrane domain. However, seven of the 38 known SNAREs, including SNAP-25, do not have a transmembrane domain. They are instead attached to the membrane via lipid modifications such as palmitoylation. The targeting of SNAREs is accomplished by altering either the composition of the C-terminal flanking amino acid residues or the length of the transmembrane domain. Replacement of the transmembrane domain with lipid anchors leads to an intermediate stage of membrane fusion where only the two contacting leaflets fuse and not the two distal leaflets of the membrane bilayer.

During membrane fusion, vesicle SNARE (v-SNARE) and target SNARE (t-SNARE) proteins on the membranes of vesicle and target cellular compartments, respectively, associate to form a SNARE complex. During membrane fusion, the SNARE complexes undergo a large conformational change that brings together the two membranes and allow them to merge. Upon the completion of membrane fusion, the two SNAREs reside on the same side of merged membrane. SNAREs are thought to be the required key components of the fusion machinery and can function independently in the absence of additional cytosolic accessory proteins. This was demonstrated by engineering flipped SNAREs, where the SNARE domains face the extracellular space rather than the cytosol (Hu et al. 2003). When cells containing v-SNAREs contact cells containing t-SNAREs, SNARE complexes form and cell-cell fusion ensues. After membrane fusion, using the energy from ATP hydrolysis, the hexameric AAA$^+$ (ATPase associated with diverse cellular activities) ATPase NSF (N-ethylmaleimide sensitive factor), together with SNAPs (soluble NSF attachment protein), disassembles the SNARE complex into its protein components, making individual SNAREs available for subsequent rounds of fusion. Large, potentially force-generating, conformational differences exist between ATP- and ADP-bound NSF. During the SNARE recycling, the NSF/SNAP/SNARE supercomplex exhibits broken symmetry, transitioning from six-fold symmetry of the NSF ATPase domains to pseudo four-fold symmetry of the SNARE complex (Zhao et al. 2015). SNAPs interact with the SNARE complex with an opposite structural twist, following an unwinding mechanism. The recycling mechanism of SNARE fusion nanomachine is a remarkable example demonstrating how ATP-fueled AAA+ family nanomachine perform works on other nanomachines to make them reusable. This paradigm might be an important avenue for the future engineering of nanomachine systems that are capable of recycling themselves.

Another large category of membrane-fusion nanomachines come from enveloped viruses including retrovirus and influenza virus. Human immunodeficiency virus type 1 (HIV-1) is a typical enveloped retrovirus that causes acquired immunodeficiency syndrome (AIDS). To enter host cells, HIV-1 utilizes a metastable, trimeric envelope glycoprotein (Env) spike as its membrane-fusion machinery to bind receptors and to fuse the viral and target cell membranes (Mao et al. 2012). During synthesis and folding in the endoplasmic reticulum of the virus-producing cell, the Env precursors (gp160) are heavily modified by N-linked glycosylation and assemble into trimers. After further glycan modification in the Golgi apparatus, the gp160 Env trimer precursors are proteolytically cleaved and directly transported to the cell surface for incorporation into virions. Each protomer composing the trimeric Env spike thus consists of a gp120 exterior subunit and a gp41 transmembrane subunit. The sequential binding of gp120 to the target cell receptors, CD4 and chemokine receptor (either CCR5 or CXCR4), allows the Env complex to transit into fusion-ready conformations. During the conformational change, gp120 subunit rotates about 45° degree around the trimer axis, whereas gp41 subunit refolds into trimeric helical bundle (Mao et al. 2013). These conformational transitions expose the hydrophobic gp41 N-terminus called the fusion peptide, promoting its insertion into the target cell membrane, and further permit the formation of a highly stable gp41 six-helix bundle that brings the viral membrane and host cell membrane together, triggering viral-cell membrane fusion. The Env spike stores its free energy for membrane fusion in its metastable trimeric assembly and is the only virus-specific component potentially accessible to neutralizing antibodies and thus has evolved a protective glycan shield and a high degree of structural variability. Other enveloped viruses also possess their own membrane-fusing machinery on their virion surfaces, such as influenza hemagglutinin on the influenza virus, which uses pH change as a trigger to mediate membrane fusion.

3.5 Information-Processing Nanomachines

A large pool of protein nanomachines process biological information that is carried by DNA and RNA (Alberts et al. 2007). Polymerase duplicates nucleic acid strands or translate DNA strands into RNA strand or vice versa (Fig. 6a). A polymerase from the thermophilic bacterium, Thermus aquaticus, called Taq, is commonly used in the polymerase chain reaction, which is a fundamental technique of DNA sequence duplication constituting the cornerstone of molecular cloning and protein engineering (Murali et al. 1998). Nucleases cut nucleic acid strands either nonspecifically or based on specific sequence. Ligases connect nucleic acid strands. Recombinases extract a portion from one strand of DNA and swap it with a portion in

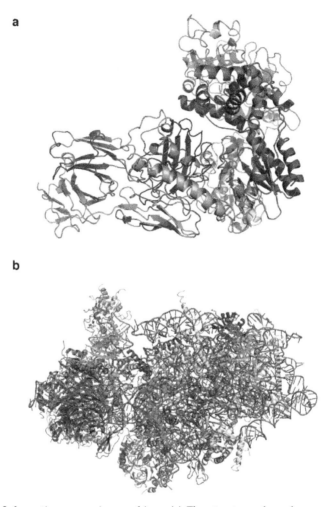

Figure 6. Information-processing machines. **(a)** The structure of a polymerase from the thermophilic bacterium, Thermus aquaticus (Taq) (PDB ID: 1BGX) that is commonly used in the polymerase chain reaction, a fundamental technique of molecular cloning (Murali et al. 1998). **(b)** The structure of eukaryotic 80S ribosome from Saccharomyces cerevisiae that is a combination of protein and RNA components and uses messenger RNA as template to synthesize protein chain (PDB ID: 4UJZ) (de Loubresse et al. 2014).

another strand. Nucleases, ligases and recombinases are very important and extremely useful in editing biological information in molecular cloning and protein recombination for functional modification and engineering. Nucleosome packages nucleic acids into compact forms for information storage. Topoisomerases topologically untangles DNA strands that are overwound, which is important for information retrieval during cell division. Splicesome edit a piece of RNA, removing a redundant loop in the middle of the strand and reconnecting two flanking segments, which is important for controlling information transfer. Ribosomes translate messenger RNA sequence to peptide chain and synthesize the proteins. These information-processing protein nanomachines essentially work around the central dogma of molecular biology, that is, "DNA makes RNA and RNA makes protein" (Alberts et al. 2007). In nearly all of these nanomachines, large conformational changes couple mechanochemical work with biological information storage, retrieval, editing, transfer, or translation. More astoundingly, these nanomachines work in concert, are spatiotemporally self-organized, and form an efficient system that maintains the life cycle of cells accurately.

The ribosome is a large and complex molecular machine, found within all living cells, that serves as the site of biological protein synthesis, translating genetic information into the 3D folds of molecular machines (Ban et al. 2000, Schluenzen et al. 2000). Ribosomes link amino acids together in the order specified by messenger RNA (mRNA) molecules. The sequence of DNA encoding for a protein may be copied many times into mRNA chains of a similar sequence. Ribosomes bind to an RNA chain and use it as a template for determining the amino acid sequence, based on which a particular protein chain is synthesized. Individual amino acids are selected and carried to the ribosome by transfer RNA (tRNA molecules), which enter one part of the ribosome and bind to the messenger RNA chain. The attached amino acids are then covalently linked together by another part of the ribosome. Once the protein chain is produced, it can fold into a specific functional three-dimensional structure. Ribosomes consist of two major components: the small ribosomal subunit (30S or 40S) that reads the mRNA, and the large subunit (50s or 60s) that joins amino acids covalently to form a polypeptide chain (Fig. 6b). These two subunits fall apart once a ribosome finishes reading an mRNA molecule. Each subunit is composed of one or more ribosomal RNA (rRNA) molecules and a variety of proteins. The ribosomes and associated molecules are also known as the translational apparatus. Ribosomes are ribozymes, because the catalytic peptidyl transferase activity that links amino acids together is performed by the ribosomal RNA. Importantly, the ribosome is a remarkable nanomachine that couples the mechanical translocation of RNA chain to the informational translation of genetic codes and to the enzymatic activity.

4. Engineering of Protein Nanomaterials

4.1 In Silico Design of Protein Nanomaterials

Computational design of protein-based nanomaterials represents an intriguing avenue for nanomaterial engineering, and can be essentially built on the techniques of *in silico* protein design. Rational design of new protein molecules aims at introducing novel function and/or behavior to a target protein structure (Richardson and Richardon 1989). Therefore, protein design is partially equivalent to protein structure-function prediction. Proteins can be designed *de novo* by fully automated sequence selection (Dahiyat and Mayo 1997, Harbury et al. 1998, Kuhlman et al. 2003) or be redesigned by making calculated variations on a known protein structure and its sequence (Wu et al. 2010). Rational protein design approaches make protein-sequence predictions that will fold to specific 3D structures. These predicted structures can then be validated experimentally at the atomic level through x-ray crystallography, electron microscopy and nuclear magnetic resonance spectroscopy (Harbury et al. 1998). In the 1970s, initial protein design approaches were based mostly on sequence redesign and composition and did not involve in the prediction of designed 3D protein structures and specific interactions between side chains at the atomic level. Recent advancement in molecular force fields, protein design algorithms, and structural bioinformatics, such as libraries of preferred side-chain conformations, have enabled the development of computational protein design tools that can predict the 3D atomic models of designed protein structures efficiently in many cases. These computational tools make thorough calculations on protein energetics, structures, interactions and flexibility, and perform exhaustive searches over enormous conformational spaces. Thanks to these developments, *in silico* rational protein design has become one of the most important tools in protein engineering.

In rational protein design, amino acid sequences are predicted so that the protein of designed sequence is expected to fold into a specific 3D structure that hopefully possesses pre-defined properties and functions. Given the functional constraints, the number of possible amino acid sequences may still grow exponentially with the size of the protein chain, and can therefore be astronomical for a large protein subunit design. Based on the energy landscape theory of protein folding, only a very small subset of them would fold reliably to a native structure. Protein design involves identifying novel sequences within this subset. Therefore, protein design is the search for sequences that have the chosen structure as a free energy minimum. In other words, a tertiary structure is designed and specified based on its predicted function, and a sequence that can fold into it is to be screened and selected. Protein design can then be translated into an optimization problem, that is, how to choose an optimized sequence that will fold to the desired structure using certain scoring criteria and energy functions. Recently, the principle of protein design was extended to predict the protein-protein interaction interfaces that may guide the self-assembly of *de novo* designed proteins into higher-order architecture and biomaterials (King et al. 2012, King et al. 2014). A general computational method for designing self-assembling protein nanomaterials was proposed, which consists of two steps: (1) symmetrical docking of protein building blocks in a target symmetric architecture, followed by (2) design of low-energy protein-protein interfaces between the building blocks to drive self-assembly. The approach was used to design a 24-subunit, 13-nm diameter complex with octahedral symmetry and a 12-subunit, 11-nm diameter complex with tetrahedral symmetry (King et al. 2012). This approach was later further generalized to the design of co-assembling of multi-component protein nanomaterials (King et al. 2014). In this case, the program design program, RosettaDesign (Leaver-Fay 2011) was used to sample the identities and configurations of the side chains near the inter-building-block interface, predicting interfacial interactions with features resembling those found in natural protein assemblies such as well-packed hydrophobic cores surrounded by polar side chains. The outcome is a pair of designed amino acid sequences, one for each building block component, which can stabilize the multi-component interface and drive assembly toward the target configuration. The approaches essentially pave the way to the design of novel protein-based molecular machines with programmable structures, dynamics and functions.

4.2 Directed Evolution of Protein Nanomaterials

In contrast to computational protein design, directed evolution represents a generic experimental approach widely used in protein engineering, which mimics the process of natural selection and genetic evolution cycles in a laboratory-adapted setting to evolve engineered proteins toward a pre-defined functional feature (MacBeath et al. 1998, Lutz 2010). Directed evolution is used both for protein engineering as an alternative to designing proteins rationally, as well as studies of fundamental evolutionary principles in a controlled, laboratory-adapted environment (Voigt et al. 2000). The directed evolution can be performed either *in vivo* or *in vitro*. It consists of three major steps in each round of selection or evolution (Dalby 2011). First, a gene is subject to iterative rounds of mutagenesis, creating a library of variants. Second, the variants are expressed in an appropriate host cells and isolating members with the desired function are isolated and selected for the next step. Third, one selected variant is amplified and used as a template for the next round. In principle, evolution takes place upon three aspects of events (Barrick et al. 2009). First, there must be variation between replication of the ancestor gene; second, the variation of offspring causes fitness differences upon which selection of certain descendants occurs; third, the feature introduced in the variation is heritable in the next round of selection. In directed evolution, a single gene is evolved in multiple rounds of mutagenesis, selection, and amplification. These steps in each round are typically repeated, using the best variant from one round as the template for the next to achieve incremental improvements of re-engineered functional features. The chance of success in a directed evolution experiment is related directly to the total library size, as evaluating more mutants increases the chances of finding one with the desired properties. Directed evolution has been widely adopted in biotechnology industry to re-engineer utility proteins to improve their enzymatic activity such as DNA polymerase frequently used in molecular cloning (Arndt and Muller 2007). The combination of directed evolution and computational *de novo* design will likely open up more opportunities to explore the universe of smart protein nanomaterials.

4.3 Reprogramming Proteins into Smart Nanomaterials

Building on the principal of rational protein design and directed evolution in protein engineering, one may also integrate other biochemical approaches in conjunction to modify the protein-protein interactions or oligomerization chemistry to program the self-assembly and functioning of the protein nanomaterials. For example, to circumvent the challenge of programming extensive non-covalent interactions to control protein self-assembly, the directionality and strength of metal coordination interactions was exploited to guide the assembly of closed, homoligomeric protein nanostructure (Brodin et al. 2012). This strategy was further extended to program protein self-assembly into one-dimensional nanotubes, two- and three-dimensional crystalline arrays. The assembly of these arrays is tunable by external stimuli, such as metal concentration and pH. In another example, an ATP-driven group II chaperonin, which resembles a barrel with a built-in lid, was reprogrammed to open and close on illumination with different wavelengths of light (Hoersh et al. 2013). By engineering photoswitchable azobenzene-based molecules into the structure, light-triggered changes in interatomic distances in the azobenzene moiety are able to drive large-scale conformational changes of the protein assembly. The different states of the assembly can be visualized with single-particle cryo-electron microscopy, and the nanocages can be used to capture and release non-native cargos. Similar strategies that switch atomic distances with light could be used to build other controllable nanoscale machines.

Small proteins or short peptides have been engineered and designed to self-assemble into various nanoconstructions. Several surfactant-like peptides undergo self-assembly to form nanotubes and nanovesicles with an average diameter of 30–50 nm, which exhibit helical symmetry (Vauthey et al. 2002). The peptide monomer contains 7–8 residues and has a hydrophilic head composed of aspartic acid and a tail of hydrophobic amino acids such as alanine, valine, or leucine. Similar self-assembling peptides were redesigned to form nanofiber scaffold (Yokoi et al. 2005). Using the mammalian visual system as a model, it was demonstrated that the nanofiber scaffold could create a permissive environment for axon regeneration at the site of an acute injury, as well as knit the brain tissue together (Ellis-Behnke et al. 2006). The peptide scaffold form a network nanofibers that are similar to the native extracellular matrix, therefore providing a 3D cell-culture environment for cell growth, migration and differentiation. A protein fiber system that is co-assembled from two peptides exhibits programmable assembly, stability and morphology (Papapostolou et al. 2007). The fibers display surface striations separated by 4.2-nm distance that matches the length expected for designed α-helical peptides precisely. The spacing of the striations can be programmed with changing the length of the peptides, demonstrating the potential programmability of protein and peptides in high-order nanomaterial self-assembly.

Engineering the interface between protein nanomachines and inorganic nanomaterials represents another important avenue for smart biomaterial development (Ishii et al. 2003). A clear understanding of interactions between proteins and inorganic nanomaterials, such as nanoparticles, solid nanopore and nanowires, could make possible the design of precise and versatile hybrid nanosystems. Although the surface properties of inorganic nanomaterials may only be modified in a limited number of ways in specific cases, the interactions between protein and inorganic nanomaterials are conceptually more programmable due to the large number of amino acid sequences that can give rise to a wide range of variation in the protein's ability to interact their counterparts. In addition, there remain options of covalently bio-conjugating proteins to the surface of inorganic nanomaterials, integrating them into programmable hybrid building blocks (Marco et al. 2010). With greater understanding of the patterns and paradigms in the interfaces between proteins and inorganic nanomaterials, the computational tools used to predict protein-protein assemblies may be adapted to program protein-inorganic-nanomaterial assemblies in the near future.

5. Applications of Protein Nanomaterials

5.1 Energy

Biological systems have evolved unique capabilities to generate, harness and store energy in most efficient ways. In recent years, biologically inspired approaches have led to breakthroughs for next generation energy devices and storage. Of particular interest are efforts to translate biological principles directly into synthetic energy systems. In nature's exquisite systems of efficient energy conversion, functionally evolved protein complexes can work as reactive sites for energy harvesting and storage, such as photosynthesis system in plants. Over the past few years, remarkable progress has been made in harnessing protein-based nanomaterials for synthetic bio-inspired energy systems (Lee et al. 2013 and references therein). The sequence-specific self-assembly and recognition properties of peptides and proteins have been used for self-assembly of unique nanostructures and for the controlled synthesis of functional nanostructures into porous electrodes. Studies on applications of engineered proteins, protein-inorganic-material hybrids and supramolecular protein structures for solar cells, bio-fuel cells, photoelectrochemical cells, solar fuel generation and lithium ion batteries demonstrate the versatility of biogenic proteins for engineering solutions to various problems in energy systems.

Future research on the energy applications of smart protein nanomaterials is expected to stress on the engineering strategies of practical energy nanodevices, by computational *de novo* design of protein assemblies, by programming protein-nanomaterials interface and other experimental approaches such as directed evolution and synthetic biology. Major breakthroughs are anticipated in the mass production of engineered protein nanomaterials and the long-term stability of these biomaterials under various operation conditions, such as highly oxidizing conditions for water oxidation. To explore these useful bioengineering principles, more diverse designer functionalities and programmability of protein building blocks should be exploited.

5.2 Nanomedicine

The molecular basis of human diseases largely involves in either protein functions alone or the interactions between proteins and other biological molecules. Proteins are either drug target or involved in the medicinal intervention. Protein nanomaterials are therefore naturally suited for a wide spectrum of applications in nanomedicine, including drug development, cancer therapy, biomedical diagnostics, tissue repair, and medical devices. First, protein nanomaterials themselves can be presented as part of treatment of diseases. Some drugs that hybridize proteins and nanoparticles are commercially available or in human clinical trials. For example, Abraxane was approved by the U.S. Food and Drug Administration (FDA) to treat breast cancer, non-small-cell lung cancer (NSCLC) and pancreatic cancer, and is a nanoparticle albumin bound paclitaxel (Green 2006). Second, biomedical diagnostics is another area where tools and devices are being developed based on protein nanomaterials. Using antibody-coated nanoparticle, the sensitivity of detecting target protein can be boosted to attomolar concentration (Nam et al. 2003). Self-assembled biocompatible nanodevices could detect, evaluate, treat diseases and be integrated as a part of health care information system that reports the diagnostic results to clinical doctors automatically. Third, protein nanomaterials are indispensible components of lab-on-a-chip technology (Yager et al. 2006). Magnetic nanoparticles, bound to a suitable antibody, are used to label specific molecules, structures or microorganisms. The nanopore protein α-hemolysin has been integrated into nanofluidic devices for single-molecule analysis and biosensing, converting strings of nucleotides or peptides directly into electronic signatures (Clarke et al. 2009, Wang et al. 2011, Rodriguez-Larrea and Bayley 2013). Fourth, protein nanomaterials may be used as part of tissue engineering to help reproduce or repair damaged tissue using suitable nanomaterial-based scaffolds and growth factors (Ellis-Behnke et al. 2006). Tissue engineering and regenerative medicine could replace conventional treatments like organ transplants or artificial implants. Protein hybrid with inorganic nanomaterials such as graphene, carbon nanotubes, molybdenum disulfide and tungsten disulfide may be used as reinforcing agents to fabricate mechanically strong biodegradable polymeric nanocomposites for bone tissue engineering applications (Lalwani et al. 2013). The addition of these nanoparticles in the polymer matrix at low concentrations (~0.2 weight %) leads to significant improvements in the compressive and flexural mechanical properties of polymeric nanocomposites. Potentially, these nanocomposites may be used as a novel, mechanically strong, lightweight composite as bone implants. Finally, how many other areas in nanomedicine that protein nanomaterials could pay a critical role is likely bound by our imagination rather than our knowledge. The rapid development of various computational and experimental tools in both protein nanomaterial engineering and molecular medicine will certainly enable many uncharted applications.

6. Perspective

Protein-based nanomaterials and nanotechnology have revolutionized many important areas in molecular biology and biomedicine, especially in the design and manipulation of proteins, their higher-order assembly and their biomedical applications at the molecular and cellular level. The marriage of molecular biology and nanotechnology opens up the possibility of manipulating atoms, molecules and structures at single-molecule and sub-molecular levels, with the potential for a wide spectrum of applications. However, the areas of protein-based nanomaterials are still in their early age of development, because our understanding of protein universe is still rather limited. Perhaps the most exciting perspective is that the protein nanomaterials will evolve into a new world of programmable nanoarchitecture and nanomachines with highly defined properties and functionality. To this end, a number of challenges remain to be addressed.

First, there are innumerous proteins whose structures and dynamics remain unknown. Even for most known protein structures, their physiological dynamics and functional interactions with other proteins remain elusive. It is not yet fully understood how other protein nanomachines existing in nature work energetically and biochemically in detail. For example, there are no high-resolution structures for the full-length complexes of bacterial flagellar rotor, making it unclear how to engineer these highly efficient, elegant nanomachines reversely. Second, the computational tools developed over the last several decades in predicting proteins structure is not yet capable of predicting sophisticated complex structures or quaternary structures, which are required to build higher-order artificial protein nanomachines and nanoarchitectures. Third, there is still lack of technology and tools to characterize the kinetics and dynamics of working complex nanomachines at the atomic level, leaving the energetic mechanism of most protein nanomachines not fully appreciated. The most recent development of cryo-electron microscopy implies such a possibility; but its technical capability awaits further growth to address this challenge. Finally, it remains to be answered how the building blocks of engineered protein nano-engines are assembled into higher-order nanoscale system with smart properties through bottom-up approaches. Addressing all these challenges will mostly require highly multidisciplinary and interdisciplinary studies that combine a wide spectrum of exploratory technologies and tools. The future technology advancement at the interface between material sciences and molecular structural biology, such as *de novo* protein prediction, single-molecule cryo-electron microscopy and super-resolution light microscopy, will hopefully provide essential tool boxes and practical solutions to design, characterize, and innovate programmable smart protein nanomaterials in greater detail, magnitude and scale.

References

Alberts, B., A. Johnson, J. Lewis, M. Raff and K. Roberts. 2007. Molecular Biology of the Cell. Garland Science, New York.

Allegretti, M., N. Klusch, D.J. Mills, J. Vonck, W. Kuhlbrandt and K.M. Davies. 2015. Horizontal membrane-intrinsic α-helices in the stator a-subunit of an F-type ATP synthase. Nature doi:10.1038/nature14185.

Alushin, G.M., V.H. Ramey, S. Pasqualato, D.A. Ball, N. Grigorieff, A. Musacchio and E. Nogales. 2010. The Ndc80 kinetochore complex forms oligomeric arrays along microtubules. Nature 467: 805–810.

Anfinsen, C.B. 1973. Principles that govern the folding of protein chains. Science 181: 223–230.

Anfinsen, C.B. and E. Haber. 1961. Studies on the reduction and re-formation of protein disulfide bonds. J. Biol. Chem. 236: 1361–1363.

Arndt, K.M. and K.M. Muller. 2007. Protein Engineering Protocols. Humana Press, Totowa, New Jersey.

Ban, N., P. Nissen, J. Hansen, P. Moore and T. Steitz. 2000. The complete atomic structure of the large ribosomal subunit at 2.4 Å resolution. Science 289: 905–20.

Barrick, J.E., D.S. Yu, S.H. Yoon, H. Jeong, T.K. Oh, D. Schneider, R.E. Lenski and J.F. Kim. 2009. Genome evolution and adaptation in a long-term experiment with *Escherichia coli*. Nature 461: 1243–1247.

Berneche, S. and B. Roux. 2001. Energetics of ion conduction through the K$^+$ channel. Nature 414: 73–77.

Bershitsky, S.Y., A.K. Tsaturyan, O.N. Bershitskaya, G.I. Mashanov, P. Brown, R. Burns and M.A. Ferenczi. 1997. Muscle force is generated by myosin heads stereospecifically attached to actin. Nature 388: 186–190.

Block, S.M. and H.C. Berg. 1984. Successive incorporation of force-generating units in the bacterial rotary motor. Nature 309: 470–473.

Branden, C. and J. Tooze. 1991. Introduction to Protein Structure. Garland, New York.

Brodin, J.D., X.I. Ambroggio, C. Tang, K.N. Parent, T.S. Baker and F.A. Tezcan. 2012. Metal-directed, chemically tunable assembly of one-, two- and three-dimensional crystalline protein arrays. Nature Chem. 4: 375–382.

Caspar, D.L.D. and A. Klug. 1962. Physical principles in the construction of regular viruses. Q. Biol. 27: 1–24.

Clarke, J., H.C. Wu, L. Jayasinghe, A. Patel, S. Reid and H. Bayley. 2009. Continuous base identification for single-molecule nanopore DNA sequencing. Nat. Nanotechnol. 4: 265–270.

Dahiyat, B.I. and S.L. Mayo. 1997. *De novo* protein design: fully automated sequence selection. Science 278: 82–87.

Dalby, P.A. 2011. Strategy and success for the directed evolution of enzymes. Curr. Opin. Struct. Biol. 21: 473–480.

de Loubresse Garreau, N., I. Prokhorova, W. Holtkamp, M.V. Rodnina, G. Yusupova and M. Yusupov. 2014. Structural basis for the inhibition of the eukaryotic ribosome. Nature 513: 517–522.

Doyle, D.A., J. Morais Cabral, R.A. Pfuetzner, A. Kuo, J.M. Gulbis, S.L. Cohen, B.T. Chait and R. MacKinnon. 1998. The structure of the potassium channel: molecular basis of K$^+$ conduction and selectivity. Science 280: 68–77.

Ellis-Behnke, R.G., Y.X. Liang, S.W. You, D.K.C. Tay, S. Zhang, K.F. So and G.E. Schneider. 2006. Nano neuro knitting: Peptide nanofiber scaffold for brain repair and axon regeneration with functional return of vision. Proc. Natl. Acad. Sci. USA 103: 5054–5059.

Fletcher, D.A. and D. Mullins. 2010. Cell mechanics and the cytoskeleton 463: 485–492.

Fuchs, E. and D.W. Cleveland. 1998. A structural scaffolding of intermediate filaments in health and disease. Science 279: 514–519.

Galland, R., P. Leduc, C. Guerin, D. Peyrade, L. Blanchoin and M. Thery. 2013. Fabrication of three-dimensional electrical connections by means of directed actin self-organization. Nat. Mater. 12: 416–421.

Gee, M.A., J.E. Heuser and R.B. Vallee. 1997. An extended microtubule-binding structure within the dynein motor domain. Nature 390: 636–639.

Gotte, L., M.G. Giro, D. Volpin and R.W. Horne. 1974. The ultrastructural organization of elastin. J. Ultrastruct. Res. 46: 23–33.

Gotte, L., D. Volpin, R.W. Horne and M. Mammi. 1976. Electron microscopy and optical diffraction of elastin. Micron 7: 95–102.

Green, M.R. 2006. Abraxane(R), a novel Cremophor(R)-free, albumin-bound particle form of paclitaxel for the treatment of advanced non-small-cell lung cancer. Annals of Oncology 17: 1263–1268.

Gunning, P.W., U. Ghoshdastider, S. Whitaker, D. Popp and R.C. Robinson. 2015. The evolution of compositionally and functionally distinct actin filaments. J. Cell Sci. doi: 10.1242/jcs.165563.

Hanukoglu, I. and E. Fuchs. 1982. The cDNA sequence of a human epidermal keratin: divergence of sequence but conservation of structure among intermediate filament proteins. Cell 31: 243–252.

Harbury, P.B., J.J. Plecs, B. Tidor, T. Alber and P.S. Kim. 1998. High-resolution protein design with backbone freedom. Science 282: 1462–7.

Hermann, H., H. Bar, L. Kreplak, S.V. Strelkov and U. Aebi. 2007. Intermediate filaments: from cell architecture to nanomechanics. Nat. Rev. Mol. Cell Biol. 8: 562–573.

Hille, B. 2001. Ion channels of excitable membranes. Sinauer Associates, Sunderland, Massachusetts.

Hoersch, D., S.H. Roh, W. Chiu and T. Kortemme. 2013. Reprogramming an ATP-driven protein machine into a light-gated nanocage. Nat. Nanotechnol. 8: 928–932.

Hu, C., M. Ahemed, T.J. Melia, T.H. Sollner, T. Mayer and J.E. Rothman. 2003. Fusion of cells by flipped SNAREs. Science 300: 1745–1749.

Hudalla, G.A., T. Sun, J.Z. Gasiorowski, H. Han, Y.F. Tian, A.S. Chong and J.H. Collier. 2014. Gradated assembly of multiple proteins into supramolecular nanomaterials. Nat. Mater. 13: 829–836.

Ishii, D., K. Kinbara, Y. Ishida, N. Ishii, M. Okochi, M. Yohda and T. Aida. 2003. Chaperonin-mediated stabilization and ATP-triggered release of semiconductor nanoparticles. Nature 423: 628–632.

Itoh, H., A. Takahashi, K. Adachi, H. Noji, R. Yasuda, M. Yoshida and K. Kinosita. 2004. Mechanically driven ATP synthesis by F_1-ATPase. Nature 427: 465–468.

Nelson, D. 2008. Lehininger Principles of Biochemistry. W.H. Freeman and Company, New York.

Jentsch, T.J., C.A. Hubner and J.C. Fuhrmann. 2004. Ion channels: Function unraveled by dysfunction. Nat. Cell Biol. 6: 1039–1047.

Jiang, Y., V. Ruta, J. Chen, A. Lee and R. MacKinnon. 2003. The principle of gating charge movement in a voltage-dependent K^+ channel. Nature 423: 42–48.

Johnson, J.E. and J.A. Speir. 2009. Desk Encyclopedia of General Virology. pp. 115–123. Academic Press, Boston.

King, N.P., W. Sheffler, M.R. Sawaya, B.S. Vollmar, J.P. Sumida, I. Andre, T. Gonen, T.O. Yeates and D. Baker. 2012. Computational design of self-assembling protein nanomaterials with atomic level accuracy. Science 336: 1171–1174.

King, N.P., J.B. Bale, W. Sheffler, D.E. McNamara, S. Gonen, T. Gonen, T.O. Yeates and D. Baker. 2014. Accurate design of co-assembling multi-component protein nanomaterials. Nature 510: 103–108.

Kon, T., T. Oyama, R. Shimo-Kon, K. Imamula, T. Shima, K. Sutoh and G. Kurisu. 2012. The 2.8 Å crystal structure of the dynein motor domain. Nature 484: 345–350.

Kuhlman, B., G. Dantas, G.C. Ireton, G. Varani, B.L. Stoddard and D. Baker. 2003. Design of a novel globular protein fold with atomic-level accuracy. Science 302: 1364–1368.

Kull, F.J., E.P. Sablin, R. Lau, R.J. Fletterick and R.D. Vale. 1996. Crystal structure of the kinesin motor domain reveals a structural similarity to myosin. Nature 380: 550–555.

Lalwani, G., A.M. Henslee, B. Farshid, L. Lin, F.K. Kasper, Y.X. Qin, A.G. Mikos and B. Sitharaman. 2013. Two-dimensional nanostructure-reinforced biodegradable polymeric nanocomposites for bone tissue engineering. Biomacromolecules 14: 900–909.

Lau, W.C. and J.L. Rubinstein. 2012. Subnanometre-resolution structure of the intact Thermus thermophiles H+-driven ATP synthase. Nature 481: 214–218.

Leaver-Fay, A., M. Tyka, S.M. Lewis, O.F. Lange, J. Thompson, R. Jacak, K. Kaufman, P.D. Renfrew, C.A. Smith, W. Sheffler, I.W. Davis, S. Cooper, A. Treuille, D.J. Mandell, F. Richter, Y.E. Ban, S.J. Fleishman, J.E. Corn, D.E. Kim, S. Lyskov, M. Berrondo, S. Mentzer, Z. Popović, J.J. Havranek, J. Karanicolas, R. Das, J. Meiler, T. Kortemme, J.J. Gray, B. Kuhlman, D. Baker and P. Bradley. 2011. ROSETTA3: an object-oriented software suite for the simulation and design of macromolecules. Methods Enzymol. 487: 545–574.

Lee, J.H., J.H. Lee, Y.J. Lee and K.T. Nam. 2013. Protein/peptide based nanomaterials for energy application. Curr. Opin. Biotechnol. 24: 599–605.

Levinthal, C. 1968. Are there pathways for protein folding. Journal de Chimie Physique et de Physico-Chimie Biologique 65: 44–45.

Lewin, B., L. Cassimeris, V.R. Lingappa and G. Plopper. 2007. Cells. Jones and Bartlett Publishers, Sudbury, Massachusetts.

Lin, S.C., Y.C. Lo and H. Wu. 2010. Helical assembly in the MyD88-IRAK4-IRAK2 complex in TLR/IL-1R signalling. Nature 465: 885–890.

Long, S.B., E.B. Campbell and R. MacKinnon. 2005a. Voltage sensor of Kv1.2: structural basis of electromechanical coupling. Science 309: 903–908.

Long, S.B., E.B. Campbell and R. MacKinnon. 2005b. Crystal structure of a mammalian voltage-dependent Shaker family K⁺ channel. Science 309: 897–903.

Lutz, S. 2010. Beyond directed evolution—semi-rational protein engineering and design. Curr. Opin. Biotechnol. 21: 734–743.

MacBeath, G., P. Kast and D. Hilvert. 1998. Redesigning enzyme topology by directed evolution. Science 279: 1958–1961.

Mao, Y., L. Wang, C. Gu, A. Herschhorn, S.H. Xiang, H. Haim, X. Yang and J. Sodroski. 2012. Subunit organization of the membrane-bound HIV-1 envelope glycoprotein trimer. Nat. Struct. Mol. Biol. 19: 893–899.

Mao, Y., L. Wang, C. Gu, A. Herschhorn, A. Desormeaux, A. Finzi, S.H. Xiang and J.G. Sodroski. 2013. Molecular architecture of the uncleaved HIV-1 envelope glycoprotein trimer. Proc. Natl. Acad. Sci. USA 110: 12438–12443.

Marco, M.D., S. Shamsuddin, K.A. Razak, A.A. Aziz, C. Devaux, E. Borghi, L. Levy and C. Sadun. 2010. Overview of the main methods used to combine proteins with nanosystems: absorption, bioconjugation and encapsulation. Int. J. Nanomedicine 5: 37–49.

Mathé, J., A. Aksimentiev, D.R. Nelson, K. Schulten and A. Meller. 2005. Orientation discrimination of single stranded DNA inside the α-hemolysin membrane channel. Proc. Natl. Acad. Sci. USA 102: 12377–12382.

Mueckler, M., C. Caruso, S.A. Baldwin, M. Panico, I. Blench, H.R. Morris, W.J. Allard, G.E. Lienhard and H.F. Lodish. 1985. Sequence and structure of a human glucose transporter. Science 229: 941–945.

Muiznieks, L.D., A.S. Weiss and F.W. Keeley. 2010. Structural disorder and dynamics of elastin. Biochem. Cell Biol. 88: 239–250.

Murail, R., D.J. Sharkey, J.L. Daiss and H.M. Murthy. 1998. Crystal structure of Taq DNA polymerase in complex with an inhibitory Fab: the Fab is directed against an intermediate in the helix-coil dynamics of the enzyme. Proc. Natl. Acad. Sci. USA 95: 12562–12567.

Nam, J.M., C.S. Thaxton and C.A. Mirkin. 2003. Nanoparticle-based bio-bar codes for the ultrasensitive detection of proteins. Science 301: 1884–1886.

Nogales, E. 2000. Structural insights into microtubule function. Annu. Rev. Biochem. 69: 277–302.

Noji, H., R. Yasuda, M. Yoshida and K. Kinosita, Jr. 1997. Direct observation of the rotation of F₁-ATPase. Nature 386: 299–302.

Papapostolou, D., A.M. Smith, E.D.T. Atkins, S.J. Oliver, M.G. Ryadnov, L.C. Serpell and D.N. Woolfson. 2007. Engineering nanoscale order into a designed protein fiber. Proc. Natl. Acad. Sci. USA 104: 10853–10858.

Pepe, A., B. Bochicchio and A.M. Tamburro. 2007. Supramolecular organization of elastin and elastin-related nanostructured biopolymers. Nanomed. 2: 203–218.

Rayment, I., W.R. Rypniewski, K. Schmidt-Base, R. Smith, D.R. Tomchick, M.M. Benning, D.A. Winkelman, G. Wesenberg and H.M. Holden. 1993. Three-dimensional structure of myosin subfragment-1: A molecular motor. Science 261: 50–58.

Reymann, A.C., J.L. Martiel, T. Cambier, L. Blanchoin, R. Boujemaa-Paterski and M. Thery. 2010. Nucleation geometry governs ordered actin networks structure. Nat. Mater. 9: 827–832.

Richardson, J.S. and D.C. Richardson. 1989. The *de novo* design of protein structures. Trends in Biochemical Sciences 14: 304–9.

Rodrigues-Larrea, D. and H. Bayley. 2013. Multistep protein unfolding during nanopore translocation. Nat. Nanotechnol. 8: 288–295.

Rondelez, Y., G. Tresset, T. Nakashima, Y. Kato-Yamada, H. Fujita, S. Takeuchi and H. Noji. 2005. Highly coupled ATP synthesis by F_1-ATPase single molecules. Nature 433: 773–777.

Ryu, W.S., R.M. Berry and H.C. Berg. 2000. Torque-generating units of the flagellar motor of *Escherichia coli* have a high duty ratio. Nature 403: 444–447.

Schluenzen, F., A. Tocilj, R. Zarivach, J. Harms, M. Gluehmann, D. Janell, A. Bashan, H. Bartels, I. Agmon, F. Franceschi and A. Yonath. 2000. Structure of functionally activated small ribosomal subunit at 3.3 Å resolution. Cell 102: 615–23.

Schnitzer, M.J. and S.M. Block. 1997. Kinesin hydrolyses one ATP per 8-nm step. Nature 388: 386–390.

Shingyoji, C., H. Higuchi, M. Yoshimura, E. Katayama and T. Yanagida. 1998. Dynein arms are oscillating force generators. Nature 393: 711–714.

Song, L., M.R. Hobaugh, C. Shustak, S. Cheley, H. Bayley and J.E. Gouaux. 1996. Structure of staphylococcal alpha-hemolysin, a heptameric transmembrane pore. Science 274: 1859–1866.

Stock, D., A.G. Leslie and J.E. Walker. 1999. Molecular architecture of the rotary motor in ATP synthase. Science 286: 1700–1705.

Stuhmer, W., F. Conti, H. Suzuki, X.D. Wang, M. Noda, N. Yahagi, H. Kubo and S. Numa. 1989. Structural parts involved in activation and inactivation of the sodium channel. Nature 339: 597–603.

Sutton, R.B., D. Fasshauer, R. Jahn and A.T. Brunger. 1998. Crystal structure of a SNARE complex involved in synaptic exocytosis at 2.4 Å resolution. Nature 395: 347–353.

Thomas, D.R., N.R. Francis, C. Xu and D.J. DeRosier. 2006. The three-dimensional structure of the flagellar rotor from a clockwise-locked mutant of Salmonella enterica serovar Typhimurium. J. Bacteriol. 188: 7039–7048.

Toyoshima, C., M. Nakasako, H. Nomura and H. Ogawa. 2000. Crystal structure of the calcium pump of sarcoplasmic reticulum at 2.6 Å resolution. Nature 405: 647–655.

Toyoshima, C. and H. Nomura. 2002. Structural changes in the calcium pump accompanying the dissociation of calcium. Nature 418: 605–611.

Urnavicius, L., K. Zhang, A.G. Diamant, C. Motz, M.A. Schlager, M. Yu, N.A. Patel, C.V. Robinson and A.P. Carter. 2015. The structure of the dynactin complex and its interaction with dynein. Science 347: 1441–1446.

Vale, R.D., T. Funatsu, D.W. Pierce, L. Romberg, Y. Harada and T. Yanagida. 1996. Direct observation of single kinesin molecules moving along microtubules. Nature 380: 451–453.

Vauthey, S., S. Santoso, H. Gong, N. Waston and S. Zhang. 2002. Molecular self-assembly of surfactant-like peptides to form nanotubes and nanovesicles. Proc. Natl. Acad. Sci. USA 99: 5355–5360.

Veesler, D., T.S. Ng, A.K. Sendamarai, B.J. Eilers, C.M. Lawrence, S.M. Lok, M.J. Young, J.E. Johnson and C.Y. Fu. 2013. Atomic structure of the 75 MDa extremophile Sulfolobus turreted icosahedral virus determined by CryoEM and X-ray crystallography. Proc. Natl. Acad. Sci. USA 110: 5504–5509.

Voigt, C.A., S. Kauffman and Z.G. Wang. 2000. Rational evolutionary design: the theory of *in vitro* protein evolution. Adv. Protein Chem. 55: 79–160.

Von der Ecken, J., M. Muller, W. Lehman, D.J. Manstein, P.A. Penczek and S. Raunser. 2015. Structure of the F-actin-tropomyosin complex. Nature 519: 114–117.

Vo-Dinh, T. 2005. Protein Nanotechnology: Protocols, Instrumentation, and Applications. Humana Press, Totowa, New Jersey.

Wang, H. and G. Oster. 1998. Energy transduction in the F_1 motor of ATP synthase. Nature 396: 279–282.

Wang, Y., D. Zhang, Q. Tan, M.X. Wang and L.Q. Gu. 2011. Nanopore-based detection of circulating microRNAs in lung cancer patients. Nat. Nanotechnol. 6: 668–674.

Weisenberg, R.C. 1972. Microtubule formation *in vitro* in solutions containing low calcium concentrations. Science 177: 1104–1105.

Whittaker, M., E.M. Wilson-Kubalek, J.E. Smith, L. Faust, R.A. Milligan and H.L. Sweeney. 1995. A 35-Å movement of smooth muscle myosin on ADP release. Nature 378: 748–751.

Wilkens, S., E. Vasilyeva and M. Forgac. 1999. Structure of the vacuolar ATPase by electron microscopy. J. Biol. Chem. 274: 31804–31810.

Wu, H. 2013. Higher-order assemblies in a new paradigm of signal transduction. Cell 153: 1–6.

Wu, X., Z.Y. Yang, Y. Li, C.M. Hogerkorp, W.R. Schief, M.S. Seaman, T. Zhou, S.D. Schmidt, L. Wu, L. Xu, N.S. Longo, K. McKee, S. O'Dell, M.K. Louder, D.L. Wycuff, Y. Feng, M. Nason, N. Doria-Rose, M. Connors, P.D. Kwong, M. Roederer, R.T. Wyatt, G.J. Nabel and J.R. Mascola. 2010. Rational design of envelope identifies broadly neutralizing human monoclonal antibodies to HIV-1. Science 329: 856–61.

Yager, P., T. Edwards, E. Fu, K. Helton, K. Nelson, M.R. Tam and B.H. Weigl. 2006. Microfluidic diagnostic technologies for global public health. Nature 442: 412–418.

Yokoi, H., T. Kinoshita and S. Zhang. 2005. Dynamic reassembly of peptide RADA16 nanofiber scaffold. Proc. Natl. Acad. Sci. USA 102: 8414–8419.

Zhang, X., L. Jin, Q. Fang, W.H. Hui and Z.H. Zhou. 2010. 3.3 Å cryo-EM structure of a nonenveloped virus reveals a priming mechanism for cell entry. Cell 141: 472–482.

Zhao, G., J.R. Perilla, E.L. Yufenyuy, X. Meng, B. Chen, J. Ning, J. Ahn, A.M. Gronenborn, K. Schulten, C. Aiken and P. Zhang. 2013. Mature HIV-1 capsid structure by cryo-electron microscopy and all-atom molecular dynamics. Nature 497: 643–646.

Zhao, M., S. Wu, Q. Zhou, S. Vivona, D.J. Cipriano, Y. Cheng and A.T. Brunger. 2015. Mechanistic insights into the recycling machine of the SNARE complex. Nature 518: 61–67.

Zhou, Y., J.H. Morais-Cabral, A. Kaufman and R. MacKinnon. 2001. Chemistry of ion coordination and hydration revealed by a K^+ channel-Fab complex at 2.0 Å resolution. Nature 414: 43–38.

Smart Materials for
Controlled Drug Release

Linfeng Chen

ABSTRACT

Smart materials involving in the applications of chemotherapeutics have been increasingly developing for several decades with the advance in materials science and nanotechnology. This chapter presents a comprehensive view of the most promising smart materials for controlled drug release. The contents are divided into three main parts based on the materials, including smart inorganic materials, smart polymer materials, and other emerging nanomaterials. These intelligent materials are designed with fascinating properties, such as biocompatibility, a long circulation time in the bloodstream, imaging, and target recognition. Especially, they could be responsive to internal or external stimulus, such as pH, temperature, light, enzymes, magnetism, and chemicals. The chapter is essential for a wide audience, including undergraduate students, graduate students, PhD students, as well as independent researchers with diverse backgrounds across chemistry, biomedical science, materials sciences, and biotechnology.

Department of Materials Science and Engineering, Technion-Israel Institute of Technology, 32000, Haifa, Israel.
Email: lfchen@tx.technion.ac.il; chlfstorm@gmail.com

1. Introduction

A disease is one of the secret powers, which could cause a disaster to the human civilization, but at the same time promote the fast development of science and technology. As one of the typical diseases, cancers have become a leading cause of death worldwide. According to the International Agency for Research on Cancer, approximately 14.1 million new cases of cancer occurred globally, which caused about 8.2 million deaths in 2012 (Stewart and Wild 2014). Cancer is defined as a group of diseases, which are featured by rapid creation of abnormal cells which can grow beyond their boundaries. In order to solve the problem, various methods including chemotherapy, radiation, surgery, palliative care, and immunotherapy are utilized, which bring serious side-effects, or can not completely defeat the disease. With the development of nanoscience and nanotechnology, controlled drug release (CDR) systems bring new hope for managing the problem.

In a broad sense, CDR refers to the strategies involving the controlling of the release event of drugs or other molecules. Specifically, smart materials are one of the critical elements. Smart nanomaterials are defined as the materials in the nanoscale size, which are sensitive to stimulus, such as pH, light, temperature, electricity, magnetic field, biomolecules, and enzymes. The smart nanomaterials applied in CDR could achieve target delivery and release of drugs triggered by the stimulus. By controlling the dose, time, and release site, the pharmacological effects are enhanced and side effects are reduced. Smart nanomaterials are exactly the development of concept "Magic Bullet" proposed by Ehrlich as early as in 1906 (Strebhardt and Ullrich 2008).

From the first report about the related release of solid to the smart materials studied in CDR, it took more than one hundred years and various kinds of materials have been employed as the formulations for CDR (Fig. 1). The evolution of CDR could be traced back to 1897 when Noyes and Whitney investigated the process of dissolution (Noyes and Whitney 1897). The breakthroughs happened in 1960s when Folkman designed a silicon rubber device which could prolong the drug release. He also achieved the release of anesthetic agents in rabbits, which was proposed to be as the planted device with controlled release of drugs (Folkman and Long 1964, Folkman et al. 1966). It wasn't until the 1970s that the nanomaterials were first proposed, followed by the first usage in the clinical in the 1980s. In the 1990s, the field of CDR came into the smart nanomaterials era (Hoffman 2008).

The evolution of CDR is dependent on the engineered materials. During the past decades, the advances of materials science have offered inspiration not only for the modification of materials, but also for numerous possibilities of new materials with potential applications in biomedicine. As the largest

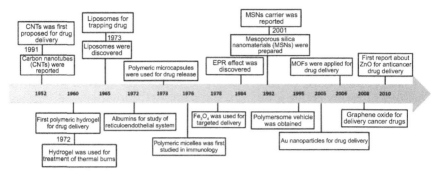

Figure 1. The development of various materials for applications in controlled drug release.

group, polymers have garnered much attention from the beginning, which is going to the present, including hydrogels, capsules, polymersomes, and micelles. In the past couple of decades, inorganic materials have received increasingly interests in the field of CDR, such as mesoporous silica nanomaterials (MSNs), iron oxide nanomaterials (IONMs), and zinc oxide (ZnO). In recent years, some emerging materials are also developed and employed in this field due to their unique properties, e.g., gold nanoparticles (AuNPs), metal-organic frameworks (MOFs), graphene and derivatives, and carbon nanotubes (CNTs). Figure 1 presents an overview of the evolution of CDR based on diverse materials.

2. Smart Inorganic Nanomaterials for CDR

2.1 Mesoporous Silica Nanoparticles

2.1.1 Introduction

In 1992, the scientists in Mobil Corporation reported the discovery of a new family of inorganic materials with uniform mesopores (pore size in the range of 2–50 nm), which has received extensive attention in the next two decades, and were employed widely in catalysis, adsorption, separation, sensors, tissue engineering, and CDR (Kresge et al. 1992, Beck et al. 1992). This kind of materials was named M41S, which could be synthesized by sol-gel chemistry method with cationic surfactants as the sacrificed templates. By altering the molar ratio of surfactant to silica source, it's able to obtain mesoporous silica with hexagonal, cubic, and lamellar structures (Fig. 2) (Kresge and Roth 2013).

The application of MSNs in CDR emerged in 2001, when Vallet-Regi used MSNs as carriers to trap and deliver anti-inflammatory drug ibuprofen (Vallet-Regi et al. 2001). As a kind of carrier materials, MSNs possess several

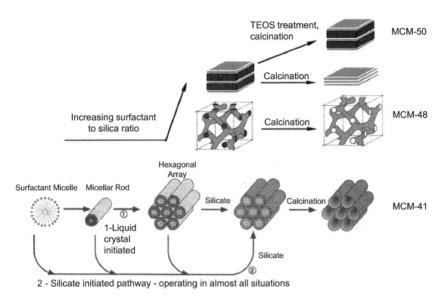

Figure 2. The mechanism for formation of mesoporous silica nanomaterials with different structures (Kresge and Roth 2013). Copyright 2013 Royal Society of Chemistry.

unique advantages. (1) The large surface area. MSNs have highly porous interior structure, which could encapsulate a distinctively high payload of drugs (200–600 mg/g silica), enhancing the chemotherapeutic efficacy (He et al. 2010). (2) Controllable particle size. MSNs could be easily synthesized by stöber method with the size of diameter from 10 nm to 1000 nm. Endocytosis efficacy is related with the particle size. Especially, particle dimension of 50–100 nm is optimal for MSNs as nanocarriers with a high efficiency of cellular uptake. (3) Tunable pore size. By changing surfactants with different chain length and addition of auxiliary organics, the mesopore size could be controlled in the range of 2–10 nm, which could be employed to encapsulate various molecules with different size. (4) Chemical stability. MSNs are stable *in vivo* and protect the payloads from bio-erosions, which is important for a carrier to deliver drugs to target sites (Borisova et al. 2011). (5) Facile surface functionalization. MSNs exhibit extensive hydroxyl groups on the surface, which could enable the decoration of functional molecules onto MSNs, and finally give rise to smart MSNs vehicles with stimuli-responsive ability. (6) Biocompatibility. Therefore, MSNs have become one of the most promising smart candidates for CDR. In the past decade, numerous smart MSNs have been designed and investigated *in vitro* and *in vivo* with stimuli-responsive ability, including pH, light, temperature, enzyme, glucose, ions, and magnetic field (Fig. 3).

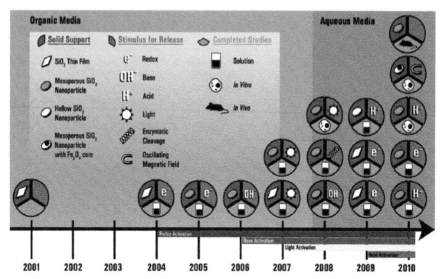

Figure 3. Controlled release systems based on mesoporous silica nanomaterials have been developing fast in the past decade (Ambrogio et al. 2011). Copyright 2011 American Chemical Society.

2.1.2 Recent Advances of Stimuli-Responsive MSNs

(1) pH-Responsive MSNs

pH-responsive MSNs for CDR are of special interests in the past decade. The development of such smart materials is based on the facts that the pHs are different in the bloodstream (pH 7.4), cancer or inflammatory tissues (pH 6.5), endosome (pH 5.5–6.0), and lysosome (pH 4.5–5.0). To design such smart vehicles, pH-sensitive components are selected and attached on the outer surface of MSNs, which could be responsive to the change of pH and release the loaded cargo. The mechanisms of triggering the flow of cargo can be divided into six main categories, including hydrolysis of chemical bonds (I) (Liu et al. 2010), disassembly of superamolecule (II) (Meng et al. 2010, Zhao et al. 2010), structure conversion (III) (Chen et al. 2011b, Xue and Findenegg 2012), dissociation of coordination bonds (IV) (Xing et al. 2012), dissolution of gate keeper (V) (Muharnmad et al. 2011), and electrostatic interactions of gate keeper (VI) (Yang et al. 2005) (Table 1).

Liu et al. reported a pH-responsive MSNs vehicle by capping AuNPs on the surface through an acid-labile acetal linker. The cargo-loaded MSNs vehicle was stable in basic condition, but allowed the escape of cargo when the pH was changed to acidic condition (e.g., 2.0 and 4.0), which was attributed to the hydrolysis of acetal linker (Liu et al. 2010). Meng et al. have deeply investigated and developed a series of pH-responsive

Table 1. The categories of pH-responsive MSNs.

Categories	Mechanisms	Examples	Ref.
I	Hydrolysis of chemical bonds	H^+	Liu et al. 2010. Copyright 2010 American Chemical Society
II	Disassembly of supramolecule	H^+	Meng et al. 2010. Copyright 2010 American Chemical Society
III	Structure conversion	H^+ HO^-	Chen et al. 2011b. Copyright 2011 Royal Society of Chemistry
IV	Dissociation of Coordination bonds	H^+	Xing et al. 2012
V	Dissolution of gate keeper	H^+	Muharnmad et al. 2011. Copyright 2011 American Chemical Society
VI	Electrostatic interaction of gate keeper	H^+ HO^-	Yang et al. 2005. Copyright 2005 American Chemical Society

MSNs vehicles based on the assembly/disassembly of supramolecular machines. For example, they modified aromatic amines (as stalk) on the outer surface of MSNs, and assembled β-cyclodextrin (β-CD, as cap) to block the pores through non-covalent bonds at pH 7.4. Lowering the pH led to the protonation of the aromatic amines, which removed the β-CD caps and allowed the release of cargo. This system was further applied to human differentiated myeloid (THP-1) and squamous carcinoma (KB-31) cell lines, which could automatically release drug in endosomal acidic condition and result in the cell apoptotic (Meng et al. 2010). In recent years, DNA molecules attract considerable attention due to their unparalleled base pairing rule, and are investigated widely in the field of CDR. More information about DNA is introduced in Chapter 2. Chen et al. demonstrated a novel MSNs vehicle by controlling the AuNPs caps through the assembly of DNA nanoswitches. The reversible structure conversion of DNA strands in alternative acidic and alkaline solution would regulate the opening/closing of pore voids (Chen et al. 2011b).

(2) Light-Responsive MSNs

Light-responsive MSNs was first reported by Mal et al. in 2003. UV light-sensitive coumarin molecules were immobilized on the surface of MSNs, which could storage and release of included organic payloads through the reversible intermolecular dimerization of coumarin derivatives (Mal et al. 2003). Since then, light-responsive MSNs have drawn increasing interests due to their potential applications in cancer therapeutics by light which is a kind of remote, and directional controllable method involving the spatial and temporal guidance (Wen et al. 2012b, Liu et al. 2013, Chen et al. 2014, Zhao et al. 2014).

Chen et al. reported an interesting UV-triggered MSNs vehicle for CDR, which was achieved by controlling the wetting property of MSNs surface coated with spiropyran (SP) and perfluorodecyltriethoxysilane (PFDTES) (Chen et al. 2014). UV exposure would cause the surface from hydrophobicity to hydrophilicity, resulting in the release of cargo. The further *in vitro* studies on EA.hy926 cells and Hela cells confirmed the feasibility of the smart vehicle for CDR triggered by light. However, the UV exposure may cause harm to living tissues and biological samples, which limits the applications of such systems to medicine. In recent years, near-infrared light (NIR) as a new method has been developing (Yang et al. 2012, Yang et al. 2013, Zhao et al. 2014), is much less damaging to normal cells and healthy tissues due to NIR, but has remarkably deeper tissue penetration (Szaciłowski et al. 2005). The generally auxiliary materials with good absorbance in NIR region include AuNPs, and upconverting nanoparticles (UCNPs). In 2012, Yang et al. reported a NIR-triggered MSNs vector fabricated with DNA attached on the outer surface and gold nanorods (GNRs) as the core. Upon NIR light irradiation, the absorbance of GNRs converted the photoenergy to heat, which causes the duplex DNA to dehybridize and unlock the pores. The *in vitro* study also showed the good feasibility of the new material regulated by NIR stimuli (Fig. 4) (Yang et al. 2012).

(3) Thermo-Responsive MSNs

As the temperature in inflammatory sites is higher than that in normal tissues, thermo-responsive MSNs for CDR are required to treat such diseases. Poly(N-isopropylacrylamide) (PNIPAM) and its derivatives are the widely investigated thermo-sensitive materials which are featured by the solubility change at the lower critical solution temperature (LCST) and hold potential applications in surface, rheology, and drug delivery. For instance, You et al. reported a thermo-responsive MSNs carrier decorated with PNIPAM on the outer surface for CDR. When the temperature was

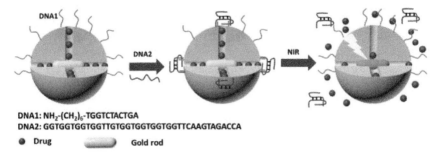

DNA1: NH₂-(CH₂)₆-TGGTCTACTGA
DNA2: GGTGGTGGTGGTTGTGGTGGTGGTGGTTCAAGTAGACCA

● Drug   〔 ▬ 〕 Gold rod

Figure 4. Scheme of near infrared light (NIR)-responsive MSNs for controlled drug release (Yang et al. 2012).

changed from 38°C to 25°C (LCST was about 32°C), PNIPAM would undergo from an insoluble state in water to a soluble and expanded state, resulting in the opening of pore voids and release of cargo (You et al. 2008). Actually, efforts are also devoted to develop novel MSNs vehicles based on other thermo-sensitive molecules in recent years. Aznar et al. found parafins assembled on MSNs surface could perform as a thermo-sensitive valve to control the access of pores. The parafins performing as a hydrophobic layer to block the pores initially would melt when the temperature increased to the melting point (in the range from 36°C to 38°C), which generated the release of included guests (Aznar et al. 2011). It should be noticed that another kind of thermo-responsive material which can be activated by light, is AuNPs. The absorbance of AuNPs in NIR could convert the photoenergy to heat, which affects the temperature in the surroundings and thus activates the thermo-sensitive switches (Yang et al. 2012). Recently, the denaturation of DNA molecules in high temperature also inspired the investigation of DNA-based MSNs vehicles (Schlossbauer et al. 2010).

(4) Enzyme-Responsive MSNs

Enzyme is a kind of vital biomolecule in organisms, which could selectively catalyze metabolic chemical reactions. Almost all metabolic chemical reactions in a biological cell need enzymes in order to occur at rates sufficient for life. Smart MSNs responsive to enzyme are critical to accomplish the delivery and selective release of cargo, which could enhance the efficacy of chemotherapy and reduce the toxicity. In addition, investigations indicate cancer cells overexpress specific receptors on the surface or contain special enzymes in cells, which provide researchers with new ideas to design enzyme-responsive MSNs for CDR (Patel et al. 2008, Bernardos et al. 2009, Chen et al. 2013). For example, hyaluronidase-1 (Hyal-1) is an enzyme major found in tumor cells. MSNs with hyaluronic acid (HA) modified on the surface are responsive to Hyal-1 (Chen et al. 2013). Here, HA acts both as a capping agent and a targeting molecule which could interact selectively with CD44 receptors overexpressed in solid tumor cells. This MSNs vector can deliver and release the anticancer drug doxorubicin (DOX) selectively inside the cancer cells. The other enzymes overexpressed in tumor/cancer cells are also utilized to design the enzyme-responsive MSNs vehicles, such as decarboxylase which plays a key role in tumor growth and inflammatory processes (Liu et al. 2011), and α-amylase and lipase which are on abnormal increase in acute pancreatitis (Park et al. 2009).

(5) Glucose-Responsive MSNs

Diabetes is a worldwide disease, which afflicted over 382 million people in 2013. Approximately 90% of them have to take medicine or insulin frequently to control the blood glucose level. This disease is still a challenge for scientists, which requires novel therapies. Glucose-responsive MSNs offer a new approach to treat diabetes instead of frequent insulin injection. The design of glucose-sensitive MSNs is similar to enzyme-responsive MSNs vectors, i.e., the unique interaction between the glucose and glucose oxidase. For example, Aznar et al. attached CD-modified-glucose oxidase (CD-GOx) as a cap to MSNs outer surface through assembly between the cyclodextrins and propylbenzimidazole groups. In the presence of glucose, glucose oxidase would catalyze glucose to gluconic acid, which induced the protonation of propylbenzimidazole, resulting in the removal of CD-GOx and the cargo flow (Aznar et al. 2013).

Zhao et al. reported a novel glucose-triggered MSNs vehicle using a different strategy (Zhao et al. 2009). The pore voids were successfully blocked by gluconic acid-modified insulin (G-Ins) through interaction with phenylboronic acid on MSNs surface. In the presence of other saccharides, such as glucose, the stronger interaction of phenylboronic acid groups with glucose would remove G-Ins and open the entrance. In addition, they incorporated cyclic adenosine monophosphate (cAMP), the molecule activating Ca^{2+} channels of cells, in the MSNs vehicle, which was controlled to release by glucose trigger. The smart MSNs vehicle was confirmed to work in rat pancreatic islet tumor, mouse liver, skin fibroblast, and human cervical cancer cells, which held promise in biological applications.

(6) Multifunctional MSNs

Smart MSNs for CDR have been developing from the initial stage with a simple function (first generation) that cargo escapes from pore voids by diffusion to the present stage with intelligent properties (second generation) that control the release of cargo by stimulus. But it's far away for the MSNs from applications in clinic because of the complex environment in biological bodies which impose multiple physiological and cellular barriers to foreign objects. In order to improve the efficacy of chemotherapy and reduce the adverse toxicity, multifunctional MSNs are highly desired for CDR. In recent years, significant efforts have been devoted to the development in this direction (Casasus et al. 2008, Angelos et al. 2009, Liu et al. 2009, Chen et al. 2011a, Wen et al. 2012a, Wu et al. 2013, Zhang et al. 2013a, Lee et al. 2014, Xiao et al. 2014, Zhou et al. 2014).

Angelos et al. designed a dual-controlled MSNs vehicle using two different types of nanomachines in tandem. The molecular nanomachine azobenzene derivatives responsive to optical signal was modified inside the mesochannels, while pH-switchable nanovalves based on cucurbit[6] uril (CB[6])/bisammonium assemblies were immobilized on outer surface of MSNs. The open and release of entrapped cargo required the input of both light and pH reduction, which behaved as an AND gate, providing a strategy for accomplishing more sophisticated levels of trigger with more precisely control (Angelos et al. 2009).

In 2013, Zhang et al. developed a multifunctional MSNs vector which could perform both the tracing and release of anticancer drugs in cells upon thermo/pH-coupling stimulus. This novel material was prepared with UCNPs $NaYF_4:Yb^{3+}/Er^{3+}$ as the cores, and MSNs as the shells coated by poly[(N-isopropylacrylamide)-co-(methyacrylic acid)] (PNIPMA-co-MAA). Upon exposure to NIR, the cells would be imaged by the bright green fluorescence from UCNPs, which could be used to trace the position of MSNs. Furthermore, under higher temperature/lower pH condition, anticancer drugs could be released from the hybrid vehicle, causing the death of Hela cells (Zhang et al. 2013b). Recently, Zhang et al. constructed a finely programmable MSNs carrier for tumor-triggered drug delivery in cancerous cells. The system was functionalized with β-CD as capping agents *via* a disulfide linker, which was subsequently coated with a targeting peptide containing Arginine-glycine-aspartic acid (Arg-Gly-Asp) (RGD) motif, peptide Pro-Leu-Gly-Val-Arg (PLGVR), and polyanion poly(aspartic acid) (PASP) (Fig. 5). The intelligent MSNs carrier could be specifically taken up by tumor cells and release encapsulated anticancer drugs because of the removal of β-CD induced by intracellular glutathione (GSH). The "cellular-uptake-shielding" multifunctional MSNs are of clinical importance (Zhang et al. 2013a).

2.1.3 Conclusions

In the past decade, MSNs have been developing quickly towards the application in clinical therapy. Diverse smart MSNs achieve anticancer drugs, genes, proteins and release them under controllable manners. And more and more multifunctional MSNs are also under increasing development in order to attain high levels of achievement in precise control in complex environment, enhancing chemotherapeutic efficacy and

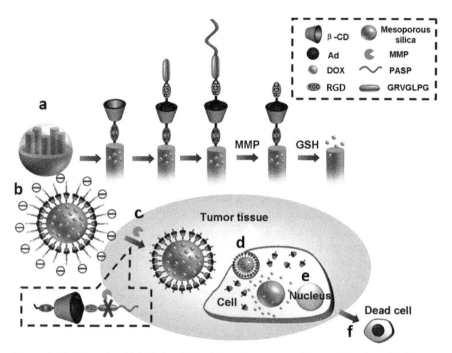

Figure 5. Multifunctional MSNs for CDR achieved by outer surface modification of β-CD cap, tumor-targeting Arg-Gly-Asp (RGD), matrix metalloproteinase substrate Pro-Leu-Gly-Val-Arg (PLGVR), and polyanion protection layer poly(aspartic acid) (PASP) (Zhang et al. 2013a). Copyright 2013 American Chemical Society.

reducing the adverse toxicity. But to develop a smart MSNs vehicle capable of delivery and release of anticancer drugs *in vivo* is still in the early stage. This task is still challenging as well as hopeful for scientists.

2.2 Iron Oxide Nanomaterials

Magnetic materials include Fe, Cobalt, Nickel, Fe_3O_4 and other metal composites. Below a critical diameter, the magnetic particles display zero coercivity, which is termed superparamagnetic. Superparamagnetic nanoparticles become magnetic in the presence of an external magnetic field, and revert to a nonmagnetic state after the removal of magnetic field. Magnetic nanoparticles are of great interests for research with wide applications in catalysis, biomedicine, magnetic resonance imaging (MRI) and data storage. As one of the most appealing candidates, IONPs have received increasing interests in the past decade, although its application in therapeutics was proposed as far back as 1978 (Senyei et al. 1978). IONPs are biodegradable and the surface could be modified with functional molecules to make it stable *in vivo* environment. IONPs less than 20 nm become super-paramagnetic when each particle consists of a single magnetic domain and thermal energy is high enough to overcome the energy barrier of magnetic flipping. IONPs perform in two ways in therapeutics. IONPs act as the vectors with both drugs or genes and functional molecules modified on the surface. The other way is IONPs are used as the core protected by polymer or inorganic shells. Both methods may accomplish the design of smart IONPs with targeting ligands and stimuli-responsive components, which could achieve the precise delivery and release of drugs under the influence of stimulus and magnetic field. In MRI, the IONPs nanoparticles generate local inhomogeneity in the magnetic field decreasing the signal, which results in the region in the body that contain SPIONs appear darker in MRI images. This information is useful in tracing the uptake of nanoparticles and monitoring the treatment of cancers.

As a kind of magnetic materials, IONPs could deliver payloads to the target site under the control of applied magnet, which is important for the treatment of diseases. Magnetic nanoparticles offer the possibility of being systematically administered but directed towards a specific target in the human body. Rao et al. reported a pH-responsive IONPs vector for transdermal delivery and release epirubicin drugs (EPI) covalently modified to IONPs through a pH-sensitive amide linker (Fig. 6).This nanomaterial is stable at the normal physiological pH with a small amount of release. But in

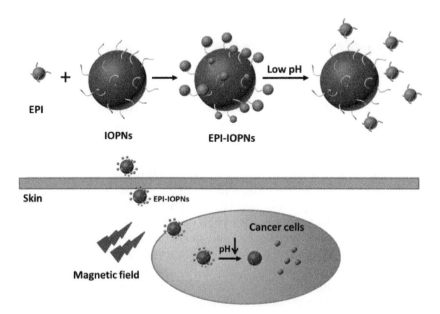

Figure 6. Superparamagnetic iron oxide nanoparticles (SIONPs) for target transdermal delivery of anticancer drugs triggered by pH (Rao et al. 2015).

pH from 6.7 to 4.5, the magnetic nanomaterial displayed fast release, which was consistent with its behavior in the *in vitro* cancer-cell experiment. A further investigation of this composite nanomaterial in human abdominal cadaver skin demonstrated the IONPs could penetrate into deeper human skin layer under the applied magnetic field. The results indicated IONPs hold promise in the application of skin cancer treatment (Rao et al. 2015).

In addition, the functionalization of IONPs on the surface brings the materials several advantages, incorporating the MRI imaging, positive targeting property, and stimuli-triggered release. For instance, IONPs were coated with redox-responsive shell of poly(ethylene glycol) (PEG) and chitosan. The materials were further modified with tumor targeting peptide chlorotoxin (CTX) and drug O^6-benzylguanine (BG) against O^6-methylguanine-DNA methyltransferase (MGMT). This multifunctional formulation was found to demonstrate redox-responsive release and trafficking of BG within human glioblastoma multiforme (GBM) cells. The *in vivo* study also showed its efficient treatment of GBM xenograft tumors by release drugs under stimulus in target sites (Stephen et al. 2014).

In summary, IONPs have shown broad potential in MRI imaging and drug delivery and release, separation *in vitro* and *in vivo*. The critical parameters for designing optimal IONPs applied in clinical applications include the size, shape, and functionalization. The nanotechnology and surface engineering strategies will further promote the development of such smart materials.

2.3 Zinc Oxide Nanostructures

Zinc oxide (ZnO) is a typical semiconductor with a wide bandgap of about 3.4 eV. It possesses good transparency and high electron mobility, which attracts wide interests in electronics. Compared with traditional organic fluorescence molecules, nanocrystal ZnO possesses advantageous photoluminescence properties, including a broad absorption, a narrow and symmetric emission band, a large Stokes shifts and weak self-adsorption, a tunable emission wavelength based on quantum size effects and high stability against photo-bleaching, which make it a promising candidate in bioimaging. As a cheap nanomaterial with low toxicity, unprotected ZnO quantum dots (QDs) are dissolved at pH 5 aqueous solutions, which make it a pH-responsive candidate for application in biomedicine.

ZnO nanostructured materials were first applied in drug delivery and release in 2010 (Yuan et al. 2010, Barick et al. 2010). Yuan et al. fabricated a blue-light emitting ZnO QDs with diameter in 2–4 nm by a chemical hydrolysis method. ZnO was combined with biodegradable chitosan (N-acetylglucosamine) conjugated with folic acid for tumor target drug delivery. DOX was trapped between ZnO QDs and polymer shells. The target delivery and uptake of ZnO QDs by cancer cells caused the decomposition of ZnO and release of DOX due to the intracellular acidity (Yuan et al. 2010). Zhang et al. also reported a pH-responsive ZnO release system. It was composed of ZnO core with a diameter of 3 nm and polyacrylamide shell which is nontoxic to animals. This material exhibited green under exposure to UV light. When internalized by cancer cells, the ZnO vectors loaded with DOX were decomposed in the acidic endosome/lysosome, resulting in the death of cancer cell (Zhang et al. 2013c).

2.4 Carbon Nanotubes

Carbon nanotubes (CNTs), first invented in 1991, belong to the fullerene family of carbon allotropes with cylindrical shape. They are a huge cylindrical large molecules consisting of a hexagonal arrangement of sp^3 hybridized carbon atoms with C-C distance about 1.4 Å. CNTs with the wall consisting of single and multiple layers of graphene sheets, are termed as single-wall carbon nanotubes (SWCNTs) and multi-wall carbon nanotubes (MWCNTs), respectively. Due to their unusual properties, CNTs are employed widely in electronics, optics and materials. CNTs possess a large surface areas in the range from a few hundred m^2/g of MWCNTs to approximately 1300 m^2/g of SWCNTs, which decreases when CNTs tend to bundle, i.e., the bundling of SWCNT decreases the special surface area to approximately 300 m^2/g. The surface of CNTs is hydrophobic, which leads to the aggregation of CNTs. In order to be dispersed in aqueous solvent for gastrointestinal absorption, blood transportation, secretion, and biocompatibility, the surface of CNTs must be capable of being wetted in solution, which could be achieved through the following approaches: surfactant-assisted dispersion, solvent dispersion, functionalization of side walls, and biomolecular dispersion (Foldvari and Bagonluri 2008). Specifically, functionalization is the most effective way, which not only modifies the wetting property, but also decreases its cytotoxicity and improves the bio-compatibility. In addition, the functionalization could design CNTs as the smart materials with functional molecules on the surface to achieve the multi-triggered release ability (Fig. 7). There are two ways working for the functionalization: through hydrophobic and π-π interactions (non-covalent interaction), and covalent bonds. For example, Singh et al. fabricated water-soluble SWCNTs functionalized with diethylenetriamine pentaacetic (DTPA) to give DTPA-SWCNTs. This material could be rapidly cleared by the blood circulation system without any toxicity. Their further investigation showed that the SWCNTs were excreted intact (Singh et al. 2006).

SWNTs have strong optical absorbance in the NIR region, which could be used to design smart release system triggered by NIR light. The

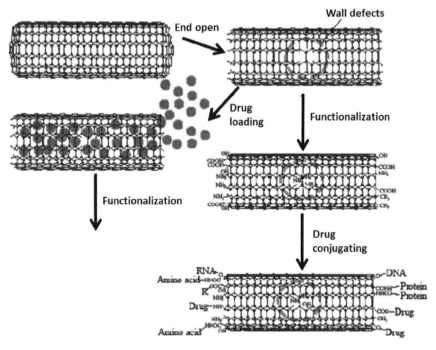

Figure 7. Illustrations of carbon nanotubes (CNTs) for drug storage and functionalization (Zhang et al. 2011). Copyright 2011 licensee Springer.

absorbance of NIR light can produce localized heat that stimulates the release of drug or genes from the nanotubes surface. Kam et al. designed a folate-modified SWNTs formulation. Selective internalization of SWNTs inside cells labeled with folate receptor tumor markers was observed. After exposure to NIR to Hela cells after Cy3-DNA-SWNT uptake, they detected the colocalization of fluorescence of Cy3-DNAs in the cell nucleus, indicating that DNAs were released from SWNTs vectors to the nucleus upon the laser stimuli (Kam et al. 2005).

The development of smart CNTs for CDR is still in the infancy. The unique property of CNTs, especially the easiness of translocation across cell membranes, makes it one of the hottest fields. However, there are many challenges existing for CNTs. The biosafety of CNTs is not very clear. Although CNTs have been used as the drug vectors to delivery anticancer drugs or genes to targeted size, and in some cases, the systems could provide the information of distribution and circling, there is a lack of research on triggered release, which is significant for the higher efficacy and lower toxicity for therapy. In the future, more smart materials based on CNTs will be designed for CDR, which hold promise in the application of medicine.

3. Smart Polymeric Materials for CDR

3.1 Polymeric Capsules

3.1.1 Preparation and Properties

Polymeric capsules (PCs) are materials with core-shell structures. The first report about capsules appeared in 1835, which became an important material with applications in pharmaceutical, cosmetic, food, adhesive and agriculture industries. The first investigation of PCs for CDR was not realized until 1970s.

Due to its easy preparation, controllable particle size from nanometer to micrometer, and facile functionalization, PCs have received considerable attention in the field of CDR. PCs are prepared by several methods, including pan coating, spray drying, centrifugal extrusion, and emulsion-based methods (Esser-Kahn et al. 2011). Each method has advantages on the control of shell wall thickness and permeability, chemical composition and mechanical integrity of the shell wall, and capsule size. Specifically, emulsion methods have drawn lots of attention, which could be divided into emulsification polymerization, layer-by-layer (LbL) assembly of polyelectrolytes, coacervation, and internal phase separation (Fig. 8) (Esser-Kahn et al. 2011). The major benefit of LbL method is their versatility, which can be utilized to fabricate PCs with various templates, giving sizes varying from a few nanometers to hundreds of micrometers, and their

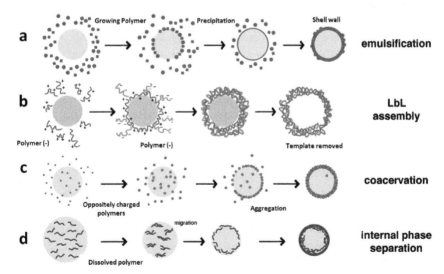

Figure 8. Schemes of emulsion strategies for preparation of polymeric capsules (Esser-Kahn et al. 2011). Copyright 2011 American Chemical Society.

chemical and mechanical properties can be precisely tailored by modulating the thickness and constitution of the shell.

The controlled release of cargo from PCs is of considerable interest in applications such as self-healing materials, nutrient preservation, fragrance release, and drug delivery. By modifying with smart molecules on the surface of capsules, or preparation of capsules with responsive molecule-conjugated polymers, the capsule vectors could encapsulate and deliver drugs and genes to target sites. Smart PCs for CDR have been developed increasingly in the past decades, including triggers as chemicals, pH, temperature, light, and magnetic field.

3.1.2 Recent Advances of Stimuli-Responsive PCs

(1) pH-Responsive PCs

As mentioned above, there exists the difference of pH between normal cells, blood circumstance cancer cells/inflammatory tissues and lysosome/ endosome, which promotes pH-responsive materials one of the most interesting issues. PCs with pH-responsive property can be fabricated by assembly of polymers modified with pH-sensitive molecules or linkers, which would induce the change of permeability of capsule shells or rapture of the shells, resulting in the release of cargo in target sites.

In 2007, Na et al. reported a pH-responsive smart PCs vehicle, which could release the loaded cargo in acidic organelle quickly. They fabricated the capsules using poly(D, L-lactic-co-glycolic acid) (PLGA) which has been widely used as a carrier material for CDR owing to its excellent biocompatibility and biodegradability (Na et al. 2007). Anticancer drugs and sodium bicarbonante ($NaHCO_3$) were loaded into the capsules. $NaHCO_3$ played an important role in the controlled release process. When the capsule vehicles were transported into an endocytic organelle of a live cell, the protons infiltrating from the compartment reacted with $NaHCO_3$ to generate CO_2 gas, which caused the shell to rupture due to the increased internal pressure. In addition, they trapped fluorescent DiO in capsule shells, which could be used to trace the release of drug inside cells (Fig. 9) (Ke et al. 2011).

(2) Chemical-Responsive PCs

Chemical signals include Ca^{2+}, Ba^{2+}, sugar, and dithiothreitol (DTT) molecules. In cancer cells, there are rich antioxidants (e.g., dihydrolipoic acid or GSH), which is usually utilized to design redox-responsive systems. Disulfide bond is the unique linkage most investigated in this field. Yan et al. demonstrated a smart PCs vector for delivery of DOX to colon cancer cells. The submicrometer PCs with size from 600 nm (pH 4.0) to 850 nm (pH 8.0) were fabricated by alternate assembly of poly(methacrylic acid) (PMAA)

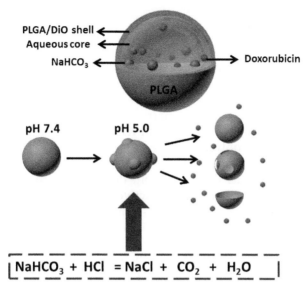

Figure 9. Scheme illustrating the capsules loaded anticancer drug and $NaHCO_3$, which could release drugs in the acidic condition (Ke et al. 2011).

and poly(vinyylpyrrolidone) (PVPON) on the SiO_2. After the cross-linking of the third group and removal of SiO_2, the capsules were obtained. Studies showed the PCs could deliver anticancer to the colon cells and cause great cytotoxicity to the cancer cells (Yan et al. 2010).

(3) Enzyme-Responsive PCs

The selectively responsive property of PCs is critical for biomedical application. Enzymes represent the typical biomolecules that are rich in living organisms and react selectively with substrate. De Geest et al. reported an enzyme-responsive polyelectrolyte PCs vehicle. The PCs were composed of poly-L-argine (PARG) and as the polycation and dextran sulfate (DEXS) as the polyanion, which were fabricated by alternate deposition of PARG and DEXS on $CaCO_3$, followed by the removal of cores. The capsules would deliver fluorescein isothiocyanate (FITC)-conjugated dextran to African green monkey kidney cells (VERO-1 cells) with the release driven by enzyme-catalyzed degradation (De Geest et al. 2006). Johnston et al. designed a novel PCs vehicle assembled by DNA molecules. The DNA molecules were engineered to contain a specific site in the sequences, which could be responsive to the restriction enzyme, resulting in the release of encapsulated proteins. This work was interesting because it provided a new strategy to design smart PCs responsive to different stimulus by programming the DNA sequences (Johnston et al. 2009).

(4) Light-Responsive PCs

Triggered release of contents based on PCs using light is appealing for a broad application. For applications in cosmetics and agriculture, capsules with UV- and visible-sensitive properties are applied because of the abundance of UV and visible light. In the fields of biomedicine, NIR-absorbing capsules are of greater interests because of the deep penetration of NIR in tissues without causing harm.

In 2012, Li et al. reported a light-triggered PCs vehicle prepared by LbL assembly of polysaccharide chitosan (CHI) as the polycation and 5-(4-aminophenyl)-10,15,20-triphenyl-porphyrin (APP) conjugated PASP (PASP-g-APP) as the polyanion. Porphyrins are a typical class of tetrapyrroles, which can yield reactive oxygen species (ROS) upon light exposure and are wide used as photosensitive chemicals. Upon light irradiation (400 nm), APP containing porphyrin moiety produced ROS to break down the polymeric chains and destroy capsule structure, subsequently causing the release of the drug (Li et al. 2012). Tao and Li et al. also reported a light-responsive capsule system composed of azo dyes, and polymers. The resulting hollow capsule displayed a sensitive response to visible light. Investigation found the photochemical reaction of the assembled hollow PCs depends strongly on the matrix. The permeability of the hollow capsule shells can be photo-controlled easily at varying irradiation time (Tao et al. 2004).

3.1.3 Conclusion

As promising candidates for medicine, PCs have been developing for decades. The unique properties, such as degradability, biocompatibility, controlled size and facile functionalization have promoted the rapid development of PCs in this field of CDR. Various kinds of smart release systems have been constructed, and tested *in vitro*, and the first commercial capsules appeared in 2001. Further efforts have to be devoted to smart capsule vehicles *in vivo* studies.

3.2 Polymeric Hydrogels

3.2.1 Introduction

A gel is a solid, jelly-like material with the properties ranging from soft to hard. It behaves like solids due to a three-dimensional cross-linked network within the liquid. Hydrogel is a kind of materials defined as a three dimensional biopolymeric network formed by chemical crosslinking by covalent bonds or physical crosslinking by non-covalent bonds, which have the tendency to retain a large amount of water. The first report on "hydrogel" appealed in 1894, which involved a colloidal gel of inorganic salts. The term hydrogel as it is known today, was first introduced by Wichterle and Lim in 1960. The absorbed property of water is attributed to the specific hydrophilic groups in the three international networks, such as $-OH$, $-CONH_2$, $-SO_3H$, $-CONH$, and $-COOR$. The soft and rubbery surface, structure, and physic-chemical properties of hydrogels mimic that of human tissue, which makes them potential candidates in biomedicine.

The advance in polymer science and technology promotes the discovery of smart biodegradable polymeric hydrogels (PHs), whose behavior would change in response to the physical or chemical triggers. The unique property of such smart PHs makes them intelligent candidates for CDR (Vashist and Ahmad 2013). The discovery of micro/nano PHs provided perspicacious means of drug delivery systems, which could be triggered by internal and external stimulus, such as pH, light, temperature, and chemicals.

3.2.2 Recent Advances of Stimuli-Responsive PHs

(1) pH-Responsive PHs

The pH-responsive *PHs* could encapsulate and release drugs or genes in different pH, which is of significance in the potential application of cancer treatment. The release process is based on the pH-hydrolyzed linker or volume expanding. In 2012, Zeng et al. fabricated a pH-sensitive poly(ethylene glycol)-poly-L-histidine hydrogel, which was used to load adeno-associated virus serotype 2 containing the green fluorescent protein gene (rAAV2-GFP). Polyhistidine (polyHis, pKa 6.10) was incorporated into the hydrogel system, and played a critical role in the controlled process. Under acidic condition the amine groups in polyHis became protonated, which caused the increase of water uptake by the hydrogel. Investigations *in vitro* indicated the rAAV2-GFP released from hydrogels in pH 6.0, which was dependent on the ratio of polyHis in Hydrogel (Zeng et al. 2012). Zan et al. recently demonstrated a dual pH-responsive PHs vehicle for CDR, the nano PHs were fabricated by host-guest interaction between adamantly (AD)-benzoic imine-conjugated poly[poly(ethylene glycol)monomethyl ether methacrylate]-co-poly(2-hydroxyethylmethacrylate) (PPEGMA-co-PHEMA) and polymer composed of DOX-hydrazone and β-CD-conjugated poly[N-(2-hydroxypropyl) methacrylamide]-co-poly(3-azidopropyl methacrylate) (PHPMA-co-PAzPMA) (PHPMA-co-PPMA-DOX-CD). The nanogels would reorganize into polymer-DOX nanoparticles with smaller size in acidic condition in tumor cells. Furthermore, the smaller polymer tides could release DOX quickly in intracellular endolysosomes at pH 5.0 due to the existence of pH-cleavable hydrozone linkage, causing greater cytotoxicity to Hela cells (Fig. 10) (Zan et al. 2014).

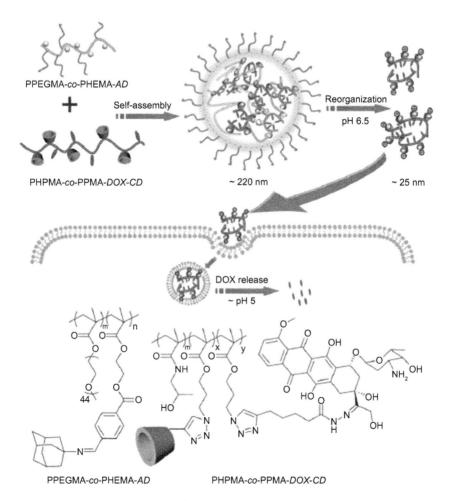

Figure 10. Polymeric nanogels formed by host-guest assembly of polymers with dual-pH stage release property (Zan et al. 2014). Copyright 2014 Royal Society of Chemistry.

(2) Light-Responsive PHs

Light is an attractive manner to control the drug delivery and release with controllable applied time, intensity and wavelength, which would achieve the remote activation of materials with relatively high spatial and temporal precision. Specifically, NIR-stimuli is more appealing because of its deep penetration into skins in the range of 700 nm to 1100 nm (Schwarz et al. 2002). Kang et al. described a NIR-controlled core-shell PHs carrier for targeted drug delivery. The PHs were obtained with Au-Ag nanorods as the cores and DNA cross-linked polymers as the shells, which was further functionalized with PEG to avoid aggregation and site-specific ligands to achieve the active tumor targeting. Upon NIR irradiation, the absorbance of Au-Ag cores would increase the temperature surroundings, leading to the melting of polymeric shells and subsequent release of guest molecules. The *in vitro* investigation further confirmed the feasibility of the core-shell PHs vehicle for targeted delivery of DOX to tumor cells with NIR light stimuli (Kang et al. 2011).

(3) Thermo-Responsive PHs

Thermo-responsive PHs can be categorized to negatively responsive PHs and positively responsive PHs depending on their temperature volume phase transition properties. PNIPAM is an outstanding thermo-sensitive polymer for fabrication of negatively temperature responsive hydrogel, which would undergo volume shrinkage above the LCST due to the phase transition from a hydrophilic coil to hydrophobic globule. PNIPAM was the polymer frequently utilized to obtain PHs for CDR. For example, Shin et al. fabricated a thermo-sensitive PHs based on PNIPAM and nanoporous silica. They incorporated PNIPAM into a porous silica host, which loaded indomethacininside. The hybrid hydrogel would exhibit a uniform release when the temperature was kept higher than 40°C, at which the PNIPAM shrank to squeeze the entrapped molecules to run out (Shin et al. 2001).

(4) Multi-Responsive PHs

Multi-responsive release PHs are capable of controlling the release of drugs under multi-stimulus, which is helpful in delivering drugs and genes in complex environment. The purpose of multi-responsive PHs is to achieve high chemotherapeutic efficacy and lowest toxicity, which is attractive for recent attention. Zhao et al. fabricated a thermo- and pH-responsive hydrogel by photo-cross-linkers. The hydrogel was composed of poly(L-glutamic acid) and PNIPAM. The investigation of hydrogel in different temperatures, pHs, and ionic strengths showed the hydrogel underwent volume shrinkage under acidic condition or at temperature above their collapse temperature, and would swell in neutral or basic media or at lower temperature. The reversible swelling/deswelling process could achieve the entrapment and release of loaded cargo. The *in vitro* study on the release of bovine serum albumin (BSA) from hydrogel at pH 6.8 and diverse temperatures indicated that hydrogels presented a slower release rate at temperature above their LCST. Additionally, the release rates of cargo from hydrogels increased sharply when pH was changed from 1.2 to 6.8 (Zhao et al. 2012). Xing et al. described a hollow nano PHs carrier consisting of a PAA and a PNIPAM network, which was fabricated by colloidal template method and loaded with isoniazid (an antibubercular drug) by controlling the equilibrium temperature. The nano PHs carrier was a dual-responsive system, and could be regulated by temperature and pH. The *in vitro* evaluation indicated the carrier was efficiently triggered to release by acidic pH (Xing et al. 2011).

3.2.3 Conclusion

Besides the smart PHs mentioned above, there are many other reports about the intelligent PHs as the candidates for CDR, including magnetic, chemicals, enzymes, and so forth. PHs have been one of the most promising materials for biomedical application, especially the emergency and development of nanoscale smart hydrogels. The further development of hydrogels depends on the advances of novel biodegradable polymers and nanotechnology. In the future, further efforts are needed in the investigation on the nanoscale PHs vectors with biodegradable/biocompatibility involving a long circulation time in the bloodstream, and with recognition of the target site.

3.3 Polymersomes

Polymersomes are a class of artificial vehicles composed of bilayered amphiphilic copolymers enclosing aqueous cores. Polymersomes are formed by the assembly of the amphiphilic copolymers with hydrophobic part contacting with each other and hydrophilic part facing water to minimize the energy (Fig. 11). They are in the size range from 50 nm to tens of micrometers. After its first reports in 1995 (Zhang and Eisenberg 1995), polymersomes have received much attention in medicine, pharmacy, and biotechnology due to their unique characteristics. Polymersomes can be prepared by dissolution of the block copolymers in an organic solvent suitable for all blockers, followed by the addition of water, which will cause the assembly of copolymers to hollow polymersome structure. Alternatively, they can also be prepared by polymer rehydration techniques. Polymers are first dissolved in organic solvent. Then, the evaporation of solvent gives the polymer film. By addition of water to the film, polymersomes will be obtained.

Polymersomes exhibit improved mechanical and chemical stability due to the higher molecular weight of copolymers. In addition, polymersomes are attractive from technological points of view. Polymersomes possess an aqueous core and hydrophobic shell, which could incorporate hydrophilic, such as drugs, enzymes, proteins, and DNA in the core and hydrophobic anticancer drugs in the shells, and are wide exploited in

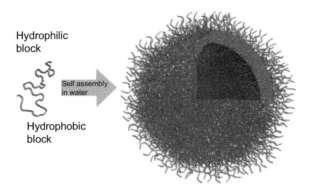

Hydrophilic
block

Hydrophobic
block

Figure 11. Structure of polymersomes (LoPresti et al. 2009). Copyright 2009 Royal Society of Chemistry.

CDR. Furthermore, as the polymersome is formed by self-assembly of amphiphilic copolymers in water, chemical methods are used to modify the copolymers to improve the hydrophilic property, which improve the interaction between polymersome and the circumstance. For instance, PEG is a famous polymer used to modify nanoparticles to improve the time of recycle. PEG containing block copolymers assemble into polymersomes with a highly hydrated and yet neutral polymer brush which has very limited interaction with proteins, which allows the PEG-polymersomes to withstand biological fluids without interacting with the immune system and reach tumor site *via* enhanced permeability and retention (EPR) effect. Most importantly, the biodegradable copolymers conjugated with active moieties and stimuli-sensitive groups will yield smart polymersomes, which could accomplish release of payloads in target sites upon triggers, such as pH, light, temperature and so on (Onaca et al. 2009, Liu et al. 2012).

For example, polymersomes formed by triblock copolymers PEG-b-(2,4,6-trimethoxybenzylidene-1,1,1-tris(hydroxymethyl)ethane methacrylate)-b-PAA (PEG-PTTMA-PAA) could be responsive to acidity stimuli. The polymersomes were reported to load anticancer drug DOX • HCl, which would be dissociated and release the encapsulated drugs when the acetals in polymersomes were hydrolyzed in acidic condition (Du et al. 2012). The amphiphilic copolymers can also be designed with multi-responsive moieties to achieve polymersomes which are responsive to multi-triggers. For instance, copolymers PEG-SS-poly(2-(diethylamino) ethyl methacrylate) (PEG-SS-PDEA) containing disulfide linkers and acetal linkers assembled to generate polymersomes, which were capable of loading and release drugs in response to acidic circumstance and reductive condition (Zhang et al. 2012).

Polymersomes for CDR has been developing for over two decades. As a promising candidate material for biomedical application, polymersomes will be attractive in the future. But at the present stage, the efforts on designing smart polymersomes are not enough. New strategies are believed to promote the development of polymersomes towards clinical tests.

3.4 Polymeric Micelles

According to IUPAC, micelle is the particle of colloidal dimensions that exists in equilibrium with the molecules or ions in solution from which it is formed. When dispersed in aqueous media with the concentration above a critical concentrate (CMC), micelles are formed by the self-assembly of amphiphilic polymers with the hydrophilic head regions in contact with the surrounding water, and the hydrophobic tails touching each other. The amphiphilic molecules could also form inverse micelles when dispersed in oil. The formation of micelles was observed as early as 1913 by McBain at the University of Bristol (McBain 1913). Small molecular amphiphililes, or surfactants, are typically composed of a hydrocarbon chain and a hydrophilic head group, which are widely used as detergents, bioactive denaturing agents, microbiocides, and repellents, but are limited due to their relatively high CMC values and cytotoxicity. Polymeric micelles (PMs) have a much lower CMC, and possess several distinct advantages, such as nanoscale size (20–100 nm), and reduced cytotoxicity of anticancer drug, improved solubility, prolonged time *in vivo* circulation time and preferential accumulation at tumor site *via* EPR, which generate great interests in PMs for biomedical applications. Some micellar anticancer drugs have been approved for clinical trials in Japan, UK, and USA (Deng et al. 2012).

Over the past decades, numerous PMs have been developed for hydrophobic drug delivery, such as Pluronics consisting of hydrophobic poly(propylene oxide) and hydrophilic poly(ethylene oxide), poly(esters), and poly(amino acids). Recently, a new sugar-based amphiphilic polymer (SBAPs) comprised of hydrophobic sugar segments and hydrophilic PEG chains was developed by Tian et al. (Tian et al. 2004), which is stable against dilution because of its extremely low CMC. With proper sugar and PEG ratio, SBAPs-based micelles could be internalized by cells, and efficiently transported to nucleus. Such micelles with particle size from 16–25 nm could evade renal clearance, penetrate tumor tissues, and avoid internalization by the reticuloendothelial system (RES), holding promise as a good vehicle for CDR (Djordjevic et al. 2008). Anticancer drug could be chemically conjugated to the hydrophobic segment of SBAPs *via* hydrazine linkage to yield drug-loaded SBAPs-micelle. The pH-cleavable property of the hydrazine linkage would lead to pH-triggered release of drugs (del Rosario et al. 2010).

However, there are some shortcomings existing in conventional PMs, which limit the advances of PMs towards *in vivo* applications. One of them is the extensive dilution of PMs causes the dissociation of the structures. Therefore, the PMs by cross-linkage have been developing in recent years, which could improve the structure stability. In addition, in order to improve the therapeutic efficacy, and control the drug release in tumor sites, stimuli-sensitive linkers or assembly units are introduced in the PMs. Li et al. (2009) reported a type of disulfide cross-linked PMs formed by dextran-lipoic acid derivatives followed by the cross-linking catalyzed by 10 mol% dithiotheitol (DTT) relative to the lipoyl units. In the presence of 10 mM DTT, the hydrolysis of disulfide led to the release of DOX from the PMs particles. The *in vitro* studies also confirmed the feasible release of DOX inside tumor cells triggered by reduction (Li et al. 2009). As the concentration of a reductive GSH in cytoplasm could be 100–1000-fold higher than that in the extracellular environments and blood pool, the redox-sensitive micelle vesicles hold promising applications in the treatment of cancers (Koo et al. 2008).

4. Other Smart Materials for CDR

4.1 Albumins

Albumin, a family of globular proteins, presents advantages of great stability *in vivo*, no toxicity and non-antigenic to the body. The main albumins investigated as vehicles are obtained mainly from egg white (ovalbumin), bovine serum (bovine serum albumin) and human serum (human serum albumin). As albumin contains charged amino acids, a significant amount of drugs can be incorporated into the albumin by electrostatic interactions. More importantly, the amino acids and carboxylic groups provide active sites for functionalization, which achieve the design of smart albumin vehicle for CDR.

Shen et al. fabricated a thermo-responsive albumin vector, which could selectively accumulate and release payload Rose Bengal (RB) in the solid tumor site with local higher temperature. The drug carrier was constructed by modification thermo-sensitive PNIPAM derivatives on the surface of albumin with RB loaded inside covalently. Below 37°C, the polymers on surface were hydrophilic and expanded, inhibiting the release of RB. When the temperature increased to 42°C, PINIPAM polymers shrank, which allowed the release of RB (Shen et al. 2008). Interestingly, albumin was recently reported to be incorporated with magnetic particles for magnetically responsive drug delivery and release (Zeybek et al. 2014). This albumin vehicle loaded with DOX displayed higher cytotoxicity to cells cancer cells than free DOX, which could be regulated under magnetic field manner.

Protein nanoparticles have a bright future in the controlled delivery of therapeutic agents due to their biodegradability, biocompatibility, and possibility of covalent derivatization with drug targeting ligands and trigger-sensitive molecules. At the present stage, developing smart albumins with precise controllability is highly required.

4.2 Liposomes

Liposome is an artificial-prepared spherical vesicle composed of phospholipid-enriched phospholipids bilayer surrounding an aqueous interior core, which is formed spontaneously when amphiphilic lipids are dispersed in water. Liposomes were first described in 1961 by British hematologists Bangham, at the Babraham Institute, in Cambridge, and soon proposed in the field of CDR due to its advantageous properties, such as biocompatibility, biodegradability and little or no antigenic reaction. The assembled core-shell structure can load hydrophobic anticancer drugs and hydrophilic molecules in the phospholipid bilayer and aqueous core, respectively.

However, the early studies found that lyposomes were rapidly cleared in the blood circulation by phagocytic cells before they accumulated at the tumor site through EPR effect. The problem was resolved after the development of PEGylation technology, which coated liposomes with hydrophilic PEG to reduce immunogenicity and prolong circulation. The long-circulating liposomes could deliver anticancer drugs to tumor sites.

In order to optimize the drug delivery in time, location, and amount, considerable efforts have been dedicated to the design and construction of new-generation of liposome-vectors, i.e., smart liposome nanomaterials, which could be triggered by stimulus, such as temperature, pH, enzymes, and light. In 2012, Mo et al. reported a multistage pH-responsive liposomal system based on zwitterionic oligopeptide liposomes (HHG2C18-L). Oligopeptide liposomes (HHG2C18-L) consist of soy phosphatidylocholine (SPC), cholesterol, and a synthetic lipid (1,5-dioctadecyl-L-glutamyl 2-histidyl-hexahdrobezoic acid, HHG2C18). The special design of HHG2C18-L would undergo multistage pH change in the process from physiological blood to cytoplasmic inside tumor cells. Typically, the negative surface change of liposomes in physiological blood would reverse to be positive in tumor environment, which was helpful for the internalization of liposomes. Then in endosome, the subsequent pH-response induced by the imidazole group of histidine facilitated the proton influx, which led to the endosomal bursting. At the same time, the hydrolysis of hexa hydrogenzoic amide yielded a stronger positive surface charge of HHG2C18-L, which was easy to accumulate at the target mitochondria by electrostatic interaction in the final stage. This intelligent liposome provided a safe and efficient strategy to deliver and release of drugs in target sites (Mo et al. 2012). Agarwal et al. reported a NIR-triggered release liposomal formulation to delivery and release DOX in tumors. The liposome incorporated with GNRs was more efficiently accumulated in the tumor site and released the chemotherapeutic DOX upon the NIR trigger, leading to the significant increase in efficacy (Agarwal et al. 2011).

Liposomes gained extensive research in the past decades, which are the first nanomedicine formulations developed from concept to clinical applications. But that's not the end. Smart liposomal release systems with targeting delivery drug and controlled release of drug under triggers are still in the early stage of development.

4.3 Graphene

Graphene is a sheet of two-dimensional carbon atoms with distinct electrical, mechanical and thermal properties, which have been employed widely in electronics, photonics, and biomedicine. Geim and coworkers first obtained the single layered graphene in 2004, which won the Nobel Prize in Physics in 2010 for the groundbreaking experiment. Now, graphene is an important new addition to the carbon family materials. Usually, the fabrication of graphene can be divided to bottom-up approaches, such as chemical vapor deposition, solvothermal and organic synthesis, and up-down approaches, such as repeated exfoliation methods. With the extremely high surface area and function ability, graphene and the derivatives, especially for GO and reduced graphene oxide (rGO), have gained lots of interests in the biomedicine since 2008. Although investigations indicated the graphene or GO displayed toxicity, the GO-derivatives exhibited no significant side effect to cells and animals in the tested dosage after coating. Specifically, the surface modification and intrinsic high NIR absorbance encourage the design of smart nanomaterials based on graphene for delivery and targeted release of drugs upon NIR trigger for disease treatment.

The 2D nanostructure of graphene with the presence of delocalized surface π electrons in plane can be utilized for effective drug loading *via* hydrophobic interactions or π-π stacking. Additionally, large surface area of graphene allows for high density bio-functionalization *via* both covalent and non-covalent surface modification, which could attain the delivery and release of drugs triggered by stimulus for treatment of cancers. Smart graphene-based nanomaterials have been applied in CDR since 2008. For example, Kurapati and Raichur demonstrated a NIR-triggered GO composites carrier for DOX release (Fig. 12a). The smart nanovector/ formulation was composed of GO and poly(allylamine hydrochlorid) (PAH) with DOX entrapped in the core. The GO exhibited strong absorbance in NIR region, which converted light to heat efficiently. Upon NIR irradiation,

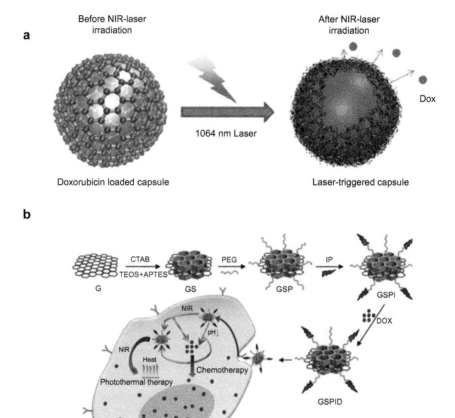

Figure 12. Graphene derivative materials for CDR. **(a)** NIR-responsive graphene oxide capsule for controlled DOX release (Kurapati and Raichur 2013), Copyright 2013 Royal Society of Chemistry. **(b)** Graphene-Fe$_3$O$_4$ composite for targeting deliver and release of anticancer drug triggered by light and pH (Wang et al. 2013). Copyright 2013 American Chemical Society.

the absorbance of light would result in the rupture of GO-PAH capsules and the release of DOX (Kurapati and Raichur 2013). Wang et al. demonstrated a multifunctional graphene nanomaterial for targeted therapy of Glioma (Fig. 12b) (Wang et al. 2013). The material was formed by mesoporous silica-coated graphene, which was further modified with glioma-targeting ligands (IP). This graphene vector possessed enlarged surface area, active target ability, and synergistic control of release, which was triggered by NIR light, and pH stimuli. Yang et al. reported as hybrid formulation formed by GO and INOPs *via* wet chemical precipitation method. The magnetic hybrids loaded with DOX can be dispersed uniformly in aqueous solution and aggregate in acidic medium. The moving of materials was under the control of magnetic field. This pH-triggered magnetic behavior of GO-Fe$_3$O$_4$ nanoparticle hybrids can be exploited potentially for CDR (Yang et al. 2009).

As emerging materials for biomedical application, graphene-based nanomaterials have attracted recent interests. Now these materials have created excitement amongst biomedical scientists. Remarkable progress in fabrication and functionalization of graphene materials has provided considerable opportunities for exploring their use in drug/gene delivery. Lots of challenges have to be faced before their clinical application, including the graphene-cell interaction, *in vivo* efficacy in animal. The future emphasis on tissue distribution, mechanisms of clearance and toxicity is required to understand their true potential.

4.4 Au nanoparticles

Au nanoparticles (AuNPs), including nanospheres, nanorods, and nanocages, are of great interests for biomedical applications due to their unique properties, e.g., the localized surface plasmon resonance and NIR absorbance. Since 2005, AuNPs have been studied widely as vectors for delivery and release of drugs for disease treatment. In addition, AuNPs could be fabricated with controlled size from 1–150 nm, which is non-toxic, and favorable for endocytosis by cells. AuNPs becoming an excellent platform to design smart vectors for CDR are also attributed to the functional versatility, which could achieve the modification on the surface of AuNPs with stimuli-sensitive molecules. The loading of drugs can be accomplished through non-covalent interaction or covalent conjugation. The application of AuNPs as promising vectors for CDR is a rapidly expanding field. In addition, the Au core in essence is non-toxic, biocompatible, and inert, which is critical for biomedical use.

In 2006, Hong et al. designed a GSH-responsive AuNPs vector (Fig. 13a). The 2 nm AuNPs were decorated with a tetra(ethylene glycol)-lyated cationic ligand (TTMA) thought Au-S bonds, which could improve the efficiency of cell uptake. A thiolated Bodipy dye (HSBOP) as the cargo was loaded by covalent interaction, which could be released in the presence of GSH (Hong et al. 2006). Xiao et al. constructed a novel AuNPs vector delivering DOX to the target site, which was then triggered to release upon NIR (Fig. 13b). The vector was composed of GNRs, complementary DNA with one pre-conjugated targeting ligand, and PEG. DOX was loaded by DNA strands. PEG modification improved the surface property and prolonged the circulation time. Upon NIR irradiation, the absorbance of AuNRs would transduce the light to heat, which denatured the DNA assembly, resulting in the release of DOX (Xiao et al. 2012).

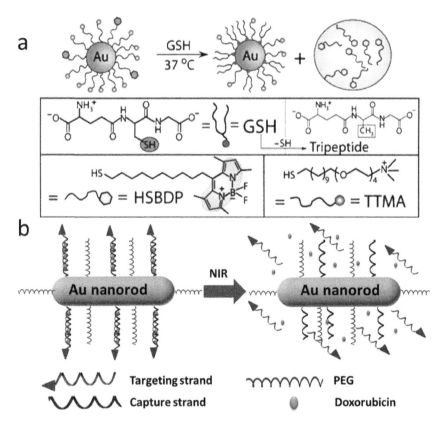

Figure 13. Smart AuNPs for CDR. **(a)** Glutathione-triggered AuNPs vehicle (Hong et al. 2006), Copyright 2006 American Chemical Society. **(b)** NIR-responsive AuNPs vehicle (Xiao et al. 2012).

AuNPs show great potential for the creation of smart vectors for biomedical applications. Their stability, tunable surface flexibility, low toxicity, and optical property offer many possibilities for further development of CDR systems. Future investigations of these systems will go forward to understand fully their interactions with the immune system, and finally construction of smart formulation and the *in vivo* studies.

4.5 Metal-Organic Framework Nanomaterials

Metal-organic frameworks (MOFs) are compounds consisting of metal ions or clusters coordinated to polydentate bridging organic ligands to form one-, two-, or three-dimensional porous structures, which has been proposed for applications since it was discovered by Robson in 1989 (Pollack et al. 1989), including catalysis, separation, gas storage, nonlinear optics and sensing. Recently, these materials have been scaled down to nanometer sizes, and become appealing for the biomedical applications due to their advantages over conventional nanomaterials. First, MOFs can be designed in diversity with different composition, shape, and pore size. MOFs are biodegradable, which could be cleared quickly by the body. In addition, the combined metal ions or organic ligands may be designed to play specific functions, such as MRI. Generally, the preparations of MOFs with homogeneous and stable structure can be divided into four methods: hydrosolvothermal, reverse-phase microemulsion, sonochemical synthesis, and nanoprecipitation.

Therapeutic agents are loaded into the nanostructures by direct incorporation or postsynthesis loading (Fig. 14). In the direct incorporation

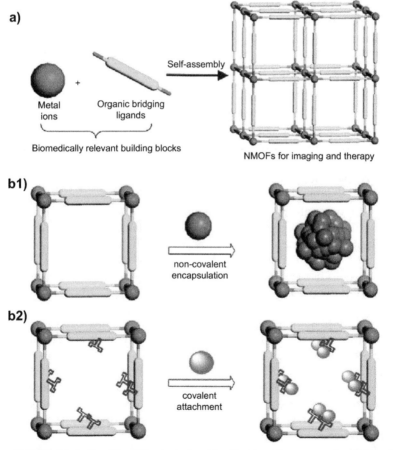

Figure 14. MOFs fabrication **(a)** and the strategies of loading drugs by non-covalent interaction **(b1)** and covalent incorporation **(b2)** (Della Rocca et al. 2011). Copyright 2011 American Chemical Society.

strategy, biomedical functional agents are the structure blocks, i.e., either the metal connection points, or the bridging ligand. In the postsynthesis loading strategy, loaded agents are attached by covalent bonding or non-covalent interactions. In addition, as the drug vector, the modification, including silica coating and polymer decoration, is important for MOFs, which could improve the stability of structure in physiological circumstance. The modification could bring the MOFs with intelligent property, which is the new promising way for treating cancer. For example, An et al. developed a MOFs vector assembled by zinc-adeninate columnar secondary units. This MOFs vector could storage cationic drug *via* cation exchange with dimethylammonium cations. Exogenous cations could triggered the release of loaded drug from the pores (An et al. 2009).

The development of smart MOFs as drug delivery vectors is in its infancy. Nonetheless, MOFs have already shown great promise as a new nanocarrier platform owing to unique properties. In future, studies should be performed on the construction of smart MOFs, which could achieve the targeted delivery of drug by triggers. At the same time, more strategies will be utilized to modify MOFs to improve the biocompatibility and prolong the circulation in blood. Although there are many reports on *in vitro* efficacy of MOFs, no systematic studies on *in vivo* efficacy have been performed yet. Such systematic *in vivo* studies are critical for optimization the performance of MOFs.

5. Promise and Challenges

Intelligent materials will bring the evolution and change in the life of human beings, including in the field of biomedical therapeutics. It's exciting to imagine the smart nanomaterials perform as nanorobots or nanovectors and carry drugs to the target sites, treating serious diseases without side effects to normal cells and tissues, such as cancers. The advances of potential materials with trigger-responsiveness for CDR are summarized in this chapter, including inorganic materials, polymeric materials, and other popular or emerging materials. This field has been developing quickly towards clinical applications with the advances of technology and materials science. Especially, some materials have been applied in clinic and commercial markets, such as liposomes and polymeric capsules. Moreover, smart materials will not only achieve the delivery and controllable release of chemotherapeutics to the target sites, but also carry genes into cells, which might express in cells and attain the gene therapy and organism improvement.

However, before the applications in clinic, many challenges have to be solved. Developing intelligent vectors with biocompatibility, nontoxicity, facile preparation, low-cost, and versatile functionalization is still a big issue.

In addition, the knowledge about the interaction of nanomaterials with cells and immunogenic system is quite limited. Finally, further investigations have to be carried out on the finely precise control of targeted delivery and release of drugs by triggers.

References

Agarwal, A., M.A. Mackey, M.A. El-Sayed and R.V. Bellamkonda. 2011. Remote triggered release of doxorubicin in tumors by synergistic application of thermosensitive liposomes and gold nanorods. Acs Nano 5: 4919–4926.

Ambrogio, M.W., C.R. Thomas, Y.-L. Zhao, J.I. Zink and J.F. Stoddartt. 2011. Mechanized silica nanoparticles: A new frontier in theranostic nanomedicine. Acc. Chem. Res. 44: 903–913.

An, J., S.J. Geib and N.L. Rosi. 2009. Cation-triggered drug release from a porous zinc-adeninate metal-organic framework. J. Am. Chem. Soc. 131: 8376–8367.

Angelos, S., Y.W. Yang, N.M. Khashab, J.F. Stoddart and J.I. Zink. 2009. Dual-controlled nanoparticles exhibiting and logic. J. Am. Chem. Soc. 131: 11344–11346.

Aznar, E., L. Mondragon, J.V. Ros-Lis, F. Sancenon, M.D. Marcos, R. Martinez-Manez, J. Soto, E. Perez-Paya and P. Amoros. 2011. Finely tuned temperature-controlled cargo release using paraffin-capped mesoporous silica nanoparticles. Angew. Chem. Int. Ed. 50: 11172–11175.

Aznar, E., R. Villalonga, C. Gimenez, F. Sancenon, M.D. Marcos, R. Martinez-Manez, P. Diez, J.M. Pingarron and P. Amoros. 2013. Glucose-triggered release using enzyme-gated mesoporous silica nanoparticles. Chem. Commun. 49: 6391–6393.

Barick, K.C., S. Nigam and D. Bahadur. 2010. Nanoscale assembly of mesoporous zno: A potential drug carrier. J. Mater. Chem. 20: 6446–6452.

Beck, J.S., J.C. Vartuli, W.J. Roth, M.E. Leonowicz, C.T. Kresge, K.D. Schmitt, C.T.W. Chu, D.H. Olson, E.W. Sheppard, S.B. Mccullen, J.B. Higgins and J.L. Schlenker. 1992. A new family of mesoporous molecular-sieves prepared with liquid-crystal templates. J. Am. Chem. Soc. 114: 10834–10843.

Bernardos, A., E. Aznar, M.D. Marcos, R. Martinez-Manez, F. Sancenon, J. Soto, J.M. Barat and P. Amoros. 2009. Enzyme-responsive controlled release using mesoporous silica supports capped with lactose. Angew. Chem. Int. Ed. 48: 5884–5887.

Borisova, D., H. Moehwald and D.G. Shchukin. 2011. Mesoporous silica nanoparticles for active corrosion protection. Acs Nano 5: 1939–1946.

Casasus, R., E. Climent, M.D. Marcos, R. Martinez-Manez, F. Sancenon, J. Soto, P. Amoros, J. Cano and E. Ruiz. 2008. Dual aperture control on pH- and anion-driven supramolecular nanoscopic hybrid gate-like ensembles. J. Am. Chem. Soc. 130: 1903–1917.

Chen, C., J. Geng, F. Pu, X. Yang, J. Ren and X. Qu. 2011a. Polyvalent nucleic acid/mesoporous silica nanoparticle conjugates: Dual stimuli-responsive vehicles for intracellular drug delivery. Angew. Chem. Int. Ed. 50: 882–886.

Chen, L., W. Wang, B. Su, Y. Wen, C. Li, Y. Zhou, M. Li, X. Shi, H. Du, Y. Song and L. Jiang. 2014. A light-responsive release platform by controlling the wetting behavior of hydrophobic surface. ACS Nano 8: 744–751.

Chen, L.F., J.C. Di, C.Y. Zhao, Y. Ma, J. Luo, Y.Q. Wen, W.G. Song, Y.L. Song and L. Jiang. 2011b. A pH-driven DNA nanoswitch for responsive controlled release. Chem. Commun. 47: 2850–2852.

Chen, Z., Z. Li, Y. Lin, M. Yin, J. Ren and X. Qu. 2013. Bioresponsive hyaluronic acid-capped mesoporous silica nanoparticles for targeted drug delivery. Chem. Eur. J. 19: 1778–1783.

De Geest, B.G., R.E. Vandenbroucke, A.M. Guenther, G.B. Sukhorukov, W.E. Hennink, N.N. Sanders, J. Demeester and S.C. De Smedt. 2006. Intracellularly degradable polyelectrolyte microcapsules. Adv. Mater. 18: 1005–1009.

del Rosario, L.S., B. Demirdirek, A. Harmon, D. Orban and K.E. Uhrich. 2010. Micellar nanocarriers assembled from doxorubicin-conjugated amphiphilic macromolecules (dox–am). Macromol. Biosci. 10: 415–423.

Della Rocca, J., D.M. Liu and W.B. Lin. 2011. Nanoscale metal-organic frameworks for biomedical imaging and drug delivery. Acc. Chem. Res. 44: 957–968.

Deng, C., Y.J. Jiang, R. Cheng, F.H. Meng and Z.Y. Zhong. 2012. Biodegradable polymeric micelles for targeted and controlled anticancer drug delivery: Promises, progress and prospects. Nano Today 7: 467–480.

Djordjevic, J., L.S. Del Rosario, Jinzhong Wang and K.E. Uhrich. 2008. Amphiphilic scorpion-like macromolecules as micellar nanocarriers. J. Bioact. Comp. Polym. 23: 532–551.

Du, Y., W. Chen, M. Zheng, F. Meng and Z. Zhong. 2012. pH-sensitive degradable chimaeric polymersomes for the intracellular release of doxorubicin hydrochloride. Biomaterials 33: 7291–7299.

Esser-Kahn, A.P., S.A. Odom, N.R. Sottos, S.R. White and J.S. Moore. 2011. Triggered release from polymer capsules. Macromolecules 44: 5539–5553.

Foldvari, M. and M. Bagonluri. 2008. Carbon nanotubes as functional excipients for nanomedicines: I. Pharmaceutical properties. Nanomedicine 4: 173–182.

Folkman, J. and D.M. Long. 1964. The use of silicone rubber as a carrier for prolonged drug therapy. Surg. Res. 4: 139–142.

Folkman, J., D.M. Long and R. Rosenbau. 1966. Silicone rubber—a new diffusion property useful for general anesthesia. Science 154: 148–149.

He, Q., J. Shi, F. Chen, M. Zhu and L. Zhang. 2010. An anticancer drug delivery system based on surfactant-templated mesoporous silica nanoparticles. Biomaterials 31: 3335–3346.

Hoffman, A.S. 2008. The origins and evolution of "controlled" drug delivery systems. J. Contrl. Release 132: 153–163.

Hong, R., G. Han, J.M. Fernandez, B.J. Kim, N.S. Forbes and V.M. Rotello. 2006. Glutathione-mediated delivery and release using monolayer protected nanoparticle carriers. J. Am. Chem. Soc. 128: 1078–1079.

Johnston, A.P.R., L. Lee, Y. Wang and F. Caruso. 2009. Controlled degradation of DNA capsules with engineered restriction-enzyme cut sites. Small 5: 1418–1421.

Kam, N.W.S., M. O'Connell, J.A. Wisdom and H. Dai. 2005. Carbon nanotubes as multifunctional biological transporters and near-infrared agents for selective cancer cell destruction. Proc. Nat. Aca. Sci. USA 102: 11600–11605.

Kang, H., A.C. Trondoli, G. Zhu, Y. Chen, Y.-J. Chang, H. Liu, Y.-F. Huang, X. Zhang and W. Tan. 2011. Near-infrared light-responsive core-shell nanogels for targeted drug delivery. Acs Nano 5: 5094–5099.

Ke, C.-J., T.-Y. Su, H.-L. Chen, H.-L. Liu, W.-L. Chiang, P.-C. Chu, Y. Xia and H.-W. Sung. 2011. Smart multifunctional hollow microspheres for the quick release of drugs in intracellular lysosomal compartments. Angew. Chem. Int. Ed. 50: 8086–8089.

Koo, A.N., H.J. Lee, S.E. Kim, J.H. Chang, C. Park, C. Kim, J.H. Park and S.C. Lee. 2008. Disulfide-cross-linked peg-poly(amino acid)s copolymer micelles for glutathione-mediated intracellular drug delivery. Chem. Commun.: 6570–6572.

Kresge, C.T., M.E. Leonowicz, W.J. Roth, J.C. Vartuli and J.S. Beck. 1992. Ordered mesoporous molecular-sieves synthesized by a liquid-crystal template mechanism. Nature 359: 710–712.

Kresge, C.T. and W.J. Roth. 2013. The discovery of mesoporous molecular sieves from the twenty year perspective. Chem. Soc. Rev. 42: 3663–3670.

Kurapati, R. and A.M. Raichur. 2013. Near-infrared light-responsive graphene oxide composite multilayer capsules: A novel route for remote controlled drug delivery. Chem. Commun. 49: 734–736.

Lee, J., H. Kim, S. Han, E. Hong, K.H. Lee and C. Kim. 2014. Stimuli-responsive conformational conversion of peptide gatekeepers for controlled release of guests from mesoporous silica nanocontainers. J. Am. Chem. Soc. 136: 12880–12883.

Li, C., Z.Y. Li, J. Zhang, K. Wang, Y.H. Gong, G.F. Luo, R.X. Zhuo and X.Z. Zhang. 2012. Porphyrin containing light-responsive capsules for controlled drug release. J. Mater. Chem. 22: 4623–4626.

Li, Y.-L., L. Zhu, Z. Liu, R. Cheng, F. Meng, J.H. Cui, S.J. Ji and Z. Zhong. 2009. Reversibly stabilized multifunctional dextran nanoparticles efficiently deliver doxorubicin into the nuclei of cancer cells. Angew. Chem. Int. Ed. 48: 9914–9918.

Liu, G.Y., C.J. Chen and J. Ji. 2012. Biocompatible and biodegradable polymersomes as delivery vehicles in biomedical applications. Soft Matter 8: 8811–8821.

Liu, J., W. Bu, L. Pan and J. Shi. 2013. Nir-triggered anticancer drug delivery by upconverting nanoparticles with integrated azobenzene-modified mesoporous silica. Angew. Chem. Int. Ed. 52: 4375–4379.

Liu, J., X. Du and X. Zhang. 2011. Enzyme-inspired controlled release of cucurbit 7 uril nanovalves by using magnetic mesoporous silica. Chem. Eur. J. 17: 810–815.

Liu, R., Y. Zhang and P. Feng. 2009. Multiresponsive supramolecular nanogated ensembles. J. Am. Chem. Soc. 131: 15128–15129.

Liu, R., Y. Zhang, X. Zhao, A. Agarwal, L.J. Mueller and P.Y. Feng. 2010. pH-responsive nanogated ensemble based on gold-capped mesoporous silica through an acid-labile acetal linker. J. Am. Chem. Soc. 132: 1500–1501.

LoPresti, C., H. Lomas, M. Massignani, T. Smart and G. Battaglia. 2009. Polymersomes: Nature inspired nanometer sized compartments. J. Mater. Chem. 19: 3576–3590.

Mal, N.K., M. Fujiwara and Y. Tanaka. 2003. Photocontrolled reversible release of guest molecules from coumarin-modified mesoporous silica. Nature 421: 350–353.

McBain, J.W. 1913. General discussion on colloids and their viscosity. Trans. Faraday Soc. 9: 99–101.

Meng, H., M. Xue, T. Xia, Y.-L. Zhao, F. Tamanoi, J.F. Stoddart, J.I. Zink and A.E. Nel. 2010. Autonomous *in vitro* anticancer drug release from mesoporous silica nanoparticles by pH-sensitive nanovalves. J. Am. Chem. Soc. 132: 12690–12697.

Mo, R., Q. Sun, J. Xue, N. Li, W. Li, C. Zhang and Q. Ping. 2012. Multistage ph-responsive liposomes for mitochondrial-targeted anticancer drug delivery. Adv. Mater. 24: 3659–3665.

Muharnmad, F., M.Y. Guo, W.X. Qi, F.X. Sun, A.F. Wang, Y.J. Guo and G.S. Zhu. 2011. pH-triggered controlled drug release from mesoporous silica nanoparticles *via* intracelluar dissolution of zno nanolids. J. Am. Chem. Soc. 133: 8778–8781.

Na, K., S. Kim, K. Park, K. Kim, D.G. Woo, I.C. Kwon, H.M. Chung and K.H. Park. 2007. Heparin/poly(l-lysine) nanoparticle-coated polymeric microspheres for stem-cell therapy. J. Am. Chem. Soc. 129: 5788–5789.

Noyes, A.A. and W.R. Whitney. 1897. The rate of solution of solid substances in their own solutions. J. Am. Chem. Soc. 19: 930–934.

Onaca, O., R. Enea, D.W. Hughes and W. Meier. 2009. Stimuli-responsive polymersomes as nanocarriers for drug and gene delivery. Macromol. Biosci. 9: 129–139.

Park, C., H. Kim, S. Kim and C. Kim. 2009. Enzyme responsive nanocontainers with cyclodextrin gatekeepers and synergistic effects in release of guests. J. Am. Chem. Soc. 131: 16614–16615.

Patel, K., S. Angelos, W.R. Dichtel, A. Coskun, Y.W. Yang, J.I. Zink and J.F. Stoddart. 2008. Enzyme-responsive snap-top covered silica nanocontainers. J. Am. Chem. Soc. 130: 2382–2383.

Pollack, S.J., P. Hsiun and P.G. Schultz. 1989. Stereospecific hydrolysis of alkyl esters by antibodies. J. Am. Chem. Soc. 111: 5961–5962.

Rao, Y.F., W. Chen, X.G. Liang, Y.Z. Huang, J. Miao, L. Liu, Y. Lou, X.G. Zhang, B. Wang, R.K. Tang, Z. Chen and X.Y. Lu. 2015. Epirubicin-loaded superparamagnetic iron-oxide nanoparticles for transdermal delivery: Cancer therapy by circumventing the skin barrier. Small 11: 239–247.

Schlossbauer, A., S. Warncke, P.M.E. Gramlich, J. Kecht, A. Manetto, T. Carell and T. Bein. 2010. A programmable DNA-based molecular valve for colloidal mesoporous silica. Angew. Chem. Int. Ed. 49: 4734–4737.

Schwarz, A., S. Stander, M. Berneburg, M. Bohm, D. Kulms, H. van Steeg, K. Grosse-Heitmeyer, J. Krutmann and T. Schwarz. 2002. Interleukin-12 suppresses ultraviolet radiation-induced apoptosis by inducing DNA repair. Nat. Cell Biol. 4: 26–31.

Senyei, A., K. Widder and G. Czerlinski. 1978. Magnetic guidance of drug—carrying microspheres. J. Appl. Phys. 49: 3578–3583.

Shen, Z.Y., G.H. Ma, T. Dobashi, Y. Maki and Z.G. Su. 2008. Preparation and characterization of thermo-responsive albumin nanospheres. Int. J. Pharm. 346: 133–142.

Shin, Y.S., J.H. Chang, J. Liu, R. Williford, Y.K. Shin and G.J. Exarhos. 2001. Hybrid nanogels for sustainable positive thermosensitive drug release. J. Contrl. Release 73: 1–6.

Singh, R., D. Pantarotto, L. Lacerda, G. Pastorin, C. Klumpp, M. Prato, A. Bianco and K. Kostarelos. 2006. Tissue biodistribution and blood clearance rates of intravenously administered carbon nanotube radiotracers. Proc. Nat. Aca. Sci. USA 103: 3357–3362.

Stephen, Z.R., F.M. Kievit, O. Veiseh, P.A. Chiarelli, C. Fang, K. Wang, S.J. Hatzinger, R.G. Ellenbogen, J.R. Silber and M. Zhang. 2014. Redox-responsive magnetic nanoparticle for targeted convection-enhanced delivery of o6-benzylguanine to brain tumors. ACS Nano 8: 10383–10395.

Stewart, B. and C.P. Wild. 2014. World cancer report 2014. Lyon: IARC Press; WHO Press.

Strebhardt, K. and A. Ullrich. 2008. Paul ehrlich's magic bullet concept: 100 years of progress. Nat. Rev. Cancer 8: 473–480.

Szaciłowski, K., W. Macyk, A. Drzewiecka-Matuszek, M. Brindell and G. Stochel. 2005. Bioinorganic photochemistry: Frontiers and mechanisms. Chem. Rev. 105: 2647–2694.

Tao, X., J.B. Li and H. Mohwald. 2004. Self-assembly, optical behavior, and permeability of a novel capsule based on an azo dye and polyelectrolytes. Chem. Eur. J. 10: 3397–3403.

Tian, L., L. Yam, N. Zhou, H. Tat and K.E. Uhrich. 2004. Amphiphilic scorpion-like macromolecules: Design, synthesis, and characterization. Macromolecules 37: 538–543.

Vallet-Regi, M., A. Ramila, R.P. del Real and J. Perez-Pariente. 2001. A new property of mcm-41: Drug delivery system. Chem. Mater. 13: 308–311.

Vashist, A. and S. Ahmad. 2013. Hydrogels: Smart materials for drug delivery. Orient. J. Chem. 29: 861–870.

Wang, Y., K. Wang, J. Zhao, X. Liu, J. Bu, X. Yan and R. Huang. 2013. Multifunctional mesoporous silica-coated graphene nanosheet used for chemo-photothermal synergistic targeted therapy of glioma. J. Am. Chem. Soc. 135: 4799–4804.

Wen, Y., L. Xu, C. Li, H. Du, L. Chen, B. Su, Z. Zhang, X. Zhang and Y. Song. 2012a. DNA-based intelligent logic controlled release systems. Chem. Commun. 48: 8410–8412.

Wen, Y.Q., L.P. Xu, W.Q. Wang, D.Y. Wang, H.W. Du and X.J. Zhang. 2012b. Highly efficient remote controlled release system based on light-driven DNA nanomachine functionalized mesoporous silica. Nanoscale 4: 4473–4476.

Wu, S., X. Huang and X. Du. 2013. Glucose- and pH-responsive controlled release of cargo from protein-gated carbohydrate-functionalized mesoporous silica nanocontainers. Angew. Chem. Int. Ed. 52: 5580–5584.

Xiao, D., H.Z. Jia, J. Zhang, C.W. Liu, R.X. Zhuo and X.Z. Zhang. 2014. A dual-responsive mesoporous silica nanoparticle for tumor-triggered targeting drug delivery. Small 10: 59159–59158.

Xiao, Z., C. Ji, J. Shi, E.M. Pridgen, J. Frieder, J. Wu and O.C. Farokhzad. 2012. DNA self-assembly of targeted near-infrared-responsive gold nanoparticles for cancer thermo-chemotherapy. Angew. Chem. Int. Ed. 51: 11853–11857.

Xing, L., H. Zheng, Y. Cao and S. Che. 2012. Coordination polymer coated mesoporous silica nanoparticles for ph-responsive drug release. Adv. Mater. 24: 6433–6437.

Xing, Z., C. Wang, J. Yan, L. Zhang, L. Li and L. Zha. 2011. Dual stimuli responsive hollow nanogels with ipn structure for temperature controlling drug loading and pH triggering drug release. Soft Matter 7: 7992–7997.

Xue, M. and G.H. Findenegg. 2012. Lysozyme as a pH-responsive valve for the controlled release of guest molecules from mesoporous silica. Langmuir 28: 17578–17584.

Yan, Y., A.P.R. Johnston, S.J. Dodds, M.M.J. Kamphuis, C. Ferguson, R.G. Parton, E.C. Nice, J.K. Heath and F. Caruso. 2010. Uptake and intracellular fate of disulfide-bonded polymer hydrogel capsules for doxorubicin delivery to colorectal cancer cells. ACS Nano 4: 2928–2936.

Yang, Q., S.H. Wang, P.W. Fan, L.F. Wang, Y. Di, K.F. Lin and F.S. Xiao. 2005. pH-responsive carrier system based on carboxylic acid modified mesoporous silica and polyelectrolyte for drug delivery. Chem. Mater. 17: 5999–6003.

Yang, X., X. Liu, Z. Liu, F. Pu, J. Ren and X. Qu. 2012. Near-infrared light-triggered, targeted drug delivery to cancer cells by aptamer gated nanovehicles. Adv. Mater. 24: 2890–2895.

Yang, X., X. Zhang, Y. Ma, Y. Huang, Y. Wang and Y. Chen. 2009. Superparamagnetic graphene oxide-fe3o4nanoparticles hybrid for controlled targeted drug carriers. J. Mater. Chem. 19: 2710–2714.

Yang, Y., B. Velmurugan, X. Liu and B. Xing. 2013. Nir photoresponsive crosslinked upconverting nanocarriers toward selective intracellular drug release. Small 9: 2937–2944.

You, Y.-Z., K.K. Kalebaila, S.L. Brock and D. Oupický. 2008. Temperature-controlled uptake and release in pnipam-modified porous silica nanoparticles. Chem. Mater. 20: 3354–3359.

Yuan, Q., S. Hein and R.D.K. Misra. 2010. New generation of chitosan-encapsulated zno quantum dots loaded with drug: Synthesis, characterization and *in vitro* drug delivery response. Acta Biomater. 6: 2732–2739.

Zan, M., J. Li, S. Luo and Z. Ge. 2014. Dual ph-triggered multistage drug delivery systems based on host-guest interaction-associated polymeric nanogels. Chem. Commun. 50: 7824–7827.

Zeng, Y.-F., S.J. Tseng, I.M. Kempson, S.F. Peng, W.T. Wu and J.R. Liu. 2012. Controlled delivery of recombinant adeno-associated virus serotype 2 using pH-sensitive poly(ethylene glycol)-poly-l-histidine hydrogels. Biomaterials 33: 9239–9245.

Zeybek, A., G. Sanli-Mohamed, G. Ak, H. Yilmaz and S.H. Sanlier. 2014. *In vitro* evaluation of doxorubicin-incorporated magnetic albumin nanospheres. Chem. Biol. Drug Des. 84: 108–115.

Zhang, J., L. Wu, F. Meng, Z. Wang, C. Deng, H. Liu and Z. Zhong. 2012. pH and reduction dual-bioresponsive polymersomes for efficient intracellular protein delivery. Langmuir 28: 2056–2065.

Zhang, J., Z.F. Yuan, Y. Wang, W.H. Chen, G.F. Luo, S.X. Cheng, R.X. Zhuo and X.Z. Zhang. 2013a. Multifunctional envelope-type mesoporous silica nanoparticles for tumor-triggered targeting drug delivery. J. Am. Chem. Soc. 135: 5068–5073.

Zhang, L. and A. Eisenberg. 1995. Multiple morphologies of "crew-cut" aggregates of polystyrene-b-poly(acrylic acid) block copolymers. Science 268: 1728–1731.

Zhang, W., Z. Zhang and Y. Zhang. 2011. The application of carbon nanotubes in target drug delivery systems for cancer therapies. Nanoscale Res. Lett. 6: 1–22.

Zhang, X., P. Yang, Y. Dai, P.a. Ma, X. Li, Z. Cheng, Z. Hou, X. Kang, C. Li and J. Lin. 2013b. Multifunctional up-converting nanocomposites with smart polymer brushes gated mesopores for cell imaging and thermo/pH dual-responsive drug controlled release. Adv. Funct. Mater. 23: 4067–4078.

Zhang, Z.Y., Y.D. Xu, Y.Y. Ma, L.L. Qiu, Y. Wang, J.L. Kong and H.M. Xiong. 2013c. Biodegradable zno@polymer coreshell nanocarriers: Ph-triggered release of doxorubicin *in vitro*. Angew. Chem. Int. Ed. 52: 4127–4131.

Zhao, C., P. He, C. Xiao, X. Gao, X. Zhuang and X. Chen. 2012. Photo-cross-linked biodegradable thermo- and pH-responsive hydrogels for controlled drug release. J. Appl. Polym. Sci. 123: 2923–2932.

Zhao, L., J. Peng, Q. Huang, C. Li, M. Chen, Y. Sun, Q. Lin, L. Zhu and F. Li. 2014. Near-infrared photoregulated drug release in living tumor tissue *via* yolk-shell upconversion nanocages. Adv. Funct. Mater. 24: 363–371.

Zhao, Y.-L., Z. Li, S. Kabehie, Y.Y. Botros, J.F. Stoddart and J.I. Zink. 2010. pH-operated nanopistons on the surfaces of mesoporous silica nanoparticles. J. Am. Chem. Soc. 132: 13016–13025.

Zhao, Y., B.G. Trewyn, Slowing, II and V.S. Lin. 2009. Mesoporous silica nanoparticle-based double drug delivery system for glucose-responsive controlled release of insulin and cyclic amp. J. Am. Chem. Soc. 131: 8398–8400.

Zhou, S., X. Du, F. Cui and X. Zhang. 2014. Multi-responsive and logic controlled release of DNA-gated mesoporous silica vehicles functionalized with intercalators for multiple delivery. Small 10: 980–988.

5

Smart Materials for Cancer Diagnosis and Treatment

Jing Huang,[1] *Haibo Cheng*[2] and *Peng Guo*[3,*]

ABSTRACT

This chapter mainly describes recent advancements in development and application of smart materials for cancer diagnosis and treatment. Cancer is a global health issue that has become one of the leading causes of morbidity and mortality around the world. In general, cancer treatment relies primarily on surgery, chemotherapy and radiation therapy. However, over 90% of cancer-related deaths are attributed to tumor metastases, which are challenging to remove by surgical resection. Additionally, chemotherapy and radiation therapy indiscriminatingly kill both cancerous and healthy cells, causing severe adverse effects. Therefore, smart materials have emerged as a novel strategy to detect and treat tumor metastases. The "smart" materials discussed in this chapter are related to the therapeutic or diagnostic nanomedicine based on active tumor

---

[1] Department of Radiology and Imaging Sciences, Emory University School of Medicine, Atlanta, Georgia 30322, United States.
   Email: ching632@gmail.com
[2] Translational Medicine Research Center, First Clinical College, Nanjing University of Chinese Medicine, Nanjing, 210023, China.
   Email: nzychb@163.com
[3] Vascular Biology Program, Boston Children's Hospital, Boston, Massachusetts 02115, United States.
* Corresponding author: peng0351@gmail.com

targeting strategy. Considering tumor heterogeneity, we will introduce and discuss smart materials and relevant cancer molecular targets based on cancer types, in the order of cancer incidence.

1. Introduction

Decades of intensive research have been devoted to translate smart and advanced materials into cancer diagnosis and treatment under interdisciplinary collaborations between oncologists, clinicians, materials scientists and biomedical engineers. In recent years, smart materials, which have mostly nanoscale structures and features within a size range from few to hundreds nanometers and composed of polymeric or inorganic materials, have emerged as a novel therapeutic tool for cancer therapy. Compared with traditional biomedical materials, smart materials demonstrate the following unique advantages, including (i) improved drug delivery efficiency, (ii) prolonged circulation time, (iii) effective protection from biodegradation, (iv) controlled or sustained release of loaded cargo, and (v) passive or active tumor targeting function. Nanoparticulate-based therapeutic/diagnostic agents, also termed as nanomedicine, play a pivotal role among all smart materials in translational oncology applications. Nanomedicines have achieved remarkable clinical successes in cancer diagnosis and treatment. For instance, several nanomedicines have been approved by United States Food and Drug Administration (U.S. FDA) as either cancer therapeutics (e.g., Abraxane and Doxil (Barenholz 2012, Green 2006)) or cancer diagnostic agents (e.g., Feridex (Barnett et al. 2007)) for clinical cancer treatment, and more than 80 nanomedicines are under investigation in preclinical/clinical trials.

Two strategies for nanomedicine to target tumor *in vivo* are passive and active targeting. For passive tumor targeting, nanomedicine in circulation systems can extravasate through the leaky vascular wall in the tumor region and accumulate preferentially in tumor tissues, which is called enhanced permeability and retention (EPR) effect (Torchilin 2011). For active tumor targeting, nanomedicine is conjugated with tumor targeting entities, such as antibodies, ligands, engineered peptides and nucleic acids, actively recognizing and binding with molecular targets that overexpress on cancer cell surface through these targeting entities. The active targeting strategy has been proven to be highly effective in animal studies.

In this chapter, we will review recent advancements in development and application of smart materials in cancer diagnosis and treatment. Because the structures and materials of smart medicine have been reviewed elaborately in the previous chapter, this chapter will focus on biomedical applications of smart materials in cancer therapy. Smart materials discussed in this chapter are defined as therapeutic or diagnostic nanomedicines based on active tumor targeting strategy. Considering the tumor heterogeneity, we will review and discuss smart materials and relevant cancer molecular targets based on cancer types, in the order of cancer incidence.

2. Background and Basic Principles

Cancer is a global health issue, which has been one of the leading causes of morbidity and mortality around the world. According to the GLOBOCAN report, for 2012, approximately 14.1 million new cases of cancer were reported globally, and caused 8.2 million deaths (Ferlay et al. 2015). In 2015, in the U.S. alone, the American Cancer Society estimates about 1.66 million new cases and 589,430 deaths of cancer (Siegel et al. 2015). In general, cancer treatment relies primarily on surgery, chemotherapy and radiation therapy. However, over 90% cancer-related deaths are attributed to tumor metastases, which are challenging to remove by surgical resection (Mehlen and Puisieux 2006). Additionally, chemotherapy and radiation therapy indiscriminatingly kill both cancerous and healthy cells, causing severe adverse effects. To understand the cancer biology that underlies the occurrence and development of human malignant cancer better, remarkable effort has been made in the past decades. Based on that understanding, tremendous novel cancer-therapeutic strategies have been developed to treat cancers, such as hormone therapy, targeted therapy and immunotherapy.

In cancer diagnosis and treatment, smart materials (as seen in Fig. 1) have a prominent characteristic as compared to conventional materials that they can recognize cancerous tissues and cells under physiological conditions, facilitate targeted attack on cancerous cells, and spare the normal healthy cells. This is extremely useful in treating patients with aggressive or advanced cancers, because many of those cancer cells have metastasized to patients' lymph nodes and other organs, which are irremovable by surgical approaches. To achieve effective targeted cancer diagnosis and therapy, there are three critical factors that must be considered during the design and engineering of smart materials: (i) Tumor specificity. Smart materials usually utilize biological molecules overexpressed on tumor cells as binding targets. These molecular targets are usually cell surface proteins, and must be absent or expressed at low levels in normal tissues and cells to avoid "off-target" binding. (ii) Overexpression level. In general, the tumor specificity of smart materials correlates positively with the overexpression level of molecular target on cancer cell surfaces. (iii) Cellular internalization. Many therapeutic and diagnostic reagent used in smart materials must be delivered into cancer cell cytoplasm to be activated and effective. To improve cellular uptake, smart materials can take advantages of ligand- or antibody-mediated endocytosis by selecting suitable molecular targets on cancer cells.

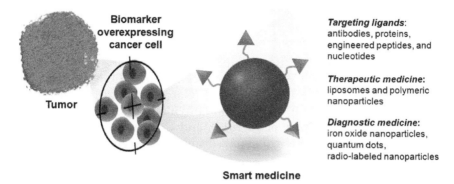

Biomarker overexpressing cancer cell

Tumor

Targeting ligands: antibodies, proteins, engineered peptides, and nucleotides

Therapeutic medicine: liposomes and polymeric nanoparticles

Diagnostic medicine: iron oxide nanoparticles, quantum dots, radio-labeled nanoparticles

Smart medicine

Figure 1. Schematic illustration of smart medicine for cancer diagnosis and treatment.

3. Smart Materials for Cancer Diagnosis and Treatment

3.1 Lung Cancer

Lung cancer is the most common cancer worldwide and the most frequent cancer among men. The mortality of lung cancer ranks as the first of overall cancer death and is over 2-fold higher than the second (liver cancer). According to GLOBOCAN report, for 2012, approximately 1.82 million new cases of lung cancer were reported globally, representing 13.0% of the total number of all new cancer cases (Ferlay et al. 2015). In the same year, lung cancer caused 1.59 million deaths, representing 19.4% of all cancer deaths (Ferlay et al. 2015). For 2015, in the U.S. alone, the American Cancer Society estimates about 221,200 new cases and 158,040 deaths of lung cancer (Siegel et al. 2015).

There are three main types of lung cancers. (i) Non-small cell lung cancer (NSCLC). NSCLC is an umbrella term for squamous cell carcinoma, adenocarcinoma and large cell carcinoma. NSCLC is the most common type of lung cancers, representing approximately 85% of all lung cancers, relatively insensitive to chemotherapy. The overall five year survival rate of NSCLC is between 11% and 17%. (ii) Small cell lung cancer (SCLC, or oat cell cancer). SCLC accounts for about 10%–15% of all lung cancers, which is generally more aggressive and metastatic than other types of lung cancers, leading to a poorer prognosis. The overall five year survival rate of SCLC is 5%. (iii) Lung carcinoid tumor (or lung neuroendocrine tumor). Lung carcinoid tumor is a less malignant lung cancer, representing fewer than 5% of all lung cancers. The overall five year survival rate of lung carcinoid tumor ranges between 50% and 90%. Tobacco smoking is usually considered as one of the major causes of lung cancer, which gives rise to about 86% of new lung cancer cases (Ganti 2006), resulting from the carcinogens, including nitrosamines and polycyclic aromatic hydrocarbons, exist in tobacco smoke. The incidence of SCLCs has been found highly correlated with tobacco smoking.

Six common targeted therapeutics for lung cancer are being used clinically (as seen in Table 1). They are Bevacizumab (Avastin) (Sandler et al. 2006), Ramucirumab (Cyramza) (Garon et al. 2012), Erlotinib (Tarceva) (Shepherd et al. 2005), Afatinib (Gilotrif) (Miller et al. 2012), Crizotinib (Xalkori) (Shaw et al. 2013), and Ceritinib (Zykadia) (Shaw et al. 2014). Bevacizumab is a humanized anti-vascular endothelial growth

Table 1. Summary of clinical targeted therapeutics for cancer therapy.

Type of Cancer	Clinical Targeted Therapeutics	References
Lung cancer	Bevacizumab (Avastin) Ramucirumab (Cyramza) Erlotinib (Tarceva) Afatinib (Gilotrif) Crizotinib (Xalkori) Ceritinib (Zykadia)	(Sandler et al. 2006) (Garon et al. 2012) (Shepherd et al. 2005) (Miller et al. 2012) (Shaw et al. 2013) (Shaw et al. 2014)
Breast cancer	Trastuzumab (Herceptin) Pertuzumab (Perjeta) Ado-trastuzumab emtansine (Kadcyla) Lapatinib (Tykerb) Palbociclib (Ibrance) Everolimus (Afinitor)	(Romond et al. 2005) (Baselga et al. 2012b) (Verma et al. 2012) (Geyer et al. 2006) (Finn et al. 2015) (Baselga et al. 2012a)
Colorectal cancer	Bevacizumab (Avastin) Ziv-aflibercept (Zaltrap) Cetuximab (Erbitux) Panitumumab (Vectibix) Regorafenib (Stivarga)	(Hurwitz et al. 2004) (Sun and Patel 2013) (Cunningham et al. 2004) (Amado et al. 2008) (Grothey et al. 2013)
Prostate cancer	Abiraterone (Zytiga) Enzalutamide (Xtandi) Sipuleucel-T (Provenge) Cabazitaxel (Jevtana)	(de Bono et al. 2011) (Cabot et al. 2012) (Kantoff et al. 2010) (de Bono et al. 2010)
Gastric cancer	Trastuzumab (Herceptin) Ramucirumab (Cyramza)	(Bang et al. 2010) (Fuchs et al. 2014)
Liver cancer	Sorafenib (Nexavar)	(Cheng et al. 2009)
Pancreatic cancer	Erlotinib (Tarceva) Sunitinib (Sutent) Everolimus (Afinitor)	(Moore et al. 2007) (Raymond et al. 2011) (Yao et al. 2011)
Brain cancer	Bevacizumab (Avastin) Everolimus (Afinitor)	(Friedman et al. 2009) (Krueger et al. 2010)
Melanoma	Vemurafenib (Zelboraf) Dabrafenib (Tafinlar) Trametinib (Mekinist)	(Chapman et al. 2011) (Hauschild et al. 2012) (Falchook et al. 2012)
Ovarian cancer	Bevacizumab (Avastin) Olaparib (Lynparza)	(Sato and Itamochi 2012) (Audeh et al. 2010)

factor (VEGF)-A monoclonal antibody. It functions as an angiogenesis inhibitor by blocking VEGF-A and inhibiting new blood vessel formation toward tumors (Sandler et al. 2006). Ramucirumab is another angiogenesis inhibitor, which is a humanized anti-VEGFR2 monoclonal antibody, and functions by neutralizing a certain type of receptor for VEGF via antibody blockade (Garon et al. 2012). Erlotinib and Afatinib are tyrosine kinase inhibitors, and function by blocking the signaling pathway of epidermal growth factor receptor (EGFR) overexpressed on lung cancer cells (Miller et al. 2012, Shepherd et al. 2005). Crizotinib and Ceritinib are small molecular inhibitors that function by targeting ALK gene mutations, which has been found in about 5% of all NSCLCs (Shaw et al. 2013, Shaw et al. 2014).

Common lung cancer molecular targets used for smart nanomedicines include EGFR, integrin $\alpha v \beta 6$, folate receptor and sigma receptor (as seen in Table 2). EGFR is a cell surface receptor for epidermal growth factor (EGF) that has been found upregulated in a variety of tumor cells, including lung cancer cells. Thus, EGFR has been widely used in smart nanomedicine for lung cancer specific affinity. In 2007, Tseng et al. reported EGF-conjugated gelatin nanoparticles could specifically bind with EGFR expressing lung cancer cells via *in vivo* aerosol administration (Tseng et al. 2008, 2007). Later in 2009, Tseng et al. further used their EGF-conjugated gelatin nanoparticles as lung tumor-targeted drug delivery systems to deliver cisplatin specifically via aerosol inhalation (Tseng et al. 2009). This EGFR-targeted constructs can inhibit the *in vivo* lung tumor growth by about 70%. In 2011, Peng et al. reported the development of single-chain variable fragment anti-EGFR antibody-conjugated heparin nanoparticles to deliver cisplatin specifically to lung tumor, in comparison with full-length EGFR antibody conjugated ones (Peng et al. 2011). Single-chain variable fragment EGFR antibody has a lower molecular weight by removing Fc regions, which showed better tumor penetration function with less non-specific interaction with Fc receptors on normal tissues.

EGFR has also been used in inorganic smart nanomedicine for cancer diagnostic and ablation applications. In 2008, Qian et al. used single-chain variable fragment anti-EGFR antibody-conjugated pegylated gold nanoparticle to facilitate a lung tumor specific detection via surface-enhanced Raman scattering (Qian et al. 2008). In 2011, Yokoyama et al. developed EGFR-targeted hybrid plasmonic magnetic nanoparticles to synergistically induce autophagy and apoptosis in NSCLC cells (Yokoyama et al. 2011). In 2013, Sadhukha et al. reported EGFR-targeted superparamagnetic iron oxide (SPIO) nanoparticles could be administered via aerosol inhalation and specifically bind with NSCLCs (Sadhukha et al. 2013). The therapeutic studies further demonstrated that the EGFR-targeted SPIO nanoparticles could facilitate efficient hyperthermia therapy and significantly inhibited *in vivo* NSCLC tumor growth.

Table 2. Summary of common molecular targets for cancer diagnosis and treatment.

Type of Cancer	Common Molecular Targets	References
Lung cancer	EGFR	(Tseng et al. 2007, 2008, 2009) (Peng et al. 2011) (Qian et al. 2008) (Yokoyama et al. 2011) (Sadhukha et al. 2013)
	Integrin αvβ6	(Sundaram et al. 2009) (Guthi et al. 2010) (Huang et al. 2009)
	Folate receptor	(Santra et al. 2009) (Yoo et al. 2012)
	Sigma receptor	(Li and Huang 2006) (Yang et al. 2012) (Zhang et al. 2013)
Breast cancer	HER2	(Sun et al. 2008) (Stuchinskaya et al. 2011) (Kikumori et al. 2009) (Lee et al. 2007) (Wu et al. 2003) (Xiao et al. 2009)
	EGFR	(Acharya et al. 2009)
	uPAR	(L. Yang et al. 2009)
	p32 receptor	(Kinsella et al. 2011)
Colorectal cancer	EGFR	(Cho et al. 2010) (Löw et al. 2011)
	VEGF	(Chen 2012)
	CD44	(Jain et al. 2010)
	Folate receptor	(Zheng et al. 2005)
Prostate cancer	PSMA	(Bagalkot et al. 2006) (Cheng et al. 2007 (Dhar et al. 2008) (Farokhzad et al. 2006, 2004) (Wang et al. 2008) (Xiao et al. 2012) (Zhang et al. 2007) (Sawant et al. 2008) (Yu et al. 2011)
	GRPR	(Steinmetz et al. 2011)
	uPAR	(Abdalla et al. 2011)
	Folate receptor	(Hattori and Maitani 2004, 2005) (Xue and Wong 2011)
	Sigma receptor	(Banerjee et al. 2004)

Table 2. contd....

Table 2. contd.

Type of Cancer	Common Molecular Targets	References
Gastric cancer	HER2	(Li et al. 2012) (Ruan et al. 2012b)
	CD44	(Chen et al. 2012)
	Folate receptor	(Pan et al. 2013)
	Transferrin receptor	(Iinuma et al. 2002)
Liver cancer	EGFR	(Klutz et al. 2011) (Liu et al. 2012)
	Folate receptor	(Fan et al. 2010) (Maeng et al. 2010) (Ji et al. 2012)
	VEGF	(Liu et al. 2011) (Wang et al. 2012) (Zhang et al. 2014)
	CD44	(Wang et al. 2012)
	ASGPR	(Gu et al. 2012) (Liang et al. 2006) (Shen et al. 2011) (Zhang et al. 2013)
	Nucleolin	(Zhang et al. 2014)
Pancreatic cancer	EGFR	(Magadala and Amiji 2008) (Patra et al. 2008) (Yang et al. 2009b) (Glazer et al. 2010) (Aggarwal et al. 2011) (Xu et al. 2014)
	CD44	(Peer and Margalit 2004)
	uPAR	(Lily Yang et al. 2009a) (Lee et al. 2013)
	Transferrin receptor	(Barth et al. 2010) (Camp et al. 2013)
Brain cancer	EGFR	(Wu et al. 2006) (Hadjipanayis et al. 2010)
	Transferrin receptor	(Zhang et al. 2003) (Gan and Feng 2010) (Ying et al. 2010)
	IL-13R	(Madhankumar et al. 2009) (Rozhkova et al. 2009)
	Nucleolin	(Reddy et al. 2006) (Gupta and Torchilin 2007)

Table 2. contd....

Table 2. contd.

Type of Cancer	Common Molecular Targets	References
Melanoma	Integrin αvβ3	(Hölig et al. 2004) (Schmieder et al. 2005) (Boles et al. 2010) (Benezra et al. 2011)
	EGFR	(El-Sayed et al. 2006)
	Sigma receptor	(Chen et al. 2010)
	Nucleolin	(Skidan et al. 2008)
	CD44	(Jin et al. 2012)
Ovarian cancer	Folate receptor	(Esmaeili et al. 2008) (Kim et al. 2009) (Kamaly et al. 2009) (Nukolova et al. 2011) (Wang et al. 2011) (Werner et al. 2011)
	Integrin αvβ3	(Zhao et al. 2009) (Xiao et al. 2012)
	Nucleolin	(Winer et al. 2010) (Perche et al. 2012)
	HER2	(Cirstoiu-Hapca et al. 2010)
	FSHR	(Zhang et al. 2009)
	EphA receptor	(Scarberry et al. 2010)
	P32 receptor	(Ren et al. 2012)

Integrin αvβ6 is a transmembrane receptor presenting on lung cancer cells, and can be targeted by engineered RGD peptides. In 2009, Sundaram et al. reported the development an RGD peptide-conjugated polymeric nanoparticle to deliver anti-VEGF intraceptor (Flt23k) specifically to lung cancer cells (Sundaram et al. 2009). In 2010, Guthi et al. developed an integrin αvβ6-targeted micelle to co-deliver magnetic resonance imaging (MRI) contrast agent iron oxide nanoparticle (IONP) and chemotherapeutic doxorubicin (Dox) specifically to lung cancer cells (Guthi et al. 2010, Huang et al. 2009). The integrin αvβ6-targeted constructs increased the cancer cell uptake by 3-fold, compared with non-specific counterparts, demonstrating potential application for image-guided, target-specific treatment of lung cancer.

Folate receptor is a cell surface glycoprotein that binds folate specifically and reduce folic acid derivatives. Folate receptor has been reported overexpressing in a variety of tumor cells including lung cancer cells, fit to be used as a molecular target for smart nanomedicine. In 2009, Santra et al. reported tumor-targeted multifunctional folate-conjugated polyacrylic acid-coated IONPs, composed of an anti-cancer drug paclitaxel (PTX), a near infrared (NIR) dye DiR and a MRI contrast agent IONP for lung cancer-targeted treatment (Santra et al. 2009). In 2012, Yoo et al. demonstrated that folic acid-conjugated, PEGylated SPIO nanoparticles could facilitate a lung tumor-specific MR imaging for *in vivo* tumor detection (Yoo et al. 2012).

Sigma receptor is another lung cancer molecular target. In 2006, Huang et al. developed sigma receptor-targeted liposome-polycation-DNA nanoparticles for a lung tumor specific antisense oligodeoxynucleotide or siRNA delivery (Li and Huang 2006). The anisamide ligand increased cellular uptake for 4 to 7-folds by the sigma receptor overexpressing cells, producing strong antisense efficacies to down-regulate survivin mRNA and protein levels in lung cancer cells. Later in 2012, Yang et al. developed a sigma receptor-targeted lipid/calcium/phosphate nanoparticle to co-deliver MDM2, c-myc and VEGF siRNAs specifically to lung metastases. The sigma receptor-targeted constructs could significantly prolong the lung tumor bearing animal survival time by 27.8% without any systematic toxicity (Yang et al. 2012). In 2013, Zhang et al. further used sigma receptor targeted lipid/calcium/phosphate nanoparticles to co-deliver siRNA targeting VEGFs and gemcitabine (Gem) monophosphate specifically to NSCLCs. A synergy of anti-angiogenesis therapy and chemotherapy has been observed, and the tumor growth was reduced effectively by 70% in orthotopic lung tumor model (Zhang et al. 2013).

3.2 Breast Cancer

Breast cancer is the second most common cancer worldwide and the most frequent cancer among women. The mortality rate of breast cancer ranks the fifth of overall cancer death and the second of cancer death in women exceeded only by lung cancer. According to the GLOBOCAN report, for 2012, approximately 1.68 million new cases of breast cancer were reported globally, representing 25% of the total number of new women's cancer cases (Ferlay et al. 2015). In the same year, breast cancer caused 522,000 deaths, representing 14.7% of women's cancer deaths. For 2015, in the U.S. alone, the American Cancer Society estimates about 234,190 new cases and 40,730 deaths of breast cancer (Siegel et al. 2015).

There are three common types of breast cancers: (i) estrogen receptor/progesterone receptor (ER/PR) positive breast cancer. ER/PR positive breast cancer is the most common type of all breast cancers, representing approximately 70% of all breast cancers. ER/PR positive breast cancer patients respond to hormone therapy, and have a five year survival rate of over 95%. (ii) Human epidermal growth factor receptor 2 (HER2) positive breast cancer. HER2 positive breast cancer is defined by the overexpression of HER2 on breast cancer cell membranes, and represents about 20% to 25% of all breast cancers. HER2 positive breast cancer patients are usually treated with HER2 targeted therapeutics, and have a five year survival rate of over 87%. (iii) Triple negative breast cancer. Triple negative breast cancer is defined by the lack of expression of ER, PR, and HER2 in breast cancer cells, and represents about 15–20% of all breast cancers. Currently there is no clinical targeted therapeutic for triple negative breast cancers. About 50% of triple negative breast cancer patients respond to adjuvant chemotherapy, but triple negative breast cancer cells are generally more proliferative and aggressive than the other types of breast cancers. Together with limited therapeutic approaches, triple negative breast cancer has a significant poorer prognosis than other types of breast cancers, which has a five year survival rate less than 74.5%.

Six targeted therapeutics are being used clinically for breast cancer (as seen in Table 1) including Trastuzumab (Herceptin) (Romond et al. 2005), Pertuzumab (Perjeta) (Baselga et al. 2012b), Ado-trastuzumab emtansine (Kadcyla) (Verma et al. 2012), Lapatinib (Tykerb) (Geyer et al. 2006), Palbociclib (Ibrance) (Finn et al. 2015) and Everolimus (Afinitor) (Baselga et al. 2012a). Trastuzumab and Pertuzumab are anti-HER2 monoclonal antibodies. They function by neutralizing HER2 signaling cascades on HER2 positive breast cancer cells via antibody blockade (Baselga et al. 2012b, Romond et al. 2005). Ado-trastuzumab emtansine is a monoclonal antibody-chemotherapeutic drug conjugate, which is used to treat advanced HER2 positive breast cancer with resistance to trastuzumab (Verma et al. 2012). Lapatinib is a dual tyrosine kinase inhibitor. It functions by interrupting both HER2 and EGFR signaling pathways in HER2 positive breast cancers (Geyer et al. 2006). Palbociclib is an aromatase inhibitor. It functions by blocking cyclin-dependent kinase (CDK) 4 and CDK6 to inhibit cancer cell proliferation in ER/PR positive breast cancers (Finn et al. 2015). Everolimus is a mammalian target of rapamycin (mTOR) inhibitor that functions by blocking mTOR in ER/PR positive breast cancers. mTOR is a protein that promotes breast cancer cells growth and division and using Everolimus to block mTOR along with a hormone therapy, can prevent tumor growth in advanced ER/PR positive breast cancer patients (Baselga et al. 2012a).

The common breast cancer molecular targets include HER2, EGFR, urokinase-type plasminogen activator receptor (uPAR), and p32 cell surface receptor (as seen in Table 2). HER2, also known as Receptor tyrosine-protein kinase erbB-2 (ERBB2), is a clinically approved biomarker and molecular target for HER2 positive breast cancers. A variety of targeted nanomedicines have been developed to target HER2 via monoclonal antibodies. For example, in 2008, Sun et al. developed a poly(d,l-lactide-co-glycolide)/montmorillonite nanoparticle decorated by Trastuzumab for targeted chemotherapy of breast cancer overexpression HER2 (Sun et al. 2008). Their findings demonstrated that the half maximal inhibitory concentration (IC50) of their targeted nanomedicine is over

13-fold more efficient than free PTX. In 2011, Stuchinskaya developed HER2-targeted PEGylated gold nanoparticles that can produce cytotoxic singlet oxygen under visible light illumination. Their findings showed that this nanoparticle can selectively target HER2 positive breast cancer to facilitate effective photodynamic therapy (PDT) (Stuchinskaya et al. 2011). In 2009, Kikumori et al. developed a HER2-targeted magnetite nanoparticle-loaded immunoliposomes to treat breast cancer via hyperthermia. It showed that hyperthermia using magnetic nanoparticle is an effective and specific therapy for breast cancer overexpressing HER2 (Kikumori et al. 2009). In 2007, Lee et al. engineered a HER2-targeted magnetism-engineered iron oxide nanoprobe that can show enhanced MRI sensitivity for the *in vivo* detection of HER2 overexpressing breast tumors (Lee et al. 2007). In 2002, Wu et al. reported that HER2 antibody-conjugated quantum dots (QDs) could be used to specifically stain actin and microtubule fibers in the cytoplasm, and to detect nuclear antigens inside the nucleus. But the application of QD for *in vivo* imaging is significantly limited by QD's safety issues (Wu et al. 2003). Single-walled carbon nanotube (SWNT) is a novel nanomaterial with potential applications in cancer detection and imaging, owing to its two unique optical properties: SWNTs have a strong Raman signal for cancer cell detection, and also demonstrate strong NIR absorbance for selective photothermal ablation of tumors. In 2009, Xiao et al. functionalized SWNTs with anti-HER2 antibodies, and used them for both detection and selective photothermal ablation of HER2 positive breast cancer cells without the need of internalization by the cells (Xiao et al. 2009).

EGFR is another common molecular target that highly expresses on breast cancer cells. In 2009, Acharya et al. reported that anti-EGFR antibody-conjugated, Rapamycin encapsulating PLGA nanoparticle can efficiently deliver anticancer drugs specifically to breast cancer cells, and significantly inhibit breast cancer cell proliferation (Acharya et al. 2009). uPAR has also been investigated as a molecular target for breast cancer. In 2009, Yang et al. explored the uPAR as a novel breast cancer molecular target and conjugated amino-terminal fragment of uPAR to IONPs for *in vivo* MR imaging of breast cancers. Their findings also showed that uPAR-targeted nanoparticles have a lower accumulation in mouse liver and spleen compared with their non-targeted counterparts (Yang et al. 2009).

Other breast cancer targets, such as p32 cell surface receptor, have also been investigated to provide high breast cancer specificity. In 2011, Kinsella et al. reported the synthesis of a LyP–1 peptide conjugated Bi_2S_3 nanoparticle as a new class of X-ray contrast agents (Kinsella et al. 2011). These NPs can produce quantitative, high fidelity CT images of breast tumors, and last for more than one week. Notably, these nanoparticles appear to undergo clearance from the mice via a fecal route during this time period. Although limiting the period of imaging, it provides a safe mechanism for nanoparticle clearance.

3.3 Colorectal Cancer

Colorectal cancer includes colon cancer and rectal cancer, and is the third most common cancer worldwide. The mortality of colorectal cancer ranks the fourth of overall cancer deaths following lung, liver and stomach cancers. According to the GLOBOCAN report, for 2012, approximately 1.36 million new cases of colorectal cancer were reported globally, representing 9.6% of the total number of all new cancer cases (Ferlay et al. 2015). In the same year, colorectal cancer caused 694,000 deaths, representing 8.5% of all cancer deaths. For 2015, in the U.S. alone, the American Cancer Society estimates about 132,700 new cases and 49,700 deaths of colorectal cancer (Siegel et al. 2015).

There are five common types of colorectal cancers: (i) Adenocarcinoma. More than 95% of colorectal cancers are adenocarcinomas originated from glands that make mucus to lubricate the inside of colon and rectum. Colon adenocarcinoma has an overall five year survival rate of 65.2%. (ii) Carcinoid tumor. Carcinoid tumors originated from specialized hormone-producing cells in the intestine, representing about 1% of cancers of the gastrointestinal tract. The five year survival rate of carcinoid cancer ranges from 65% to 90% based on carcinoid tumor location. (iii) Gastrointestinal stromal tumor (GIST). GIST originates from specialized cells in the wall of the colon called the interstitial cells of Cajal, representing less than 1% of all gastrointestinal tumors. But it is the most common mesenchymal tumors of the gastrointestinal (GI) tract. GISTs have a five year survival rate of 35%. (iv) Lymphoma. Primary colorectal lymphoma is a rare tumor of the GI tract, and the most common variety of colonic lymphoma is non-Hodgkin's lymphoma (NHL). The five year survival rate of primary colorectal lymphoma patient is approximately 40%. (v) Colorectal sarcoma. Colorectal sarcoma is also a rare GI tract cancer originating from muscle and connective tissue in the wall of the colon and rectum. The five year survival rate of colorectal sarcoma patient is approximately 51%.

There are five common clinical targeted therapeutics for colorectal cancer (as seen in Table 1). They are Bevacizumab (Avastin) (Hurwitz et al. 2004), Ziv-aflibercept (Zaltrap) (Sun and Patel 2013), Cetuximab (Erbitux) (Cunningham et al. 2004), Panitumumab (Vectibix) (Amado et al. 2008) and Regorafenib (Stivarga) (Grothey et al. 2013). Bevacizumab is anti-VEGF-A monoclonal antibody, and its therapeutic function is efficiently inhibiting tumor angiogenesis by blocking VEGF-A signaling pathways (Hurwitz et al. 2004). See more information about Bevacizumab in lung cancer section. Ziv-aflibercept is a recombinant fusion protein with VEGF-binding portions (Sun and Patel 2013). Cetuximab and Panitumumab are both anti-EGFR monoclonal antibodies that function by neutralizing EGFR signaling pathways in EGFR overexpressing cells via antibody blockade (Amado et al. 2008, Cunningham et al. 2004). Regorafenib is a kinase inhibitor, and can block several important kinase proteins that promote tumor growth and angiogenesis (Grothey et al. 2013).

Common colorectal cancer molecular targets include EGFR, VEGF, CD44, and folate receptor (as seen in Table 2). EGFR is a colorectal cancer target protein highly overexpressed in about 80.0% human colorectal tumor tissues. In 2010, Cho et al. developed Cetuximab-conjugated magneto-fluorescent silica nanoparticles and used them for *in vivo* colon cancer targeting and MR imaging. Their findings indicated that the potential application of Cetuximab as a nanomedicine targeting ligand for the detection of EGFR-expressing colon cancer using *in vivo* imaging approaches (Cho et al. 2010). In 2011, Löw et al. reported the development of the Cetuximab-modified human serum albumin (HSA) nanoparticles as a colon cancer-targeted drug delivery carrier (Löw et al. 2011). Their nanoparticles can significantly improve cellular binding and intracellular accumulation with EGFR overexpressing colon cancer cells.

VEGF is a colorectal cancer target detected in about 64.0% human colorectal tumor tissues. In 2012, Hsieh et al. reported the development of anti-VEGF antibody-conjugated SPIO nanoparticles for colon cancer specific MR imaging (Hsieh et al. 2012). Their findings demonstrated that VEGF is an effective colon cancer target for *in vivo* tumor targeting and efficient accumulation of nanoparticles in tumor tissues after systemic delivery in a colon cancer animal model.

CD44 is detected in 54% of colorectal tumor tissues. In 2010, Jain et al. developed a hyaluronic acid-coupled chitosan nanoparticles bearing oxaliplatin for targeted treatment of colon tumors (Jain et al. 2010). They demonstrated using hyaluronic acid to target CD44 expressing colon tumor in murine model which can achieve high local drug concentration within the colonic tumors with prolonged exposure time, indicating a potential for enhanced antitumor efficacy with low systematic toxicity.

Folate receptor was detected positive in about 33% of primaries and 44% of metastases of colorectal tumor tissues. In 2005, Zheng et al. developed folic acid-conjugated lipoprotein-based nanoplatform to targeted folate receptor overexpressing prostate cancer cells and their findings indicated that folic acid conjugation to the Lys side-chain amino groups can block binding to the normal LDL receptor and reroute the resulting conjugate to cancer cells through their folate receptors (Zheng et al. 2005). In 2010, Yang et al. demonstrated that folic acid-conjugated chitosan nanoparticles enhanced protoporphyrin IX accumulation in colorectal cancer cells, as an ideal vector for colorectal-specific delivery of 5-ALA for fluorescent endoscopic detection (Yang et al. 2010).

3.4 Prostate Cancer

Prostate cancer is the fourth most common cancer worldwide and the second most frequent cancer among men. The mortality of prostate cancer ranks the eighth of overall cancer death and the fifth of cancer death in men. According to the GLOBOCAN report, in 2012, approximately 1.11 million new cases of prostate cancer were reported globally, representing 7.9% of the total number of new men's cancer cases (Ferlay et al. 2015). In the same year, prostate cancer caused 307,000 deaths, representing 6.6% of men's cancer deaths. In 2015, in the U.S. alone, the American Cancer Society estimates about 220,800 new cases and 27,540 deaths of prostate cancer (Siegel et al. 2015).

Prostate adenocarcinoma is most common type of prostate cancers, representing over 95% of all prostate cancers. Prostate adenocarcinoma is defined as a glandular cancer in the male reproductive system. The relative five year survival rate of prostate adenocarcinoma is almost 100%.

There are four common clinical targeted therapeutics for prostate cancer (as seen in Table 1). They are Abiraterone (Zytiga) (de Bono et al. 2011), Enzalutamide (Xtandi) (Cabot et al. 2012), Sipuleucel-T (Provenge) (Kantoff et al. 2010), and Cabazitaxel (Jevtana) (de Bono et al. 2010). Abiraterone is a small molecular inhibitor used in combination with prednisone in HRPC. It functions by inhibiting cytochrome P17 to reduce androgen biosynthesis (de Bono et al. 2011). Enzalutamide is a small molecular androgen receptor antagonist. It was approved by U.S. FDA in 2012 to treat castration-resistant prostate cancer (Cabot et al. 2012). Sipuleucel-T is a cell-based cancer immunotherapy for prostate cancer. It functions by programming a patient's own immune system with prostate cancer cell antigens to amplify the immunoresponse to metastatic prostate tumors. Sipuleucel-T was approved by U.S. FDA in 2010 to treat asymptomatic or minimally symptomatic metastatic HRPC (Kantoff et al. 2010). Cabazitaxel is a small molecular inhibitor derived from a natural taxoid. It was approved by U.S. FDA in 2010 for targeted treatment of hormone-refractory prostate cancer. Cabazitaxel functions by stabilizing microtubules to suppress nuclear translocation of the androgen receptor in prostate cancer cells, which eventually leads to cell apoptosis (de Bono et al. 2010).

Common molecular targets for prostate cancer includes: prostate-specific membrane antigen (PSMA), gastrin releasing peptide receptor (GRPR), uPAR, folate receptor, and sigma receptor (as seen in Table 2). PSMA is the most commonly used prostate cancer molecular target for smart nanomedicine. Farokhzad et al. conducted a series of studies on PSMA aptamer-conjugated polymeric (PLG-PEG and PLGA-PEG) nanoparticles to facilitate specific delivery of therapeutics to prostate tumors *in vivo*

(Bagalkot et al. 2006, Cheng et al. 2007, Dhar et al. 2008, Farokhzad et al. 2006, 2004, Wang et al. 2008, Xiao et al. 2012, Zhang et al. 2007). In 2008 Sawant et al. conjugated prostate cell-specific monoclonal antibody 5D4 (mAb 5D4) on long-circulating liposomes loaded with Dox to facilitate prostate tumor-targeted therapy, and proved its significantly enhanced cytotoxicity toward these cells compared with the non-targeted Dox-liposomes *in vitro* (Sawant et al. 2008). In 2011, Yu et al. reported the development of PSMA aptamer-conjugated SPIO nanoparticles to facilitate image-guided prostate cancer therapy (Yu et al. 2011). These prepared nanoparticles can detect prostate tumor *in vivo* by MR imaging and selectively deliver therapeutics to the tumor tissue simultaneously.

GRPR has been found overexpressing in a number of cancer cells including prostate cancer. In 2011, Steinmetz et al. reported the synthesis and characterization of NIR cowpea mosaic virus particles targeting GRPRs that are over-expressed in human prostate cancers (Steinmetz et al. 2011). They observed significantly increased tumor accumulation in human prostate tumor xenografts on the chicken chorioallantoic membrane model using intravital imaging.

uPAR has also been investigated as a prostate cancer molecular target. In 2011, Abdalla et al. engineered a multifunctional nanoscale delivery vehicle targeting uPAR on prostate cancer cells for specific MR imaging and NIR imaging (Abdalla et al. 2011). They also demonstrated uPAR-targeted noscapine-loaded nanoparticles enhanced intracellular noscapine accumulation by the approximately 6-fold stronger inhibitory effect on PC-3 prostate cancer cell growth compared to free noscapine.

Folate receptor is also a prostate cancer molecular target, and folate has been explored as a smart nanomedicine targeting ligand. In 2004, Hattori and Maitani reported the development of folate-conjugated cationic lipid nanoparticles to selectively deliver DNA to LNCaP prostate cancer cells, suggesting that such nanoparticles are potentially targeted vectors to prostate cancer for gene delivery (Hattori and Maitani 2004). Later in 2005, Hattori and Maitani further used folate-conjugated nanoparticles to mediate suicide gene therapy in human prostate cancer (Hattori and Maitani 2005). In 2011, Xue and Wong developed folate-conjugated, survivin siRNA encapsulating solid lipid-PEI hybrid nanoparticles to facilitate prostate tumor specific treatment (Xue and Wong 2011). They found that the extended release and toxicity reduction features of lipid nanoparticles are particularly suitable for the treatment of chronic disease conditions. Sigma receptor has also been found overexpressing in prostate cancer cells. In 2004, Banerjee et al. engineered anisamide-targeted stealth liposomes to specifically deliver Dox to human prostate cancer cells *in vivo* (Banerjee et al. 2004).

3.5 Gastric Cancer

Gastric cancer is the fifth most common cancer worldwide. The mortality of gastric cancer ranks the third of overall cancer death exceeded only by lung and liver cancers. According to the GLOBOCAN report, in 2012, approximately 951,000 new cases of gastric cancer were reported globally, representing 6.8% of the total number of overall cancer cases (Ferlay et al. 2015). In the same year, gastric cancer caused 723,000 deaths, representing 8.8% of overall cancer deaths. In 2015, in the U.S. alone, the American Cancer Society estimates about 24,590 new cases and 10,720 deaths of gastric cancer (Siegel et al. 2015).

There are four common types of gastric cancers: (i) Gastric adenocarcinoma. Gastric adenocarcinoma is most common type of gastric cancers, representing about 90% to 95% of gastric cancers. It is defined as cancers developing from the cells that form the innermost lining of the stomach. The five year survival rate of gastric adenocarcinoma is approximately 29%. (ii) Gastric lymphoma. Gastric lymphoma is the second most common type of gastric cancers, representing about 5% of gastric cancers. It is defined as cancers developed in the immune system tissues in stomach walls. The overall five year survival rate of gastric lymphoma ranges between 56% and 91% depending on the stage of disease and grade of lesion. (iii) GIST. GIST is rare stomach cancer developed from interstitial cells in the wall of stomach. The incidence of GIST in the U.S. alone is estimated at 4,000 to 5,000 per year. The overall five year survival rate of GIST was estimated to be about 76%. (iv) Gastric carcinoid tumor. Gastric carcinoid tumor is defined as cancers developed in hormone-making cells of the stomach. It is also a rare gastric cancer comprising only 3% of all gastric cancers. The overall five year survival rate of gastric carcinoid tumor is about 75%.

There are two common clinical targeted therapeutics for gastric cancer (as seen in Table 1). They are Trastuzumab (Herceptin) (Bang et al. 2010) and Ramucirumab (Cyramza) (Fuchs et al. 2014). Trastuzumab is a human monoclonal HER2 antibody. See more introduction of trastuzumab in breast cancer section. Trastuzumab, in combination with cisplatin and a fluoropyrimidine (either capecitabine or 5-fluorouracil), was approved by U.S. FDA in 2010 to treat gastric or gastro-esophageal junction adenocarcinoma (Bang et al. 2010). Ramucirumab is a human monoclonal VEGFR antibody. See more introduction of Ramucirumab in lung cancer section. It was approved by U.S. FDA in 2014 to treat gastric or gastro-esophageal junction adenocarcinoma as single agent or in combination with PTX (Fuchs et al. 2014).

Common gastric cancer molecular targets include HER2, CD44, folate receptor, and transferrin receptor (as seen in Table 2). HER2 is a common molecular target for gastric cancers, and has been investigated as targeting site for therapeutic and diagnostic nanomedicine. In 2012, Li et al. synthesized anti-HER2 antibody Fab fragment-conjugated immunomicelles to deliver Dox into gastric tumors *in vivo* (Li et al. 2012). In 2012, Ruan et al. further developed anti-HER2 monoclonal antibody-conjugated RNase-A-associated CdTe QDs for targeted fluorescent imaging and therapy of gastric cancer (Ruan et al. 2012b).

CD44 has been found overexpressed in gastric cancer tissues and cells, revealed its potential as an effective molecular target for gastric cancer. In 2012, Chen et al. engineered CD44v6 single-chain variable fragment-conjugated IONPsas a gastric cancer-specific MRI contrast agent and siRNA delivery nanocarrier (Chen et al. 2012). Folate receptor has also been used for gastric cancer targeting sites. In 2013, Pan et al. successfully synthesized folic acid-conjugated NaYbF4 upconversion nanocrystals for gastric tumor specific NIR imaging of *in vivo* gastric cancer tissues. Transferrin receptor was also studied as a gastric cancer target (Pan et al. 2013). In 2002 Iinuma et al. prepared transferrin-polyethylene glycol liposome to facilitate gastric tumor specific delivery of cisplatin (Iinuma et al. 2002). They found that transferrin receptor-targeted liposomes can achieve high liposome and cisplatin levels in ascites and showed a prolonged residence time in the peripheral circulation. In comparison, uptakes of transferrin receptor-targeted PEGylated liposomes in the liver and spleen were also significantly lower than those of non-specific liposomes.

In addition to molecular targets, macrophages and mesenchymal stem cells have also been studied as cell-based targeting ligands for gastric cancer specific drug delivery and tumor detection. In 2010, Matsui et al. prepared oligomannose-coated liposomes, and demonstrated these liposomes could be effectively uptaken by mouse peritoneal macrophages to carry anticancer drugs to omental milky spots known as initial metastatic sites in the peritoneal cavity in mice (Matsui et al. 2010). Their findings provide a new strategy to use human macrophages as a cellular vehicle targeting peritoneal micrometastases in the omentum of gastric cancer patients. In 2012, Ruan et al. investigated to use fluorescent magnetic nanoparticle-labeled mesenchymal stem cells to find gastric tumor and facilitate tumor-targeted imaging and hyperthermia therapy (Ruan et al. 2012a).

3.6 Liver Cancer

Liver cancer is the sixth most common cancer worldwide. The mortality of liver cancer ranks the second of overall cancer death exceeded only by lung cancer. According to the GLOBOCAN report, for 2012, approximately 782,000 new cases of liver cancer were reported globally, representing 5.6% of the total number of overall cancer cases (Ferlay et al. 2015). In the same year, liver cancer caused 745,000 deaths, representing 8.8% of overall cancer deaths. For 2015, in the U.S. alone, the American Cancer Society estimates about 35,660 new cases and 24,550 deaths of liver cancer (Siegel et al. 2015).

There are four common types of liver cancers: (i) Hepatocellular carcinoma (HCC). HCC is the most common type of liver cancers, representing approximately 80% of all liver cancers. The incidence of HCC strongly correlates with prevalence of hepatitis B, hepatitis C and chronic liver diseases. The five year survival rate of HCC is below 12%. (ii) Intrahepatic cholangiocarcinoma. Intrahepatic cholangiocarcinoma is also called bile duct cancer, representing about 10–20% of liver cancers. It is defined as cancers developed in small bile duts within the liver. The five year survival rate of intrahepatic cholangiocarcinoma ranges between 2% to 15%. (iii) Angiosarcoma and hemangiosarcoma. They are rare liver cancers, and defined as cancers developed in cells lining the blood vessels in the liver. The incidence of angiosarcoma and hemangiosarcoma is strongly associated with exposure to vinyl chloride or thorium dioxide. (iv) Hepatoblastoma. Hepatoblastoma is a rare liver cancer that mostly develops in children under 4 years old.

Currently, there is one common clinical targeted therapeutic for liver cancer, which is Sorafenib (Nexavar) (Cheng et al. 2009) (as seen in Table 1). Sorafenib a small molecular kinase inhibitor approved for the treatment of advanced HCC. It functions by blocking several tyrosine protein kinases (VEGFR and PDGFR) and Raf kinases (more avidly C-Raf than B-Raf), and inhabiting new blood vessel formation toward tumors (Cheng et al. 2009).

Common liver cancer molecular targets include EGFR, folate receptor, VEGF, CD44, asialoglycoprotein receptor (ASGPR), and nucleolin (as seen in Table 2). EGFR has been investigated as a smart nanomedicine molecular target for liver cancer. In 2011, Klutz et al. prepared PEI/PEG polyplexes conjugated with synthetic peptide GE11 as an EGFR-specific ligand to facilitate a sodium iodide symporter gene delivery specifically to HCC (Klutz et al. 2011). Their studies indicated that systemic NIS gene transfer using polyplexes coupled with an EGFR-targeting ligand is capable of inducing tumor-specific iodide uptake, which represents a promising innovative strategy for systemic sodium iodide symporter gene therapy in metastatic cancers. In 2012, Liu et al. developed EGFR-targeted therapeutic gene delivery nanocomplex for liver cancer-targeted therapy. Targeting ligand YC21 (an oligopeptide composed of 21 amino acid units) was coupled with the nanovector composed of β-cyclodextrin and low-molecular-weight polyethylenimine (PEI, Mw 600) to form the EGFR-targeted gene vector (termed as YPCs) (Liu et al. 2012). The reported gene vectors possessed the highly efficient gene delivery capability to the EGFR expressing liver cancer cells.

Folate receptor is another important liver cancer target for smart nanomedicine. In 2010, Fan et al. reported the development of folate-chitosan micellar nanoparticles to co-deliver Pyrrolidinedithiocarbamate (PDTC) and Dox for liver cancer targeted therapy (Fan et al. 2010). Their findings showed further that co-delivery of PDTC and DOX can overcome the multidrug resistance (MDR) of DOX during folate receptor-mediated endocytosis process. In 2010, Maeng et al. developed a folate receptor-targeted, Dox-loaded SPIO nanoparticles for chemotherapy and MR imaging in liver cancer (Maeng et al. 2010). In 2012, Ji et al. prepared folic acid-conjugated, chitosan modified SWNTs for targeted delivery of Dox to liver cancer cells. Their Dox loaded SWNTs were approved to effectively inhibit the liver tumor growth in nude mice compared with free Dox (Ji et al. 2012).

Other common cancer targets such as VEGF, CD44, and nucleolin, have also been explored as liver cancer molecular targets for smart nanomedicine (Liu et al. 2011, Wang et al. 2012, Zhang et al. 2014). In 2011, Liu et al. developed anti-VEGF antibody-conjugated Gadolinium-loaded polymeric nanoparticles modified as multifunctional MRI contrast agents to detect liver cancer (Liu et al. 2011). During *in vivo* studies, the VEGF-targeted nanoparticles showed significantly signal intensity enhancement at the tumor site compared with non-specific nanoparticles in liver tumor bearing mice. VEGF-targeted nanoparticles also prolonged the MR imaging time from less than an hour to 12 h compared with free Gadolinium. In 2012, Wang et al. reported the development of CD44-targeted liposomal nanoparticles to deliver Dox or a triple fusion gene containing the herpes simplex virus truncated thymidine kinase, renilla luciferase and red fluorescent protein (Wang et al. 2012). Engineered nanoparticles successfully demonstrated time intensive preclinical steps involved in molecular target identification, validation, and characterization by dual molecular imaging. In 2014, Zhang et al. reported the synthesis of mesoporous silica nanoparticles (MSNs) for liver tumor-targeted triplex therapy (Zhang et al. 2014). They immobilized CytC onto the MSNs as sealing agent for redox-responsive intracellular drug delivery. AS1411 aptamer was further tailored onto MSNs for liver tumor specificity. Dox was eventually loaded on the triplex nanoparticles for chemotherapy. Their results displayed a promising triplex anti-tumor triplex therapy via the loading DOX, gatekeeper of CytC and AS1411 aptamer.

ASGPR has been found overexpressing in liver cancerous tissues, but not in normal tissues (Gu et al. 2012, Liang et al. 2006, Shen et al. 2011, Zhang et al. 2013). Thus it has been explored as a liver cancer specific antigen and molecular target. In 2006, Liang et al. engineered galactosamine-conjugated, PTX-loaded poly(γ-glutamic acid)-poly(lactide) nanoparticles as a targeted drug delivery system for the treatment of liver cancer

(Liang et al. 2006). Galactosamine-conjugated on nanoparticle surface can specifically increase the *in vivo* tumor accumulation of nanoparticles due to the specific interaction with hepatoma tumor via ligand-receptor recognition. In 2011, Shen et al. developed a galactosamine-conjugated, dox encapsulating, albumin nanoparticles for targeted liver cancer therapy (Shen et al. 2011). Their results implied that galactosamine-conjugated on the nanoparticle surface can have specific interaction with HepG2 liver cancer cells via the recognition between galactosamine and ASGPR. In 2012, Gu et al. coated a layer of PEGylated-galactose onto the external surface of Dox-loaded MSNs, and demonstrated targeted Dox delivery to liver cancer cells (Gu et al. 2012). In 2013, Zhang et al. developed actosylated gramicidin-containing lipid nanoparticles to deliver anti-microRNA-155 to HCC cells. MiR-155 is an oncomiR frequently elevated in HCC (Zhang et al. 2013).

3.7 Pancreatic Cancer

Pancreatic cancer is one of the most deadly cancers, and the mortality of pancreatic cancer ranks the seventh of overall cancer death. According to the GLOBOCAN report, for 2012, approximately 338,000 million new cases of pancreatic cancer were reported globally, representing 2.4% of the total number of overall cancer cases (Ferlay et al. 2015). In the same year, pancreatic cancer caused 331,000 deaths, representing 4.0% of overall cancer deaths. For 2015, in the U.S. alone, the American Cancer Society estimates about 48,960 new cases and 40,560 deaths of pancreatic cancer (Siegel et al. 2015).

There are two common types of pancreatic cancers: (i) Pancreatic exocrine tumor. Pancreatic exocrine tumor is an umbrella term for pancreatic adenocarcinoma, solid pseudopapillary neoplasms, ampullary cancer and other less common types of pancreatic cancers. Pancreatic exocrine tumor is the most common type of pancreatic cancers. Amongst pancreatic exocrine tumors, pancreatic adenocarcinoma develops in the ducts of pancreas, and comprises approximately 95% of pancreatic exocrine tumors. The overall five year survival rate of pancreatic exocrine tumors ranges between 1% and 14% depending on the stage of disease. (ii) Pancreatic endocrine tumor. Pancreatic endocrine tumor is also called pancreatic neuroendocrine tumor or islet cell tumor, and is an umbrella term for pancreatic functioning tumors, non-functioning tumors, and carcinoid tumors. Pancreatic exocrine tumors are uncommon and only comprise 4% of all pancreatic cancers. The overall five year survival rate of pancreatic endocrine tumors is about 16%.

There are three common clinical targeted therapeutics approved to treat pancreatic cancer (as seen in Table 1). They are Erlotinib (Tarceva) (Moore et al. 2007), Sunitinib (Sutent) (Raymond et al. 2011), and Everolimus

(Afinitor) (Yao et al. 2011). Erlotinib is a small molecular tyrosine kinase inhibitor. Erlotinib was approved by U.S. FDA in 2005 in combination with Gem to treat locally advanced, unresectable, or metastatic pancreatic cancer (exocrine pancreatic cancers) (Moore et al. 2007). Sunitinib is small molecular multitargeted receptor tyrosine kinase inhibitor. Sunitinib was approved by U.S. FDA in 2005 to treat progressive neuroendocrine cancerous tumors (Raymond et al. 2011). Everolimus is a small molecular inhibitor of mTOR. It functions by blocking the mTOR intracellular signaling pathways of pancreatic cancer cells to inhibit tumor proliferation, growth, and angiogenesis. In 2011, Everolimus was approved by U.S. FDA to treat progressive or metastatic pancreatic neuroendocrine tumors (Yao et al. 2011).

Common pancreatic cancer molecular targets include EGFR, CD44, uPAR, and transferrin receptor (as seen in Table 2). EGFR is the most common pancreatic cancer molecular target used for smart nanomedicine. In 2008, Magadala et al. developed EGFR-targeted gelatin nanoparticles for pancreatic tumor specific DNA delivery (Magadala and Amiji 2008). In the same year, Patra et al. developed a Cetuximab-conjugated, Gem-loaded gold nanoparticle to facilitate pancreatic tumor specific drug delivery *in vivo* (Patra et al. 2008). Cetuximab conjugated on the nanoparticle surface can induce about 5 to 10-fold increases in both EGFR expressing cell uptake and *in vivo* tumor accumulation, suggesting EGFR is a promising nanomedicine target for pancreatic cancer. In 2009, Yang et al. investigated single chain anti-EGFR antibody as a pancreatic cancer targeting ligand for QDs and IONPs (Yang et al. 2009b). Their results demonstrated that single-chain anti-EGFR antibody conjugated QD and IONPs can efficiently target and accumulate in orthotopic pancreatic tumors *in vivo*, and facilitate a pancreatic cancer-targeted fluorescent imaging (with EGFR-targeted QDs) and MR imaging (with EGFR-targeted IONPs). In 2010, Glazer et al. applied Cetuximab-conjugated gold nanoparticles in noninvasive radiofrequency radiation to investigate human pancreatic xenograft-targeted destruction in a murine model (Glazer et al. 2010). Their results showed pancreatic tumor can be noninvasively ablated by intracellular hyperthermia with targeted gold nanoparticles in a RF field. The tumor specificity of anti-EGFR antibodies can minimize the hyperthermia toxicity to normal tissues under systemic therapy and whole-body RF field exposure. In 2011, Aggarwal et al. prepared an anti-EGFR monoclonal antibody-conjugated PLGA nanoparticle to deliver Gem specifically to pancreatic cancer cells *in vitro* (Aggarwal et al. 2011). In 2014, Xu et al. developed EGFR-targeted, redox-responsive thiolated gelatin nanoparticles to co-deliver wt-p53 expressing plasmid DNA and Gem specifically to pancreatic tumors (Xu et al. 2014). Their therapeutic results showed that these pancreatic cancer-targeted drug delivery nanoparticles can produce gene/drug synergistic effects that improve the therapeutic performance of the delivery system significantly as compared to the gene or drug alone.

CD44 is another pancreatic cancer molecular target. In 2004, Peer and Margalit reported the synthesis of hyaluronan-conjugated liposomal Dox to facilitate a pancreatic tumor specific Dox delivery (Peer and Margalit 2004). Their therapeutic results showed that hyaluronan targeting with CD44 receptor on pancreatic cancer cells can induce an over 3.5-fold increase of tumor accumulation compared with nonspecific counterparts. In 2014, Nigam et al. developed a hyaluronic acid-functionalized and green fluorescent graphene quantum dot-labeled HSA nanoparticle, and applied it for pancreatic cancer specific drug delivery and bioimaging (Nigam et al. 2014).

uPAR has been characterized as a novel and important pancreatic cancer biomarker and therapeutic target. In 2009, Yang and Mao et al. developed an uPAR-targeted dual-modality molecular imaging nanoparticulate probe comprising NIR dye for fluorescent imaging and IONPs for MR imaging, and applied this dual-modality molecular imaging probe to facilitate a pancreatic tumor specific detection study *in vivo* (Yang et al. 2009a). In 2013, Lee et al. further developed an uPAR-targeted IONPs carrying chemotherapeutic Gem to simultaneously treat uPAR-expressing pancreatic tumor and stromal cells (Lee et al. 2013).

Transferrin receptor has also been used as nanomedicine targeting ligand for pancreatic cancer. In 2010, Barth et al. functionalized fluorescent calcium phosphosilicate nanoparticles with anti-CD71 antibody to investigate the enhancement of drug delivery, targeting, and imaging of pancreatic cancer tumors (Barth et al. 2010). In 2013, Camp et al. reported the development of transferrin receptor-targeted immunoliposome to specifically deliver p53 gene to metastatic pancreatic tumors (Camp et al. 2013). From their *in vivo* therapeutic results, in combination with Gem therapy, this transferrin receptor-targeted, p53 gene encapsulating immunoliposome can sensitize the metastatic pancreatic tumor to respond conventional Gem therapy.

3.8 Brain Cancer

Brain cancer accounted for approximately 257,000 new cases of cancer worldwide in 2012, representing 1.8% of the total number of all new cancer cases, according to the GLOBOCAN report (Ferlay et al. 2015). In the same year, brain cancer caused 189,000 deaths, representing 2.3% of all cancer deaths. For 2015, in the U.S. alone, the American Cancer Society estimates about 22,850 new cases and 15,320 deaths of brain cancer (Siegel et al. 2015).

There are four common types of brain cancers: (i) Glioma. Glioma is an umbrella term for glioblastoma, astrocytoma, oligodendroglioma, and ependymoma. Gliomas are defined as cancers developed in brain glial cells, and represent about 30% of all brain tumors. The five year survival rate of gliomas strongly depends on the type of tumor: glioblastoma

(4%–17%), astrocytomas (10%–65%), oligodendrogliomas (38%–85%), and ependymomas (85%–91%). (ii) Meningioma. Meningioma is defined as cancers developed in the meninges surrounding the brain and spinal cord, representing about 30% of primary brain and spinal cord tumors. The overall five year survival rate of meningiomas is over 67%. (iii) Medulloblastoma. Medulloblastoma is defined as cancers developed from neuroectodermal cells in the cerebellum. Medulloblastomas are highly malignant tumors and occur more frequently in children than adults. The overall five year survival rate of medulloblastomas is 69%. (iv) Schwannoma. Schwannoma is also called neurilemmoma, and defined as the brain cancers developed from Schwann cells surrounding the cranial nerves and other nerves. It represents about 8% of all brain tumors. The overall five year survival rate of schwannoma ranges between 37.6% and 65.7%.

There are two common clinical targeted therapeutics for brain cancer (as seen in Table 1). They are Bevacizumab (Avastin) (Friedman et al. 2009) and Everolimus (Afinitor) (Krueger et al. 2010). Bevacizumab is humanized anti-VEGF-A monoclonal antibody. Bevacizumab was approved by U.S. FDA to treat glioblastoma (Friedman et al. 2009). Everolimus is a small molecular inhibitor of mTOR. Everolimus was approved by U.S. FDA in 2012 to treat a rare brain tumor called subependymal giant cell astrocytoma (Krueger et al. 2010).

Common brain cancer molecular targets include EGFR, transferrin receptor, interleukin 13 receptor (IL-13R), and nucleolin (as seen in Table 2). EGFR is a common brain cancer target used for nanomedicine. In 2006, Wu et al. developed a Cetuximab-conjugated polyamidoamine dendrimer to facilitate a brain tumor specific delivery of cytotoxic drug methotrexate (Wu et al. 2006). In 2010, Hadjipanayis et al. reported the development of EGFRvIII antibody-conjugated IONPs for MR imaging-guided convection-enhanced delivery and targeted therapy of glioblastoma (Hadjipanayis et al. 2010). They used EGFRvIII antibody as both therapeutic and targeting agent to selectively bind and neutralize the EGFR deletion mutant (EGFRvIII) present on human glioblastoma multiforme cells, and their EGFRvIII targeted IONPs achieved a significant inhibition in glioblastoma cell growth with no toxicity in human astrocytes.

Transferrin receptor is another common brain cancer target. In 2003, Zhang et al. developed anti-transferrin receptor antibody-conjugated, PEGylated immunoliposomes as a brain tumor-targeted gene delivery system (Zhang et al. 2003). Their therapeutic results showed that their transferrin receptor-targeted liposomes can inhibit luciferase gene expression in the brain cancer by 90% for at least 5 days after a single intravenous injection. In 2010, Gan and Feng reported the development of transferrin-conjugated nanoparticles of poly(lactide)-d-α-Tocopheryl polyethylene glycol succinate diblock copolymer (PLA-TPGS) (Gan and Feng 2010). Their findings demonstrated that these transferrin-conjugated

PLA-TPGS nanoparticles can readily co-deliver imaging agents and chemotherapeutics across the blood-brain barrier (BBB) for brain cancer-targeted therapy. In 2010, Ying et al. developed a p-aminophenyl-alpha-D-manno-pyranoside and transferrin dual-targeting daunorubicin liposome, which was designed to improve transporting the drug across the BBB and then target brain glioma (Ying et al. 2010). Their therapeutic studies showed their dual targeting liposomes can improve the therapeutic efficacy of brain glioma *in vitro* and in animals.

IL-13R has been found selectively overexpressed by human glioblastoma tumors. In 2009, Madhankumar et al. reported the development of an IL-13R-targeted liposomal Dox, and they applied this immunoliposome to facilitate a human glioblastoma tumor specific drug delivery (Madhankumar et al. 2009). In 2009, Rozhkova et al. modified 5 nm TiO2 nanoparticles tethered through a DOPAC linker to the anti-IL-13R antibody. Their findings demostrated the visible light-induced photo-toxicity of the TiO2-nanobio hybrid toward human brain cancer (Rozhkova et al. 2009). The photo-toxicity was mediated by reactive oxygen species (ROS) that initiate programmed death of cancer cells.

Nucleolin is another common molecular target for brain cancer. In 2006, Reddy et al. reported the development of a F3 peptide-conjugated Photofrin- and IONP–encapsulated polymeric nanoparticle to facilitate a brain cancer-targeted PDT and MR imaging (Reddy et al. 2006). In their *in vivo* MR imaging measurement, F3 peptides can prolong the nanoparticle half life in tumor from 39 min to 120 min, and increase the tumor accumulation of nanoparticle by about 2-fold compared with non-specific constructs. Their PDT results further showed a significant improvement in survival rate when compared with animals who received PDT after administration of non-targeted nanoparticles or free Photofrin. In 2007, Gupta and Torchilin developed monoclonal antibody 2C5-modified Dox-loaded liposomes to enhance therapeutic activity against intracranial human brain tumor xenograft in nude mice (Gupta and Torchilin 2007). Their findings revealed that monoclonal antibody 2C5 can guide the drug delivery liposome specifically to brain tumors, and improve the therapeutic outcome of breast tumor in an intracranial model *in vivo*.

3.9 Melanoma

Melanoma is a severe skin cancer with high malignance that initiates in the melanocytes. Although comprising less than 2% of all skin cancer cases, melanoma causes over 90% of skin cancer deaths due to its malignancy. The overall five year survival rate of melanoma is about 91%. In worldwide, melanoma accounted for approximately 232,000 new cases in 2012, which represents 1.6% of the total number of all new cancer cases, according to the

GLOBOCAN report (Ferlay et al. 2015). In the same year, melanoma caused 55,000 deaths, representing 0.7% of all cancer deaths. In 2015, in the U.S. alone, the American Cancer Society estimates about 73,870 new cases and 9,940 deaths by melanoma (Siegel et al. 2015). The incidence of melanoma in Caucasians is more than 20-fold higher than that of African Americans, and the overall incidence of melanoma (all races) has been increasing for the past 30 years (Ferlay et al. 2015).

Three targeted therapeutics are used to treat melanoma in clinic right now (as seen in Table 1), including Vemurafenib (Zelboraf) (Chapman et al. 2011), Dabrafenib (Tafinlar) (Hauschild et al. 2012), and Trametinib (Mekinist) (Falchook et al. 2012). Among them, Vemurafenib and Dabrafenib are small molecular inhibitors of BRAF gene, approved by U.S. FDA to treat advanced melanomas (Chapman et al. 2011, Hauschild et al. 2012). Trametinib is a small molecular inhibitor that functions by blocking MEK1 and MEK2 (Falchook et al. 2012), which was approved by U.S. FDA in 2013 to treat metastatic melanoma.

Prevailing molecular targets of melanoma include integrin $\alpha v\beta 3$, EGFR, sigma receptor, nucleolin, and CD44 (as seen in Table 2). Integrin $\alpha v\beta 3$ is the most common melanoma target for smart nanomedicine. In 2004, Hölig et al. developed an RGD peptide-conjugated liposome to facilitate melanoma-targeted drug delivery (Hölig et al. 2004). In 2005, Schmieder et al. used integrin $\alpha v\beta 3$-targeted SPIO nanoparticles to detect sparse integrin $\alpha v\beta$ expression on neovasculature induced by nascent melanoma xenografts by contrast enhanced MR imaging (Schmieder et al. 2005). Their results showed integrin $\alpha v\beta 3$-targeted SPIO nanoparticles can enhance MRI contrast by 173% compared with non-specific counterparts. The reported nanoparticles can be applied noninvasively detecting very small regions of angiogenesis associated with nascent melanoma tumors, and early phenotype/stage melanomas in a clinical setting. In 2010, Boles et al. prepared both Robo4- and integrin $\alpha v\beta 3$-targeted SPIO nanoparticles for MR angiogenesis imaging of melanoma in a mouse model (Boles et al. 2010). MR angiogenesis mapping results indicated Robo4 expression generally colocalized with integrin $\alpha v\beta 3$ expression. It was found that integrin $\alpha v\beta 3$ expression was more detectable by MR than Robo4 in melanoma. In 2011, Benezra et al. reported the development of an RGD peptide-conjugated multimodal silica nanoparticle to facilitate sensitive, real-time detection and imaging of lymphatic drainage patterns, particle clearance rates, nodal metastases, and differential tumor burden in a large animal model of melanoma (Benezra et al. 2011).

EGFR, another widely used cancer target, has been found to effectively target melanomas. In 2006, El-Sayed et al. developed anti-EGFR antibody-conjugated gold nanoparticles for a selective laser photo-thermal therapy of epithelial carcinoma (El-Sayed et al. 2006). Their findings indicated less than half of the laser energy was required to kill the malignant cells than that of benign cells after incubating with anti-EGFR antibody-conjugated gold nanoparticles.

Other common cancer biomarkers such as nucleolin, sigma receptor, and CD44 have also been investigated for melanoma-targeted nanomedicine. In 2008, Skidan et al. developed nucleolin-targeted 5, 10, 15, 20-tetraphenylporphin (TPP)-loaded PEG-PE micelles for melanoma specific PDT (Skidan et al. 2008). Compared with free TPP, non-specific TPP-loaded PEG-PE micelles and nucleolin-targeted, TPP-loaded PEG-PE2C5-immunomicelles can generate approximately 3.5- and 7.5-fold inhibitory effects on tumor weights, respectively. In 2010, Chen et al. reported the development of sigma receptor-targeted solid lipid nanoparticle to systemically deliver c-Myc siRNA into the cytoplasm of B16F10 murine melanoma cells, indicating effective therapeutic effect for melanoma using targeted c-Myc siRNA loaded nanoparticles (Chen et al. 2010). In 2012, Jin et al. reported a hyaluronic acid-ceramide nanoparticle for the targeted delivery of Dox to melanoma *in vivo* (Jin et al. 2012). The results showed Dox carried in CD44-targeted nanoparticles were efficiently uptaken by the B16F10 cells through CD44 receptor-mediated endocytosis, producing a significant inhibitory effect on tumor growth without any serious changes in body weight in melanoma bearing mouse models.

In addition to molecular targets, neural progenitor cells have also been investigated as cell-based melanoma targeting ligands. Neural progenitor cells are known for their tumor-tropic functions, and have been investigated previously as cell delivery vehicles to many solid tumors. In 2010, Rachakatla et al. investigated tumor-tropic neural progenitor cells as cell delivery vehicles to achieve preferential accumulation of IONPs within a mouse model of melanoma (Rachakatla et al. 2010). After neural progenitor cell-mediated IONPs were delivered and accumulated to subcutaneous melanomas, an alternating current magnetic field was applied to ablate the melanoma tumors via magnetic nanoparticle-mediated hyperthermia.

3.10 Ovarian Cancer

Ovarian cancer is the seventh most common in women. The mortality of ovarian cancer ranks the eighth of all women's cancer-related deaths. According to the GLOBOCAN report, for 2012, approximately 239,000 new cases of ovarian cancer were reported globally, representing 3.6% of the total number of women's cancer cases (Ferlay et al. 2015). In the same year, ovarian cancer caused 152,000 deaths, representing 4.3% of women's cancer deaths. For 2015, in the U.S. alone, the American Cancer Society estimates about 21,290 new cases and 14,180 deaths of ovarian cancer (Siegel et al. 2015).

There are three common types of ovarian cancers: (i) Epithelial ovarian tumor. Epithelial ovarian tumor is defined as cancers developed from epithelial cells in the outer surface of ovary. It is the most common type of ovarian cancers, and represents 85% to 90% of all ovarian cancers. The five year survival rate of epithelial ovarian tumor ranges between 17% and 94%, depending on the stage of disease. (ii) Ovarian germ cell tumor. Ovarian germ cell tumor is defined as cancers developed from the germ (egg) cells in the ovary. It is uncommon, and only accounts for less than 2% of all ovarian cancers. The five year survival rate of ovarian germ cell tumor ranges between 69% and 98%, depending on the stage of disease. (iii) Ovarian stromal tumor. Ovarian stromal tumor is defined as cancers developed from the stromal cells in the connective tissues of ovary. It is a rare ovarian cancer that accounts for only about 1% of all ovarian cancers. The five year survival rate of ovarian stromal tumor ranges between 35% and 95%, depending on the stage of disease.

Two common targeted therapeutics are used for ovarian cancer (as seen in Table 1). They are Bevacizumab (Avastin) (Sato and Itamochi 2012) and Olaparib (Lynparza) (Audeh et al. 2010). Bevacizumab is a human monoclonal anti-VEGF antibody. More information is supplied in lung cancer section. Bevacizumab was approved by U.S. FDA in combination with chemotherapy to treat epithelial ovarian tumors (Sato and Itamochi 2012). Olaparib is a small molecular inhibitor of poly ADP ribose polymerase (PARP) (Audeh et al. 2010). It functions by blocking the PARP signaling transduction to inhibit tumor growth. Olaparib was approved by U.S. FDA in 2014 as monotherapy to treat BRCA mutated advanced ovarian cancers.

Common ovarian cancer molecular targets include folate receptor, integrin $\alpha v\beta 3$, nucleolin, HER2, follicle-stimulating hormone receptor (FSHR), EphA receptor, and P32 receptor (as seen in Table 2). Folate receptor is the most common ovarian cancer target used for smart nanomedicine. In 2008, Esmaeili et al. reported the synthesis of folate receptor-targeted Docetaxel (DTX) encapsulating PLGA-PEG nanoparticles to improve therapeutic efficacy in human ovarian cancer cells that express folate receptors (Esmaeili et al. 2008). In 2009, Kim et al. reported the development of folate receptor-targeted, Dox-loaded polymeric micelles to improve Dox efficacy to drug-resistant ovarian cancer (Kim et al. 2009). The folate receptor-targeted constructs can effectively suppress the growth of existing MDR ovarian tumors in mice for at least 50 days. In 2009, Kamaly et al. developed folate receptor-targeted bimodal paramagnetic and fluorescent liposomes to facilitate ovarian tumor specific fluorescent and MR imaging (Kamaly et al. 2009). Their *in vivo* studies revealed that folate receptor-targeted liposome had a 4-fold higher tumor accumulation compared with non-specific liposomes, indicating folate receptor is an effective ovarian cancer target for nanomedicine. In 2011, Nukolova et al.

developed folic acid-conjugated, diblock copolymer poly(ethylene oxide)-b-poly(methacrylic acid) nanogels to deliver chemotherapeutics, cisplatin or Dox specifically to human ovarian tumors *in vivo* (Nukolova et al. 2011). In 2011, Wang et al. prepared folic acid-conjugated polyglycerol-grafted Fe3O4@SiO2 nanoparticles for ovarian cancer specific MRI (Wang et al. 2011). It was found that folic acid molecules conjugated on nanoparticles can introduce approximately a 10-fold increase on ovarian cancer cellular uptake compared with non-specific counterparts. In 2011, Werner et al. developed folate-targeted PLGA–lecithin–PEG core–shell nanoparticles to facilitate co-delivery of chemotherapeutic PTX and therapeutic radioisotope yittrium-90 specifically to ovarian cancer peritoneal metastasis (Werner et al. 2011).

Integrin αvβ3 is another common ovarian cancer target for nanomedicine. In 2009, Zhao et al. developed RGD peptide-modified liposomes to deliver PTX specifically to human ovarian carcinoma. The therapeutic results showed that their RGD peptide-modified liposomes produced a 3-fold lower IC50 than non-specific counterparts in SKOV3 ovarian cancer cells (Zhao et al. 2009). The *in vivo* experiment further confirmed that RGD peptide-modified liposomes can significantly inhibited ovarian tumor growth without significant change in the animal body weight. In 2012, Xiao et al. developed OA02 peptide-conjugated PTX-loaded micellar nanoparticles to improve the ovarian tumor specificity and facilitate targeted treatment for ovarian cancer *in vivo* (Xiao et al. 2012). It was revealed that OA02 peptide can guide nanoparticles toward ovarian cancer cells and generated a 6-fold increase of *in vitro* cellular uptake, and a 1.7-fold increase of *in vivo* tumor accumulation, compared with non-specific counterparts.

Nucleolin has also been investigated as an ovarian cancer target for smart nanomedicine. In 2010, Winer et al. developed F3-targeted cisplatin-hydrogel nanoparticles with high specificity to both human ovarian tumor cells and tumor-associated endothelial cells (Winer et al. 2010). These nucleolin-targeted constructs primarily bound tumor-associated endothelial cells and generated significant vascular necrosis, in consistent with anti-angiogenesis effects. In 2012, Perche et al. evaluated the tumor specificity and resulting toxicity of their constructs in an ovarian cancer cell spheroid model using 2C5 antibody conjugated, doxorubicin-loaded PEG-PE micelles (Perche et al. 2012). Similarly, HER2 is also overexpressed in some ovarian cancer cells and can be used to target ovarian tumors. In 2010, Cirstoiu-Hapca et al. reported HER2 antibody conjugated, PTX-loaded PLA nanoparticles improving PTX efficacy in the treatment of disseminated ovarian cancer overexpressing HER2 (Cirstoiu-Hapca et al. 2010).

Several other proteins and genes (e.g., FSHR, EphA receptor, and P32) overexpress in ovarian cancer cells, and have been reported to guide nanomedicine to target ovarian tumors effectively *in vivo*. In 2009, Zhang et al. investigated FSHR as a molecular target for ovarian cancer, and developed FSHR-targeted PTX-loaded nanoparticles to facilitate specific drug delivery to ovarian tumor *in vivo* (Zhang et al. 2009). In 2010, Scarberry et al. developed ephrin-A1 mimetic peptide-conjugated SPIO nanoparticles to remove ovarian cancer cells from human ascites fluid selectively using magnetic separation, in order to inhibit ovarian cancer metastases (Scarberry et al. 2010). In 2012, Ren et al. identified ID4 as an essential oncogene in human ovarian cancer, and developed a p32-targeted, tumor-penetrating nanocomplex to deliver ID4 siRNA specifically to human ovarian cancer cells (Ren et al. 2012). Their therapeutic results showed that intravenously administered p32-targeted constructs can facilitate 80–90% of ID4 mRNA knockdown and 82% of tumor growth inhibition.

4. Summary

In conclusion, smart materials, especially nanomedicine, are currently one of the leading directions in cancer research. It offers a broad range of advantages to improve the quality of clinical prognosis. Furthermore, smart materials can serve as multifunctional drug delivery systems that exhibit exceptional therapeutic properties over traditional chemotherapeutics. Meanwhile, owing to the unique optical/magnetic properties, many inorganic smart nanomedicines can be used as novel molecular imaging contrast agents to improve early detection and diagnosis of cancer. Combining smart nanomedicine's advantages, we believe that, in the near future, tumor-targeting nanotechnology will revolutionize the field of cancer research by significantly enhancing tumor detection accuracy, therapeutic sensitivity and efficiency, and subsequently improve cancer prognosis outcomes.

References

Abdalla, M.O., P. Karna, H.K. Sajja, H. Mao, C. Yates, T. Turner and R. Aneja. 2011. Enhanced noscapine delivery using uPAR-targeted optical-MR imaging trackable nanoparticles for prostate cancer therapy. J. Control. Release Off. J. Control. Release Soc. 149: 314–322.

Acharya, S., F. Dilnawaz and S.K. Sahoo. 2009. Targeted epidermal growth factor receptor nanoparticle bioconjugates for breast cancer therapy. Biomaterials 30: 5737–5750.

Aggarwal, S., S. Yadav and S. Gupta. 2011. EGFR targeted PLGA nanoparticles using gemcitabine for treatment of pancreatic cancer. J. Biomed. Nanotechnol. 7: 137–138.

Amado, R.G., M. Wolf, M. Peeters, E. Van Cutsem, S. Siena, D.J. Freeman, T. Juan, R. Sikorski, S. Suggs, R. Radinsky, S.D. Patterson and D.D. Chang. 2008. Wild-Type KRAS is required for panitumumab efficacy in patients with metastatic colorectal cancer. J. Clin. Oncol. 26: 1626–1634.

Audeh, M.W., J. Carmichael, R.T. Penson, M. Friedlander, B. Powell, K.M. Bell-McGuinn, C. Scott, J.N. Weitzel, A. Oaknin, N. Loman, K. Lu, R.K. Schmutzler, U. Matulonis,

M. Wickens and A. Tutt. 2010. Oral poly(ADP-ribose) polymerase inhibitor olaparib in patients with BRCA1 or BRCA2 mutations and recurrent ovarian cancer: a proof-of-concept trial. The Lancet 376: 245–251.

Bagalkot, V., O.C. Farokhzad, R. Langer and S. Jon. 2006. An aptamer-doxorubicin physical conjugate as a novel targeted drug-delivery platform. Angew. Chem. Int. Ed Engl. 45: 8149–8152.

Banerjee, R., P. Tyagi, S. Li and L. Huang. 2004. Anisamide-targeted stealth liposomes: a potent carrier for targeting doxorubicin to human prostate cancer cells. Int. J. Cancer J. Int. Cancer 112: 693–700.

Bang, Y.J., E. Van Cutsem, A. Feyereislova, H.C. Chung, L. Shen, A. Sawaki, F. Lordick, A. Ohtsu, Y. Omuro, T. Satoh, G. Aprile, E. Kulikov, J. Hill, M. Lehle, J. Rüschoff and Y.-K. Kang. 2010. Trastuzumab in combination with chemotherapy versus chemotherapy alone for treatment of HER2-positive advanced gastric or gastro-oesophageal junction cancer (ToGA): a phase 3, open-label, randomised controlled trial. The Lancet 376: 687–697.

Barenholz, Y. 2012. Doxil®—The first FDA-approved nano-drug: Lessons learned. J. Controlled Release 160: 117–134.

Barnett, B.P., A. Arepally, P.V. Karmarkar, D. Qian, W.D. Gilson, P. Walczak, V. Howland, L. Lawler, C. Lauzon, M. Stuber, D.L. Kraitchman and J.W.M. Bulte. 2007. Magnetic resonance–guided, real-time targeted delivery and imaging of magnetocapsules immunoprotecting pancreatic islet cells. Nat. Med. 13: 986–991.

Barth, B.M., R. Sharma, E.I. Altinoğlu, T.T. Morgan, S.S. Shanmugavelandy, J.M. Kaiser, C. McGovern, G.L. Matters, J.P. Smith, M. Kester and J.H. Adair. 2010. Bioconjugation of calcium phosphosilicate composite nanoparticles for selective targeting of human breast and pancreatic cancers *in vivo*. ACS Nano 4: 1279–1287.

Baselga, J., M. Campone, M. Piccart, H.A. Burris, H.S. Rugo, T. Sahmoud, S. Noguchi, M. Gnant, K.I. Pritchard, F. Lebrun, J.T. Beck, Y. Ito, D. Yardley, I. Deleu, A. Perez, T. Bachelot, L. Vittori, Z. Xu, P. Mukhopadhyay, D. Lebwohl and G.N. Hortobagyi. 2012a. Everolimus in postmenopausal hormone-receptor–positive advanced breast cancer. N. Engl. J. Med. 366: 520–529.

Baselga, J., J. Cortés, S.B. Kim, S.A. Im, R. Hegg, Y.H. Im, L. Roman, J.L. Pedrini, T. Pienkowski, A. Knott, E. Clark, M.C. Benyunes, G. Ross and S.M. Swain. 2012b. Pertuzumab plus trastuzumab plus docetaxel for metastatic breast cancer. N. Engl. J. Med. 366: 109–119.

Benezra, M., O. Penate-Medina, P.B. Zanzonico, D. Schaer, H. Ow, A. Burns, E. DeStanchina, V. Longo, E. Herz, S. Iyer, J. Wolchok, S.M. Larson, U. Wiesner and M.S. Bradbury. 2011. Multimodal silica nanoparticles are effective cancer-targeted probes in a model of human melanoma. J. Clin. Invest. 121: 2768–2780.

Boles, K.S., A.H. Schmieder, A.W. Koch, R.A.D. Carano, Y. Wu, S.D. Caruthers, R.K. Tong, S. Stawicki, G. Hu, M.J. Scott, H. Zhang, B.A. Reynolds, S.A. Wickline and G.M. Lanza. 2010. MR angiogenesis imaging with Robo4- vs. alphaVbeta3-targeted nanoparticles in a B16/F10 mouse melanoma model. FASEB J. 24: 4262–4270.

Cabot, R.C., N.L. Harris, E.S. Rosenberg, J.A.O. Shepard, A.M. Cort, S.H. Ebeling, E.K. McDonald, H.I. Scher, K. Fizazi, F. Saad, M.E. Taplin, C.N. Sternberg, K. Miller, R. de Wit, P. Mulders, K.N. Chi, N.D. Shore, A.J. Armstrong, T.W. Flaig, A. Fléchon, P. Mainwaring, M. Fleming, J.D. Hainsworth, M. Hirmand, B. Selby, L. Seely and J.S. de Bono. 2012. Increased survival with enzalutamide in prostate cancer after chemotherapy. N. Engl. J. Med. 367: 1187–1197.

Camp, E.R., C. Wang, E.C. Little, P.M. Watson, K.F. Pirollo, A. Rait, D.J. Cole, E.H. Chang and D.K. Watson. 2013. Transferrin receptor targeting nanomedicine delivering wild-type p53 gene sensitizes pancreatic cancer to gemcitabine therapy. Cancer Gene Ther. 20: 222–228.

Chapman, P.B., A. Hauschild, C. Robert, J.B. Haanen, P. Ascierto, J. Larkin, R. Dummer, C. Garbe, A. Testori, M. Maio, D. Hogg, P. Lorigan, C. Lebbe, T. Jouary, D. Schadendorf, A. Ribas, S.J. O'Day, J.A. Sosman, J.M. Kirkwood, A.M.M. Eggermont, B. Dreno, K. Nolop, J. Li, B. Nelson, J. Hou, R.J. Lee, K.T. Flaherty and G.A. McArthur. 2011.

Improved survival with vemurafenib in melanoma with BRAF V600E mutation. N. Engl. J. Med. 364: 2507–2516.

Cheng, A.L., Y.K. Kang, Z. Chen, C.J. Tsao, S. Qin, J.S. Kim, R. Luo, J. Feng, S. Ye, T.S. Yang, J. Xu, Y. Sun, H. Liang, J. Liu, J. Wang, W.Y. Tak, H. Pan, K. Burock, J. Zou, D. Voliotis and Z. Guan. 2009. Efficacy and safety of sorafenib in patients in the Asia-Pacific region with advanced hepatocellular carcinoma: a phase III randomised, double-blind, placebo-controlled trial. Lancet Oncol. 10: 25–34.

Cheng, J., B.A. Teply, I. Sherifi, J. Sung, G. Luther, F.X. Gu, E. Levy-Nissenbaum, A.F. Radovic-Moreno, R. Langer and O.C. Farokhzad. 2007. Formulation of functionalized PLGA-PEG nanoparticles for *in vivo* targeted drug delivery. Biomaterials 28: 869–876.

Chen, Y., S.R. Bathula, Q. Yang and L. Huang. 2010. Targeted nanoparticles deliver siRNA to melanoma. J. Invest. Dermatol. 130: 2790–2798.

Chen, Y., W. Wang, G. Lian, C. Qian, L. Wang, L. Zeng, C. Liao, B. Liang, B. Huang, K. Huang and X. Shuai. 2012. Development of an MRI-visible nonviral vector for siRNA delivery targeting gastric cancer. Int. J. Nanomedicine 7: 359–368.

Cho, Y.S., T.J. Yoon, E.S. Jang, K. Soo Hong, S. Young Lee, O. Ran Kim, C. Park, Y.J. Kim, G.C. Yi and K. Chang. 2010. Cetuximab-conjugated magneto-fluorescent silica nanoparticles for *in vivo* colon cancer targeting and imaging. Cancer Lett. 299: 63–71.

Cirstoiu-Hapca, A., F. Buchegger, N. Lange, L. Bossy, R. Gurny and F. Delie. 2010. Benefit of anti-HER2-coated paclitaxel-loaded immuno-nanoparticles in the treatment of disseminated ovarian cancer: Therapeutic efficacy and biodistribution in mice. J. Control. Release Off. J. Control. Release Soc. 144: 324–331.

Cunningham, D., Y. Humblet, S. Siena, D. Khayat, H. Bleiberg, A. Santoro, D. Bets, M. Mueser, A. Harstrick, C. Verslype, I. Chau and E. Van Cutsem. 2004. Cetuximab monotherapy and cetuximab plus irinotecan in irinotecan-refractory metastatic colorectal cancer. N. Engl. J. Med. 351: 337–345.

De Bono, J.S., C.J. Logothetis, A. Molina, K. Fizazi, S. North, L. Chu, K.N. Chi, R.J. Jones, O.B. Goodman, F. Saad, J.N. Staffurth, P. Mainwaring, S. Harland, T.W. Flaig, T.E. Hutson, T. Cheng, H. Patterson, J.D. Hainsworth, C.J. Ryan, C.N. Sternberg, S.L. Ellard, A. Fléchon, M. Saleh, M. Scholz, E. Efstathiou, A. Zivi, D. Bianchini, Y. Loriot, N. Chieffo, T. Kheoh, C.M. Haqq and H.I. Scher. 2011. Abiraterone and increased survival in metastatic prostate cancer. N. Engl. J. Med. 364: 1995–2005.

De Bono, J.S., S. Oudard, M. Ozguroglu, S. Hansen, J.P. Machiels, I. Kocak, G. Gravis, I. Bodrogi, M.J. Mackenzie, L. Shen, M. Roessner, S. Gupta and A.O. Sartor. 2010. Prednisone plus cabazitaxel or mitoxantrone for metastatic castration-resistant prostate cancer progressing after docetaxel treatment: a randomised open-label trial. The Lancet 376: 1147–1154.

Dhar, S., F.X. Gu, R. Langer, O.C. Farokhzad and S.J. Lippard. 2008. Targeted delivery of cisplatin to prostate cancer cells by aptamer functionalized Pt(IV) prodrug-PLGA-PEG nanoparticles. Proc. Natl. Acad. Sci. U.S.A. 105: 17356–17361.

El-Sayed, I.H., X. Huang and M.A. El-Sayed. 2006. Selective laser photo-thermal therapy of epithelial carcinoma using anti-EGFR antibody conjugated gold nanoparticles. Cancer Lett. 239: 129–135.

Esmaeili, F., M.H. Ghahremani, S.N. Ostad, F. Atyabi, M. Seyedabadi, M.R. Malekshahi, M. Amini and R. Dinarvand. 2008. Folate-receptor-targeted delivery of docetaxel nanoparticles prepared by PLGA-PEG-folate conjugate. J. Drug Target. 16: 415–423.

Falchook, G.S., K.D. Lewis, J.R. Infante, M.S. Gordon, N.J. Vogelzang, D.J. DeMarini, P. Sun, C. Moy, S.A. Szabo, L.T. Roadcap, V.G. Peddareddigari, P.F. Lebowitz, N.T. Le, H.A. Burris, W.A. Messersmith, P.J. O'Dwyer, K.B. Kim, K. Flaherty, J.C. Bendell, R. Gonzalez, R. Kurzrock and L.A. Fecher. 2012. Activity of the oral MEK inhibitor trametinib in patients with advanced melanoma: a phase 1 dose-escalation trial. Lancet Oncol. 13: 782–789.

Fan, L., F. Li, H. Zhang, Y. Wang, C. Cheng, X. Li, C.H. Gu, Q. Yang, H. Wu and S. Zhang. 2010. Co-delivery of PDTC and doxorubicin by multifunctional micellar nanoparticles to achieve active targeted drug delivery and overcome multidrug resistance. Biomaterials 31: 5634–5642.

Farokhzad, O.C., J. Cheng, B.A. Teply, I. Sherifi, S. Jon, P.W. Kantoff, J.P. Richie and R. Langer. 2006. Targeted nanoparticle-aptamer bioconjugates for cancer chemotherapy *in vivo*. Proc. Natl. Acad. Sci. U.S.A. 103: 6315–6320.

Farokhzad, O.C., S. Jon, A. Khademhosseini, T.N.T. Tran, D.A. Lavan and R. Langer. 2004. Nanoparticle-aptamer bioconjugates: a new approach for targeting prostate cancer cells. Cancer Res. 64: 7668–7672.

Ferlay, J., I. Soerjomataram, R. Dikshit, S. Eser, C. Mathers, M. Rebelo, D.M. Parkin, D. Forman and F. Bray. 2015. Cancer incidence and mortality worldwide: Sources, methods and major patterns in GLOBOCAN 2012: Globocan 2012. Int. J. Cancer 136: E359–E386.

Finn, R.S., J.P. Crown, I. Lang, K. Boer, I.M. Bondarenko, S.O. Kulyk, J. Ettl, R. Patel, T. Pinter, M. Schmidt, Y. Shparyk, A.R. Thummala, N.L. Voytko, C. Fowst, X. Huang, S.T. Kim, S. Randolph and D.J. Slamon. 2015. The cyclin-dependent kinase 4/6 inhibitor palbociclib in combination with letrozole versus letrozole alone as first-line treatment of oestrogen receptor-positive, HER2-negative, advanced breast cancer (PALOMA-1/TRIO-18): a randomised phase 2 study. Lancet Oncol. 16: 25–35.

Friedman, H.S., M.D. Prados, P.Y. Wen, T. Mikkelsen, D. Schiff, L.E. Abrey, W.K.A. Yung, N. Paleologos, M.K. Nicholas, R. Jensen, J. Vredenburgh, J. Huang, M. Zheng and T. Cloughesy. 2009. Bevacizumab alone and in combination with irinotecan in recurrent glioblastoma. J. Clin. Oncol. 27: 4733–4740.

Fuchs, C.S., J. Tomasek, C.J. Yong, F. Dumitru, R. Passalacqua, C. Goswami, H. Safran, L.V. dos Santos, G. Aprile, D.R. Ferry, B. Melichar, M. Tehfe, E. Topuzov, J.R. Zalcberg, I. Chau, W. Campbell, C. Sivanandan, J. Pikiel, M. Koshiji, Y. Hsu, A.M. Liepa, L. Gao, J.D. Schwartz and J. Tabernero. 2014. Ramucirumab monotherapy for previously treated advanced gastric or gastro-oesophageal junction adenocarcinoma (REGARD): an international, randomised, multicentre, placebo-controlled, phase 3 trial. The Lancet 383: 31–39.

Gan, C.W. and S.-S. Feng. 2010. Transferrin-conjugated nanoparticles of poly(lactide)-D-alpha-tocopheryl polyethylene glycol succinate diblock copolymer for targeted drug delivery across the blood-brain barrier. Biomaterials 31: 7748–7757.

Ganti, A.K. 2006. Hormone replacement therapy is associated with decreased survival in women with lung cancer. J. Clin. Oncol. 24: 59–63.

Garon, E.B., D. Cao, E. Alexandris, W.J. John, S. Yurasov and M. Perol. 2012. A randomized, double-blind, phase III study of docetaxel and ramucirumab versus docetaxel and placebo in the treatment of stage IV non–small-cell lung cancer after disease progression after 1 previous platinum-based therapy (REVEL): treatment rationale and study design. Clin. Lung Cancer 13: 505–509.

Geyer, C.E., J. Forster, D. Lindquist, S. Chan, C.G. Romieu, T. Pienkowski, A. Jagiello-Gruszfeld, J. Crown, A. Chan, B. Kaufman, D. Skarlos, M. Campone, N. Davidson, M. Berger, C. Oliva, S.D. Rubin, S. Stein and D. Cameron. 2006. Lapatinib plus capecitabine for HER2-positive advanced breast cancer. N. Engl. J. Med. 355: 2733–2743.

Glazer, E.S., C. Zhu, K.L. Massey, C.S. Thompson, W.D. Kaluarachchi, A.N. Hamir and S.A. Curley. 2010. Noninvasive radiofrequency field destruction of pancreatic adenocarcinoma xenografts treated with targeted gold nanoparticles. Clin. Cancer Res. Off. J. Am. Assoc. Cancer Res. 16: 5712–5721.

Green, M.R. 2006. Abraxane(R), a novel Cremophor(R)-free, albumin-bound particle form of paclitaxel for the treatment of advanced non-small-cell lung cancer. Ann. Oncol. 17: 1263–1268.

Grothey, A., E.V. Cutsem, A. Sobrero, S. Siena, A. Falcone, M. Ychou, Y. Humblet, O. Bouché, L. Mineur, C. Barone, A. Adenis, J. Tabernero, T. Yoshino, H.J. Lenz, R.M. Goldberg, D.J. Sargent, F. Cihon, L. Cupit, A. Wagner and D. Laurent. 2013. Regorafenib monotherapy for previously treated metastatic colorectal cancer (CORRECT): an international, multicentre, randomised, placebo-controlled, phase 3 trial. The Lancet 381: 303–312.

Gu, J., S. Su, M. Zhu, Y. Li, W. Zhao, Y. Duan and J. Shi. 2012. Targeted doxorubicin delivery to liver cancer cells by PEGylated mesoporous silica nanoparticles with a pH-dependent release profile. Microporous Mesoporous Mater. 161: 160–167.

Gupta, B. and V.P. Torchilin. 2007. Monoclonal antibody 2C5-modified doxorubicin-loaded liposomes with significantly enhanced therapeutic activity against intracranial human brain U-87 MG tumor xenografts in nude mice. Cancer Immunol. Immunother. CII 56: 1215–1223.

Guthi, J.S., S.G. Yang, G. Huang, S. Li, C. Khemtong, C.W. Kessinger, M. Peyton, J.D. Minna, K.C. Brown and J. Gao. 2010. MRI-Visible micellar nanomedicine for targeted drug delivery to lung cancer cells. Mol. Pharm. 7: 32–40.

Hadjipanayis, C.G., R. Machaidze, M. Kaluzova, L. Wang, A.J. Schuette, H. Chen, X. Wu and H. Mao. 2010. EGFRvIII antibody-conjugated iron oxide nanoparticles for magnetic resonance imaging-guided convection-enhanced delivery and targeted therapy of glioblastoma. Cancer Res. 70: 6303–6312.

Hattori, Y. and Y. Maitani. 2005. Folate-linked nanoparticle-mediated suicide gene therapy in human prostate cancer and nasopharyngeal cancer with herpes simplex virus thymidine kinase. Cancer Gene Ther. 12: 796–809.

Hattori, Y. and Y. Maitani. 2004. Enhanced *in vitro* DNA transfection efficiency by novel folate-linked nanoparticles in human prostate cancer and oral cancer. J. Control. Release Off. J. Control. Release Soc. 97: 173–183.

Hauschild, A., J.J. Grob, L.V. Demidov, T. Jouary, R. Gutzmer, M. Millward, P. Rutkowski, C.U. Blank, W.H. Miller, E. Kaempgen, S. Martín-Algarra, B. Karaszewska, C. Mauch, V. Chiarion-Sileni, A.M. Martin, S. Swann, P. Haney, B. Mirakhur, M.E. Guckert, V. Goodman and P.B. Chapman. 2012. Dabrafenib in BRAF-mutated metastatic melanoma: a multicentre, open-label, phase 3 randomised controlled trial. The Lancet 380: 358–365.

Hölig, P., M. Bach, T. Völkel, T. Nahde, S. Hoffmann, R. Müller and R.E. Kontermann. 2004. Novel RGD lipopeptides for the targeting of liposomes to integrin-expressing endothelial and melanoma cells. Protein Eng. Des. Sel. PEDS 17: 433–441.

Hsieh, W.J., C.J. Liang, J.J. Chieh, S.H. Wang, I.R. Lai, J.H. Chen, F.H. Chang, W.K. Tseng, S.Y. Yang, C.C. Wang and Y.L. Chen. 2012. *In vivo* tumor targeting and imaging with anti-vascular endothelial growth factor antibody-conjugated dextran-coated iron oxide nanoparticles. Int. J. Nanomedicine 2833.

Huang, G., C. Zhang, S. Li, C. Khemtong, S.G. Yang, R. Tian, J.D. Minna, K.C. Brown and J. Gao. 2009. A novel strategy for surface modification of superparamagnetic iron oxide nanoparticles for lung cancer imaging. J. Mater. Chem. 19: 6367–6372.

Hurwitz, H., L. Fehrenbacher, W. Novotny, T. Cartwright, J. Hainsworth, W. Heim, J. Berlin, A. Baron, S. Griffing, E. Holmgren, N. Ferrara, G. Fyfe, B. Rogers, R. Ross and F. Kabbinavar. 2004. Bevacizumab plus irinotecan, fluorouracil, and leucovorin for metastatic colorectal cancer. N. Engl. J. Med. 350: 2335–2342.

Iinuma, H., K. Maruyama, K. Okinaga, K. Sasaki, T. Sekine, O. Ishida, N. Ogiwara, K. Johkura and Y. Yonemura. 2002. Intracellular targeting therapy of cisplatin-encapsulated transferrin-polyethylene glycol liposome on peritoneal dissemination of gastric cancer. Int. J. Cancer J. Int. Cancer 99: 130–137.

Jain, A., S.K. Jain, N. Ganesh, J. Barve and A.M. Beg. 2010. Design and development of ligand-appended polysaccharidic nanoparticles for the delivery of oxaliplatin in colorectal cancer. Nanomedicine Nanotechnol. Biol. Med. 6: 179–190.

Jin, Y.J., U. Termsarasab, S.H. Ko, J.S. Shim, S. Chong, S.J. Chung, C.K. Shim, H.J. Cho and D.D. Kim. 2012. Hyaluronic acid derivative-based self-assembled nanoparticles for the treatment of melanoma. Pharm. Res. 29: 3443–3454.

Ji, Z., G. Lin, Q. Lu, L. Meng, X. Shen, L. Dong, C. Fu and X. Zhang. 2012. Targeted therapy of SMMC-7721 liver cancer *in vitro* and *in vivo* with carbon nanotubes based drug delivery system. J. Colloid Interface Sci. 365: 143–149.

Kamaly, N., T. Kalber, M. Thanou, J.D. Bell and A.D. Miller. 2009. Folate receptor targeted bimodal liposomes for tumor magnetic resonance imaging. Bioconjug. Chem. 20: 648–655.

Kantoff, P.W., C.S. Higano, N.D. Shore, E.R. Berger, E.J. Small, D.F. Penson, C.H. Redfern, A.C. Ferrari, R. Dreicer, R.B. Sims, Y. Xu, M.W. Frohlich and P.F. Schellhammer. 2010.

Sipuleucel-T immunotherapy for castration-resistant prostate cancer. N. Engl. J. Med. 363: 411–422.

Kikumori, T., T. Kobayashi, M. Sawaki and T. Imai. 2009. Anti-cancer effect of hyperthermia on breast cancer by magnetite nanoparticle-loaded anti-HER2 immunoliposomes. Breast Cancer Res. Treat. 113: 435–441.

Kim, D., Z.G. Gao, E.S. Lee and Y.H. Bae. 2009. *In vivo* evaluation of doxorubicin-loaded polymeric micelles targeting folate receptors and early endosomal pH in drug-resistant ovarian cancer. Mol. Pharm. 6: 1353–1362.

Kinsella, J.M., R.E. Jimenez, P.P. Karmali, A.M. Rush, V.R. Kotamraju, N.C. Gianneschi, E. Ruoslahti, D. Stupack and M.J. Sailor. 2011. X-Ray computed tomography imaging of breast cancer by using targeted peptide-labeled bismuth sulfide nanoparticles. Angew. Chem. Int. Ed. 50: 12308–12311.

Klutz, K., D. Schaffert, M.J. Willhauck, G.K. Grünwald, R. Haase, N. Wunderlich, C. Zach, F.J. Gildehaus, R. Senekowitsch-Schmidtke, B. Göke, E. Wagner, M. Ogris and C. Spitzweg. 2011. Epidermal growth factor receptor-targeted (131)I-therapy of liver cancer following systemic delivery of the sodium iodide symporter gene. Mol. Ther. J. Am. Soc. Gene Ther. 19: 676–685.

Krueger, D.A., M.M. Care, K. Holland, K. Agricola, C. Tudor, P. Mangeshkar, K.A. Wilson, A. Byars, T. Sahmoud and D.N. Franz. 2010. Everolimus for subependymal giant-cell astrocytomas in tuberous sclerosis. N. Engl. J. Med. 363: 1801–1811.

Lee, G.Y., W.P. Qian, L. Wang, Y.A. Wang, C.A. Staley, M. Satpathy, S. Nie, H. Mao and L. Yang. 2013. Theranostic nanoparticles with controlled release of gemcitabine for targeted therapy and MRI of pancreatic cancer. ACS Nano 7: 2078–2089.

Lee, J.H., Y.M. Huh, Y. Jun, J. Seo, J. Jang, H.T. Song, S. Kim, E.J. Cho, H.G. Yoon, J.S. Suh and J. Cheon. 2007. Artificially engineered magnetic nanoparticles for ultra-sensitive molecular imaging. Nat. Med. 13: 95–99.

Liang, H.F., C.T. Chen, S.C. Chen, A.R. Kulkarni, Y.L. Chiu, M.C. Chen and H.W. Sung. 2006. Paclitaxel-loaded poly(gamma-glutamic acid)-poly(lactide) nanoparticles as a targeted drug delivery system for the treatment of liver cancer. Biomaterials 27: 2051–2059.

Li, S.D. and L. Huang. 2006. Targeted delivery of antisense oligodeoxynucleotide and small interference RNA into lung cancer cells. Mol. Pharm. 3: 579–588.

Liu, M., Z.H. Li, F.J. Xu, L.H. Lai, Q.Q. Wang, G.P. Tang and W.T. Yang. 2012. An oligopeptide ligand-mediated therapeutic gene nanocomplex for liver cancer-targeted therapy. Biomaterials 33: 2240–2250.

Liu, Y., Z. Chen, C. Liu, D. Yu, Z. Lu and N. Zhang. 2011. Gadolinium-loaded polymeric nanoparticles modified with Anti-VEGF as multifunctional MRI contrast agents for the diagnosis of liver cancer. Biomaterials 32: 5167–5176.

Li, W., H. Zhao, W. Qian, H. Li, L. Zhang, Z. Ye, G. Zhang, M. Xia, J. Li, J. Gao, B. Li, G. Kou, J. Dai, H. Wang and Y. Guo. 2012. Chemotherapy for gastric cancer by finely tailoring anti-Her2 anchored dual targeting immunomicelles. Biomaterials 33: 5349–5362.

Löw, K., M. Wacker, S. Wagner, K. Langer and H. von Briesen. 2011. Targeted human serum albumin nanoparticles for specific uptake in EGFR-Expressing colon carcinoma cells. Nanomedicine Nanotechnol. Biol. Med. 7: 454–463.

Madhankumar, A.B., B. Slagle-Webb, X. Wang, Q.X. Yang, D.A. Antonetti, P.A. Miller, J.M. Sheehan and J.R. Connor. 2009. Efficacy of interleukin-13 receptor-targeted liposomal doxorubicin in the intracranial brain tumor model. Mol. Cancer Ther. 8: 648–654.

Maeng, J.H., D.H. Lee, K.H. Jung, Y.H. Bae, I.S. Park, S. Jeong, Y.S. Jeon, C.K. Shim, W. Kim, J. Kim, J. Lee, Y.M. Lee, J.H. Kim, W.H. Kim and S.S. Hong. 2010. Multifunctional doxorubicin loaded superparamagnetic iron oxide nanoparticles for chemotherapy and magnetic resonance imaging in liver cancer. Biomaterials 31: 4995–5006.

Magadala, P. and M. Amiji. 2008. Epidermal growth factor receptor-targeted gelatin-based engineered nanocarriers for DNA delivery and transfection in human pancreatic cancer cells. AAPS J. 10: 565–576.

Matsui, M., Y. Shimizu, Y. Kodera, E. Kondo, Y. Ikehara and H. Nakanishi. 2010. Targeted delivery of oligomannose-coated liposome to the omental micrometastasis by peritoneal macrophages from patients with gastric cancer. Cancer Sci. 101: 1670–1677.

Mehlen, P. and A. Puisieux. 2006. Metastasis: a question of life or death. Nat. Rev. Cancer 6: 449–458.

Miller, V.A., V. Hirsh, J. Cadranel, Y.M. Chen, K. Park, S.W. Kim, C. Zhou, W.C. Su, M. Wang, Y. Sun, D.S. Heo, L. Crino, E.H. Tan, T.Y. Chao, M. Shahidi, X.J. Cong, R.M. Lorence and J.C.H. Yang. 2012. Afatinib versus placebo for patients with advanced, metastatic non-small-cell lung cancer after failure of erlotinib, gefitinib, or both, and one or two lines of chemotherapy (LUX-Lung 1): a phase 2b/3 randomised trial. Lancet Oncol. 13: 528–538.

Moore, M.J., D. Goldstein, J. Hamm, A. Figer, J.R. Hecht, S. Gallinger, H.J. Au, P. Murawa, D. Walde, R.A. Wolff, D. Campos, R. Lim, K. Ding, G. Clark, T. Voskoglou-Nomikos, M. Ptasynski and W. Parulekar. 2007. Erlotinib plus gemcitabine compared with gemcitabine alone in patients with advanced pancreatic cancer: a phase III trial of the national cancer institute of canada clinical trials group. J. Clin. Oncol. 25: 1960–1966.

Nigam, P., S. Waghmode, M. Louis, S. Wangnoo, P. Chavan and D. Sarkar. 2014. Graphene quantum dots conjugated albumin nanoparticles for targeted drug delivery and imaging of pancreatic cancer. J. Mater. Chem. B 2, 3190.

Nukolova, N.V., H.S. Oberoi, S.M. Cohen, A.V. Kabanov and T.K. Bronich. 2011. Folate-decorated nanogels for targeted therapy of ovarian cancer. Biomaterials 32: 5417–5426.

Pan, L., M. He, J. Ma, W. Tang, G. Gao, R. He, H. Su and D. Cui. 2013. Phase and size controllable synthesis of NaYbF4 nanocrystals in oleic acid/ionic liquid two-phase system for targeted fluorescent imaging of gastric cancer. Theranostics 3: 210–222.

Patra, C.R., R. Bhattacharya, E. Wang, A. Katarya, J.S. Lau, S. Dutta, M. Muders, S. Wang, S.A. Buhrow, S.L. Safgren, M.J. Yaszemski, J.M. Reid, M.M. Ames, P. Mukherjee and D. Mukhopadhyay. 2008. Targeted delivery of gemcitabine to pancreatic adenocarcinoma using cetuximab as a targeting agent. Cancer Res. 68: 1970–1978.

Peer, D. and R. Margalit. 2004. Tumor-targeted hyaluronan nanoliposomes increase the antitumor activity of liposomal Doxorubicin in syngeneic and human xenograft mouse tumor models. Neoplasia N. Y. N. 6: 343–353.

Peng, X.H., Y. Wang, D. Huang, Y. Wang, H.J. Shin, Z. Chen, M.B. Spewak, H. Mao, X. Wang, Y. Wang, Z. Chen (Georgia), S. Nie and D.M. Shin. 2011. Targeted delivery of cisplatin to lung cancer using ScFvEGFR-Heparin-Cisplatin nanoparticles. ACS Nano 5: 9480–9493.

Perche, F., N.R. Patel and V.P. Torchilin. 2012. Accumulation and toxicity of antibody-targeted doxorubicin-loaded PEG-PE micelles in ovarian cancer cell spheroid model. J. Control. Release Off. J. Control. Release Soc. 164: 95–102.

Qian, X., X.H. Peng, D.O. Ansari, Q. Yin-Goen, G.Z. Chen, D.M. Shin, L. Yang, A.N. Young, M.D. Wang and S. Nie. 2008. *In vivo* tumor targeting and spectroscopic detection with surface-enhanced Raman nanoparticle tags. Nat. Biotechnol. 26: 83–90.

Rachakatla, R.S., S. Balivada, G.M. Seo, C.B. Myers, H. Wang, T.N. Samarakoon, R. Dani, M. Pyle, F.O. Kroh, B. Walker, X. Leaym, O.B. Koper, V. Chikan, S.H. Bossmann, M. Tamura and D.L. Troyer. 2010. Attenuation of mouse melanoma by A/C magnetic field after delivery of bi-magnetic nanoparticles by neural progenitor cells. ACS Nano 4: 7093–7104.

Raymond, E., L. Dahan, J.L. Raoul, Y.J. Bang, I. Borbath, C. Lombard-Bohas, J. Valle, P. Metrakos, D. Smith, A. Vinik, J.S. Chen, D. Hörsch, P. Hammel, B. Wiedenmann, E. Van Cutsem, S. Patyna, D.R. Lu, C. Blanckmeister, R. Chao and P. Ruszniewski. 2011. Sunitinib Malate for the treatment of pancreatic neuroendocrine tumors. N. Engl. J. Med. 364: 501–513.

Reddy, G.R., M.S. Bhojani, P. McConville, J. Moody, B.A. Moffat, D.E. Hall, G. Kim, Y.E.L. Koo, M.J. Woolliscroft, J.V. Sugai, T.D. Johnson, M.A. Philbert, R. Kopelman, A. Rehemtulla and B.D. Ross. 2006. Vascular targeted nanoparticles for imaging and treatment of brain tumors. Clin. Cancer Res. Off. J. Am. Assoc. Cancer Res. 12: 6677–6686.

Ren, Y., H.W. Cheung, G. von Maltzhan, A. Agrawal, G.S. Cowley, B.A. Weir, J.S. Boehm, P. Tamayo, A.M. Karst, J.F. Liu, M.S. Hirsch, J.P. Mesirov, R. Drapkin, D.E. Root, J. Lo, V. Fogal, E. Ruoslahti, W.C. Hahn and S.N. Bhatia. 2012. Targeted tumor-penetrating siRNA nanocomplexes for credentialing the ovarian cancer oncogene ID4. Sci. Transl. Med. 4: 147ra112.

Romond, E.H., E.A. Perez, J. Bryant, V.J. Suman, C.E. Geyer, N.E. Davidson, E. Tan-Chiu, S. Martino, S. Paik, P.A. Kaufman, S.M. Swain, T.M. Pisansky, L. Fehrenbacher, L.A. Kutteh, V.G. Vogel, D.W. Visscher, G. Yothers, R.B. Jenkins, A.M. Brown, S.R. Dakhil, E.P. Mamounas, W.L. Lingle, P.M. Klein, J.N. Ingle and N. Wolmark. 2005. Trastuzumab plus adjuvant chemotherapy for operable HER2-positive breast cancer. N. Engl. J. Med. 353: 1673–1684.

Rozhkova, E.A., I. Ulasov, B. Lai, N.M. Dimitrijevic, M.S. Lesniak and T. Rajh. 2009. A high-performance nanobio photocatalyst for targeted brain cancer therapy. Nano Lett. 9: 3337–3342.

Ruan, J., J. Ji, H. Song, Q. Qian, K. Wang, C. Wang and D. Cui. 2012a. Fluorescent magnetic nanoparticle-labeled mesenchymal stem cells for targeted imaging and hyperthermia therapy of *in vivo* gastric cancer. Nanoscale Res. Lett. 7: 309.

Ruan, J., H. Song, Q. Qian, C. Li, K. Wang, C. Bao and D. Cui. 2012b. HER2 monoclonal antibody conjugated RNase-A-associated CdTe quantum dots for targeted imaging and therapy of gastric cancer. Biomaterials 33: 7093–7102.

Sadhukha, T., T.S. Wiedmann and J. Panyam. 2013. Inhalable magnetic nanoparticles for targeted hyperthermia in lung cancer therapy. Biomaterials 34: 5163–5171.

Sandler, A., R. Gray, M.C. Perry, J. Brahmer, J.H. Schiller, A. Dowlati, R. Lilenbaum and D.H. Johnson. 2006. Paclitaxel–Carboplatin alone or with bevacizumab for non–small-cell lung cancer. N. Engl. J. Med. 355: 2542–2550.

Santra, S., C. Kaittanis, J. Grimm and J.M. Perez. 2009. Drug/Dye-Loaded, Multifunctional Iron Oxide Nanoparticles for Combined Targeted Cancer Therapy and Dual Optical/ Magnetic Resonance Imaging. Small 5: 1862–1868.

Sato, S. and H. Itamochi. 2012. Bevacizumab and ovarian cancer: Curr. Opin. Obstet. Gynecol. 24: 8–13.

Sawant, R.M., M.B. Cohen, V.P. Torchilin and O.W. Rokhlin. 2008. Prostate cancer-specific monoclonal antibody 5D4 significantly enhances the cytotoxicity of doxorubicin-loaded liposomes against target cells *in vitro*. J. Drug Target. 16: 601–604.

Scarberry, K.E., E.B. Dickerson, Z.J. Zhang, B.B. Benigno and J.F. McDonald. 2010. Selective removal of ovarian cancer cells from human ascites fluid using magnetic nanoparticles. Nanomedicine Nanotechnol. Biol. Med. 6: 399–408.

Schmieder, A.H., P.M. Winter, S.D. Caruthers, T.D. Harris, T.A. Williams, J.S. Allen, E.K. Lacy, H. Zhang, M.J. Scott, G. Hu, J.D. Robertson, S.A. Wickline and G.M. Lanza. 2005. Molecular MR imaging of melanoma angiogenesis with alphanubeta3-targeted paramagnetic nanoparticles. Magn. Reson. Med. 53: 621–627.

Shaw, A.T., D.W. Kim, R. Mehra, D.S.W. Tan, E. Felip, L.Q.M. Chow, D.R. Camidge, J. Vansteenkiste, S. Sharma, T. De Pas, G.J. Riely, B.J. Solomon, J. Wolf, M. Thomas, M. Schuler, G. Liu, A. Santoro, Y.Y. Lau, M. Goldwasser, A.L. Boral and J.A. Engelman. 2014. Ceritinib in *ALK*—rearranged non–small-cell lung cancer. N. Engl. J. Med. 370: 1189–1197.

Shaw, A.T., D.W. Kim, K. Nakagawa, T. Seto, L. Crinó, M.J. Ahn, T. De Pas, B. Besse, B.J. Solomon, F. Blackhall, Y.L. Wu, M. Thomas, K.J. O'Byrne, D. Moro-Sibilot, D.R. Camidge, T. Mok, V. Hirsh, G.J. Riely, S. Iyer, V. Tassell, A. Polli, K.D. Wilner and P.A. Jänne. 2013. Crizotinib versus chemotherapy in advanced *ALK*—positive lung cancer. N. Engl. J. Med. 368: 2385–2394.

Shen, Z., W. Wei, H. Tanaka, K. Kohama, G. Ma, T. Dobashi, Y. Maki, H. Wang, J. Bi and S. Dai. 2011. A galactosamine-mediated drug delivery carrier for targeted liver cancer therapy. Pharmacol. Res. Off. J. Ital. Pharmacol. Soc. 64: 410–419.

Shepherd, F.A., J. Rodrigues Pereira, T. Ciuleanu, E.H. Tan, V. Hirsh, S. Thongprasert, D. Campos, S. Maoleekoonpiroj, M. Smylie, R. Martins, M. van Kooten, M. Dediu, B. Findlay, D. Tu, D. Johnston, A. Bezjak, G. Clark, P. Santabárbara and L. Seymour. 2005. Erlotinib in previously treated non–small-cell lung cancer. N. Engl. J. Med. 353: 123–132.

Siegel, R.L., K.D. Miller and A. Jemal. 2015. Cancer statistics, 2015: Cancer Statistics, 2015. CA. Cancer J. Clin. 65: 5–29.

Skidan, I., P. Dholakia and V. Torchilin. 2008. Photodynamic therapy of experimental B-16 melanoma in mice with tumor-targeted 5,10,15,20-tetraphenylporphin-loaded PEG-PE micelles. J. Drug Target. 16: 486–493.

Steinmetz, N.F., A.L. Ablack, J.L. Hickey, J. Ablack, B. Manocha, J.S. Mymryk, L.G. Luyt and J.D. Lewis. 2011. Intravital imaging of human prostate cancer using viral nanoparticles targeted to gastrin-releasing Peptide receptors. Small Weinh. Bergstr. Ger. 7: 1664–1672.

Stuchinskaya, T., M. Moreno, M.J. Cook, D.R. Edwards and D.A. Russell. 2011. Targeted photodynamic therapy of breast cancer cells using antibody–phthalocyanine–gold nanoparticle conjugates. Photochem. Photobiol. Sci. 10: 822.

Sun, B., B. Ranganathan and S.S. Feng. 2008. Multifunctional poly(D,L-lactide-co-glycolide)/montmorillonite (PLGA/MMT) nanoparticles decorated by Trastuzumab for targeted chemotherapy of breast cancer. Biomaterials 29: 475–486.

Sundaram, S., R. Trivedi, C. Durairaj, R. Ramesh, B.K. Ambati and U.B. Kompella. 2009. Targeted drug and gene delivery systems for lung cancer therapy. Clin. Cancer Res. Off. J. Am. Assoc. Cancer Res. 15: 7299–7308.

Sun, W. and A. Patel. 2013. Ziv-aflibercept in metastatic colorectal cancer. Biol. Targets Ther. 13.

Torchilin, V. 2011. Tumor delivery of macromolecular drugs based on the EPR effect. Adv. Drug Deliv. Rev. 63: 131–135.

Tseng, C.L., W.Y. Su, K.C. Yen, K.C. Yang and F.H. Lin. 2009. The use of biotinylated-EGF-modified gelatin nanoparticle carrier to enhance cisplatin accumulation in cancerous lungs via inhalation. Biomaterials 30: 3476–3485.

Tseng, C.L., T.W. Wang, G.C. Dong, S. Yueh-Hsiu Wu, T.H. Young, M.J. Shieh, P.J. Lou and F.H. Lin. 2007. Development of gelatin nanoparticles with biotinylated EGF conjugation for lung cancer targeting. Biomaterials 28: 3996–4005.

Tseng, C.L., S.Y.H. Wu, W.H. Wang, C.L. Peng, F.H. Lin, C.C. Lin, T.H. Young and M.J. Shieh. 2008. Targeting efficiency and biodistribution of biotinylated-EGF-conjugated gelatin nanoparticles administered via aerosol delivery in nude mice with lung cancer. Biomaterials 29: 3014–3022.

Verma, S., D. Miles, L. Gianni, I.E. Krop, M. Welslau, J. Baselga, M. Pegram, D.Y. Oh, V. Diéras, E. Guardino, L. Fang, M.W. Lu, S. Olsen and K. Blackwell. 2012. Trastuzumab emtansine for HER2-positive advanced breast cancer. N. Engl. J. Med. 367: 1783–1791.

Wang, A.Z., V. Bagalkot, C.C. Vasilliou, F. Gu, F. Alexis, L. Zhang, M. Shaikh, K. Yuet, M.J. Cima, R. Langer, P.W. Kantoff, N.H. Bander, S. Jon and O.C. Farokhzad. 2008. Superparamagnetic iron oxide nanoparticle-aptamer bioconjugates for combined prostate cancer imaging and therapy. ChemMedChem. 3: 1311–1315.

Wang, L., K.G. Neoh, E.T. Kang and B. Shuter. 2011. Multifunctional polyglycerol-grafted Fe3O4@SiO2 nanoparticles for targeting ovarian cancer cells. Biomaterials 32: 2166–2173.

Wang, L., W. Su, Z. Liu, M. Zhou, S. Chen, Y. Chen, D. Lu, Y. Liu, Y. Fan, Y. Zheng, Z. Han, D. Kong, J.C. Wu, R. Xiang and Z. Li. 2012. CD44 antibody-targeted liposomal nanoparticles for molecular imaging and therapy of hepatocellular carcinoma. Biomaterials 33: 5107–5114.

Werner, M.E., S. Karve, R. Sukumar, N.D. Cummings, J.A. Copp, R.C. Chen, T. Zhang and A.Z. Wang. 2011. Folate-targeted nanoparticle delivery of chemo- and radiotherapeutics for the treatment of ovarian cancer peritoneal metastasis. Biomaterials 32: 8548–8554.

Winer, I., S. Wang, Y.E.K. Lee, Y.E.K. Lee, W. Fan, Y. Gong, D. Burgos-Ojeda, G. Spahlinger, R. Kopelman and R.J. Buckanovich. 2010. F3-targeted cisplatin-hydrogel nanoparticles as an effective therapeutic that targets both murine and human ovarian tumor endothelial cells *in vivo*. Cancer Res. 70: 8674–8683.

Wu, G., R.F. Barth, W. Yang, S. Kawabata, L. Zhang and K. Green-Church. 2006. Targeted delivery of methotrexate to epidermal growth factor receptor-positive brain tumors by means of cetuximab (IMC-C225) dendrimer bioconjugates. Mol. Cancer Ther. 5: 52–59.

Wu, X., H. Liu, J. Liu, K.N. Haley, J.A. Treadway, J.P. Larson, N. Ge, F. Peale and M.P. Bruchez. 2003. Immunofluorescent labeling of cancer marker Her2 and other cellular targets with semiconductor quantum dots. Nat. Biotechnol. 21: 41–46.

Xiao, K., Y. Li, J.S. Lee, A.M. Gonik, T. Dong, G. Fung, E. Sanchez, L. Xing, H.R. Cheng, J. Luo and K.S. Lam. 2012. "OA02" peptide facilitates the precise targeting of paclitaxel-loaded micellar nanoparticles to ovarian cancer *in vivo*. Cancer Res. 72: 2100–2110.

Xiao, Y., X. Gao, O. Taratula, S. Treado, A. Urbas, R.D. Holbrook, R.E. Cavicchi, C.T. Avedisian, S. Mitra, R. Savla, P.D. Wagner, S. Srivastava and H. He. 2009. Anti-HER2 IgY antibody-functionalized single-walled carbon nanotubes for detection and selective destruction of breast cancer cells. BMC Cancer 9: 351.

Xiao, Z., E. Levy-Nissenbaum, F. Alexis, A. Lupták, B.A. Teply, J.M. Chan, J. Shi, E. Digga, J. Cheng, R. Langer and O.C. Farokhzad. 2012. Engineering of targeted nanoparticles for cancer therapy using internalizing aptamers isolated by cell-uptake selection. ACS Nano 6: 696–704.

Xue, H.Y. and H.L. Wong. 2011. Solid lipid-PEI hybrid nanocarrier: an integrated approach to provide extended, targeted, and safer siRNA therapy of prostate cancer in an all-in-one manner. ACS Nano 5: 7034–7047.

Xu, J., A. Singh and M.M. Amiji. 2014. Redox-responsive targeted gelatin nanoparticles for delivery of combination wt-p53 expressing plasmid DNA and gemcitabine in the treatment of pancreatic cancer. BMC Cancer 14: 75.

Yang, L., H. Mao, Z. Cao, Y.A. Wang, X. Peng, X. Wang, H.K. Sajja, L. Wang, H. Duan, C. Ni, C.A. Staley, W.C. Wood, X. Gao and S. Nie. 2009a. Molecular imaging of pancreatic cancer in an animal model using targeted multifunctional nanoparticles. Gastroenterology 136: 1514–1525.e2.

Yang, L., H. Mao, Y.A. Wang, Z. Cao, X. Peng, X. Wang, H. Duan, C. Ni, Q. Yuan, G. Adams, M.Q. Smith, W.C. Wood, X. Gao and S. Nie. 2009b. Single chain epidermal growth factor receptor antibody conjugated nanoparticles for *in vivo* tumor targeting and imaging. Small Weinh. Bergstr. Ger. 5: 235–243.

Yang, L., X.-H. Peng, Y.A. Wang, X. Wang, Z. Cao, C. Ni, P. Karna, X. Zhang, W.C. Wood, X. Gao, S. Nie and H. Mao. 2009. Receptor-targeted nanoparticles for *in vivo* imaging of breast cancer. Clin. Cancer Res. 15: 4722–4732.

Yang, S.J., F.H. Lin, K.C. Tsai, M.F. Wei, H.M. Tsai, J.M. Wong and M.J. Shieh. 2010. Folic acid-conjugated chitosan nanoparticles enhanced protoporphyrin IX accumulation in colorectal cancer cells. Bioconjug. Chem. 21: 679–689.

Yang, Y., J. Li, F. Liu and L. Huang. 2012. Systemic delivery of siRNA via LCP nanoparticle efficiently inhibits lung metastasis. Mol. Ther. 20: 609–615.

Yao, J.C., M.H. Shah, T. Ito, C.L. Bohas, E.M. Wolin, E. Van Cutsem, T.J. Hobday, T. Okusaka, J. Capdevila, E.G.E. de Vries, P. Tomassetti, M.E. Pavel, S. Hoosen, T. Haas, J. Lincy, D. Lebwohl and K. Öberg. 2011. Everolimus for advanced pancreatic neuroendocrine tumors. N. Engl. J. Med. 364: 514–523.

Ying, X., H. Wen, W.L. Lu, J. Du, J. Guo, W. Tian, Y. Men, Y. Zhang, R.J. Li, T.Y. Yang, D.W. Shang, J.N. Lou, L.R. Zhang and Q. Zhang. 2010. Dual-targeting daunorubicin liposomes improve the therapeutic efficacy of brain glioma in animals. J. Control. Release 141: 183–192.

Yokoyama, T., J. Tam, S. Kuroda, A.W. Scott, J. Aaron, T. Larson, M. Shanker, A.M. Correa, S. Kondo, J.A. Roth, K. Sokolov and R. Ramesh. 2011. EGFR-targeted hybrid plasmonic magnetic nanoparticles synergistically induce autophagy and apoptosis in non-small cell lung cancer cells. PLoS ONE 6: e25507.

Yoo, M.K., I.K. Park, H.T. Lim, S.J. Lee, H.L. Jiang, Y.K. Kim, Y.J. Choi, M.H. Cho and C.S. Cho. 2012. Folate–PEG–superparamagnetic iron oxide nanoparticles for lung cancer imaging. Acta Biomater. 8: 3005–3013.

Yu, M.K., D. Kim, I.H. Lee, J.S. So, Y.Y. Jeong and S. Jon. 2011. Image-guided prostate cancer therapy using aptamer-functionalized thermally cross-linked superparamagnetic iron oxide nanoparticles. Small Weinh. Bergstr. Ger. 7: 2241–2249.

Zhang, B., Z. Luo, J. Liu, X. Ding, J. Li and K. Cai. 2014. Cytochrome c end-capped mesoporous silica nanoparticles as redox-responsive drug delivery vehicles for liver tumor-targeted triplex therapy *in vitro* and *in vivo*. J. Control. Release Off. J. Control. Release Soc. 192: 192–201.

Zhang, L., A.F. Radovic-Moreno, F. Alexis, F.X. Gu, P.A. Basto, V. Bagalkot, S. Jon, R.S. Langer and O.C. Farokhzad. 2007. Co-delivery of hydrophobic and hydrophilic drugs from nanoparticle-aptamer bioconjugates. ChemMedChem. 2: 1268–1271.

Zhang, M., X. Zhou, B. Wang, B.C. Yung, L.J. Lee, K. Ghoshal and R.J. Lee. 2013. Lactosylated gramicidin-based lipid nanoparticles (Lac-GLN) for targeted delivery of anti-miR-155 to hepatocellular carcinoma. J. Control. Release.

Zhang, X., J. Chen, Y. Zheng, X. Gao, Y. Kang, J. Liu, M. Cheng, H. Sun and C. Xu. 2009. Follicle-stimulating hormone peptide can facilitate paclitaxel nanoparticles to target ovarian carcinoma *in vivo*. Cancer Res. 69: 6506–6514.

Zhang, Y., R.J. Boado and W.M. Pardridge. 2003. *In vivo* knockdown of gene expression in brain cancer with intravenous RNAi in adult rats. J. Gene Med. 5: 1039–1045.

Zhang, Y., N.M. Schwerbrock, A.B. Rogers, W.Y. Kim and L. Huang. 2013. Codelivery of VEGF siRNA and gemcitabine monophosphate in a single nanoparticle formulation for effective treatment of NSCLC. Mol. Ther. 21: 1559–1569.

Zhao, H., J.C. Wang, Q.S. Sun, C.L. Luo and Q. Zhang. 2009. RGD-based strategies for improving antitumor activity of paclitaxel-loaded liposomes in nude mice xenografted with human ovarian cancer. J. Drug Target. 17: 10–18.

Zheng, G., J. Chen, H. Li and J.D. Glickson. 2005. Rerouting lipoprotein nanoparticles to selected alternate receptors for the targeted delivery of cancer diagnostic and therapeutic agents. Proc. Natl. Acad. Sci. 102: 17757–17762.

6

Smart Materials for Fluorescence Sensing

Fan Zhang and *Haibing Li**

ABSTRACT

Smart materials have attracted increasing interest in them because of their special properties and wide applications. Quantum dots (QDs), in particular, emerged as one of the most promising classes of smart inorganic nanomaterials which have superior tunable optical properties including size-controlled fluorescence, high fluorescence quantum yields, large effective stokes shifts, broad excitation range with narrow symmetric photoluminescence spectra and so on. These properties enable QDs become a focus at the leading edge of the rapidly developing field of nanotechnology for bioanalysis and function materials, which have broadly applied in energy conversion and storage, optoelectronic devices, sensor applications, photocatalysis and more. However, to exploit these properties fully, functionalization of the QD surface is critical and essential. In general, important factors considered when functionalizing QDs, are the maintenance of the original physical properties of the QDs, and in many situations, the development of a flexible surface chemistry of the QDs. Introduction of an organic ligand onto the surface of nanoparticles affords not only the stability of

Key Laboratory of Pesticide and Chemical Biology (CCNU), Ministry of Education, College of Chemistry, Central China Normal University.
Email: zhangfan@mails.ccnu.edu.cn
* Corresponding author: lhbing@mail.ccnu.edu.cn

these nanoentities in different solvents but also the desired surface functionality. Therefore, given the tunable cavity of macrocyclic molecules and hydrophobic properties, many efforts have been made in combining the host-guest chemistry on the surface modification and functionalization of the QDs, which changed their luminescent properties. In this chapter we will focus on the design and fabrication of macrocyclic molecules (crownether, porphyrin, cyclodextrin and calixarene) coated QDs as smart sensors towards metal ions and small organic molecules. The necessary theoretical background to understand the general properties of QDs are provided. Different methods of surface modification to obtain the smart sensing materials and most common processes affecting the fluorescence sensing are described. And finally the summary of this chapter and future perspectives will be discussed. The overview information should be of interest to researchers in areas relevant to chemistry sensor, biological analysis, physics, nanotechnology and diagnostics.

1. Introduction

Smart nanomaterials such as quantum dots (QDs), gold nanoparticles (AuNPs), silver nanoparticles (AgNPs), silica nanoparticles (SiNPs) with unique size-dependent properties in the field of nanotechnology have attracted increasing interest for bioanalytical applications in recent years (Yao et al. 2014). Quantum dots (QDs) as a unique smart semiconductor nanoparticle has become a notable star in the development of optical labels for labelling and sensing and have been broadly applied in fluorescence assay or imaging for biological and medical researches due to their superior physical and chemical properties (Zrazhevskiy et al. 2010). They were first discovered by Alexey Ekimov in 1981 in a glass matrix and then in colloidal solutions by Louis E. Brus in 1985 (Ekimov and Onushchenko 1982). QDs are semiconductor nanocrystals composed of elements from groups II–VI (e.g., Cd, Zn, Se, and Te) or III–V (e.g., In, P, and As) in the periodic table, for example CdSe, CdTe, HgTe, PbS, PbSe, PbTe, InAs, InP, and GaAs (Klostranec and Chan 2006). They are small enough and range from 2 to 10 nanometers in diameter (physical size smaller than the exciton Bohr radius) to exhibit quantum mechanical properties, such as the quantum-size effect, surface effect, macroscopic quantum tunneling effect and so forth (Alivisatos 2004). These quantum effects would make it invalidate the governing rules at the macroscopic level and generate both high signal-to-noise ratio and signal amplification (Bau et al. 2011). Additionally, for their excitons are confined in all three spatial dimensions, the electrons are quantized to certain energies, similar to that of a small molecule, thus giving QDs excellent properties that is not achievable in bulk materials (Alivisatos 1996). With size variable photoluminescence, QDs are promising alternatives to organic dyes for their outstanding optical properties and fluorescence-based applications. These properties include high quantum yields, long fluorescence lifetime, large extinction coefficients, pronounced photostability, and more importantly, broad absorption with narrow symmetric photoluminescence spectra (full-width at half-maximum ~25–40 nm) spanning the UV to near-infrared region (NIR) (Resch-Genger et al. 2008, Esteve-Turrillas and Abad-Fuentes 2013). All these properties allow the excitation of QDs of various sizes at a common wavelength manifold while imaging, in parallel, the fluorescence of the different QDs. As a consequence, QDs are suitable for multiplexing analysis and they can be simultaneously excited by a single light source when multiple colors are acquired together (Giepmans et al. 2005).

Since biocompatible QDs were first described in 1998 (Bruchez et al. 1998), they have been utilized in many biological assays and functionalization materials. The efficient fluorescence and stability of QDs improve sensitivity and prolong lifetime in their use as optical labels, which has greatly prompted the basic and applied studies in life sciences (Smith et al. 2006). These include nucleic acid detection, gene expression analysis, detection of proteins and enzymatic activities, diagnosis of metabolic processes, cell biology and animal imaging (Somers et al. 2007, Medintz et al. 2005). Particularly, as is described, DNA/QDs conjugates have been used as fluorescence probes in analyzing human metaphase chromosomes for the fluorescence *in situ* hybridization (FISH) assays (Fletcher 1999). In such assays, genomic DNA is denatured and then allowed to hybridize with a fluorescent-labeled DNA sequence to visualize the presence or absence of specific DNA sequences in the chromosomes. The results showed that QD labeled DNA gave a higher signal-to-noise ratio and was more photostable than commonly used fluorophores, such as FITC and Texas Red. Another focus of this field was use of QDs that conjugated with biomolecules (e.g., antibodies) in FRET based assays and biosensors. For instance, by the application of a CdSe/ZnS QD conjugated to maltose binding protein (MBP) allowed binding of either maltose or a quenching molecule, where maltose was the target analyte, provided the basis for the development of several FRET-based transduction strategies (Medintz et al. 2003, Medintz et al. 2005). Besides the robust application of biological assay, QDs also showed great promise with high brightness and photostability for smart materials applications as well as universal platforms for engineering of multifunctional nanodevices. Researchers have studied applications of quantum dots in transistors, solar cells, LEDs, diode lasers and so forth, which have a great deal of practical applications and higher commercial value (Gill et al. 2008, Ai et al. 2013).

The applications described above rely mainly on fluorescence spectroscopy of QDs, which is one of the most promising and sensitive analytical techniques that have played and continue to play significant roles in modern research. Therefore, in order to achieve the purpose of rapid analysis and sensitive detection, many efforts have been focused on developing new methods to obtain smart QDs with surface modification and functionalization to make QDs more soluble and stable in aqueous media. Substantial progress has also been made in the utilization of host-guest interaction in analytical science for the capability of recognizing analytes with high specificity (Dickert and Haunschild 1993). Especially, supramolecular chemistry can be utilized for sensing technology with macrocyclic compounds like crown ether, porphyrin cyclodextrine and calixarene that are capable of forming molecular cavities to include analytes because of the high preorganization of these structures (Lehn et al. 1996). Recently, the molecular recognitions of host-guest interactions extended to surfaces for sensor applications have drawn considerable attention (Beulen et al. 2000, Kitano et al. 2000, Chen et al. 2004, Wu et al. 2009). A large number of chemical sensing of ions and small molecules with QDs by analyte-induced changes in photoluminescence have been reported. For example, Chen and Rosenzweig demonstrated luminescent quantum dots probes for Cu^{2+} and Zn^{2+} ions by utilizing functionalized CdS QDs capped with different organic ligands in aqueous medium (Chen and Rosenzweig 2002); Sanz-Medel and co-workers reported CdSe quantum dots modified with 2-mercaptoethane sulfonate as selective fluorescent probes for the determination of free cyanide (Jin et al. 2005); Thioaniline ligand and aza-crown ether—functionalized CdSe/ZnS QDs not only used to detect pH changes or ions, but also developed as a smart logic gate construct (Kaur et al. 2010). And more recently, calixarene-modified CdTe QDs were reported as fluorescence sensors for organic pollutants by Li's group. In this chapter the emphasis will be on the process of design and fabrication of macrocyclic molecules-coated quantum dots which can be acted as smart optical labels for fluorescence sensing (Scheme 1). Furthermore, the versatility and potential application will be outlined in this field.

2. Theory and Background

QDs as a smart photo-electrochemically active material is composed mainly of semiconductor cores, which often coated with one or more shell(s) consisting of semiconductor material (e.g., CdSe/ZnS core/shell or CdSe/CdS/ZnS core/shell/shell QDs) (Fig. 1) (Murray et al. 2000, Medintz et al. 2005). Because of surface defects and the surrounding medium, the shell passivates the QD core from quenching effects and increases the photoluminescence quantum yields of QDs (Esteve-Turrillas and Abad-

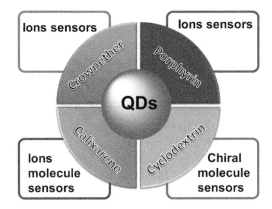

Scheme 1. Graphical abstract of macrocyclic molecules coated quantum dots for fluorescence sensing.

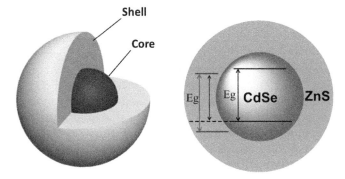

Figure 1. Schematic illustration of the quantum dots structure.

Fuentes 2013). Owing to the suitable lattice parameters and a relatively small band gap or higher band-gap energy between the valence band and conduction band, QDs can be behaving like insulators at ambient conditions and exhibiting electrical conductivity only under external stimulation (Kamat 2007). The optical properties of QDs can be described by conventional semiconductor physics and quantum mechanics. When a semiconductor is optically or electrically excited, static electrons (electrons located in the valence band) become mobile (electrons located in the conduction band) within the semiconductor matrix and after a certain period of time the electrons and holes recombine (Li et al. 2012). The quantum confinement effect will occur only when the size of the nanostructure is on the order of the exciton Bohr radius (< 10 nm) (Brus 1986), and the excitation and emission peaks of QDs can be modulated easily by changing the nanoparticles diameters or engineering the QD core–shell structures, which results in wide ultraviolet–visible (UV–vis) absorption spectra, and narrow and symmetric emission bands (Medintz et al. 2003) (Fig. 2).

So far, fluorescence-quenching and enhancement effects account for the main mechanisms of QDs as optical sensors. The mechanism of quantum dots fluorescent quenching is usually explained by electron transfer (ET) and fluorescence resonance energy transfer (FRET). Electron and energy transfer processes can therefore be designed to switch the luminescence of QDs in response to molecular recognition events, providing extremely sensitive probes or efficiently sensitize the electrodes for solar cell applications. The electron transfer (ET) quenching of photoexcited QDs is a versatile useful photophysical mechanism to follow spatially-restricted close interactions between electron donor–acceptor sites. It may occur over long distances and be associated with major dipole moment changes, making the process particularly sensitive to the microenvironment of the QDs (Freeman and Willner 2012). Thus, it can be expected that electrostatic interaction between the donor and the acceptor moieties with different charge (i.e., positive charge or negative charge) will change the photophysical properties of the functionalized QDs.

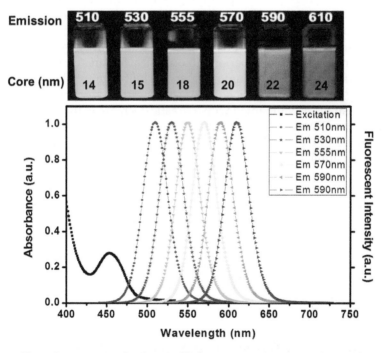

Figure 2. Photo demonstrating the size-tunable fluorescence properties and spectral range of QD dispersions plotted. Reproduced with permission from ref. (Medintz et al. 2005) Copyright © 2005 Nature Publishing Group.

Fluorescence resonance energy transfer (FRET) is a distance-dependent physical process by which energy is transferred to non-radiative from an excited molecular fluorophore (the donor) to another fluorophore (the acceptor) by means of intermolecular long-range dipole–dipole coupling (Bagalkot et al. 2007). It mainly influenced by three factors: the distance between the donor and the acceptor, the extent of spectral overlap between the donor emission and acceptor absorption spectrum and the relative orientation of the donor emission dipole moment and acceptor absorption moment. As is required, the distance between the donor and the acceptor is smaller than a critical radius, known as the Förster radius. This returns the donor to its electronic ground state, and emission may then occur from the acceptor center. QDs can be served both as FRET donors and acceptors, which have been treated in several comprehensive reviews in the recent literature. As donors, QDs can be combined with a large variety of acceptors (e.g., organic dyes or fluorescent proteins). Using QDs as acceptors is less common because of their broad excitation spectra, which will cause QD excitation at almost any wavelength (independent of the donor). This will lead to many QD acceptors in excited states, which is very counter productive for FRET (the acceptor must be in the ground state).

The photophysical properties of QDs can be controlled by their nanocrystal core sizes, the shell thickness, and the composition of the semiconductor materials of cores and shell(s) and partly by their surface ligands (Green 2010). The superior photophysical features of semiconductor QDs (high fluorescence quantum yields and stability against photobleaching) are usually observed in organic solvents, and their introduction into aqueous media is usually accompanied with a drastic decrease in the luminescence yields of the QDs (Gao et al. 2005). Thereafter, one of the main problems encountered in the development of QD-based systems is the modification of suitable capping ligands on the surface of QDs to improve the selectivity, sensitivity and biocompatibility of these systems with enhanced brightness and stability. To address this issue, various strategies have proposed which including (a) attaching a certain metal-specific ligand based on an electron transfer process; (b) exchanging of the ligands with more thiol groups by host-guest chemistry on the surfaces of the particles; (c) capped the surface with amphiphilic molecules or comb like polymers possessed a hydrophilic backbone and hydrophobic side chains. As a result, QDs are intrinsically soluble in non-polar media and suitable surface modification will be performed.

3. Quantum Dots-Based Smart Sensing Systems

Luminescence based transduction has been the primary approach for developing QDs based smart optical sensors of small molecules and ions, as demonstrated by numerous works summarized in the literature (Galian and de la Guardia 2009, Frasco and Chaniotakis 2009, Callan et al. 2007). As the luminescence of QDs is very sensitive to their surface states, fluorescence transduction is based on the principle that chemical or physical interactions occurring at the surface of the QDs, change the efficiency of the radiative recombination, leading to photoluminescence activation or quenching.

Macrocyclic molecules with their unique structure and properties have been applied extensively, particularly in molecular recognition, materials, supramolecular self-assembly, catalysis, etc. (Miao et al. 2012, Mao et al. 2012, Mao and Li 2013, Zhao et al. 2013, Homden and Redshaw 2008). Recently, the functionalization of nanoparticle surfaces with macrocyclic molecules in well-defined host-guest interactions has drawn considerable attention (Tshikhudo et al. 2005, Li et al. 2006, Sawicki and Cier 2006, Leyton et al. 2004). The combination of excellent optical properties of QDs and the molecular recognition ability of host molecules is an active line of research, which has contributed to creating sophisticated sensors based on modifying the QD surface using specific ligands. Here, we summarized some of recent efforts designing macrocyclic molecules-coated QDs and their application as smart fluorescent chemosensors.

3.1 Crown-ether-Coated Quantum Dots for Fluorescence Sensing

Surface modification of the QDs by attaching a certain metal-specific ligand could be an attractive approach to achieve specific response of the QDs to metal ions. Crown ethers as a class of heterocyclic chemical compounds that consist of a ring containing several ether groups, is a good candidate. Recently, based on the well-known binding ability of crown ethers to metals, crown-coated nanoparticles have been designed as selectively ion sensors (Ho et al. 2009, Lin et al. 2002, Lin et al. 2005). The most common crown ethers are oligomers of ethylene oxide, the repeating unit being ethyleneoxy, i.e., $-CH_2CH_2O-$, which strongly bind certain cations, forming stable chelation complexes (Izatt et al. 1991, Gokel et al. 2004, Rurack et al. 2000). The denticity of polyether influences the affinity of the crown ether for various cations. For example, 18-crown-6 has high affinity for potassium cations, 15-crown-5 has high affinity for sodium cations, and 12-crown-4 has high affinity for lithium cations. In addition, based on a sandwich complex of 15-crown-5/K^+/15-crown-5, 15-crown-5 derivatives have also been reported to recognize alkali metal ions and they have the tendency to bring the donor and acceptor molecules into close proximity (Toupance et al. 1997, Flink et al. 1998, Lin et al. 2006). When one or more of the oxygen donoratoms are replaced with nitrogen atoms, aza-crown macrocyclic compounds generated. Particularly, in contrast to crown ethers, aza-crown ethers bind the guest ions using both nitrogen and oxygen donors, and they have specific complex selectivity and stability for heavy or noble metal ions (Izaatl 1985, Pond et al. 2004).

By a method of ligand exchange and fluorescence resonance energy transfer (FRET) mechanism, two different sizes (3.2 nm and 5.6 nm) of 15-crown-5 modified CdSe/ZnS QDs sensors were synthesized (Fig. 3) (Chen et al. 2006). Upon addition of K^+, two neighboring CdSe/ZnS QDs were bridged by a sandwich complex of 15-crown-5 and K^+, which resulted in the different sized QDs coming close enough together to engage in energy transfer. In the process, QD in the size of 3.2 nm (emission at 545 nm) served as the energy donor and a 5.6 nm (emission 635 nm) particle act as the energy acceptor. Quantitative analysis was realized through a ratiometric response, the emissions at 545 nm and 635 nm decreasing and increasing respectively. This recognition scheme sparked a broad spectrum of interest to fabricate an intelligent switchable sensor due to its great versatility and flexibility for future applications.

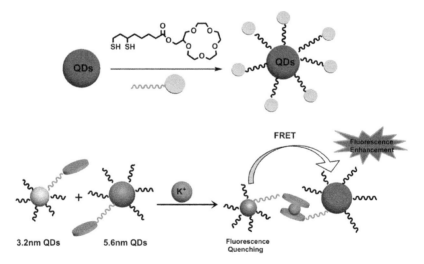

Figure 3. 15-Crown-5 functionalized CdSe/ZnS QDs for potassium ion recognition in a sandwich model.

Except the fluorescence metal ion sensor based on the mechanism of energy transfer between QDs in different size, there are also fluorescent sensor based on energy transfer between QDs and organic dyes and other nanoparticles. Recently, Lin et al. reported a ratiometric fluorescent sensor for K^+ ions based on the mechanism of fluorescence resonance energy transfer (FRET) between the synthesized 15-crown-5-ether capped CdSe/ZnS quantum dots and 15-crown-5-ether attached rhodamine B in pH 8.3 buffer solution (Fig. 4) (Lee et al. 2015). In this design, the synthesized

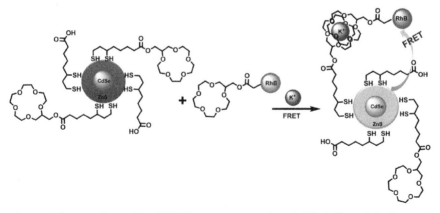

Figure 4. Schematic illustration of FRET between crown ether modified QDs and rhodamine B.

QDs (ex: 515 nm, em: 530 nm) and the crown ether attached rhodamine B (ex: 530 nm, em: 575 nm) was used as an energy donor and acceptor respectively. In the presence of potassium ions, the QD units and RhB units in aqueous solution formed the QD-RhB conjugate by the interaction between two 15-crown-5-ethers and one potassium ion. Subsequently, the fluorescence spectra displayed a decrease at 530 nm and an increase in 575 nm respectively due to the FRET from QDs units to RhB units. The fluorescent sensor showed high selectivity for potassium ions compared with other metal ions with a LOD of 4.3×10^{-6} M. This water soluble ratio metric sensor system can act as an excellent FRET probe for sensing applications especially in biological systems.

Kim and Park reported a fluorescent chemosensor for metal ions composited of a pair of aza-crown ether acridinedione-functionalized quantum dots (ACEADD-QDs) and aza-crown ether acridinedione-functionalized gold nanorods (ACEADD-GNRs) (Velu et al. 2012). The ACEADD-QDs showed two emissions at 430 nm from the acridinedione moiety and 775 nm from the CdTeSe quantum dots moiety. In acetonitrile, the emission at 430 nm was suppressed due to the photoinduced electron transfer from aza-crown ether to the acridinedione moiety. Upon the presence of Ca^{2+} or Mg^{2+}, the ACEADD-GNRs and ACEADD-QDs formed a sandwich complex induced by the metal ion. As a result, the near-infrared fluorescence of QDs was quenched effectively by the gold nanoparticles due to the nanometal surface energy transfer. The fluorescence spectra showed an increased in 430 nm and a decrease in 775 nm. The sensor displayed high selectivity towards Ca^{2+} and Mg^{2+} ions among other metal ions and provided a robust and sensitive method to detecting Ca^{2+} and Mg^{2+} with dual fluorescence emissions.

In addition, based on the mechanism of electron transfer (ET), the introduction of 1, 10-diaza-18-crown-6 to CdS:Mn/ZnS QDs has been used to specifically sense Cd^{2+} ions (Banerjee et al. 2008). The detection was based on an electron transfer process between the QDs and the ligands, and subsequent blocking of the electron transfer pathways upon exposure to Cd^{2+} ions owing to the complex formation between Cd^{2+} and 1, 10-diaza-18-crown-6. The switching on the QD emission allowed the detection of low concentrations of Cd^{2+} ions. The covalent linking of aza-macrocyclic compounds (1, 4, 7-triazacyclononane, 1, 4, 7, 10-tetraazacyclododecane, and 1, 4, 8, 11-tetraazacyclo tetradecane) on the surface of QDs has resulted in the development of a new family of zinc ions nanosensors based on the similar mechanism (Ruedas-Rama and Hall 2008). Zinc ions at a concentration lower than 2.4 µM zinc ions could be detected via fluorescence enhancement.

3.2 Porphyrin-Coated Quantum Dots for Fluorescence Sensing

Porphyrins are heterocyclic macrocycles characterized by the presence of modified pyrrole subunits interconnected at their α-carbon atoms via methine bridges. The existence of a variety of commercially available native and functionalized porphyrin structures makes them ideal building blocks for the design of electrochemical and optical sensing systems (Vlascici et al. 2008). Besides covalent bonding to organic QD shell, porphyrins can be linked directly to QD surface via coordination with metal atoms of the QD core/shell. Porphyrins can bind strongly to the surfaces of semiconductor substrates such as CdS and CdSe and their optical properties can be profoundly influenced by the presence of small gaseous molecules such as dioxygen and nitric oxide (Chrysochoos 1992, Isarov and Chrysochoos 1997, Isarov and Chrysochoos 1998, Ivanisevic and Ellis 1999). Ivanisevic and Ellis (Ivanisevic and Ellis 2000) demonstrated that films of trivalent metalloporphyrins (Fe or Mn as the metal) deposited onto single-crystal CdSe substrates could serve as transducers toward oxygen with a detection limit of approximately 0.1 atm. In addition, fluorescence quenching of CdSe QDs in the presence of trivalent metalloporphyrins, MTPPCl (TPP is tetraphenyl-porphyrin; M is Mn, Fe, Co), was developed as an NO sensor. The electron transfer from QDs to the porphyrin's aromatic system caused fluorescence quenching. When NO was added to this assembly, restoration of luminescence was observed. The authors proposed that the formation of a nitrosyl adduct of metalloporphyrin caused the ligand to donate additional electron density to the bulk of the semiconductor, thereby shrinking the depletion region and enhancing the photoluminescence intensity (Ivanisevic et al. 2000). Because these changes are readily reversible, such MTPPCl films have the potential to serve as online detectors for NO.

By the strategy of supramolecular assembly, a quantum dot (QD) associated to palladium (II) porphyrins have been developed to detect oxygen (O_2) in organic solvents (Lemon et al. 2013). In the system, the QD acted as the two-photon antenna of NIR (700–1000 nm) excitation, and the QD emission was quenched in the presence of a surface-bound Pd porphyrin via a FRET mechanism. The insensitivity of the QD emission to O_2 afforded an internal reference to establish a ratiometric O_2 response. Owing to the insensitivity of the QD to O_2, a ratiometric signal transduction mechanism may be established. Because of superior spectral overlap and efficient surface binding, the FRET efficiency in these systems was 67–94%. This result demonstrated that QD-palladium porphyrin conjugates may be used for oxygen sensing over physiological oxygen ranges.

Ivanisevic et al. reported the formation of nano-assemblies of CdSe/ZnS QDs and pyridyl-substituted porphyrins. The coordination bonding of the pyridyl group with the ZnS shell of the CdSe/ZnS nanoparticles gave rise to a strong complex formation accompanied by fluorescence quenching of CdSe/ZnS QDs. This quenching was explained partially by FRET from CdSe/ZnS nanoparticles to porphyrins (Ellis et al. 1997). Recently, a novel hybrid structure for the direct sensing of zinc ions based on CdSe QDs functionalized with tetrapyridyl-substituted porphyrin was developed by Chaniotakis' group (Fig. 5) (Frasco et al. 2010). The pyridyl-substituted porphyrins were conjugated on the surface of CdSe QDs through one or two pyridyl nitrogen atoms, while at the same time they preserved the zinc recognition capabilities of the porphyrin, relying on the nitrogens from the pyrrole or pyridyl rings, depending on the orientation of the macrocycle. Upon coordination with zinc ions, this porphyrin capping was shown to strongly contribute to the increase in the fluorescence efficiency of CdSe, via an activating interaction with the surface of the QDs. The detection limit of this nanosensor was about 0.5 μM.

Recently, Renganathan's group reported a quantum dots–cationic porphyrin nanohybrid sensor for double stranded DNA (dsDNA) (Vaishnavi and Renganathan 2014). In this sensor, the thioglycolic acid capped CdTe QDs (CdTe-QD TGA) possessed negative charge showed strong fluorescence. Meso-tetrakis (4-N-methylpyridyl) porphyrin (TMPyP), cationic porphyrin, was readily assembled on the surface of CdTe-QDs TGA through electrostatic interaction. The fluorescence of CdTe-QDs was quenched drastically by the cationic porphyrin through the photoinduced electron transfer (PET) process. Upon the addition of target DNA, the fluorescence of QDs restored sharply due to the planar cationic porphyrin intercalate or externally bind with DNA and the PET process from the QDs to porphyrin was blocked. The sensor showed high selectivity and sensitivity for double strand DNA by tracing "on-off-on" fluorescence signals utilizing fluorescence and synchronous fluorescence measurements. The increasing

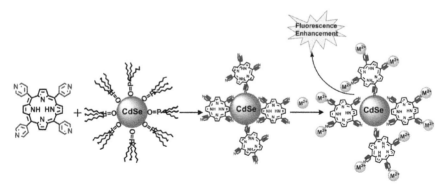

Figure 5. Porphyrin functionalized CdSe QDs for direct fluorescent sensing of metal ions.

fluorescence intensity was proportional to the concentration of calf thymus DNA in the range of 6.5×10^{-9} M to 29.6×10^{-8} M.

Furthermore, Patra's group has fabricated a fluorescence switch by alloy $(Cd_{1-x}Zn_xS)$ quantum dot (QD) in the presence of porphyrin and cucurbituril (Mandal et al. 2013). The assemblies of $Cd_{1-x}Zn_xS$ QD, 5-(4-aminophenyl)-10, 15, 20-triphenyl-21 H, 23 H-porphyrin (APTPP), have been prepared by electrostatically attaching the negatively charged QD with positively charged APTPP. The drastic photoluminescence (PL) quenching and the shortening of decay time of alloy QD in the presence of porphyrin indicated the efficient energy transfer from QD to porphyrin. Furthermore, in the presence of cucurbituril, cucurbituril acted as a receptor to bind the porphyrin (quencher) and restored the luminescence of the QD by preventing the energy transfer from QD to porphyrin. The turn off/on fluorescence of luminescent QD opened a new opportunity for designing a new optical-based sensor for bioapplications.

3.3 Cyclodextrins-Coated Quantum Dots for Fluorescence Sensing

Expanding the applications of modified QDs to develop fluorescent sensors in water media is a topic of current interest. There have been many reports of chemical sensing of ions and small molecules with QDs by analyte-induced changes in photoluminescence. Cyclodextrins (CDs) as the most widely used macrocyclic molecule are cyclic oligosaccharides that consist of six, seven, or eight glucopyranose units in α, β, and γ forms, respectively, have attracted great interest in supramolecular chemistry (Kuwabara et al. 2002, Fragoso et al. 2002, Haider et al. 2003, Stanier et al. 2002). They are well known for forming inclusion complexes with various guest molecules because of their special molecular structure—hydrophobic internal cavity and hydrophilic external surface, which made it widely developed in different sensors and separation matrices (Szejtli 1998). Since cyclodextrins are chiral, different chromatographic cyclodextrin-based chiral separation processes were accomplished. Therefore, by the method of chemical cross-link or self-assembly, cyclodextrin capped QDs can be acquired and used in different nanodevices.

Thiol groups are known to have a great affinity to the nanoparticles surface. Thiolated cyclodextrin have been widely used in the modification of metal nanoparticles (Alvarez et al. 2000, Liu et al. 1998, Nelles et al. 1996, Rojas et al. 1995). Palaniappan et al. have prepared water-soluble, perthiolated β-CD-modified CdS QDs (Palaniappan et al. 2004) and monothiolated β-CD-capped CdSe/CdS QDs (Palaniappan et al. 2006) by a one-pot approach. The surface-immobilized β-CDs retain the capability of engaging molecular recognition in aqueous solutions. These receptor-modified QDs have been successfully employed as a proof-of-concept system to control the analyte-induced fluorescence change of QDs selectively and reversibly by introducing host-guest chemistry on the surface of these particles (Fig. 6). The addition of ferrocene derivatives to the system produced fluorescence quenching by a photo-induced electron transfer mechanism, but, after the addition of adamantine molecules, a high luminescent response was observed by replacement of ferrocene with adamantine. This is an interesting example of the capability of the analytes to induce a smart controlled "on-off-on" fluorescence switching behavior in aqueous solution. In addition, this system can also be applied to redox-active organic molecules (e.g., quinone derivatives). Benzoquinone enhances the fluorescence, but the mechanism is still uncovered. The addition of ferrocene produced a decrease in the fluorescence response owing to the replacement of benzoquinone molecules, which should have a high affinity for the receptor.

Most recently, Ji and co-workers contributed an alkaline phosphatase (ALP) activity detection system based on quenching effect of the enzyme substrate and product on the β-CD-functionalized CdTe QDs (Jia et al. 2010). The CdTe QDs functionalized with GSH was covalently tethered to *p*-aminophenyl boronic acid (APBA). The CD units were linked to QDs via the covalently coupling between the boronic acid and secondary vicinal hydroxyl groups of the sugar units. In aqueous solution, the water-soluble β-CD and QDs exhibited highly fluorescence in the presence of *p*-nitrophenyl phosphate. However, the alkaline phosphatase catalytic product, *p*-nitrophenol (NP), included in the cavity of β-CD quenched the fluorescence of QDs effectively through electron transfer. The fluorescent senor provided a simple, rapid and sensitive method for the detection of ALP activity.

By a simple and convenient sonochemical method, a water-soluble CdSe/ZnS QDs using CDs as surface-coating agents have been developed.

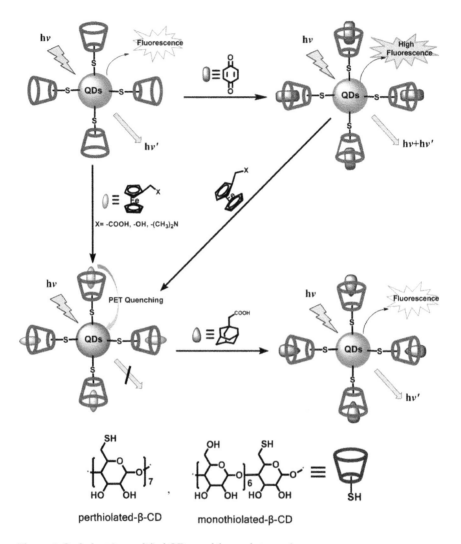

Figure 6. Cyclodextrin modified QDs used for analyte sensing.

The formation of a tri-n-octylphosphine oxide (TOPO) complex with CD transferred QDs from organic solutions to aqueous media has been developed. The n-CD-QDs (n = α, β, γ) have a high level of emission efficiency in aqueous solution (the quantum yields were about 27–45%). The fluorescence enhancement observed after the interaction of β-CD-capped QDs with polycyclic aromatic hydrocarbons (PAHs) which can be used for their analytical determination with a detection limit of 1.6×10^{-8} M (Han and Li 2008). The enhancement of the QD fluorescence after addition of anthracene was due to the generation of a new and efficient radiative path involving the bound anthracene and/or from the suppression of a nonradiative process.

By changing CD coating, n-CD-QDs allow highly sensitive determination of phenols via fluorescence intensity quenching (Fig. 7) (Li and Han 2008). The α-CD-QDs and β-CDQDs were sensitive toward *p*-nitrophenol and 1-naphthol, with corresponding detection limits of 7.92×10^{-9} M and 4.83×10^{-9} M, respectively. The quenching luminescence of the CD-QDs was attributed to the fact that the phenol molecules competed with TOPO to form inclusion complexes with CD and destabilized the TOPO-CD complexes, which lead to CDs being peeled off the surface of QDs and the fluorescence of QDs was quenched by water. The surface-tagged CD should have an important role in recognition of the phenols. Because of the fixed size of the CD cavity, only guest molecules of the appropriate size and structure can be included in the cavity. The most compatible size and steric arrangement with the CD cavity result in the observed selectivity of structural isomers (*o*-nitrophenol, *m*-nitrophenol, *p*-nitrophenol, 1-naphthol and 2-naphthol).

Moreover, the chiral cavities of CDs have been found to be very suitable for enantioselective analysis of chiral compounds (Risley and Strege 2000, Schumacher et al. 2003, Han et al. 2005). It has been demonstrated that the CD-QDs can carry out highly enantioselective fluorescent recognition of amino acids (Han and Li 2008). Within a certain concentration range, one

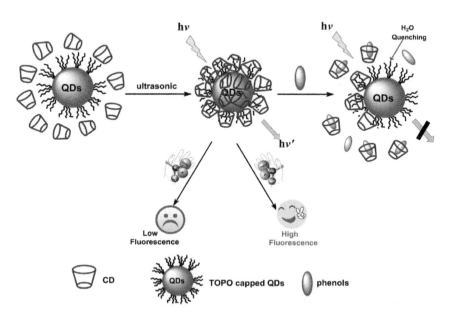

Figure 7. Synthesis of cyclodextrin functionalized CdSe/ZnS QDs for phenol and chiral recognition.

enantiomer of the chiral amino acids can increase the fluorescence intensity of the CD-QDs, whereas the other enantiomer scarcely influences the fluorescence. Such unusually high enantioselective responses make these CD-QDs very attractive as fluorescent sensors in determining enantiomeric compounds.

Additionally, CD can also affect the interactions between the organic molecules and the QDs. It has been reported that photoinduced electron transfer between 1-anilino-8-naphthalene and CdS colloidal systems increased 12.5 times in the presence of β-CD (Yang et al. 2001). In addition, supramolecular nano-sensitizers combining of CdTe QDs and CDs were reported to fluorescence quenching analysis method to PAHs (Qu and Li 2009). In the presence of CDs, PAHs embedded into the hydrophobic cavity of CDs to form the water soluble host-guest complex CD/PAHs, which resulted in PAHs easily interacting with hydrophilic CdTe QDs. The generation of a new and efficient nonradiative path involving the bound PAHs and/or from the suppression of a radiative process caused the fluorescence quenching.

The fluorescence resonance energy transfer (FRET) mechanism is central to the design of luminescent sensors with organic fluorophores (Medintz et al. 2003, Zhang et al. 2005, Algar and Krull 2009). CD-capped QDs was designed as energy donor and organic dyes such as Nile Red (Rakshit and Vasudevan 2008) as the energy acceptor. The inclusion of an organic dye within the cavities of the capping CD to form a noncovalent CD/dye supramolecular assembly was critical to FRET process, as the host-guest complexes of CD/dye resulted in QDs and dye coming close enough together to engage in energy transfer. More recently, Willner's group reported FRET-based competitive assay using β-CD-modified CdSe/ZnS QDs as sensors and chiroselective sensors (Fig. 8) (Freeman et al. 2009). The FRET between the QDs and Rhodamine B incorporated in the β-CD receptor sites was used for the competitive analysis of adamantine carboxylic acid and *p*-hydroxytoluene. Also, the dye-incorporated β-CD-modified QDs could be used for the chiral selective discrimination between D, L-phenylalanine and D, L-tyrosine by optical method.

3.4 Calixarene-Coated Quantum Dots for Fluorescence Sensing

Calixarenes as the third best host molecules after crown ethers and cyclodextrins, are cavity-shaped cyclic phenol molecules capable of forming host-guest complexes with a variety of inorganic and organic guests, such as ions and/or neutral molecules. Considering the excellent host ability of calixarenes, it has been widely used as surface coating agents combined with QDs for the development of nanosensors (Bott et al. 1986, Arduini et al. 2005, Arduini et al. 1992, Li et al. 2005, Tshikhudo et al. 2005). For example,

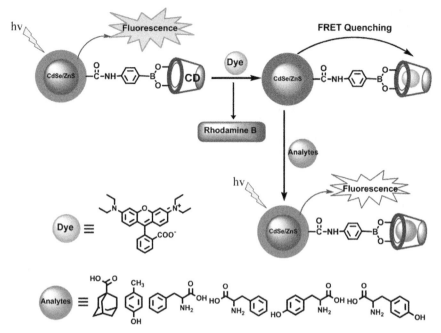

Figure 8. Sensing substrates by a competitive FRET assay using cyclodextrin-modified QDs with receptor-bound rhodamine B.

sulfur calixarene modified CdSe/ZnS QDs has been developed as a sensor for Hg^{2+} based on a selective fluorescence quenching (Li et al. 2007). This system showed a detection limit of 15 nM and the influence of other metal ions was very weak,with the exception of Pb^{2+} at higher concentrations, which produced a measurable quenching of QD fluorescence.

Menon's group developed an optical nanoprobe for menadione (VK3) based on calixarene capped ZnS quantum dots (Joshi et al. 2012). The ZnS QDs was capped by *p*-sulfonatocalix[4]arene. Upon the addition of menadione, the *p*-sulfonatocalix[4]arene included menadione through host-guest interaction. As a consequence, the fluorescence of QDs was quenched with the increased concentration of menadione. The fluorescence response to menadione showed a good linear relationship in the range of 5 $\times 10^{-9}$–1×10^{-6} M with a LOD value of 80 nM. This ultrasensitive probe was sucessfully determined menadione in pharmaceutical commercial samples with almost 100% recovery.

Also amphiphilic calixarene derivatives have been used as a capping ligand for QDs, giving rise to water soluble QDs (Jin et al. 2005, Jin et al. 2006, Basilio et al. 2013). The calixerene forms a bilayer structure with TOPO molecules surrounding CdSe/ZnS QDs by hydrophobic interaction. The optical properties of CdSe/ZnS QDs could be controlled by changing the surface coating layer with different oligomer sizes of calixarene derivatives without changing the particle sizes. The use of p-sulfonatocalix[4] arene-coated CdSe/ZnS QDs was reported for the detection of acetylcholine (ACh) based on the fluorescence quenching observed after binding the ammonium cation of the acethylcholine with the calixarene (Fig. 9) (Jin et al. 2005). ACh molecules interacted with the semiconductor surface to reduce the core electron–hole recombination, which resulted in the observed fluorescence quenching. The fluorescence quenching of the QDs is quite selective to acetylcholine among the anionic and neutral neurotransmitters (L-glutamic acid and GABA).

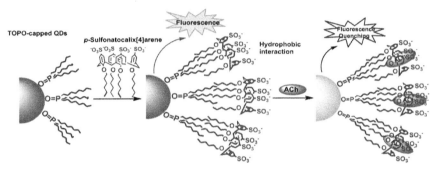

Figure 9. Synthesis of water-soluble CdSe/ZnS QDs by surface coating with p-sulfonatocalix[4] arene and their application as acetylcholine (ACh) sensors.

In the above cases of calixarene modifiers, the interfacial process is driven by the hydrophobic van der Waals interactions between the primary alkane of the stabilizing ligand like TOPO and the secondary calixarene layer, resulting in bilayer structures. It is undesirable that the loose bilayer structures based on the weak interaction of van der Waals are not quite stable. Therefore, it is a challenge to construct stable calixarene modified QDs.

It is known that QDs often need to be coated with inert materials to improve the chemical and photochemical stabilities, especially in aqueous systems. Silica is a good option for the inert material, because not only does it impede the diffusion of charge carriers generated upon photoexcitation as well as the diffusion of oxygen from the environment, but it also prevents the coated NPs from coagulating in aqueous dispersions (Algar et al. 2011). By using the well-known Stober method, Liz-Marza´n and co-workers have succeeded in coating aqueous CdS QDs stabilized by sodium citrate to obtain core-shell structured SiO_2/CdS particles. More recently, Nann and co-workers further demonstrated that the silica coating on organic-soluble QDs by the Stober process could also lead to well-defined structures (Resch-Genger et al. 2008). It is reasonable to believe that the sol-gel technique in the Stober method will be a powerful tool to introduce artificial receptor moieties onto the surface of silica sphere.

Consequently, highly luminescent calix[4]arene-coated CdTe/SiO$_2$ QDs prepared via sol-gel technique in aqueous media were employed as a highly sensitive luminescence probe for optical recognition of methomyl by Li and Qu (Li and Qu 2007). Enhancement of the luminescence emitted by the synthesized nanoparticles allows the determination of methomyl at concentrations as low as 0.08 µM. Moreover, the authors demonstrated that calix[4,7]arene-coated CdTe/SiO$_2$ QDs could be applied as selective fluorescent probes for the determination of PAHs by changing the surface coating layer with different oligomer sizes of calixarenes via enhancement of the response of the fluorescence intensity of QDs (Fig. 10) (Li and Qu 2007). The calix[4]arene-coated QDs turned out to be sensitive to the presence of anthracene (detection limit 2.45 × 10^{-8} M); meanwhile the calix[7]arene-coated QDs were sensitive to the presence of pyrene (detection limit 2.94 × 10^{-8} M). A luminescence intensity enhancement mechanism was proposed

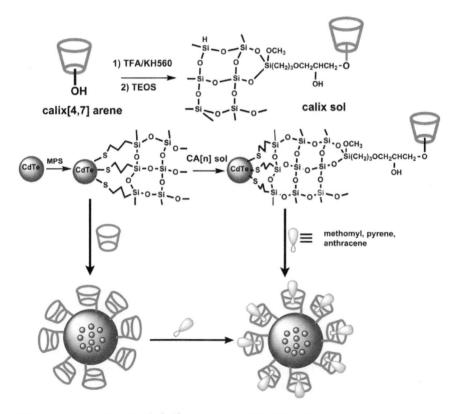

Figure 10. Preparation of calix[4,7]arene-coated CdTe/SiO$_2$ QDs via sol-gel technique and their application as pesticide and PAH sensors.

in which the analytes induced ordered orientation and/or enhanced the conformational rigidity of the surface substituent and suppressed the quenching path to the medium by effective core protection and thus increase the luminescence intensity.

p-Sulfonatocalixarene could be used as a stable ligand to prepared water-soluble nanoparticles via chemical bonding (Ben-Ishay and Gedanken 2007, Chen et al. 2007). A recent application of highly fluorescent *p*-sulfonatocalix[4,6]arene-coated QDs as probes for amino acids (e.g., methionine and phenylalanine) was reported with enhanced fluorescence and a detection limit of 3–4 μM. Structural changes of the host molecules around the QDs seemed to play an important role in the fluorescence changes. Increased rigidity of the system could suppress nonradiative paths and protect the core (Wang et al. 2008). Simple and rapid optical switch detection systems for pesticides (Qu et al. 2009) and amino acids (Li and Wang 2008) combining *p*-sulfonatocalix[4]arene as the specific detecting switch and QDs as the signal reporter were established by Li's group.

In consideration of tuning the cavity size of calixarene can be included different size targets, large cavity calixarene for macromolecule with host-guest interaction has also been studied. Recently, Simonet's group exploited a calix[8]arene Coated CdSe/ZnS quantum dots optical nanosensor for C60 (Carrillo-Carrion et al. 2011). The calix[8]arene-coated CdSe/ZnS QDs showed intensive fluorescence emission with high quantum yield. The hydrophobic pockets of the calixarene molecules interacted with the aliphatic chains of the TOPO present on the surface of QDs. When C60 added to the solution, the calix[8]arene adapted conical conformation to accommodate the guest in the cavity in an induced-fit manner. As a result, the quencher C60 quenched the fluorescence of calix[8]arene-coated CdSe/ZnS QDs effectively. Based on this sensing mechanism, the developed method could be applicable for determination of C60 in river water samples.

4. Perspective of the Future Research

This chapter highlighted some of the recent developments of macrocyclic molecules functionalized inorganic QDs as a smart optical nanomaterial in the application of fluorescent sensor. Owing to host-guest interaction of macrocyclic molecules, the modification and functionalized of QDs with macrocyclic molecules is quite simple, rapid and efficient to develop as the smart sensor. The macrocyclic molecules not only can serve as ligands to stabilize the photophysics properties of QDs but also can afford the recognition units for a special analyze. The combination of QDs and host molecule provided a flexible and versatile method to determine

DNA, protein and other biomolecules. Highly fluorescent and selective nanosensors were achieved due to a synergic manner of the typical optical properties of QDs and selective recognition properties of the host molecules.

Among the various sensing mechanisms including electron transfer, energy transfer, electrostatic interaction, QDs based FRET assay has obtained broad attention and can be developed as an effective way to fabricate smart materials. As FRET is sensitive to molecular rearrangements on the 1–10 nm range, researchers have long used this photophysical process to monitor intracellular interactions and binding events (Miyawaki 2003). Due to the FRET assay is usually showed with "turn on" or dual emission fluorescence response, the FRET assay can be utilized in selective and sensitive detection in complex environment with minimal interference. Recent years, QDs as FRET donor has also been reported (Willard et al. 2001). Compared with other fluorescent nanomaterials and organic fluorescent dyes, one unique advantage is obvious: QDs donor emission could be size-tuned to improve spectral overlap with a particular FRET acceptor which can improve the FRET efficiency. Thereafter, macrocyclic molecules functionalized QDs based on FRET mechanism will be one of the most promising methods for fluorescence-based biosensing.

With the development of chemosensors based on macrocyclic molecules-coated QDs gradually to mature, the use of QDs functionalized with host molecules as smart sensors in fluorescence imaging will be the focal point in future's work, and increasing trend on converging QDs with other nanotechnology and biotechnologies can be more and more favored. The emission of QDs can easily be tuned to near-infrared region which is suitable for deep tissue imaging. Furthermore, the small size of encapsulated QDs (20–30 nm) indicated the QDs can be loaded into cells readily through endocytosis, electroporation, direct microinjection. Peculiarly, cellular components labelled with QDs including nucleus, mitochondria has also been reported (Derfus et al. 2004, Guo et al. 2003). Despite the widely application of QDs in imaging *in vitro* and *in vivo*, the use of QDs with host molecule for fluorescence imaging is rarely reported. With the merits of QDs and host-guest interaction, we envision that smart fluorescent sensors composited of QDs and macrocyclic molecules will provide a powerful tool to study inorganic and organic species in biological process with fluorescence imaging *in vitro* and *in vivo*.

References

Ai, X., Y. Wang, X. Hou, L. Yang, C. Zheng and L. Wu. 2013. Advanced oxidation using Fe_3O_4 magnetic nanoparticles and its application in mercury speciation analysis by high performance liquid chromatography-cold vapor generation atomic fluorescence spectrometry. Analyst. 138: 3494–3501.

Algar, W.R. and U.J. Krull. 2009. Toward a multiplexed solid-phase nucleic acid hybridization assay using quantum dots as donors in fluorescence resonance energy transfer. Anal. Chem. 81: 4113–4120.

Algar, W.R., K. Susumu, J.B. Delehanty and I.L. Medintz. 2011. Semiconductor quantum dots in bioanalysis: crossing the valley of death. Anal. Chem. 83: 8826–8837.

Alivisatos, A.P. 1996. Perspectives on the physical chemistry of semiconductor nanocrystals. J. Phys. Chem. 100: 13226–13239.

Alivisatos, A.P. 2004. The use of nanocrystals in biological detection. Nat. Biotechnol. 22: 47–52.

Alvarez, J., J. Liu, E. Roman and A.E. Kaifer. 2000. Water-soluble platinum and palladium nanoparticles modified with thiolated β-cyclodextrin. Chem. Commun. 13: 1151–1152.

Arduini, A., A. Casnati, M. Fabbi, P. Minari, A. Pochini, A.R. Sicuri and R. Ungaro. 1992. New shapes for selective molecular recognition from calixarenes. Supramol. Chem. 371: 31–50.

Arduini, A., D. Demuru, A. Pochini and A. Secchi. 2005. Recognition of quaternary ammonium cations by calix[4]arene derivatives supported on gold nanoparticles. Chem. Commun. 645–647.

Bagalkot, V., L. Zhang, E. Levy-Nissenbaum, S. Jon, P.W. Kantoff, R. Langer and O.C. Farokhzad. 2007. Quantum dot–aptamer conjugates for synchronous cancer imaging, therapy, and sensing of drug delivery based on bi-fluorescence resonance energy transfer. Nano Lett. 7: 3065–3070.

Banerjee, S., S. Kar and S. Santra. 2008. A simple strategy for quantum dot assisted selective detection of cadmium ions. Chem. Commun. 3037–3039.

Basilio, N., V. Francisco and L. Garcia-Rio. 2013. Aggregation of *p*-sulfonatocalixarene-based amphiphiles and supra-amphiphiles. Int. J. Mol. Sci. 14: 3140–3157.

Bau, L., P. Tecilla and F. Mancin. 2011. Sensing with fluorescent nanoparticles. Nanoscale. 3: 121–133.

Ben-Ishay, M.L. and A. Gedanken. 2007. Difference in the bonding scheme of calix[6]arene and *p*-sulfonic calix[6]arene to nanoparticles of Fe_2O_3 and Fe_3O_4. Langmuir. 23: 5238–5242.

Beulen, M.W.J., J. Bugler, M.R. Jong, B. Lammerink, J. Huskens, H. Schonherr, G.J. Vancso, B.A. Boukamp, H. Wieder, A. Offenhauser, W. Knoll, F.C.J.M. Veggel and D.N. Reinhoudt. 2000. Host-guest interactions at self-assembled monolayers of cyclodextrins on gold. Chem. Eur. J. 6: 1176–1183.

Bott, S.G., A.W. Coleman and J.L. Atwood. 1986. Inclusion of both cation and neutral molecule by a calixarene. Structure of the [*p*- tert-butylmethoxycalix[4]arene-sodium-toluene]+ cation. J. Am. Chem. Soc. 108: 1709–1710.

Bruchez, M., M. Moronne, P. Gin, S. Weis and A.P. Alivisatos. 1998. Semiconductor nanocrystals as fluorescent biological labels. Science 281: 2013–2016.

Brus, L.E. 1986. Electronic wave functions in semiconductor clusters: experiment and theory. J. Phys. Chem. 90: 2555–2560.

Callan, J.F., A.P.D.E. Silva, R.C. Mulrooney and B.M.C. Caughan. 2007. Luminescent sensing with quantum dots. J. Incl. Phen. Macr. Chem. 58: 257–262.

Carrillo-Carrion, C., B. Lendl, B.M. Simonet and M. Valcarcel. 2011. Calix[8]arene coated CdSe/ZnS quantum dots as C60-nanosensor. Anal. Chem. 83: 8093–8100.

Chen, C., C. Cheng, C. Lai, P. Wu, K. Wu, P. Chou, Y. Chou and H. Chiu. 2006. Potassium ion recognition by 15-crown-5 functionalized CdSe/ZnS quantum dots in H_2O. Chem. Commun. 263–265.

Chen, M., G.W. Diao and X.M. Zhou. 2007. Formation of nanoclusters of cuprous oxide with bridge linker of *p*-sulfanated calix[8]arene host and its mechanism. Nanotechnology 18: 275606–275615.

Chen, Y. and Z. Rosenzweig. 2002. Luminescent CdS quantum dots as selective ion probes. Anal. Chem. 74: 5132–5138.

Chen, Y., I.A. Banerjee, L. Yu, R. Djalali and H. Matsui. 2004. Attachment of ferrocene nanotubes on beta-cyclodextrin self-assembled monolayers with molecular recognitions. Langmuir. 20: 8409–8413.

Chrysochoos, J. 1992. Recombination luminescence quenching of nonstoichiometric cadmium sulfide clusters by ZnTPP. J. Phys. Chem. 96: 2868–2873.

Derfus, A.M., W.C.W. Chan and S.N. Bhatia. 2004. Intracellular delivery of quantum dots for live cell labeling and organelle tracking. Adv. Mater. 16: 961–966.

Dickert, F.L. and A. Haunschild. 1993. Sensor materials for solvent vapor detection-donor-acceptor and host-guest interactions. Adv. Mater. 5: 887–895.

Ekimov, A.I. and A.A. Onushchenko. 1982. Quantum size effect in the optical-spectra of semiconductor micro-crystals. Soviet Physics Semiconductors-USSR. 16: 775–778.

Ellis, A.B., R.J. Brainard, K.D. Kepler, D.E. Moore, E.J. Winder, T.F. Kuech and G.C. Lisensky. 1997. Modulation of the photoluminescence of semiconductors by surface adduct formation: an application of inorganic photochemsitry to chemical sensing. J. Chem. Educ. 74: 680–684.

Esteve-Turrillas, F.A. and A. Abad-Fuentes. 2013. Applications of quantum dots as probes in immunosensing of small-sized analytes. Biosens. Bioelectron. 41: 12–29.

Fletcher, J.A. 1999. DNA *in situ* hybridization as an adjunct in tumor diagnosis. Am. J. Clin. Pathol. 112: S11–18.

Flink, S., B.A. Boukamp, A. van den Berg, F.C.J.M. van Veggel and D.N. Reinhoudt. 1998. Electrochemical detection of electrochemically inactive cations by self-assembled monolayers of crown ethers. J. Am. Chem. Soc. 120: 4652–4657.

Fragoso, A., J. Caballero, E. Almirall, R. Villalonga and R. Cao. 2002. Immobilization of adamantane-modified cytochrome c at electrode surfaces through supramolecular interactions. Langmuir. 18: 5051–5054.

Frasco, M.F. and N. Chaniotakis. 2009. Semiconductor quantum dots in chemical sensors and biosensors. Sensors 9: 7266–7286.

Frasco, M.F., V. Vamvakaki and N. Chaniotakis. 2010. Porphyrin decorated CdSe quantum dots for direct fluorescent sensing of metal ions. J. Nanopart. Res. 12: 1449–1458.

Freeman, R. and I. Willner. 2012. Optical molecular sensing with semiconductor quantum dots (QDs). Chem. Soc. Rev. 41: 4067–4085.

Freeman, R., T. Finder, L. Bahshi and I. Willner. 2009. Beta-cyclodextrin-modified CdSe/ZnS quantum dots for sensing and chiroselective analysis. Nano. Lett. 9: 2073–2076.

Galian, R.E. and M. de la Guardia. 2009. The use of quantum dots in organic chemistry. TrAC Trends Anal. Chem. 28: 279–291.

Gao, X.H., L.L. Yang, J.A. Petros, F.F. Marshal, J.W. Simons and S.M. Nie. 2005. *In vivo* molecular and cellular imaging with quantum dots. Curr. Opin. Biotechnol. 16: 63–72.

Giepmans, B.N., T.J. Deerinck, B.L. Smarr, Y.Z. Jones and M.H. Ellisman. 2005. Correlated light and electron microscopic imaging of multiple endogenous proteins using Quantum dots. Nat. Methods 2: 743–749.

Gill, R., M. Zayats and I. Willner. 2008. Semiconductor quantum dots for bioanalysis. Angew. Chem. Int. Ed. 47: 7602–7625.

Gokel, G.W., W.M. Leevy and M.E. Weber. 2004. Crown ethers: sensors for ions and molecular scaffolds for materials and biological models. Chem. Rev. 104: 2723–2750.

Green, M. 2010. The nature of quantum dot capping ligands. J. Mater. Chem. 20: 5797–5809.

Guo, W., J.J. Li, Y.A. Wang and X. Peng. 2003. Conjugation chemistry and bioapplications of semiconductor box nanocrystals prepared via dendrimer bridging. Chem. Mater. 15: 3125–3133.

Haider, J.M., R.M. Williams, L.D. Cola and Z. Pikramenou. 2003. Vectorial control of energy-transfer processes in metallocyclodextrin heterometallic assemblies. Angew. Chem. Int. Ed. 42: 1830–1833.

Han, C. and H. Li. 2008. Novel beta-cyclodextrin modified quantum dots as fluorescent probes for polycyclic aromatic hydrocarbons. Chin. Chem. Lett. 19: 215–218.

Han, C. and H. Li. 2008. Chiral recognition of amino acids based on cyclodextrin-capped quantum dots. Small. 4: 1344–1350.

Han, X., Q. Zhong, D. Yue, C. Della, R.C.N. Larock and D.W. Armstrong. 2005. Charge-transfer interaction of iodine with some polyamidoamines. Chromatographia 61: 205–211.

Homden, D.M. and C. Redshaw. 2008. The use of calixarenes in metal-based catalysis. Chem. Rev. 108: 5086–5130.

Ho, M., J. Hsieh, C. Lai, H. Peng, C. Kang, I. Wu, C. Lai, Y. Chen and P. Chou. 2009. 15-Crown-5 functionalized Au nanoparticles synthesized via single molecule exchange on silica nanoparticles: its application to probe 15-crown-5/K$^+$/15-crown-5 "sandwiches" as linking mechanisms. J. Phys. Chem. C. 113: 1686–1693.

Isarov, A.V. and J. Chrysochoos. 1997. Optical and photochemical properties of nonstoichiometric cadmium sulfide nanoparticles: surface modification with copper (II) ions. Langmuir. 13: 3142–3149.

Isarov, A.V. and J. Chrysochoos. 1998. Interfacial electron transfer from nonstoichiometric cadmium sulfide nanoparticles to free and complexed copper(II) ions in 2-propanol. Proc. Indian Acad. Sci., Chem. Sci. 110: 277–295.

Ivanisevic, A. and A.B. Ellis. 1999. Photoluminescent properties of cadmium selenide in contact with solutions and films of metalloporphyrins. Evidence for semiconductor-mediated adduct formation of oxygen with metalloporphyrins at room temperature. J. Phys. Chem. B. 103: 1914–1919.

Ivanisevic, A. and A.B. Ellis. 2000. Linker-enhanced binding of metalloporphyrins to cadmium selenide and implications for oxygen detection. Langmuir. 16: 7852–7858.

Ivanisevic, A., M.F. Reynolds, J.N. Burstyn and A.B. Ellis. 2000. Photoluminescent properties of cadmium selenide in contact with solutions and films of metalloporphyrins: nitric oxide sensing and evidence for the aversion of an analyte to a buried semiconductor–film interface. J. Am. Chem. Soc. 122: 3731–3738.

Izaatl, R.M. 1985. Thermodynamic and kinetic data for cation-macrocycle interactiont. Chem. Rev. 85: 271–339.

Izatt, R.M., K. Pawlak, J.S. Bradshaw and R.L. Bruening. 1991. Thermodynamic and kinetic data for macrocycle interaction with cations and anions. Chem. Rev. 91: 1721–2085.

Jia, L., J.P. Xu, D. Li, S.P. Pang, Y. Fang, Z.G. Song and J. Ji. 2010. Fluorescence detection of alkaline phosphatase activity with β-cyclodextrin-modified quantum dots. Chem. Commun. 46: 7166–7168.

Jin, T., F. Fujii, H. Sakata, T. Mamoru and M. Kinjo. 2005. Amphiphilic *p*-sulfonatocalix[4] arene-coated CdSe/ZnS quantum dots for the optical detection of the neurotransmitter acetylcholine. Chem. Commun. 4300–4302.

Jin, T., F. Fujii, H. Sakata, M. Tamura and M. Kinjo. 2005. Calixarene-coated water-soluble CdSe-ZnS semiconductor quantum dots that are highly fluorescent and stable in aqueous solution. Chem. Commun. 2829–2831.

Jin, T., F. Fujii, E. Yamada, Y. Nodasaka and M. Kinjo. 2006. Control of the optical properties of quantum dots by surface coating with calix[n]arene carboxylic acids. J. Am. Chem. Soc. 128: 9288–9289.

Jin, W.J., M.T. Fernandez-Arguelles, J.M. Costa-Fernandez, R. Pereiro and A. Sanz-Medel. 2005. Photoactivated luminescent CdSe quantum dots as sensitive cyanide probes in aqueous solutions. Chem. Commun. 21: 883–885.

Joshi, K.V., B.K. Joshi, A. Pandya, P.G. Sutariya and S.K. Menon. 2012. Calixarene capped ZnS quantum dots as an optical nanoprobe for detection and determination of menadione. Analyst. 137: 4647–4650.

Kamat, P.V. 2007. Meeting the clean energy demand: nanostructure architectures for solar energy conversion. J. Phys. Chem. C. 111: 2834–2860.

Kaur, N., N. Singh, B. McCaughan and J.F. Callan. 2010. AND molecular logic using semiconductor quantum dots. Sens. Actuators B. 144: 88–91.

Kitano, H., Y. Taira and H. Yamamoto. 2000. Inclusion of phthalate esters by a self-assembled monolayer of thiolated cyclodextrin on a gold electrode. Anal. Chem. 72: 2976–2980.

Klostranec, J.M. and W.C.W. Chan. 2006. Quantum dots in biological and biomedical research: recent progress and present challenges. Adv. Mater. 18: 1953–1964.

Kuwabara, T., T. Aoyagi, M. Takamura, A. Matsushita, A. Nakamura and A. Ueno. 2002. Heterodimerization of dye-modified cyclodextrins with native cyclodextrins. J. Org. Chem. 67: 720–725.

Lee, H.L., N. Dhenadhayalan and K.C. Lin. 2015. Metal ion induced fluorescence resonance energy transfer between crown ether functionalized quantum dots and rhodamine B: selectivity of K^+ ion. RSC Adv. 5: 4926–4933.

Lehn, J.M., J.L. Atwood, J.E.D. Davies, D.D. MacNicol and F. Vogtle. 1996. Molecular recognition: receptors for cationic guests. Comprehensive Supramolecular Chemistry 1: 831.

Lemon, C.M., E. Karnas, M.G. Bawendi and D.G. Nocera. 2013. Two-photon oxygen sensing with quantum dot-porphyrin conjugates. Inorg. Chem. 52: 10394–10406.

Leyton, P., S. Sanchez-Cortes, J. Garcia-Ramos and M. Campos-Vallette. 2004. Selective molecular recognition of polycyclic aromatic hydrocarbons (PAHs) on calix[4]arene-functionalized Ag nanoparticles by surface-enhanced raman scattering. J. Phys. Chem B. 108: 17484–17490.

Li, H. and C. Han. 2008. Sonochemical synthesis of cyclodextrin-coated quantum dots for optical detection of pollutant phenols in water. Chem. Mater. 20: 6053–6059.

Li, H. and F. Qu. 2007. Selective inclusion of polycyclic aromatic hydrocarbons (PAHs) on calixarene coated silica nanospheres englobed with CdTe nanocrystals. J. Mater. Chem. 17: 3536–3544.

Li, H. and F. Qu. 2007. Synthesis of CdTe quantum dots in sol-gel-derived composite silica spheres coated with calix[4]arene as luminescent probes for pesticides. Chem. Mater. 19: 4148–4154.

Li, H. and X. Wang. 2008. Tuning the fluorescence response of surface modified CdSe quantum dots between tyrosine and cysteine by addition of *p*-sulfonato-calix[4]arene. Photochem. Photobiol. Sci. 7: 694–699.

Li, H., Y. Chen, Z. Zeng, C. Xie and X. Yang. 2005. p-tert-Butylcalix[4]arene-1, 3-bis(allyloxyethoxy) ether coated capillaries for open-tubular electrochromato-graphy. Anal. Sci. 6: 717–720.

Li, H., Y. Zhang, X. Wang, D. Xiong and Y. Bai. 2007. Calixarene capped quantum dots as luminescent probes for Hg^{2+} ions. Mater. Lett. 61: 1474–1477.

Li, L., X. Sun, Y. Yang, N. Guan and F. Zhang. 2006. Synthesis of anatase TiO_2 nanoparticles with beta-cyclodextrin as a supramolecular shell. Chem. Asian J. 1: 664–668.

Lin, S., S. Wu and C. Chen. 2006. A simple strategy for prompt visual sensing by gold nanoparticles: general applications of interparticle hydrogen bonds. Angew. Chem. Int. Ed. 45: 4948–4951.

Lin, S.Y., S.W. Liu, C.M. Lin and C. Chen. 2002. Recognition of potassium ion in water by 15-crown-5 functionalized gold nanoparticles. Anal. Chem. 74: 330–335.

Lin, S.Y., C.H. Chen, M.C. Lin and H.F. Hsu. 2005. A cooperative effect of bifunctionalized nanoparticles on recognition: sensing alkali ions by crown and carboxylate moieties in aqueous media. Anal. Chem. 77: 4821–4828.

Liu, J., R.L. Xu and A.E. Kaifer. 1998. *In situ* modification of the surface of gold colloidal particles. Preparation of cyclodextrin-based rotaxanes supported on gold nanospheres. Langmuir. 14: 7337–7339.

Li, Y., J. Zhou, C. Liu and H. Li. 2012. Composite quantum dots detect Cd(II) in living cells in a fluorescence "turning on" mode. J. Mater. Chem. 22: 2507–2511.

Mandal, S., M. Rahaman, S. Sadhu, S.K. Nayak and A. Patra. 2013. Fluorescence switching of quantum dot in quantum dot-porphyrin-cucurbit [7] uril assemblies. J. Phys. Chem. C. 117: 3069–3077.

Mao, X.W., D.M. Tian and H.B. Li. 2012. *p*-Sulfonated calix[6]arene modified graphene as a highly selective fluorescent probe for L-carnitine. Chem. Commun. 48: 4851–4853.

Mao, X.W. and H.B. Li. 2013. Chiral imaging in living cells with a functionalized graphene oxide. J. Mater. Chem. B. 1: 4267–4272.

Medintz, I.L., A.R. Clapp, H. Mattoussi, E.R. Goldman, B. Fisher and J.M. Mauro. 2003. Self-assembled nanoscale biosensors based on quantum dot FRET donors. Nat. Mater. 2: 630–638.

Medintz, I.L., H.T. Uyeda, E.R. Goldman and H. Mattoussi. 2005. Quantum dot bioconjugates for imaging, labelling and sensing. Nat. Mater. 4: 435–446.

Miao, F.J., J. Zhou, D.M. Tian and H.B. Li. 2012. Enantioselective recognition of mandelic acid with (R)-1,1-Bi-2-naphthol-linked calix[4]arene via fluorescence and dynamic light scattering. Org. Lett. 14: 3572–3575.

Miyawaki, A. 2003. Visualization of the spatial and temporal dynamics of intracellular signaling. Dev. Cell. 4: 295–305.

Murray, C.B., C.R. Kagan and M.G. Bawendi. 2000. Synthesis and characterization of monodisperse nanocrystals and close-packed nanocrystal assemblies. Annu. Rev. Mater. Sci. 30: 545–610.

Nelles, G., M. Weisser, R. Back, P. Wohlfart, G. Wenz and S. Mittler-Neher. 1996. Controlled orientation of cyclodextrin derivatives immobilized on gold surfaces. J. Am. Chem. Soc. 118: 5039–5046.

Palaniappan, K., C. Xue, G. Arumugam, S.A. Hackney and J. Liu. 2006. Water-soluble, cyclodextrin-modified CdSe-CdS core-shell structured quantum dots. Chem. Mater. 18: 1275–1280.

Palaniappan, K., S.A. Hackney and J. Liu. 2004. Supramolecular control of complexation-induced fluorescence change of water-soluble, beta-cyclodextrin-modified CdS quantum dots. Chem. Commun. 2704–2705.

Pond, S.J.K., O. Tsutsumi, M. Rumi, O. Kwon, E. Zojer, J. Brédas, S.R. Marder and J.W. Perry. 2004. Metal-ion sensing fluorophores with large two-photon absorption cross sections: aza-crown ether substituted donor-acceptor-donor distyryl-benzenes. J. Am. Chem. Soc. 126: 9291–9306.

Qu, F. and H. Li. 2009. Selective molecular recognition of polycyclic aromatic hydrocarbons using CdTe quantum dots with cyclodextrin as supramolecular nano-sensitizers in water. Sens. Act. B. 135: 499–505.

Qu, F., X. Zhou, J. Xu, H. Li and G. Xie. 2009. Luminescence switching of CdTe quantum dots in presence of *p*-sulfonatocalix[4]arene to detect pesticides in aqueous solution. Talanta. 78: 1359–1363.

Rakshit, S. and S. Vasudevan. 2008. Resonance energy transfer from beta-cyclodextrin-capped ZnO:MgO nanocrystals to included Nile Red guest molecules in aqueous media. ACS Nano. 2: 1473–1479.

Resch-Genger, U., M. Grabolle, S. Cavaliere-Jaricot, R. Nitschke and T. Nann. 2008. Quantum dots versus organic dyes as fluorescent labels. Nat. Methods 5: 763–775.

Risley, D.S. and M.A. Strege. 2000. Chiral separations of polar compounds by hydrophilic interaction chromatography with evaporative light scattering detection. Anal. Chem. 72: 1736–1739.

Rojas, M.T., R. Koniger, J.F. Stoddart and A.E. Kaifer. 1995. Supported monolayers containing preformed binding sites. Synthesis and interfacial binding properties of a thiolated p-cyclodextrin derivative. J. Am. Chem. Soc. 117: 336–343.

Ruedas-Rama, M.J. and E.A.H. Hall. 2008. Azamacrocycle activated quantum dot for zinc ion detection. Anal. Chem. 80: 8260–8268.

Rurack, K., W. Rettig and U. Resch-Genger. 2000. Unusually high cation-induced fluorescence enhancement of a structrally simple intrinsic fluoroionophore with a donor-acceptor-donor constitution. Chem. Commun. 407–408.

Sawicki, R. and L. Cier. 2006. Evaluation of mesoporous cyclodextrin-silica nanocomposites for the removal of pesticides from aqueous media. Environ. Sci. Technol. 40: 1978–1983.

Schumacher, D.D., C.R. Mitchell, T.L. Xiao, R.V. Rozhkov, R.C. Larock and D.W. Armstrong. 2003. Cyclodextrin-based liquid chromatographic enantiomeric separation of chiral dihydrofurocoumarins, an emerging class of medicinal compounds. J. Chromatogr. A 1011: 37–47.

Smith, A.M., G. Ruan, M.N. Rhyner and S. Nie. 2006. Engineering luminescent quantum dots for *in vivo* molecular and cellular imaging. Ann. Biomed. Eng. 34: 3–14.

Szejtli, J. 1998. Introduction and general overview of cyclodextrin chemistry. Chem. Rev. 98: 1743–1753.

Somers, R.C., M.G. Bawendi and D.G. Nocera. 2007. CdSe nanocrystal based chem-/bio-sensors. Chem. Soc. Rev. 36: 579–591.

Stanier, C.A., S.J. Alderman, T.D.W. Claridge and H.L. Anderson. 2002. Unidirectional photoinduced shuttling in a rotaxane with a symmetric stilbene dumbbell. Angew. Chem. Int. Ed. 41: 1769–1772.

Toupance, T., H. Benoit, D. Sarazin and J. Simon. 1997. Ionoelectronics. pillar like aggregates formed via highly nonlinear complexation processes. A light-scattering study. J. Am. Chem. Soc. 119: 9191–9197.

Tshikhudo, R., D. Demuru, Z. Wang, M. Brust, A. Secchi, A. Arduini and A. Pochini. 2005. Molecular recognition by calix[4]arene modified gold nanoparticles in aqueous solution. Angew. Chem. Int. Ed. 44: 2913–2916.

Vaishnavi, E. and R. Renganathan. 2014. "Turn-on-off-on" fluorescence switching of quantum dots–cationic porphyrin nanohybrid: a sensor for DNA. Analyst. 139: 225–234.

Velu, R., N. Won, J. Kwag, S. Jung, J. Hur, S. Kim and N. Park. 2012. Metal ion-induced dual fluorescent change for aza-crown ether acridinedione-functionalized gold nanorods and quantum dots. New J. Chem. 36: 1725–1728.

Vlascici, D., E.F. Cosmas, E.M. Pica, V. Cosma, O. Bizerea, G. Mihailescu and L. Olenic. 2008. Free base porphyrins as ionophores for heavy metal sensors. Sensors 8: 4995–5004.

Wang, X., J. Wu, F. Li and H. Li. 2008. Synthesis of water-soluble CdSe quantum dots by ligand exchange with p-sulfonatocalix[n]arene (n = 4, 6) as fluorescent probes for amino acids. Nanotechnology 19: 205501–205508.

Willard, D.M., L.L. Carillo, J. Jung and A. Van Orden. 2001. CdSe-ZnS quantum dots as resonance energy transfer donors in a model protein-protein binding assay. Nano Lett. 1: 469–474.

Wu, Y., F. Zuo, Z. Zheng, X. Ding and Y. Peng. 2009. A novel approach to molecular recognition surface of magnetic nanoparticles based on host-guest effect. Nanoscale. Res. Lett. 4: 738–747.

Yang, J., J. Xiang, C. Chen, D. Lu and G. Xu. 2001. Effect of beta-cyclodextrin on the photoinduced charge transfer in sodium 1-anilino-8-naphthalene sulfate (ANS)/CdS colloidal system. J. Colloid Interface Sci. 240: 425–431.

Yao, J., M. Yang and Y. Duan. 2014. Chemistry, biology, and medicine of fluorescent nanomaterials and related systems: new insights into biosensing, bioimaging, genomics, diagnostics, and therapy. Chem. Rev. 114: 6130–6178.

Yildiz, I., M. Tomasulo and F.M. Raymo. 2006. A mechanism to signal receptor-substrate interactions with luminescent quantum dots. Proc. Natl. Acad. Sci. USA 103: 11457–11460.

Zhang, C., H. Yeh, M.T. Kuroki and T. Wang. 2005. Single-quantum-dot-based DNA nanosensor. Nat. Mater. 4: 826–831.

Zhao, H.Y., J.Y. Zhan, Z.L. Zou, F.J. Miao, H. Chen, L. Zhang, X.L. Cao, D.M. Tian and H.B. Li. 2013. Novel 1,3-alternatethiacalix[4]arenes: click synthesis, silver ion-binding and self-assembly. RSC Advances 3: 1029–1032.

Zrazhevskiy, P., M. Sena and X.H. Gao. 2010. Designing multifunctional quantum dots for bioimaging, detection and drug delivery. Chem. Soc. Rev. 39: 4326–4354.

7

Smart Materials for Controlled Droplet Motion

Lin Wang[1,2] and *Hao Bai*[1,*]

ABSTRACT

Designing smart materials to control droplet motion has attracted many ongoing studies. Smart materials outperform traditional materials by their reverse tunable wettability, comprehensive gradients and stimuli-responsibility. Nature always serves as a rich source of inspirations for developing smart materials. Some plants and animals in particular have evolved intriguing wetting properties which enable them to control droplet motion on their surfaces for helping their daily lives, such as spider silk, cactus, desert beetles, and seabirds. On the other hand, engineering synthetic materials that mimic these properties to control droplet motion are of great importance in many fields such as heat transfer, water collection, and anti-icing. During the last few decades, the secrets of these creatures have been gradually revealed, and great progress has been made in developing smart surface for controlled droplet motion. In this chapter, we will go over some basic concepts and theories about surface wettability and droplet motion. Then, we will summarize some prime examples of biological surfaces and their synthetic analogues that can control droplet motion. In the end applications of such surfaces and future perspectives in this specific field will be discussed.

[1] State Key Laboratory of Chemical Engineering, College of Chemical and Biological Engineering, Zhejiang University, Hangzhou 310027, China.
[2] Department of Materials Science and Engineering, The Pennsylvania State University, Park, N-316, Millennium Science Complex, State College, PA, 16803, USA.
* Corresponding author: hbai@zju.edu.cn

1. Introduction

Droplet motion on a solid surface is an important and interesting topic (Grunze 1999, Wasan et al. 2001) in both fundamental research and real applications such as microfluidics (Gau et al. 1999, Zahner et al. 2011), filtration, water collection (Bai et al. 2011, Bai et al. 2014), and condensers. For example, microfluidic systems involving self-powered droplets would be more energy-favorable. Condensers with smart surfaces where condensed droplets can be removed by their spontaneous motion would be more efficient in heat exchange (Daniel et al. 2001). Water collectors could even collect very tiny droplets which usually evaporate into air again after hitting on the surface by controllably driving them towards each other and forming larger droplets (Ju et al. 2012, Parker and Lawrence 2001, Zheng et al. 2010). As a result, there are always pressing needs for developing smart surfaces for controlled droplet motion.

Such intriguing property as droplet motion is actually quite a common phenomenon in our daily lives like 'tears of wine' on a wine glass. From a scientific point of view, this is a typical process of liquid motion driven by surface tension gradient, also known as Marangoni effect (Marangoni et al. 1872, Scriven and Sternling 1960). When alcohol in the wine evaporates at the meniscus on the glass wall, the alcohol concentration decreases and lowers the surface tension of the water/alcohol mixture. Such a gradient of surface tension, as a driving force, fuels up the liquid motion on the glass wall until it falls back into the bulk wine because of gravity and leaves the 'tears' behind.

There are also many examples in natural surfaces that favor droplet motion. As water droplets may cause infection and block stomata on plant leaves, it is highly desirable to remove them from the leaf surfaces. To this end, many plant leaves, such as lotus leaf (Barthlott and Neinhuis 1997) and rice leaf (Feng et al. 2002), have evolved to be capable of driving droplet motion. For similar reasons, insects such as the butterfly (Zheng et al. 2007) and the cicada (Lee et al. 2004) have very powerful wings to shed water droplets directionally and efficiently. On the other hand, some plants and animals have applied the same principles to collect water for their living by basically driving tiny water droplets in a controllable manner to form larger ones that have less evaporation loss. Various examples, including desert beetles' back with hydrophilic/hydrophobic patterns (Parker and Lawrence 2001), spider silks with periodic variation in radius (Zheng et al. 2010), shorebirds with long beaks (Prakash et al. 2008), and cactus with grooved spines (Ju et al. 2012), have been deeply investigated. The secrets of these creatures have been revealed gradually during the last decades.

With the designing strategies and principles discovered in the natural surfaces, great efforts have also been devoted to mimic these properties synthetically and create artificial smart surfaces for controlled droplet motion. Indeed, it has been shown to be a successful strategy to follow the designing principles found in natural materials to engineer smart surfaces. For example, fibers that can drive submillimeter-size water droplets directionally can be developed by mimicking the periodic spindle-knot structure of spider silk (Zheng et al. 2010). Surfaces that shed off droplets directionally can be developed by mimicking the ratchet structure of butterfly wings (Guo et al. 2012, Liu et al. 2014). In terms of materials system, such smart surfaces can be created with polymer, metal, ceramic and their composites. In terms of dimensions, there are one dimensional bioinspired fibers as well as two dimensional patterned or ratchet surfaces that can control droplet motion. In terms of driving forces, there are chemical, thermal, light, and structural gradients working individually or collaboratively to drive droplet motion. In terms of applications, for controlled droplet motion such smart surfaces have already been demonstrated beneficial for such as heat transfer (Daniel et al. 2001), water collection (Bai et al. 2011), chemical reactions (Gilet et al. 2009, Gilet et al. 2010), and materials assembling (Garrod et al. 2007).

Despite all these achievements, there still remain many challenges for controlled droplet motion. For example, how to involve more driving forces to make droplets move faster, how to drive droplets with a broad spectrum of size and surface tension, how to broaden the application of such smart surfaces for controlled droplet motion in harsh environments, are some of those challenges.

In this chapter, we will go over some basic concepts and theories about surface wettability and droplet motion. Then, we will summarize some prime examples of biological surfaces and their synthetic analogues that can control droplet motion. In the end, applications of such surfaces and future perspectives in this specific field will be discussed.

2. Basic Theories About Droplet Motion

The static contact angle (CA) (θ) of a droplet on a solid surface is determined by the solid-liquid (γ_{sl}), solid-gas (γ_{sg}), and liquid-gas (γ_{lg}) interfacial energies following Young's equation (Fig. 1a):

$$cos\theta = \frac{\gamma_{sg} - \gamma_{sl}}{\gamma_{lg}}$$

Thermodynamically, if there is a difference in contact angle at opposite sides of a droplet, then the droplet will tend to move to the more wettable

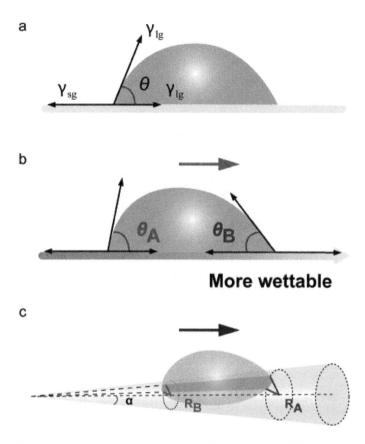

Figure 1. Basic theories about droplet motion. **(a)** Contact angle and Young's equation. **(b)** Asymmetric contact angles on anisotropic wettability surface. θ_A represents advancing contact angle and θ_R represents receding angle. **(c)** Water droplet moves on conical fibers owing to the Laplace pressure difference.

side to minimize the energy of the whole system (Young 1805). According to Young's equation, this contact angle difference might come from local variations in any of the three interfacial energies (γ_{sl}, γ_{sg}, γ_{lg}). For example, consider a liquid droplet placed on a solid surface that has a gradient in its surface free energy, i.e., γ_{sg} is lower on Side A than on Side B (Fig. 1b). If the droplet moves from A to B, the higher energy solid-gas interface (B) will be replaced by lower energy solid-gas interface (A). As a result, the energy of the whole system can be reduced. In other words, the droplet motion towards more wettable surface (higher surface free energy, lower contact angle) is energy-favorable. Such a difference in contact angle on the two sides of a droplet generates a driving force ($F_{Driving}$) which can be described as:

$$F_{Driving} \sim \pi R_0 \gamma (\cos \theta_B - \cos \theta_A)$$

Where R_0 is the radius of the droplet, γ is the surface tension. More generally, πR_0 could be replaced by the local length of contact line (De Gennes et al. 2004).

In another case, even though the solid surface has a constant free energy, i.e., γ_{sg} is the same all over the surface, the droplet can also be imbalanced if it has different surface tensions (γ_{lg}). This could be achieved, for example, by placing the droplet in a thermal gradient, as the surface tension of a liquid usually decreases with temperature (Mettu and Chaudhury 2008, Yarin et al. 2002). In this way, the equilibrium contact angle could be different on the two sides of a droplet and also act as a driving force for droplet motion.

For droplets on a solid surface with topological gradient, such as wedge patterned surface, cone fibers, and tubes, there is another driving force arising from the gradient of Laplace pressure within the droplet (Bico and Quéré 2002, Lorenceau and Quéré 2004). For example, for a droplet sitting on the outer surface of a conical fiber, the driving force of Laplace pressure gradient ($F_{Laplace}$) can be described as (Fig. 1c):

$$F_{Laplace} \sim -\int_{R_B}^{R_A} \frac{2\gamma}{(R + R_0)^2} \sin \alpha \, dz$$

where R_A and R_B are the local radius of conical fiber, R_0 is the drop radius, α is the half apex-angle, γ is the surface tension (Bico and Quéré 2002, Lorenceau and Quéré 2004).

All these driving forces arising from either chemical, thermal, light or topological gradient can act solely or collaboratively to drive droplet motion. However, having a driving force between the opposite sides of a droplet is not sufficient enough for droplet motion. The driving force should at least be able to conquer the resistance force such as contact angle hysteresis arising from surface inhomogeneity. The $F_{Hysteresis}$ can be described as:

$$F_{Hysteresis} \sim \pi R_0 \gamma \left(\cos \theta_{Receding} - \cos \theta_{Advancing}\right)$$

where γ is the surface tension, R_0 is the droplet radius, $\theta_{Receding}$ and $\theta_{Advancing}$ are the receding and advancing contact angles of the droplet (De Gennes et al. 2004).

Guided by these basic principles, chemical (Bain et al. 1994, Chaudhury and Whitesides 1992, Domingues dos Santos and Ondarçuhu 1995, Sumino et al. 2005), thermal (Mettu and Chaudhury 2008, Yarin et al. 2002), light and topological gradients (Ichimura et al. 2000) have been used to drive liquid droplets in the past decades. This was experimentally achieved by Chaudhury and Whitesides (Chaudhury and Whitesides 1992) in

1992, where a water droplet (1 to 2 microliters) was driven uphill on a decyltrichlorosilane modified silicone substrate with a gradient of water contact angle changing from 97° to 25°. Another method to construct a chemical gradient is the so called 'reactive flow' (Bain et al. 1994), where liquid droplets containing surface reactive agents are used. When these agents, e.g., fluorinated fatty acid, react with the solid surface, the surface free energy of the modified area is altered compared to the unmodified area. Thus, the surface free energy difference at the two opposite sides of the liquid droplet induces a net force to make it move (Bain et al. 1994, Domingues dos Santos and Ondarçuhu 1995, Sumino et al. 2005). These liquid droplets can move at a velocity of up to 10 cm/s. As surface tension of a liquid decreases with temperature, a thermal gradient could be another kind of driving force. When further combining a thermal gradient with the chemical gradient, Daniel et al. showed that the condensed water droplets could be largely accelerated to a really high velocity of about 1 m/s (Daniel et al. 2001). Mettu and Chaudhury investigated the motion of droplets on a thermal gradient surface, aided by vibration (Mettu and Chaudhury 2008). The thermal gradient induced motion was also observed along fibers, reported by Yarin et al. (Yarin et al. 2002). In spite of chemical and thermal gradients, light can make liquid droplets move. When a surface is modified with light responsive molecules, e.g., azobenzene, the light induced molecular conformation change will affect the wettability of the surface. Ichimura et al. (Ichimura et al. 2000) created a surface covered by a photoisomerizable monolayer. By asymmetrically photoirradiating the surface, a surface free energy gradient was realized, making liquid droplets move under the manipulation of light.

In brief, in order to drive a droplet move, we need to maximize driving force and minimize resistance. It is also noteworthy that a smaller droplet is usually more difficult to move, as there is not enough imbalanced force at the two sides because of its small size.

3. Controlled Droplet Motion Phenomena in Nature

In nature, some creatures have developed micro- and nano-structured surfaces after millions of years of evolution, which have the ability of driving water droplets directionally, such as butterfly wings (Zheng et al. 2007), spider silks (Zheng et al. 2010), shorebird's beaks (Prakash et al. 2008), and rice leaves (Feng et al. 2002). This gives inspirations in designing bioinspired interfacial materials with multiscale structures to drive liquid droplets in a controllable manner. Combining surface structure design and surface free energy gradients, smart surfaces can be engineered for controlled droplet motion.

3.1 Butterfly Wing

The superhydrophobic wing of the butterfly *Morpho aega* is a good example of directional driving of water droplets (Fig. 2a) (Zheng et al. 2007). The wings are composed of overlapping micrometer scales and fine lamella nano-stripes on the scales. These structures arranged in a direction-dependent way, which causes water droplets on the wings to roll off easily along the scales in the outward direction of the central axis of their body and are strongly pinned in the opposite direction. This phenomenon has been investigated in detail in the view point of anisotropic hysteresis on ratchet superhydrophobic surfaces (Kusumaatmaja and Yeomans 2009). The authors reported that the easy direction for the drops to move was different, when they were pushed across the surface, depending on whether the droplets contacted the surface in suspended or collapsed states. This result is consistent with the anisotropic moving behavior of the superparamagnetic liquid drop (Zhang et al. 2009) and the Leidenfrost drop (Linke et al. 2006) on saw tooth surfaces. Besides superhydrophobic surfaces, Extrand (Extrand 2007) provided a more general discussion on

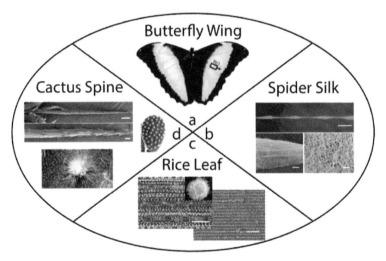

Figure 2. Controlled droplet motion phenomena in nature. **(a)** Butterfly wing. Along the radial outward (RO) direction, the sliding angle of water droplets is much smaller than other directions. Scale bar represents 100 μm and 100 nm, respectively. Reprinted with permission from ref. (Liu et al. 2014). Copyright 2014 American Chemical Society. **(b)** SEM images of spider silk: periodic spindle-knots and joints. Scale bar represents 50 μm, 2 μm and 500 nm, respectively. Reprinted with permission from ref. (Zheng et al. 2010). Copyright 2010 Nature Publishing Group. **(c)** Optical and SEM images of a rice leaf and a patterned aligned carbon nanotube film. Scale bar represents 50 μm and 100 μm, respectively. Reprinted with permission from ref. (Feng et al. 2002). Copyright 2002 John Wiley and Sons. **(d)** Multiple scale structures on cactus spines. Scale bar represents 100 μm and 20 μm, respectively. Reprinted with permission from ref. (Ju et al. 2012). Copyright 2012 Nature Publishing Group.

the retention force when the liquid drop tends to move along the saw tooth structure. It is the ratio of the retention forces in the opposite moving directions that determines the preferred direction for the liquid drop to move. While the motion can be directed by the sawtooth structures, an external force overcoming the retention force is needed to initiate it. Buguin et al. (Buguin et al. 2002) utilized varies physical means, like electric field and vibration, to power the liquid drop to move between two sawtooth structured surfaces. Most recently, Demirel's group fabricated an anisotropic nanofilm with bioinspired structures that has unidirectional wetting properties (Malvadkar et al. 2010). The nanofilm, which is composed of an array of poly (*p*-xylylene) nanorods, displays an asymmetric nanoscale roughness. Since the roughness is at nanoscale, smaller liquid drops can be driven directionally and the energy dissipated by the vibration during the moving process can be reduced. Additionally, the difference of the retention forces in the pin and release directions is higher on this nanofilm, compared to the surface with asymmetric structural features at microscale.

3.2 Desert Beetle's Back

In the Namib Desert, some beetles have evolved a special strategy of collecting water from foggy wind by creating wettability patterns on their back, hydrophilic humps surrounded by wax-covered hydrophobic region (Hamilton and Seely 1976, Parker and Lawrence 2001). Whenever there is a morning fog, the tenebrionid beetle *Stenocara* sp. usually lifts up its back facing the direction of the wind. When droplets carried by the wind hit the beetle's back, they can be trapped on the hydrophilic region because of its high adhesion rather than being blown away or disappearing in the wind again. Gradually, such droplets cover the hydrophilic humps completely, but because of the restriction of the hydrophobic background, they grow rapidly in the ratio of mass and contact area. Until the growing droplet reaches a critical size, it can no longer be blown away by the wind, but conquers the adhesion on the hydrophilic hump and roll off because of gravity. In such a water collection strategy, the directional motion of the droplet from hydrophobic to hydrophilic area and rolling off from hydrophilic area are both critical for the collection efficiency.

3.3 Rice Leaf

While water droplets roll freely in all directions on the lotus leaf, they roll asymmetrically on a rice leaf. This is because the micropapillae on a rice leaf arrange in an orderly parallel to the leaf edge and randomly in other directions (Fig. 2c). The three phase contact line of the liquid droplet is thus different in the parallel and perpendicular direction, which engenders a preferred direction for the drop to move. The role of asymmetrical structures in causing asymmetrical wettability of liquid drops was demonstrated on an aligned carbon nanotube film (Sun et al. 2003). And a two-step phase-separation micromolding method was used to fabricate artificial rice leaf that showed even better asymmetrical wettability (Gao et al. 2009). This indicates that topological structure of the surface have a critical effect on controlled droplet motion.

3.4 Seabird's Beak

Some shorebirds feed themselves by driving the prey-containing liquid droplets between their long thin beaks (Prakash et al. 2008). Their beaks with an opening angle are typical asymmetrical structures which make the liquid drops move differently in the mouthward and the reverse direction. Through repeatedly opening and closing the beaks, the liquid drops are driven mouthward in a ratcheting manner. This motion is related to the wettability of beaks, the opening and closing angles, the fine structures on the inner face of the beaks. It is noteworthy that the shorebirds drive liquid drops taking advantage of the CA hysteresis, which is usually considered to be a resistance in drop motion (Quéré 2008). Another example is in the tapered tubes (Renvoise et al. 2009). Because of the radius variation in the tube, the liquid drops can be driven towards the taper side by the Laplace pressure gradient.

3.5 Spider Silk

Spider silk is famous for its outstanding mechanical properties and has aroused much attention in the field of biomimetic research, trying to get artificial fibers with similar properties (Porter and Vollrath 2009). Despite the mechanical properties, its ability to capture water from fog has been investigated recently (Fig. 2b) (Zheng et al. 2010). The capture silk of the cribellate spider *Uloborus walckenaerius* is composed of two main-axis fibers covered by nanofibers. These nanofibers shrink to form a periodic spindle-knots and joint structure, with random and align nanofibers on these two positions respectively. These structural features combine two driving forces together: Laplace pressure gradient induced from spindle-knot shape and surface energy gradient induced from random and align nanofibers. As a result, the submillimeter-sized droplets that condense on the silk are directionally driven from the joint to the spindle-knot position. Before this work, Quéré et al. (Lorenceau and Quéré 2004) reported that a silicone oil droplet with volumes of 0.2–1 mm³ could move toward the side with larger radius on a smooth conical copper fiber. They also anticipated that the CA hysteresis would hinder the motion of smaller liquid droplets if the Laplace pressure gradient is the only driving force (De Gennes et al. 2004). Indeed, the spider silk gives some inspirations to achieve this goal by combining different driving forces together through surface structure design. The successive biomimetic research shows that artificial fibers can not only reproduce the ability of natural silks in driving liquid droplets directionally, but also the moving direction of these droplets can be controlled, through designing the surface chemical compositions and roughness on the fiber surfaces.

3.6 Cactus Spine

Many Cactaceae species can survive in extremely arid environment. They all have a well-developed water management strategy that can minimize water loss. Some of the species are even smarter, collecting fog droplets as an alternative water source. Take cactus *O. microdasys* as a typical system, it has been revealed that the continuous fog collection system of cactus relies greatly on the multiple scale structures of their spines (Fig. 2d) (Ju et al. 2012). The spine can be divided by three regions, i.e., oriented barbs, gradient grooves, and belt-structured trichomes, from the tip to the base. Water droplets in the fog first hit uniformly all over the spine. Then, those water droplets on the small barbs first move directionally toward the spine because of Laplace pressure gradient arising from the conical shape of the barb. The grooves on the barb also help the movement. In addition, larger droplets on the spine can be driven further by the conical spine from tip to base. Again, the energy barrier for droplet motion is reduced by grooved structure on the spine. The driving force is so strong that the droplet can be driven upwards even against gravity. Actually, water droplets can be driven from tip to base of the spine regardless of its orientation. At the last step, the capillary system formed by the belt-structured trichomes and root of the spine absorbs water droplets into the cactus body quickly. The complex system of cactus has utilized multiple driving forces to control droplet motion in every step so as to increase its water collection efficiency. This is a very inspiring example for developing a materials system for continuous and controlled droplet motion.

4. Bioinspired Surfaces for Controlled Droplet Motion

4.1 One Dimensional Materials

Asymmetric droplet motion on conical fibers was first reported by Lorenceau and Quéré in 2004 (Lorenceau and Quéré 2004). It was shown that silicone oil droplets with volume of 0.2–1 mm^3 could spontaneously move to the region of lower curvature. The driving force solely comes from a difference of Laplace pressure between the two sides of the conical fibers. Stemming from the same mechanism, liquid droplets could also move spontaneously toward the tip in an asymmetric tube (Bico and Quéré 2002). For smaller droplets (tens of micrometers), however, it is much more challenging to drive them (Lorenceau and Quéré 2004) because only one driving force (either from curvature or surface energy gradient) is inadequate to conquer the contact angle hysteresis. As mentioned in Section 3.5, *Cribellate* spiders utilize the periodic spindle-knots along their silks, coupling with the surface energy gradient between spindle-knots and joints, to drive tiny water droplets to move directionally (Zheng et al. 2010).

Inspired by spider silks, a lot of research has been conducted to explore the method of preparing conical fibers (Bai et al. 2010, Bai et al. 2011a, Bai et al. 2011b, Chen et al. 2012, Chen et al. 2013, Chen et al. 2014, Feng et al. 2013, Feng et al. 2013, Hou et al. 2013, Song et al. 2014, Xue et al. 2014, Zheng et al. 2010). Dip-coating method is the one that has been studied most systemically (Bai et al. 2011, Zheng et al. 2010). A uniform bare nylon fiber was immersed into a polymer solution, poly (methyl methacrylate) (PMMA) dissolved in N, N-dimethylformamide (DMF). When the fiber is drawn out horizontally, the fiber surface will be covered by a cylindrical PMMA film. This film then spontaneously breaks up into polymer droplets owing to the Rayleigh instability (Bai et al. 2010, Strutt and Rayleigh 1878). After the evaporation of solvent, periodic PMMA spindle-knots form on the nylon fibers (Fig. 4). By controlling the solution viscosity, solution surface tension, and fiber drawing-out velocity, geometry parameters of spindle-knots could be manipulated thus the abilities of collecting water could be controlled (Fig. 10b and Fig. 10c).

In the spider silk case, as mentioned, the surface energy gradients also play a key role in driving droplet motion. Instead of changing the size of spindle-knots, a lot of research has focused on introducing the surface energy gradient on conical fibers (Cao et al. 2014, Ju et al. 2013, Li et al. 2013). Water tends to propagate to the region with higher surface energy, and also higher wettability. By planting polymers with different wettability on the spindle-knots and joints, the direction of water motion could be changed arbitrarily (Bai et al. 2010). That is if the spindle-knots were covered by smooth hydrophobic polystyrene (PS), and smooth hydrophilic PMMA on joints, although the curvature gradient works against the surface energy gradient, water droplet (31.3 to 55.5 μm in size) could still move away from the hydrophobic spindle-knots. Likely, if the spindle-knots are covered by rough hydrophobic PS, and joints are covered by smooth hydrophilic PMMA, water droplet could move toward the hydrophobic PS quickly (362.5 to 544.7 μm·s⁻¹) (Fig. 3).

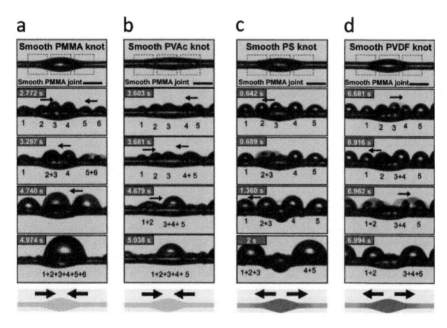

Figure 3. Optical photos of moving water droplets on artificial spider silks. On surfaces with different wettability, water droplets move toward different directions: either towards the knots on **(a)** PMMA and **(b)** poly (vinylacetate) (PVAc) surfaces, or away from the knots on **(c)** PS and **(d)** poly (vinylidene fluoride) (PVDF) surfaces. Scale bar is 50 μm. Reprinted with permission from ref. (Bai et al. 2010). Copyright 2010 John Wiley and Sons.

Instead of utilizing static surface energy gradient to drive water droplets, some surfaces could be transformed *in situ* from hydrophilic to hydrophobic to drive water droplets by introducing external stimulus, e.g., temperature changing or UV light (Feng et al. 2013, Hou et al. 2013). By changing the configuration of stimulus-responsive molecules, the wettability of surface could be inversed. Spindle-knots covered by copolymer poly (methyl methacrylate)-b-poly (N-isopropylacrylamide) (PMMA-b-PNIPAAm) which is temperature sensitive could drive water droplets move either towards or away from spindle-knots at different temperature. A surface covered by PMMA-b-PNIPAAm is more wettable below the lower critical solution temperature (LCST, ~ 32°C) and less

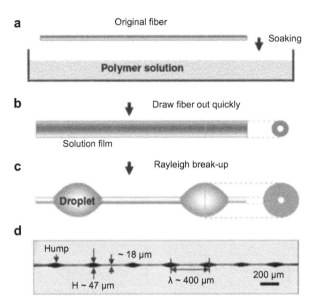

Figure 4. Illustration of dip-coating method for preparing artificial spider silks. **(a)** The original fiber is a nylon fiber with a diameter ~ 18 μm. PMMA is dissolved into DMF to serve as polymer solution. **(b)** A uniform solution film covers the nylon fiber after drawing out quickly. **(c)** The solution film breaks into periodic droplets due to Rayleigh Instability. **(d)** Optical photos of the prepared artificial spider silks with periodic humps and joints upon it. Reprinted with permission from ref. (Tian et al. 2011). Copyright 2011 John Wiley and Sons.

wettable above the LCST. Water droplets move toward spindle-knots at 25°C but move away from them at 40°C in the same humidity. Similarly, photosensitive azobenzene polymer is responsive to UV light. A surface modified by it could be changed from less hydrophilic to hydrophilic owing to the molecular configuration transformation under UV light.

On a single spindle-knot, water droplet's motion could be controlled arbitrarily. However, on a fiber with uniform spindle-knots, droplets can move only between two adjacent spindle-knots. While, on a fiber with non-uniform spindle-knots, droplets could move for a longer directional distance. Through multiple dip-coating treatments, the artificial spider silk with non-uniform spindle-knots could be fabricated. Multilevel sized spindle-knots were achieved: the main, the satellite, and the sub-satellite spindle-knots (Fig. 5a and Fig. 5b). When this fiber was exposed in a fog flow, tiny water droplets first deposited everywhere on the fiber and directionally moved toward spindle-knots to form larger droplets. The droplets on sub-satellite would coalesce to the droplet on satellite then finally to the droplet on main spindle-knot (Fig. 5a and Fig. 5b). The increasing capillary force originating from size gradient contributes to droplets' movement from sub-satellite to satellite to main spindle-knots. Following the same

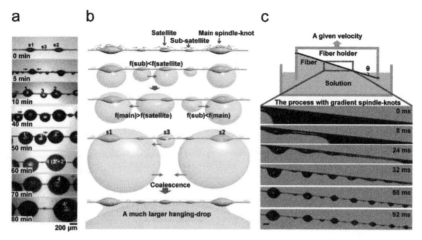

Figure 5. Transportation of water droplets on multi-level structure gradient artificial spider silks. **(a)** Optical photos of moving water droplets on this type of artificial spider silks. **(b)** Detailed process of water droplets merging. **(a) (b)** are reprinted with permission from ref. (Chen et al. 2012). Copyright 2012 Royal Society of Chemistry. **(c)** Method illustration. Scale bar is 200 μm. Reprinted with permission from ref. (Chen et al. 2013). Copyright 2013 Nature Publishing Group.

mechanism, plotting a series of spindle-knots with increasing size on a fiber through tilt angle dip-coating method (Fig. 5c) could drive water droplet move for a much longer distance (~ 5.00 mm, velocity of 0.1–0.22 m·s⁻¹) (Fig. 5c) than on the fiber with uniform spindle-knots.

Vertical dip-coating method could only prepare a spindle-knot fiber with a finite length. However, the horizontal fluid-coating method is capable of large scale fabrication (Fig. 6a) (Bai et al. 2011). A motor was used to drag a bare nylon fiber horizontally pass through a polymer tank at a well-defined speed. Coating in this way, the nylon fiber could be continuously covered by a uniform cylindrical polymer film. The same as in dip-coating method, this film then spontaneously broke up into polymer droplets owing to the Rayleigh instability and the periodic spindle-knots were formed (Fig. 6a).

Apart from dip coating method, the coaxial electrospinning method is also able to fabricate spindle-knot fiber (Fig. 6b) (Dong et al. 2012, Tian et al. 2011). Outer solutions was dilute PMMA solution and inner solution was concentrated PS solution. The viscous PS solution stretched into liquid thread at a high-voltage electric field. The PMMA solution flowed out and stuck on the PS thread forming a film. Then it broke into discontinuous liquid droplets, after the solvent evaporating, periodic spindle-knots formed on the PS fiber. A microfluidic system consisting of digital and programmable flow controller was also reported to be used to continuously fabricate spindle-knot fiber (Kang et al. 2011). The shape and micro-/nano-structures of the fibers could be well defined by manipulating

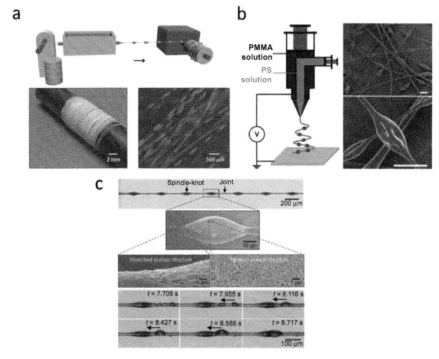

Figure 6. Methods for preparing artificial spider silks. **(a)** A continuous coating method for large scale fabrication. Reprinted with permission from ref. (Bai et al. 2011). Copyright 2011 John Wiley and Sons. **(b)** Electrospun method, scale bars represent 10 μm. Reprinted with permission from ref. (Dong et al. 2012). Copyright 2012 John Wiley and Sons. **(c)** The SEM and optical images of one type of artificial spider silks. Reprinted with permission from ref. (Zheng et al. 2010). Copyright 2010 Nature Publishing Group.

the valve operation of two channels with different fluids. In order to achieve porous spindle-knotted fibers, an alginate solution containing salt was introduced at a high feeding rate into one of the channels and then the salt was removed by dissolution. All the methods mentioned above could fabricate uniform spindle-knots fibers. They all have the ability to move water droplets directionally from joints toward spindle-knots (Fig. 6c).

Another great example of one dimensional water droplet transportation in nature is the cactus spine system (Ju et al. 2012). As aforementioned in Section 3.6, each spine is featured by oriented conical barbs on the tip, gradient grooves in the middle and belt-structured trichomes around the base. These features working together provide both the Laplace pressure gradient and surface energy gradient and then intensively overcome the CA hysteresis to transport fog droplets deposited upon it from tip to base continuously.

By mimicking the mechanism of cactus spine, a kind of artificial cactus spine (ACS) was prepared via gradient electrochemical corrosion followed by gradient chemical modification (Fig. 7a) (Ju et al. 2013). In specific, conical copper wires served as a substrate and they could be achieved by lifting and lowering the electrolyte container repeatedly and controlling other corrosion parameters. Their apex angles (2α) could be controlled finely as 9°, 11°, 13°, 15°, and were much close to that of cactus spine (~ 12.3 ± 1.6°). Hydrophilic compound and hydrophobic compound were assembled onto the base and tip of conical copper surface respectively to

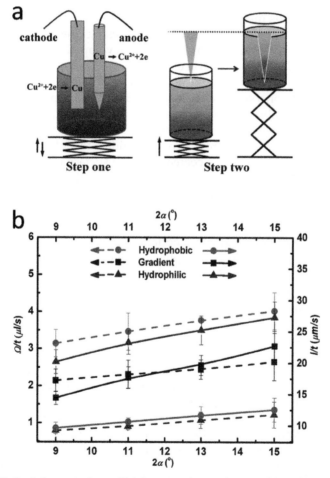

Figure 7. Methods for preparing artificial cactus spines and comparision of water collecting efficiency on different surfaces. **(a)** Method illustration of preparing conical copper wires and introducing a chemical gradient. **(b)** Efficiency comparision of different artificial cactus spines. Reprinted with permission from ref. (Ju et al. 2013). Copyright 2013 John Wiley and Sons.

form a unidirectional wettability gradient. Hydrophobic surface is more suitable for droplet volume growing (growth rate of about 3.27 $\mu L \cdot s^{-1}$), because micrometer-sized water droplets in a fog flow are more readily to jump away when they collide with a water film than with a dry surface, which curbs further deposition and increasing of drop volume. Whereas, the hydrophilic surface is more suitable for droplet transportation (velocity of about 26.03 $\mu m \cdot s^{-1}$). With a wettability gradient, ACS could both grow droplets quickly (growth rate of about 2.08 $\mu L \cdot s^{-1}$) and transport them more efficiently (velocity of about 20.05 $\mu m \cdot s^{-1}$) (Fig. 7b). X-ray etching method has also been used to prepare cactus spine-like conical tips (Moon et al. 2005).

These methods mentioned above, however, are restricted to fabricate limited amount of samples within a single process (Cao et al. 2014). Methods capable of large scale fabrication have been developed recently (Cao et al. 2014, Ju et al. 2013, Ju et al. 2014, Li et al. 2013). One of them is a method that combines mechanical lithography and mold replica technology to fabricate conical tip arrays on a substrate (Li et al. 2013). A piece of polyethylene (PE) slice served as a substrate and was pricked under a jet dispensing system which was attached by a commercial conical needle to form the hole arrays. Polydimethylsiloxane (PDMS) is a good candidate to replicate fine features of mold. So when PDMS was coated onto the surface of hole arrays and then lifted away after hardening, it replicated exactly the morphology of the commercial conical needle, which meant the parameter of conical tip could be manipulated precisely by choosing different conical needles. This process was highly efficient in scalable preparation of conical tip arrays which not only showed a great potential in collecting water droplets from fog, but in separating tiny oil droplets (smaller than 10 μm) from water (Ju et al. 2014, Li et al. 2013). More detailed information about its application will be discussed in Section 5.4.

Another approach for large scale fabrication of conical tip arrays is a modified magnetic particle-assisted molding (MPAM) method (Cao et al. 2014). Mixture of PDMS and magnetic particles on flat plane could form disordered conical tip arrays with the assistance of external magnetic field. In order to achieve ordered conical tip arrays, the substrate plane needs to be modified with geometric patterns. Once the magnetic particles (MPs) were filled into the geometric patterns, the magnetic field would selectively impact the patterns owing to the higher magnetic conductivity, subsequently springing the ordered conical tip arrays. After solidification by IR irradiation, these conical tip arrays attained a fixed morphology. The weight ratio of PDMS to MPs strongly affect the geometry of conical tips. The height and apex angle of conical tip was roughly 1.7 mm and 14°, respectively, when PDMS was fixed at 67 wt.%. As expected, this conical tip arrays also showed a high efficiency in fog collecting (Cao et al. 2014).

The seabird is also an expert in transporting tiny water droplets one dimensionally with its beaks (Prakash et al. 2008). It utilizes a different mechanism for asymmetric water droplet transportation. Instead of maintaining a static curvature like those on cones, the seabird changes the opening angle of their beaks dynamically to achieve asymmetric motion of water droplets, by which they could move their drinking water or preys into mouth. Although silicone oil droplets could move spontaneously toward narrower gap between non-parallel plates (Bouasse 1924, Prakash et al. 2008), when water took the place of the silicone oil, no droplet motion arose. The driving force is counteracted by an adhesive force that comes from contact angle hysteresis. Unlike silicone oil, water cannot wet the solid completely, which will lead to a difference between advancing (θ_A) and receding (θ_R) contact angles. During cyclically opening and closing the seabird breaks, the trailing edge and leading edge of droplets move asynchronously. When it is closing, both edges tend to move outward, but the leading edge always moves first, because the contact angle near the leading edge reaches advancing angle θ_A first, above which contact line motion happens. Conversely, when it is opening, both edges tend to move inwards, but the trailing edge always moves first, because the contact angle near trailing edge reaches the receding angle θ_R first. The droplet could only move a finite distance during one circle, however, by optimally tuning the droplet volume and beak geometry, the droplet could arrive to the buccal cavity of seabird in three circles (Rubega and Obst 1993).

In order to mimic seabird beaks, Prakash et al. built a mechanical wedgelike beak with stainless steel whose surface was polished by Buehler Metadidiamond slurry (Prakash et al. 2008). Its opening and closing was controlled precisely by a motorized micrometer stage that was connected to a computer. The opening and closing angles need to be carefully controlled, because for a given volume droplet, if the opening angle exceeds the maximum angle, the droplet will break; if the opening angle is lower than minimum angle, the droplet will split. More importantly, with the fixed droplet volume 1.5 μL, the most efficient transportation regime could be achieved (3 cycles) by turning the opening angles. That is, a droplet could be transported from beak tip to the base within 3 cycles, which is interestingly consistent with the data in the wild (Rubega and Obst 1993). This mechanical beak could still transport water droplet when it was put vertically. The force of gravity was canceled by adhesive force due to the contact angle hysteresis. This vertically regime is much more like the real way in which the seabird works.

4.2 Surface with Patterned Wettability

Namib Desert beetles, *Stenocara* sp. beetle, as mentioned in Section 3.2, could use the hydrophobic micrometer patterns on their hydrophilic back to collect tiny water droplets from the fog (Parker and Lawrence 2001). Inspired by desert beetles, fabricating a surface with patterned wettability is a highly desirable approach to drive water droplets towards specific areas. Park et al. achieved this by a simple way that is embedding hydrophilic glass spheres into waxed slides (Parker and Lawrence 2001). Their results showed that if the glass spheres were plotted in square arrays, more water could be collected from the artificial mist than in random way.

To mimic *Stenocara* sp. beetle's back, many methods have been developed to prepare patterned wettability surface, such as microcontact printing (Drelich et al. 1994, Lopez et al. 1993), polyelectrolyte multilayers (Zhai et al. 2006), and plasma polymers (Garrod et al. 2007).

Zhai et al. assembled a patterned wettability surface by selectively delivering polyelectrolytes onto a superhydrophobic surface. Aqueous solution is not capable of wetting superhydrophobic surface. However, it has been reported that by introducing surfactants into an aqueous solution, the superhydrophobic surface became wettable, since the surfactants reduced the surface tension of the solution (Mohammadi et al. 2004, Soeno et al. 2004). A mixture of water (surface tension: 72.2 mJ/m^2) with a low surface tension organic solvent, 2-propanol (surface tension: 21.7 mJ/m^2) was used to dissolve poly (fluorescein isothio-cyanate allylamine hydrochloride) (FITC-PAH) in Zhai's case. Their superhydrophobic surface consisted of rough microporous poly (allylamine hydrochloride) (PAH)/poly (acrylic acid) (PAA) (bottom layer), PAH/silica nanoparticles (PAH/SiO$_2$) (middle layer) and a hydrophobic network of semi-fluorosilane molecules (top layer). Then when the FITC-PAH water/2-propanol solution was micropipeted onto this superhydrophobic surface, the polyelectrolyte, FITC-PAH, could pass through the semi-fluorosilane network, then formed electrostatic bonds with the underlying PAH or the silica nanoparticles. Meanwhile, part of the FITC-PAH remains on the surface, changing local area into hydrophilic. Some other chemicals were also used by Zhai to mimic the desert beetle's back, such as poly-(acrylic acid) (PAA), methylene blue (a positively charged dye) and rose bengal (a negatively charged dye). Figure 8a shows the behavior of small water droplets on a superhydrophobic surface decorated by PAA patterns. Although Zhai et al. developed a valid method for preparing patterned superhydrophilic and superhydrophobic surfaces, they did not investigate the water collection efficiency of their samples.

A similar method was developed by Dorrer and Rühe to prepare the hydrophilic patterned superhydrophobic surfaces (Dorrer and Rühe 2008). By repeatedly dispensing defined volume of hydrophilic polymer solution onto a superhydrophobic background, after the solvent evaporating on a hotplate (80°C), a hydrophilic bump was left. Several polymer solutions were used: poly (dimethyl acrylamide) (PDMAA) in ethanol,

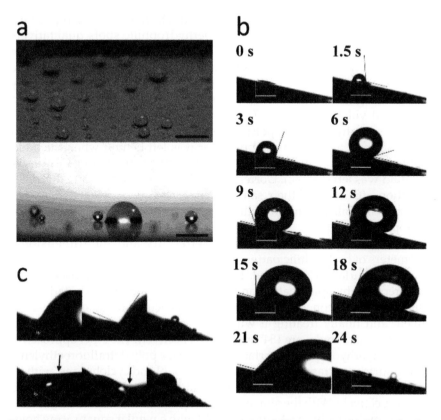

Figure 8. Behavior of water droplets on patterned wettability surfaces. **(a)** Water droplets accumulated on superhydrophobic spots. Scale bars represent 5 mm (top) 750 μm (bottom). Reprinted with permission from ref. (Zhai et al. 2006). Copyright 2006 American Chemical Society. **(b)** Optical photos of a growing water droplet. **(c)**, On a less hydrophilic spot (top), the contact line receded into spot, and no water was left when the adult droplet rolled off. (The transparent tiny drop was newly deposited from the fog.) On a superhydrophilic spot (bottom), the contact line pined and left a part of the adult droplet when it rolled off. The white lines in both **(b)** **(c)** represent spot diameter of 2 mm. **(b)** **(c)** are reprinted with permission from ref. (Dorrer and Rühe 2008). Copyright 2008 American Chemical Society.

poly (styrene) (PS) in toluene, and poly (heptadecafluorodecylacrylate) (PFA) in 1,1,2-trichlorotrifluoroethane. The receding angles of those polymer solutions on the superhydrophobic background are zero, therefore the bump diameter could be retained constant during evaporation. The superhydrophobic background was prepared by planting a layer of PFA on a roughed silicon surface. An alternative approach to make hydrophilic patterns is to selectively remove the PFA. An aluminum mask with circular holes was used to do this when the sample was exposed under UV light. Then patterned spots on the roughed silicon surface, which is a superhydrophilic surface, will be uncovered. The authors investigated the development of droplets on those superhydrophilic spots qualitatively under a fog flow. The smaller droplets initially appeared everywhere on the surface. Then they continuously coalesced with each other and began to run over the surface until finally were stuck at superhydrophilic spots. They kept growing and then rolled off the surface when their volume exceeded the critical volume (Fig. 8b). The dewetting mechanism is different for different superhydrophilic spots. On more hydrophilic spots, bare roughed silicon surface, a small amount of water was left behind when the adult droplets rolled off. On a relative less hydrophilic spot, PFA or PS bumps, no liquid was left (Fig. 8c). Dorrer and Rühe did not investigate the water collection efficiency of their samples either, but their results suggested that the wettability of hydrophilic patterns could influence the behavior of droplets upon them.

Garrod et al. studied how the wettability and the dimensional parameters of hydrophilic patterns affect the water collection efficiency of a patterned surface (Garrod et al. 2007). The superhydrophobic background was prepared by spin-coating an 8% (w/v) solution of polybutadiene in toluene onto a silicon wafer and annealing in vacuum to remove remained solvent and finally treating it with CF_4 gas plasma. The advancing and receding contact angles are 154° and 152°, respectively. The other approach to obtain a superhydrophobic surface was to treat a poly (tetrafluoroethylene) (PTFE) surface, ultrasonic cleaned in propan-2-ol and cyclohexane, with O_2 gas plasma. The advancing contact angle and receding angle were 152° and 151°, respectively. With the aim to selectively deposit hydrophilic layer onto the superhydrophobic background, a grid with a regular square array holes was embossed on it. After the plasma deposition of hydrophilic layer, the grid was removed and a patterned hydrophilic superhydrophobic surface was achieved.

The wettability of patterns could be regulated by depositing different hydrophilic layers, such as poly (4-vinyl pyridine) (4-VP), poly (glycidyl methacrylate) (GMA), 4-vinyl aniline (4-VA), bromoethyl-acrylate (BEA), maleic anhydride (MA), 3-vinylbenzaldehyde (3-VBAL). The patterned surface was fixed on a 45° tilted stage and put under a fog flow for 2 hours to collect water. The water collection efficiency could be evaluated via comparing the weight of collected water. Attributing to its lowest contact angle, 4-VP was found to be the most efficient, no matter on plasma-fluorinated polybutadiene or O_2 plasma etched PTFE surfaces. The results also demonstrated that purely hydrophilic or purely hydrophobic surfaces performed worse than patterned surfaces.

The dimensional parameters of hydrophilic patterns could be adjusted by using different grid, whose holes were drilled by a machine. The patterns with a diameter of 500 μm and a center to center distance of 1000 μm were the optimal patterns. Either increasing or decreasing the spot size, water collection became less. These results are well consistent with the design of desert beetle's back, on which the nature favorable diameter is 500 μm and the center to center distance is 0.5–1.5 mm (Parker and Lawrence 2001).

Those results indicated that a maximum efficiency of water collection on patterned surface could be achieved by regulating both the wettability and the dimensional parameters of hydrophilic areas. Recently, Bai et al. reported that the shape of patterns also affected the efficiency significantly (Bai et al. 2014). As discussed above, spider silks take advantage of both wettability (surface energy) and shape gradients to directionally collect water from fog. Therefore, by replacing the circular spots with star-shaped patterns, the wedgelike tips introduce a shape gradient (or Laplace pressure gradient) to the superhydrophilic patterned superhydrophobic surfaces (Fig. 9a). As these two integrated gradients work together, smaller

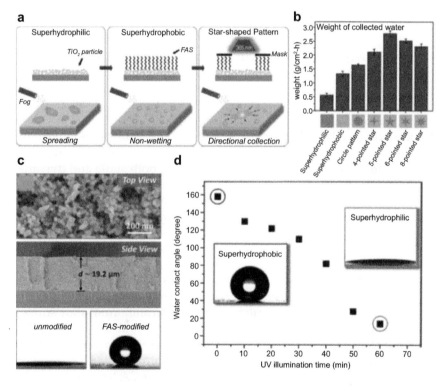

Figure 9. Preparation method and performance of star-shape patterned surfaces. **(a)** Fabrication procedures. **(b)** Collected water by various surfaces. **(c)** SEM images of TiO$_2$ layers and the behaviors of water droplets on modified/unmodified TiO$_2$ surface. **(d)** FAS layer could be removed by UV light and switch the superhydrophobic surface back into superhydrophilic surface. Reprinted with permission from ref. (Bai et al. 2014). Copyright 2014 John Wiley and Sons.

droplets could be transported to the superhydrophilic patterns more quickly and therefore resulting in a faster collection cycle. Surfaces with 4-, 5-, 6-, 8-pointed star-shaped patterns all performed better than the one with circle-shaped patterns (about 1.65 g/cm·2h). Particularly, the one with 5-pointed star-shaped patterns performed best (about 2.78 g/cm·2h) (Fig. 9b). To fabricate a star-shaped superhydrophilic patterned surface, TiO_2 slurry was first deposited onto a glass substrate via spin-coating to form a superhydrophobic background; then, heptadecafl uorodecyltrimethoxysilane (FAS) was deposited onto the porous TiO_2 film (particle size of about 50 nm, thickness of about 19.2 μm, Fig. 9c) through chemical vapor deposition method; finally, a photo mask was used to selectively photocatalytic decompose FAS (Zhang et al. 2007), the size of holes on photo mask was well defined (Fig. 9a). The wettability of patterned areas could be finely controlled by adjusting the UV lightening time (Fig. 9d), which means this method is much flexible in designing artificial dessert beetle's back.

4.3 Surface with Asymmetric Structure

Asymmetric structure on surfaces could lead to an anisotropic water droplets wetting behavior. Mentioned in Section 3.1 (Zheng et al. 2007), butterfly wing is embossed with overlapping micrometer scales and these scales are covered by fine lamella nano-stripes. These direction-dependent ratchet structures gives the butterfly wing an ability to drive water droplets to move along radial outward (RO) direction in both static and dynamic states (Liu et al. 2014). Even when the butterfly wing was put upright, the water droplet could not move against RO direction, but ran away immediately along RO direction when the butterfly wing was tilted 9°. If the butterfly wing was put horizontally on a vibrating stage (vibration mode was 35 Hz, 3 mm), fog droplets moved readily along RO direction (Liu et al. 2014).

Constructing an artificial butterfly wing is highly desirable, because it could provide a surface to drive water droplet move asymmetrically. Zhang et al. fabricated the microscale ratchet structure on aluminum alloy surface by a mechanical method using a machine (Zhang et al. 2009). To endow superhydrophobicity on this surface, polystyrene (PS) microparticles which have a coating of nanostructures were coated onto it. This PS coating on a flat substrate has a contact angle of 150.3° and a CA hysteresis of 3.3°. They put a superparamagnetic microdroplet (M-droplets) to characterize the asymmetric wettability of this artificial butterfly wing. When the ratchet structure was given, M-droplet moved if the intensity of magnetic field exceeded a threshold to overcome the surface retention. M-droplets moved much easier along the ratchet pointing direction (acceleration about

1.8 m/s²) than against the ratchet pointing direction (acceleration about 1 m/s²), which was consistent with butterfly wing.

Malvadkar et al. outlined a preparation method that produced a nanofilm on a substrate to achieve asymmetric wetting (Malvadkar et al. 2010). The nanofilm was consisted of tilted nanorod arrays with a diameter about 150 nm. Poly (p-xylylene) (parylene or PPX) was used to construct nanorods via a bottom-up vapor-phase technique and oblique angle polymerization (OAP). OAP vaporized and decomposed a p-cyclophane precursor to create a diradical vapor flux through a commercial system. The flux was set at a defined angle respect to the substrate, resulting in anisotropic growing of structures during polymerization. As expected, the nanofilm showed an obvious anisotropic wetting behavior. Although water contact angles were as high as 120°, water droplets could adhere to the surface even when the nanofilm was tilted vertically with nanorods pointing upward or inverted. However, when the nanofilm surface was tilted vertically with the nanorods pointing downward, the droplets released themselves immediately.

Asymmetric structured surface could also work at high temperature of 350°C, a temperature widely used in hot wetting studies (Bai et al. 2010, Tian et al. 2011). Liu et al. prepared a surface with tilted nanowires, which was superhydrophilic at ambient temperature, but superhydrophobic at 350°C (Liu et al. 2014). By subtly tuning the length of tilted nanowires and chemical characteristic of surface, directional water droplet transportation at high temperature could be realized. The authors used off-cut (OC) Si wafers as substrates and etched them by an etchant solution (HF = 4.6 mol/L and AgNO₃ = 0.02 mol/L) in a polytetrafluoroethene (PTFE) container. Since etching always happened along a ⟨100⟩ direction for single-crystal silicon wafers, the tilted angle of nanowires could be tuned by adjusting the angle of ⟨100⟩ direction respecting to top plane (Ma et al. 2013).

These butterfly wings inspired asymmetric structured surface all achieved the anisotropic wetting. It is easier to drive water droplets to move along the specific direction, toward which tilted nano-wires/walls point, than the other direction. The driving force comes from either gravity (Malvadkar et al. 2010, Zheng et al. 2007) or magnetic field (Zhang et al. 2009). Water droplets could move spontaneously on asymmetric structured surface, even at a high temperature of 350°C (Liu et al. 2014).

5. Application of Controlled Droplet Motion

5.1 Heat Transfer

Condensing saturated water steam on heat exchanger involves many important applications in industry. A widely-used method is utilizing a cold metallic plate to capture a cold water stream from a superheated vapor. The heat released from the condensation is conducted to the cold fluid via the metallic wall. However, the thin insulating liquid film that sticks on the metallic wall troubles this process. Although some methods have been designed to eliminate the insulating water film, they are far from efficient enough (Bejan 1993). Daniel et al. applied a surface with wettability gradient onto the heat exchanger top to avoid the insulating water film and to accelerate heat exchange (Daniel et al. 2001). The surface with a wettability gradient was prepared by diffusion-controlled silanization. A small drop of silane was held above the center of a clean silicon surface. The silane evaporated from the drop and diffused radically on the silicon surface. On this surface, whose center was hydrophobic and the edge was hydrophilic, condensed water droplets moved rapidly towards the edge. The energy driving the rapid motion of these water droplets came from coalescence and wettability gradient.

To test the heat transfer ability, they built a heat exchange system with the radically wettability gradient surface. When a saturated steam (100°C) was passed over the gradient surface, condensed droplets were continuously transported away the surface center by the wettability gradient. Meanwhile, heat flux was continuously conducted through the exchanger. Heat transfer coefficients of two types of surface were compared: filmwise condensation on a horizontal unmodified surface and dropwise condensation on the gradient surface. The gradient surface outperformed the unmodified surface significantly. The gradient wettability surface could also be used on a large scale as discussed by Daniel. A surface with a periodic gradient can overcome its size limit and benefit the rapid condensed water droplets motion continuously.

5.2 Water Collection

Harvesting water from air is a vital way to solve water storage issue, particularly in extremely arid areas where have the access to adequate fog flow (Ju et al. 2012, Parker and Lawrence 2001, Zheng et al. 2010). Desert beetles (Parker and Lawrence 2001), spider silks (Zheng et al. 2010) and cactus spines (Ju et al. 2012) all evolved the wettability patterned surfaces to capture tiny water droplets from the misty air. Therefore, by mimicking these biological surfaces, many artificial surfaces have been designed to collect water efficiently from fog flow.

Spider silk is a great example for scientists to study and mimic. Introduced previously in this chapter, the driving force for water droplet directional motion on spider silk stems from both surface energy gradient and shape gradient. Hao et al. reported that the size of the spindle-knots, which provided the shape gradient, would affect the water collection efficiency significantly (Bai et al. 2014). They fabricated a series of bio-inspired artificial spider silks (BASs) with different size of spindle-knots through a dip-coating method (Fig. 4). There were four kinds of BASs with the spindle-knots volume decrease from 0.12 nL, 0.08 nL, 0.03 nL, to 0 nL (smooth fiber) (Fig. 10b). These BASs were put under a fog flow (velocity of about 75 cm/s) and the water collection process was recorded using charge-coupled device (CCD) camera. The total volumes of all the water droplets on BASs at different time were plotted in a diagram (Fig. 10c). Water droplets kept growing on the spindle-knots under fog flow (Fig. 10a). The BAS with the largest spindle-knots could collect about 35 nL of water within 12 s. While the smooth fiber could collect only about 3 nL of water during the same time. Clearly, the BASs with larger spindle-knots

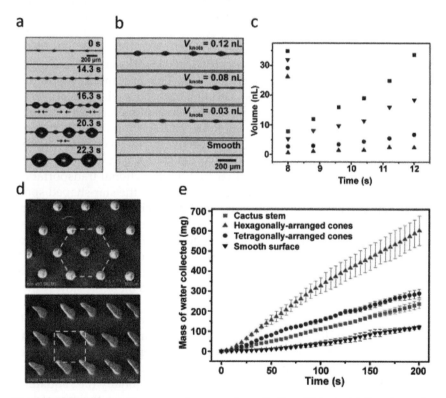

Figure 10. Effect of geometry parameters on water collection efficiency. **(a)** Growing water droplets on artificial spider silks. **(b)** Fibers with different sized knots. **(c)** Volume of collected water was plotted corresponding to time. The four kinds of dots in **(c)** correspond to the four types of fibers in **(b)** from the top down respectively, e.g., the square dots represent the fiber with V_{knots} = 0.12 nL. **(a) (b) (c)** are reprinted with permission from ref. (Bai et al. 2011). Copyright 2011 John Wiley and Sons. **(d)** Hexagonally and tetragonally arranged cone arrays on artificial cactus surfaces. **(e)** Mass of collected water on various artificial cactus surfaces. **(d) (e)** are reprinted with permission from ref. (Ju et al. 2014). Copyright 2014 John Wiley and Sons.

are more capable of collecting water than the smaller ones. The spindle-knots contribute water collection process by fuelling the transportation of condensed water droplets toward specific sites and releasing the original area for new condensation. Therefore, when this acceleration was intensified by larger spindle-knots, a more efficient collection cycle achieved.

The desert beetle's back also inspired many studies on fabricating artificial patterned wettability surfaces. There are many factors that need to be considered when designing such surfaces. The arrangement of spot arrays could be either hexagonal or square, which was studied by Parker (Parker and Lawrence 2001). Although nature chooses to use hexagonal arrays on the desert beetle's back, square arrays collected more water in Parker's experiments. And the following researches all chose to use square arrays (Bai et al. 2014, Garrod et al. 2007). Garrod et al. systemically studied the effect of the spot size and the spacing distance on the water collection efficiency. They designed a series of arrays with the spot size of 100 μm and center-to-center distance of 200 μm (100/200), 500/1000 and 2000/4000. Results showed that the combination of 500/1000 worked best in collecting water for two hours under a given fog flow. This explained why nature chooses to use 500/1000 on the desert beetle's back. Following this principle, Hao et al. (Hao et al. 2014) also prepared a series of patterned wettability surfaces with square arrays. The spot size and spacing distance in their experiment was 500 μm and 1000 μm, respectively. However, they replaced the circle spots with star-shaped spots in order to utilize the tips to generate a Laplace pressure gradient to further fuel the directional transportation of water droplets. Their results indicated that all of the star-shaped spots performed better than the circle spots. But increasing the number of tips did not guarantee an improvement on water collection efficiency. 5-pointed star-shaped spots had the highest efficiency, but 6, 8-pointed decreased. Actually, superhydrophilic area not only helped to capture fog droplets, it curbed droplet from running away. More tips means larger adhesion area, which will intensify the pining of collected water and therefore slows down the collection process.

It is reported that cactus spine is an expert in directional water droplets transportation, which helps the cactus survive in the desert (Ju et al. 2012). Ju et al. mimicked the cactus spine and prepared several types of artificial cactus surfaces (ACSs) (Ju et al. 2014). The cone arrays upon these surfaces were arranged either hexagonally or tetragonally (i.e., squarely) (Fig. 10d). To test the efficiency of water collection on these surfaces, smooth, tetragonally-arranged, hexagonally-arranged, cactus stem surfaces (same size of 1 cm by 1 cm) were put under a horizontal fog flow (velocity of about 125 cm/s), and the mass of collected water was recorded real-timely. Within 200 seconds, surface with hexagonally arranged cones collected the most water, while smooth surface collected the least (Fig. 10e). That is, surface with hexagonally arranged cones was more efficient in fog harvesting than surface with tetragonally arranged cones and the smooth surface. Interestingly, this result is consistent with the design of nature on the desert beetle's back but not with Parker's experiments. In the case of the cactus, the water collection efficiency scales with the turbulence of fog flow field. In the hexagonally arranged cone arrays, differing from the tetragonally arranged cone arrays, fog droplets can not only deposit on the windward side, but also on the leeward side of the cones, which contributes to a more efficient water collection.

5.3 Anti-Icing

Hydrophobicity and icephobicity play an important role in cold environment devices (Ryerson 2011). Reports indicated that butterfly wings not only repel water at ambient temperature, but also delay ice forming at subzero temperature (Guo et al. 2012, Zheng et al. 2007). In order to mimic the ratchet structure of butterfly wings, Guo et al. designed a mechanical working and crystal growth method (Guo et al. 2012). The microratchets were characterized with a periodicity of about 300 µm and the height of about 80 µm. ZnO nanohairs were grown upon the microratchets surface to build the hierarchical structures (MN-surface). Microratchets surfaces without nanohairs (M-surface), surfaces with only nanohairs (N-surface) and smooth surface (S-surface) were prepared as reference surfaces. FAS was deposited onto all these surfaces to modify them with low surface energy.

The icephobicity was measured by delay time (DT), the time taken for reference water droplets to freeze or become non-transparent. Reference water droplets on M-surface was the first one to become non-transparent after DT ~ 30.5 s. This happened on S-surface after DT ~ 1260 s and on N-surface after DT ~ 1740 s. However, on the MN-surface the reference droplets was transparent until DT was about 7220 s. Additionally, it was harder for ice crystals to form on the edges of the MN-surface than on N-, M-, and S-surfaces. Clearly, the microratchets surface with nanohairs performs best in anti-icing.

5.4 Oil/Water Separation

Separating oil/water mixture is still challenging in petroleum industry and oil spill accidents. The behavior of micro-sized oil droplets in water is much similar as micro-sized water droplets in air. Therefore, same as fog harvesting process, the micro-sized oil droplets could move directionally in water. By putting a surface with arrays of artificial cactus spines into oil/water mixture, high efficient and continuous oil/water separation could be achieved (Li et al. 2013). Li et al. used a PDMS replication and programmable mechanical drilling method to prepare their samples. Four types of surfaces were made: cone arrays with tetragonal or hexagonal arrangements of either smooth or rough surface.

When the arrays were fixed downward-sloping at an appropriate angle, the micro-sized oil droplets could continuously move from the spine tips to bases and coalesce into larger droplets because of the Laplace pressure gradient. The collected oil droplets could be moved away by the pump, which releases the arrays surface for a new cycle of collection and achieved a high throughput. The efficiency of oil/water separation could be as highest as 98.9 ± 0.4% on rough hexagonal arrays, and was 96.2 ± 1.9% on smooth hexagonal arrays, 95.6 ± 0.8% on rough tetragonal arrays, 93.7 ± 2.3% on tetragonal square arrays, respectively. Clearly, the rough surfaces worked better than smooth surfaces. The rough surfaces have divided three-phase contact lines (TCLs). This division raised the local radius of TCLs, and further enlarged the Laplace pressure gradient. Hence, the rough surfaced arrays were more efficient as the results showed.

Compared to traditional materials used for oil/water separation, such as membrane-based separating materials (Feng et al. 2004, Xue et al. 2011), bulk absorbing materials (Hayase et al. 2013), and hygro-responsive membrane (Kota et al. 2012), the artificial cactus spine arrays not only avoids the fouling but also achieves a high throughout.

6. Perspective

As discussed in the previous sections, great achievements have been made during the last decades to create bioinspired smart materials for controlled droplet motion. In terms of energy conversion, many external driving forces such as chemical, thermal, light and topological gradients, have been utilized to achieve droplet motion. Moreover, various kinds of surface structures have been created on multiple dimensional materials systems. More importantly, such smart materials capable of driving liquid droplets in a controllable manner have shown great potential in a wide range of applications, particularly in those areas related to energy, health and environment. Despite all these achievements, however, there still remain many challenges such as driving smaller droplets that are of greater importance in microfluidic devices but have relatively higher resistance owing to contact angle hysteresis, precisely control droplet moving direction, and combine more driving forces to drive droplets faster and in a longer distance. In addition, there is always a pressing need for building smart surfaces that are able to control droplet motion to achieve more complex missions in the fields of anti-corrosion, anti-icing, high efficient heat exchange, etc.

Can we make current smart materials even smarter? Where can we acquire more inspirations? Learning from nature is absolutely one feasible route. By just learning from the designing principles of several typical plant and animal models, we have already achieved a great success in building artificial smart surfaces for controlled droplet motion. There are still plenty of opportunities if we take nature as a rich source for inspiration. Indeed, after billions of years of evolution, natural surfaces are so smart and successful as to suit various kinds of living environments. They are made with regular materials at ambient conditions and have fine structures carefully assembled at multiple length scales ranging from nano- to macroscopic. In addition, they are always self-repairable, energy-efficient, specific for applications and multifunctional. All these intriguing properties of natural surfaces have opened up new opportunities for making artificial smart surfaces. Scientists and engineers can always learn novel ideas and concepts from nature to guide them in fabricating functional materials and developing smart engineering systems. There are always three questions for bioinspired research, in particular for bioinspired smart surfaces for controlled droplet motion.

Can we clearly describe the structural features of natural surfaces and summarize their design motifs? The answer is 'Yes' with the aid of modern equipment. Powerful tools such as scanning electron microscopy and high-speed cameras have shown the possibility to observe the fine structures of natural surfaces, and monitor the detailed process of droplet motion on such surfaces. Together with simulation studies, the mechanism for controlled droplet motion on natural surfaces can be thoroughly recovered. It is worth notifying that current research often deals with droplet motion on one dimensional or two dimensional natural surfaces, like spider silks and butterfly wings, leaving three dimensional structure seldom investigated. Indeed, droplet motion or more generally, liquid transport is crucial for the living of creatures.

Can we mimic the natural system in a reasonable time and cost? It is still quite challenging. For example, mimicking the fine structure of cactus spine involves engineering materials in three dimensions. However, it is not completely impossible. Techniques such as three dimensional printing or two photon polymerization can precisely control the local composition and architecture of materials, which are very promising to design smart surfaces with more complicated structures for controlled droplet motion.

Can we do better than natural system? The answer is, it depends. While natural surfaces are made with a fairly limited selection of materials, mostly polymers, artificial smart materials can be made with a wider range of materials, including ceramics, metals, polymers, and their composites. Therefore, more kinds of surfaces can be developed for specific applications. For example, condensers that work at organic solvent or high temperature environment. On the other hand, the finely assembled structures at multiple scales found in natural surfaces, have not yet to be successfully replicated in artificial analogues. Also, some aspects of natural surfaces are still challenging to mimic, such as self-healing capability and fast responsiveness.

Looking down the road, future efforts should be made to: (i) thoroughly investigate unique structural features of natural surfaces; (ii) unravel the relationship between their structures and controlled droplet motion; (iii) develop various methods to mimic structural features found in natural surfaces with all kinds of materials; (iv) incorporate more functions on the as-prepared smart surfaces and broaden their applications. In brief, the next generation of smart surfaces for controlled droplet motion would be designed and fabricated based on further mimicking natural surfaces on both structures and properties. This approach will bring various kinds of even 'smarter' materials for controlled droplet motion.

References

Bai, H., X. Tian, Y. Zheng, J. Ju, Y. Zhao and L. Jiang. 2010. Direction controlled driving of tiny water drops on bioinspired artificial spider silks. Advanced Materials 22: 5521–5525.

Bai, H., J. Ju, R. Sun, Y. Chen, Y. Zheng and L. Jiang. 2011a. Controlled fabrication and water collection ability of bioinspired artificial spider silks. Advanced Materials 23: 3708–3711.

Bai, H., R. Sun, J. Ju, X. Yao, Y. Zheng and L. Jiang. 2011b. Large-scale fabrication of bioinspired fibers for directional water collection. Small 7: 3429–3433.

Bai, H., L. Wang, J. Ju, R. Sun, Y. Zheng and L. Jiang. 2014. Efficient water collection on integrative bioinspired surfaces with star-shaped wettability patterns. Advanced Materials 26: 5025–5030.

Bain, C.D., G.D. Burnett-Hall and R.R. Montgomerie. 1994. Rapid motion of liquid drops. Nature 372: 414–415.

Barthlott, W. and C. Neinhuis. 1997. Purity of the sacred lotus, or escape from contamination in biological surfaces. Planta 202: 1–8.

Bejan, A. 1993. Heat Transfer. Wiley, New York.

Bico, J. and D. Quéré. 2002. Self-propelling slugs. Journal of Fluid Mechanics 467: 101–127.

Bouasse, H. 1924. Capillarité phénomènes superficiels. Paris: Delagrave.

Buguin, A., L. Talini and P. Silberzan. 2002. Ratchet-like topological structures for the control of microdrops. Applied Physics A: Materials Science & Processing 75: 207–212.

Cao, M., J. Ju, K. Li, S. Dou, K. Liu and L. Jiang. 2014. Facile and large-scale fabrication of a cactus-inspired continuous fog collector. Advanced Functional Materials 24: 3235–3240.

Chaudhury, M.K. and G.M. Whitesides. 1992. How to make water run uphill. Science 256: 1539–1541.

Chen, Y., L. Wang, Y. Xue, Y. Zheng and L. Jiang. 2012. Bioinspired spindle-knotted fibers with a strong water-collecting ability from a humid environment. Soft Matter 8: 11450.

Chen, Y., L. Wang, Y. Xue, L. Jiang and Y. Zheng. 2013. Bioinspired tilt-angle fabricated structure gradient fibers: micro-drops fast transport in a long-distance. Scientific Reports 3: 2927.

Chen, Y., J. He, L. Wang, Y. Xue, Y. Zheng and L. Jiang. 2014. Excellent bead-on-string silkworm silk with drop capturing abilities. Journal of Materials Chemistry A 2: 1230.

Daniel, S., M.K. Chaudhury and J.C. Chen. 2001. Fast drop movements resulting from the phase change on a gradient Surface. Science 291: 633–636.

De Gennes, P.-G., F. Brochard-Wyart and D. Quéré. 2004. Capillarity and wetting phenomena: drops, bubbles, pearls, waves. Springer Science & Business Media, New York.

Domingues dos Santos, F. and T. Ondarçuhu. 1995. Free-running droplets. Physical Review Letters 75: 2972–2975.

Dong, H., N. Wang, L. Wang, H. Bai, J. Wu, Y. Zheng, Y. Zhao and L. Jiang. 2012. Bioinspired Electrospun Knotted Microfibers for Fog Harvesting. Chemphyschem 13: 1153–1156.

Dorrer, C. and J.r. Rühe. 2008. Mimicking the stenocara beetle–dewetting of drops from a patterned superhydrophobic surface. Langmuir 24: 6154–6158.

Drelich, J., J.D. Miller, A. Kumar and G.M. Whitesides. 1994. Wetting characteristics of liquid drops at heterogeneous surfaces. Colloids and Surfaces A: Physicochemical and Engineering Aspects 93: 1–13.

Extrand, C.W. 2007. Retention forces of a liquid slug in a rough capillary tube with symmetric or asymmetric features. Langmuir 23: 1867–1871.

Feng, L., S. Li, Y. Li, H. Li, L. Zhang, J. Zhai, Y. Song, B. Liu, L. Jiang and D. Zhu. 2002. Super-hydrophobic surfaces: from natural to artificial. Advanced Materials 14: 1857–1860.

Feng, L., Z. Zhang, Z. Mai, Y. Ma, B. Liu, L. Jiang and D. Zhu. 2004. A super-hydrophobic and super-oleophilic coating mesh film for the separation of oil and water. Angewandte Chemie International Edition 43: 2012–2014.

Feng, S., Y. Hou, Y. Chen, Y. Xue, Y. Zheng and L. Jiang. 2013. Water-assisted fabrication of porous bead-on-string fibers. Journal of Materials Chemistry A 1: 8363.

Feng, S., Y. Hou, Y. Xue, L. Gao, L. Jiang and Y. Zheng. 2013. Photo-controlled water gathering on bio-inspired fibers. Soft Matter 9: 9294.

Gao, J., Y.L. Liu, H.P. Xu, Z.Q. Wang and X. Zhang. 2009. Mimicking biological structured surfaces by phase-separation micromolding. Langmuir 25: 4365–4369.

Garrod, R., L. Harris, W. Schofield, J. McGettrick, L. Ward, D. Teare and J. Badyal. 2007. Mimicking a stenocara beetle's back for microcondensation using plasmachemical patterned superhydrophobic-superhydrophilic surfaces. Langmuir 23: 689–693.

Gau, H., S. Herminghaus, P. Lenz and R. Lipowsky. 1999. Liquid morphologies on structured surfaces: from microchannels to microchips. Science 283: 46–49.

Gilet, T., D. Terwagne and N. Vandewalle. 2009. Digital microfluidics on a wire. Applied Physics Letters 95: 014106.

Gilet, T., D. Terwagne and N. Vandewalle. 2010. Droplets sliding on fibres. The European Physical Journal E: Soft Matter and Biological Physics 31: 253–262.

Grunze, M. 1999. Driven liquids. Science 283: 41–42.

Guo, P., Y. Zheng, C. Liu, J. Ju and L. Jiang. 2012. Directional shedding-off of water on natural/bio-mimetic taper-ratchet array surfaces. Soft Matter 8: 1770.

Guo, P., Y. Zheng, M. Wen, C. Song, Y. Lin and L. Jiang. 2012. Icephobic/anti-icing properties of micro/nanostructured surfaces. Advanced Materials 24: 2642–2648.

Hamilton, W.J. and M.K. Seely. 1976. Fog basking by the Namib Desert beetle, Onymacris unguicularis. Nature 262: 284–285.

Hayase, G., K. Kanamori, M. Fukuchi, H. Kaji and K. Nakanishi. 2013. Facile Synthesis of marshmallow-like macroporous gels usable under harsh conditions for the separation of oil and water. Angewandte Chemie International Edition 52: 1986–1989.

Hou, Y., L. Gao, S. Feng, Y. Chen, Y. Xue, L. Jiang and Y. Zheng. 2013. Temperature-triggered directional motion of tiny water droplets on bioinspired fibers in humidity. Chemical Communications 49: 5253–5255.

Ichimura, K., S.-K. Oh and M. Nakagawa. 2000. Light-driven motion of liquids on a photoresponsive surface. Science 288: 1624–1626.

Ju, J., H. Bai, Y.M. Zheng, T.Y. Zhao, R.C. Fang and L. Jiang. 2012. A multi-structural and multi-functional integrated fog collection system in cactus. Nature Communications 3: 1247.

Ju, J., K. Xiao, X. Yao, H. Bai and L. Jiang. 2013. Bioinspired conical copper wire with gradient wettability for continuous and efficient fog collection. Advanced Materials 25: 5937–5942.

Ju, J., X. Yao, S. Yang, L. Wang, R. Sun, Y. He and L. Jiang. 2014. Cactus stem inspired cone-arrayed surfaces for efficient fog collection. Advanced Functional Materials 24: 6933–6938.

Kang, E., G.S. Jeong, Y.Y. Choi, K.H. Lee, A. Khademhosseini and S.-H. Lee. 2011. Digitally tunable physicochemical coding of material composition and topography in continuous microfibres. Nature Materials 10: 877–883.

Kota, A.K., G. Kwon, W. Choi, J.M. Mabry and A. Tuteja. 2012. Hygro-responsive membranes for effective oil-water separation. Nature Communications 3: 1025.

Kusumaatmaja, H. and J.M. Yeomans. 2009. Anisotropic hysteresis on ratcheted superhydrophobic surfaces. Soft Matter 5: 2704–2707.

Lee, W., M.-K. Jin, W.-C. Yoo and J.-K. Lee. 2004. Nanostructuring of a polymeric substrate with well-defined nanometer-scale topography and tailored surface wettability. Langmuir 20: 7665–7669.

Li, K., J. Ju, Z. Xue, J. Ma, L. Feng, S. Gao and L. Jiang. 2013. Structured cone arrays for continuous and effective collection of micron-sized oil droplets from water. Nature Communications 4: 2276.

Linke, H., B. Aleman, L. Melling, M. Taormina, M. Francis, C. Dow-Hygelund, V. Narayanan, R. Taylor and A. Stout. 2006. Self-propelled Leidenfrost droplets. Physical Review Letters 96: 154502.

Liu, C., J. Ju, J. Ma, Y. Zheng and L. Jiang. 2014. Directional drop transport achieved on high-temperature anisotropic wetting surfaces. Advanced Materials 26: 6086–6091.

Liu, C., J. Ju, Y. Zheng and L. Jiang. 2014. Asymmetric ratchet effect for directional transport of fog drops on static and dynamic butterfly wings. ACS Nano 8: 1321–1329.

Lopez, G.P., H.A. Biebuyck, C.D. Frisbie and G.M. Whitesides. 1993. Imaging of features on surfaces by condensation figures. Science 260: 647–649.

Lorenceau, L. and D. Quéré. 2004. Drops on a conical wire. Journal of Fluid Mechanics 510: 29–45.

Ma, J., L. Wen, Z. Dong, T. Zhang, S. Wang and L. Jiang. 2013. Aligned silicon nanowires with fine-tunable tilting angles by metal-assisted chemical etching on off-cut wafers. Physica Status Solidi (RRL)—Rapid Research Letters 7: 655–658.

Malvadkar, N.A., M.J. Hancock, K. Sekeroglu, W.J. Dressick and M.C. Demirel. 2010. An engineered anisotropic nanofilm with unidirectional wetting properties. Nature Materials 9: 1023–1028.

Marangoni, C., P. Stefanelli and R. Liceo. 1872. Monografia sulle bolle liquide. Nuovo Cim 7-8: 301–356.

Mettu, S. and M.K. Chaudhury. 2008. Motion of drops on a surface induced by thermal gradient and vibration. Langmuir 24: 10833–10837.

Mohammadi, R., J. Wassink and A. Amirfazli. 2004. Effect of surfactants on wetting of super-hydrophobic surfaces. Langmuir 20(22): 9657–9662.

Moon, S.J., S.S. Lee, H. Lee and T. Kwon. 2005. Fabrication of microneedle array using LIGA and hot embossing process. Microsystem Technologies 11: 311–318.

Parker, A.R. and C.R. Lawrence. 2001. Water capture by a desert beetle. Nature 414: 33–34.

Porter, D. and F. Vollrath. 2009. Silk as a biomimetic ideal for structural polymers. Advanced Materials 21(4): 487–492.

Prakash, M., D. Quéré and J.W.M. Bush. 2008. Surface tension transport of prey by feeding shorebirds: the capillary ratchet. Science 320: 931–934.

Quéré, D. 2008. Wetting and roughness. Annual Review of Materials Research 38: 71–99.

Renvoise, P., J.W.M. Bush, M. Prakash and D. Quéré. 2009. Drop propulsion in tapered tubes. Europhysics Letters 86: 64003.

Rubega, M.A. and B.S. Obst. 1993. Surface-tension feeding in phalaropes: discovery of a novel feeding mechanism. The Auk 1993: 169–178.

Ryerson, C.C. 2011. Ice protection of offshore platforms. Cold Regions Science and Technology 65: 97–110.

Scriven, L.E. and C.V. Sternling. 1960. The Marangoni effects. Nature 187: 186–188.

Soeno, T., K. Inokuchi and S. Shiratori. 2004. Ultra-water-repellent surface: fabrication of complicated structure of SiO_2 nanoparticles by electrostatic self-assembled films. Applied Surface Science 237: 539–543.

Song, C., L. Zhao, W. Zhou, M. Zhang and Y. Zheng. 2014. Bioinspired wet-assembly fibers: from nanofragments to microhumps on string in mist. Journal of Materials Chemistry A 2: 9465.

Strutt, J.W. and L. Rayleigh. 1878. On the instability of jets. Proc. London Math. Soc. 10: 4–13.

Sumino, Y., N. Magome, T. Hamada and K. Yoshikawa. 2005. Self-running droplet: emergence of regular motion from nonequilibrium noise. Physical Review Letters 94: 068301.

Sun, T., G.J. Wang, H. Liu, L. Feng, L. Jiang and D.B. Zhu. 2003. Control over the wettability of an aligned carbon nanotube film. Journal of the American Chemical Society 125: 14996–14997.

Tian, X., H. Bai, Y. Zheng and L. Jiang. 2011. Bio-inspired heterostructured bead-on-string fibers that respond to environmental wetting. Advanced Functional Materials 21: 1398–1402.

Tian, X., Y. Chen, Y. Zheng, H. Bai and L. Jiang. 2011. Controlling water capture of bioinspired fibers with hump structures. Advanced Materials 23: 5486–5491.

Wasan, D.T., A.D. Nikolov and H. Brenner. 2001. Fluid dynamics-droplets speeding on surfaces. Science 291: 605–606.

Xue, Y., Y. Chen, T. Wang, L. Jiang and Y. Zheng. 2014. Directional size-triggered microdroplet target transport on gradient-step fibers. Journal of Materials Chemistry A 2: 7156–7160.

Xue, Z., S. Wang, L. Lin, L. Chen, M. Liu, L. Feng and L. Jiang. 2011. A novel superhydrophilic and underwater superoleophobic hydrogel-coated mesh for oil/water separation. Advanced Materials 23: 4270–4273.

Yarin, A.L., W. Liu and D.H. Reneker. 2002. Motion of droplets along thin fibers with temperature gradient. J. Applied Physics 91: 4751–4760.

Young, T. 1805. An essay on the cohesion of fluids. Philosophical Transactions of the Royal Society of London 1805: 65–87.

Zahner, D., J. Abagat, F. Svec, J.M. Fréchet and P.A. Levkin. 2011. A facile approach to superhydrophilic–superhydrophobic patterns in porous polymer films. Advanced Materials 23: 3030–3034.

Zhai, L., M.C. Berg, F.Ç. Cebeci, Y. Kim, J.M. Milwid, M.F. Rubner and R.E. Cohen. 2006. Patterned superhydrophobic surfaces: toward a synthetic mimic of the namib desert beetle. Nano Letters 6: 1213–1217.

Zhang, J., Z. Cheng, Y. Zheng and L. Jiang. 2009. Ratchet-induced anisotropic behavior of superparamagnetic microdroplet. Applied Physics Letters 94: 144104.

Zhang, X., H. Kono, Z. Liu, S. Nishimoto, D.A. Tryk, T. Murakami, H. Sakai, M. Abe and A. Fujishima. 2007. A transparent and photo-patternable superhydrophobic film. Chemical Communications 46: 4949–4951.

Zheng, Y., X. Gao and L. Jiang. 2007. Directional adhesion of superhydrophobic butterfly wings. Soft Matter 3: 178–182.

Zheng, Y., H. Bai, Z. Huang, X. Tian, F.-Q. Nie, Y. Zhao, J. Zhai and L. Jiang. 2010. Directional water collection on wetted spider silk. Nature 463: 640–643.

Advanced Materials for Thermoelectric Applications

Lianjun Wang,[1,*] *Qihao Zhang*[2] *and Wan Jiang*[3]

ABSTRACT

Thermal to electrical energy conversion, through thermoelectric materials, has been proposed to be much more efficient in improving the existing energy efficiency, so as to meet the demand for energy conservation and environment protection. The strongly-linked physical parameters of thermoelectric materials, however, restrict the improvement of thermoelectric conversion efficiency, which always requires tradeoffs between various material properties. Designing smart materials is one of the important and effective ways to develop advanced functional materials, which plays a crucial role in developing materials for thermoelectric applications as well. This chapter will summarize the progress and applications that have been made in recent years in developing advanced bulk nanostructured thermoelectric materials with a high dimensionless

[1] State Key Laboratory for Modification of Chemical Fibers and Polymer Materials, Donghua University, 2999 North Renmin Road, Songjiang, Shanghai 201620, People's Republic of China.

[2] State Key Laboratory for Modification of Chemical Fibers and Polymer Materials, Donghua University, 2999 North Renmin Road, Songjiang, Shanghai 201620, People's Republic of China.
  Email: zhangqihao@student.sic.ac.cn

[3] School of Material Science and Engineering, Jingdezhen Ceramic Institute, Jingdezhen 333000, China.
  Email: wanjiang@dhu.edu.cn

* Corresponding author: wanglj@dhu.edu.cn

figure of merit and the related fabrication processes. The chapter will provide detailed information on some typical bulk nanostructured thermoelectric materials, including Bi_2Te_3-based alloys, SiGe alloys, $CoSb_3$-based skutterudites, PbTe and its alloys, Half-Heusler compounds, etc. Finally, current applications of bulk nanostructured thermoelectric materials will be reviewed and an outlook of new directions in this field will be provided.

1. Introduction

Energy is universally recognized as the physical basis of human life as well as the necessary demand of social development. However, traditional non-renewable energy has become increasingly exhausted with the ongoing of human consumption and the industrial activities, of which the discovered oil reserves, as is predicted, can only last for another 50 years at the current consumption rate. Exploitation of advanced, cleaner, more sustainable energy sources and improvement of the existing energy efficiency, therefore, are imminent. Among all viable technologies for addressing this issue, thermoelectric (TE) energy converters that can easily achieve the mutual conversion between thermal energy and electric energy have attracted increasing attention as promising alternative environmental friendly applications. This is because they have advantages of solid-state operation, no mechanical moving parts, no releasing of greenhouse gases, compact in size, light in weight, being powered by direct current, being silent, reliable, and versatile (Rowe 1995, Rowe 2006). The solid-state TE devices can transform heat into electric power using the Seebeck effect and, conversely, change electrical energy into thermal energy for cooling or heating using the Peltier effect. This has been utilized or is at the trial stage in fields like military, aerospace, industrial, transportation tools, medical services, electronic, detecting temperature and measuring facilities, meeting the demand for energy conservation and environment protection (Snyder and Toberer 2008).

TE materials as the core component of TE devices, are decisive to the efficiency of conversion. The larger the value of ZT, the higher is the conversion efficiency. A suitable high-performance TE material should have a ZT of ≥ 3 for practical purposes, as a TE power conversion device with ZT = 3 operating between 500 and 30°C would yield about 50% of the Carnot efficiency (Yang and Caillat 2006). The ZT value of TE materials is defined as ZT = $(\alpha^2\sigma/\kappa)T$, where α is the Seebeck coefficient, σ is the electric conductivity, T is the absolute temperature and κ is the thermal conductivity that consists of three parts ($\kappa = \kappa_c + \kappa_l + \kappa_{bi}$, where κ_c, κ_l and κ_{bi} represent the carrier thermal conductivity, lattice thermal conductivity and bipolar thermal conductivity, respectively) (Venkatasubramanian et al. 2001). It is obvious that an ideal TE material should combine a large α to create a potential difference, a high σ to minimize Joule heating with a low κ to maintain a large temperature gradient between the hot and cold sides. However, nature does not provide many materials with these properties, since metals possess both high σ and κ, while glass or ceramic insulates heat and electricity. In principle, the parameters (α, σ and κ) are strongly linked to each other via band structures and scattering mechanisms. Therefore, producing the "BEST" TE material becomes a challenging task, which requires tradeoffs between various material properties. Designing smart materials is one of the important and effective ways to develop advanced functional materials, which plays a crucial role in developing materials for thermoelectric applications as well.

Figure 1 presents the ZT_{max} values of many typical TE materials as a function of years (Chen 2014). TE materials were initially studied on metals, which display considerably low Seebeck coefficients of a few tens of $\mu V \cdot K^{-1}$. In the middle of the 20th century, interest turned to use semiconductors as TE materials due to their high Seebeck coefficients and heat conduction dominated by phonon transport, regardless of small ratios of electrical to thermal conductivity (Rowe 1995, Rowe 2006). As a result, bismuth telluride (Bi_2Te_3) was discovered by Goldsmid in 1954 with the selection rules of choosing materials with high Seebeck coefficients and high atomic weights (Goldsmid and Douglas 1954). Soon afterwards, Ioffe in 1956 demonstrated that the ratio of electrical to thermal conductivity

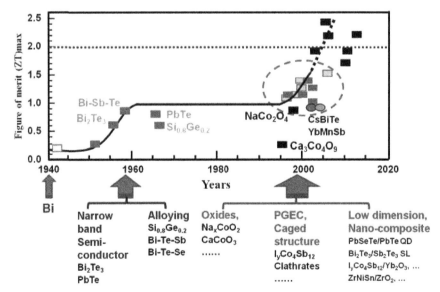

Figure 1. The ZT_{max} values of typical TE materials as a function of years. Adapted with permission from ref. (Chen 2014), 33rd ICT Nashville, USA.

could be enhanced if the TE material is alloyed with an isomorphous element or compound (Ioffe 1956). He suggested that the disturbances in the short-range order would scatter phonons, but the preservation of long-range order would prevent scattering of electrons and holes, resulting in reduced lattice thermal conductivity without affecting carrier mobility. This brought about an extensive study of the TE performance of various semiconductor alloy systems over a wide range of temperatures. Examples are Bi_2Te_3 and its alloys such as $Bi_xSb_{2-x}Te_3$ for room temperature, PbTe and its alloys like PbTe-PbSe for moderate temperature, and the $Si_{0.8}Ge_{0.2}$ alloy for high temperature (Minnich et al. 2009). The major approach to enhance the TE properties in bulk materials was carried out by tuning or doping techniques to reduce the lattice thermal conductivity with no effect on the electrical conductivity. It was quickly realized that Bi_2Te_3 alloyed with Sb_2Te_3 and Bi_2Se_3 could allow for the fine tuning of the carrier concentration along with a reduction in lattice thermal conductivity (Snyder and Toberer 2008). The most commonly studied p-type compositions are near $(Sb_{0.8}Bi_{0.2})_2Te_3$ whereas n-type compositions are close to $Bi_2(Te_{0.8}Se_{0.2})_3$, of which peak ZT values are typically in the range of 0.8 to 1.1. From 1960s to 1990s, however, the progress in improving ZT was nearly stagnant and the value of maximum ZT remained essentially around 1 with the $(Bi_{1-x}Sb_x)_2(Se_{1-y}Te_y)_3$ alloy family. During this period, the TE field received little attention from the worldwide scientific research community and research funding in this area dwindled (Dresselhaus et al. 2007). Fortunately, the situation in TE research picked up in the 1990s due to several new conceptual developments and renewed interest from several research funding agencies. Since Hicks and Dresselhaus proposed the concept that quantum confinement of electrons and holes in low-dimensional materials could dramatically increase ZT over 1 by independently changing $\alpha^2\sigma$ in 1993, intense research into developing nanostructured TE materials was ignited (Hicks and Dresselhaus 1993). Over the past two decades, two different approaches have been developed to search for the advanced materials for TE applications: one is developing new families of bulk TE materials with complex crystal structures, and the other is synthesizing and using low-dimensional TE systems. The maximum value of ZT has constantly reached new highs for different TE systems, reaching 1.4 in a p-type nanocrystalline BiSbTe bulk alloy (Poudel et al. 2008), 1.5 for "phonon-liquid electron-crystal" structured $Cu_{2-x}Se$ (Liu et al. 2012), 1.7 for $Cu_{2-x}S$ (He et al. 2014), 2.2 for bulk TE materials with all-scale hierarchical architectures (PbTe + 4 mol% SrTe + 2 mol% Na) (Biswas et al. 2012), and 2.6 for SnSe single crystals measured along the b axis of the room-temperature orthorhombic unit cell (Zhao et al. 2014). The TE materials have thus developed into a huge family, including different materials from semimetals, semiconductor, ceramic, oxides to polymers, containing various crystalline forms from monocrystals, polycrystals to nanocomposite and covering varying dimensions from bulk, film, wires to clusters (Alam and Ramakrishna 2013). With the soaring interests in TE materials, there are some outstanding review articles on the different types, including fundamental phenomena of physics terms contributing to ZT value (Minnich et al. 2009, Shakouri 2011, Pichanusakorn and Bandaru 2010), nanostructures modifying the electron and phonon transport (Medlin and Snyder 2009, Lan et al. 2010, Li et al. 2010, Vineis et al. 2010, Wan et al. 2010), advances in TE nanocomposites (Alam and Ramakrishna 2013, Chen et al. 2012, He et al. 2013, Zhao et al. 2014), development in low-dimensional TE materials (Dresselhaus et al. 2007), and so on (Liu et al. 2012, Han et al. 2014).

Although a high ZT has been reported in nanostructured materials such as superlattices (Venkatasubramanian et al. 2001), quantum dots (Harman et al. 2002) and nanowires (Hochbaum et al. 2008, Boukai et al. 2008), many of these materials are not practical for large-scale commercial use because they are fabricated by atomic layer deposition processes such as molecular beam epitaxy, making them slow and expensive to fabricate and restricting the amount of material that can be produced (Minnich et al. 2009). Another type of nanostructured material, known as a bulk nanostructured material, is a kind of advanced material which is fabricated using a bulk process rather than a nanofabrication process. This advanced material has the important advantage of being able to be produced in large quantities and in a form that is compatible with commercially available devices. So far, of all the nanostructured materials only bulk nanostructured materials have been produced in enough quantity to be used in practice. In this chapter, we will summarize the developments made in the last two decades in developing advanced bulk nanostructured TE materials with a high dimensionless figure of merit and the related fabrication processes. The chapter will then provide detailed information on some selected advanced bulk nanostructured TE materials, including Bi_2Te_3-based alloys, PbTe and its alloys, Skutterudites, Half-Heusler compounds and SiGe alloys. Finally, we will introduce current applications of the advanced bulk nanostructured TE materials and provide an outlook of new directions in this field.

2. Fabrication of Advanced Bulk Nanostructured Thermoelectric Materials

There are two advanced approaches to the synthesis of advanced bulk nanostructured TE material. One is the bottom-up strategy, by which TE nanopowders are usually fabricated first via various methods and then assembled into dense bulk form by hot press or spark plasma sintering. The other is an *in situ* strategy, or top-down strategy. In this case, nanophases are formed *in situ* from the bulk matrix. Such a nanocomposite has an improved homogeneity and clean interfaces. More importantly, the nanostructures in *in situ* composites are more thermodynamically stable than those in *ex situ* nanocomposites by a bottom-up method. However, the latter has significant grain growth during thermal consolidation, weakening the effect of nanostructures (Zhu et al. 2010).

2.1 Ball-Milling and Hot-Pressing Method

In recent years, researchers have increasingly turned to nontraditional processing technologies in order to obtain novel TE compositions with highly refined microstructures. Among these techniques, mechanical alloying (MA) has emerged as a leading alternative to equilibrium melt-growth processes (Rowe 2006, Koch et al. 2010). The MA technique has not only relative convenience and minimal requirement for complex equipment, but also exhibits many of the characteristics of rapid solidification thanks to the solid-state nature of the process. Ball-milling is an effective top-down industrial approach to obtain fine particles. Conventional ball-milling has been employed to create large quantities of fine particles with the size of one to several micrometers while high-energy ball-milling, developed in the 1970s, can produce particles as small as several nanometers (Pierrat et al. 1997, Kim et al. 2000). It has become a popular method because of its simplicity, applicability to essentially all classes of material and the possibility of easy scale-up (Wang and Qin 2003, Papageorgiou et al. 2010, Kuo et al. 2010, Ioannou et al. 2012, Kanatzia et al. 2013). In this technique, elemental powders or intermetallic compounds are inserted into a milling vial with some balls. The vial is then loaded into a high energy ball mill. The constant agitation of the ball mill causes the powders to be subjected to a series of impact collisions between the powder and balls. As a result, the powders are constantly cold-welded and fractured, leading to the formation of nanostructured domains. Details on the ball milling process, different types of mills, kinetics, thermodynamics and related information can be found in Ref. (Suryanarayana 2001).

Hot-pressing was used to prepare PbSe composites as early as 1960 and was then employed to produce SiGe polycrystalline materials for NASA space missions (Lan et al. 2010). Currently, the combination of ball-milling and hot-pressing has become one mature and powerful method to create nanograined bulk TE materials. In this technique, raw materials, either crystalline ingots or individual elements, are ground into nanoparticles using a ball milling machine in a short time without involving the tedious melting and slow cooling for crystal growth. The resulting mixture is then hot pressed at an appropriate temperature and pressure. If the pressing temperature and duration are adjusted correctly, the nanoparticles will fuse together but leave the interface between each particle intact, creating a fully dense solid material with nanosize grains, as shown in Fig. 2 (Minnich et al. 2009). This technique, which does not involve melting and quenching and thus avoids crystallization during quenching, allows the manufacture of bulk amorphous samples with higher dimensions as compared to those prepared by the melt/quenching technique (Hubert et al. 2013).

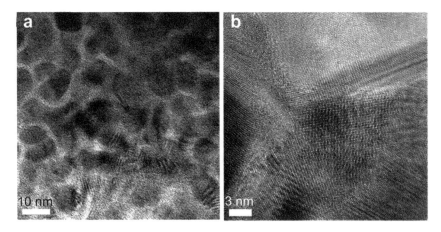

Figure 2. (a) TEM and **(b)** HRTEM images of a $Si_{80}Ge_{20}$ nanocomposite. Reproduced with permission from ref. (Wang et al. 2008), copyright © AIP Publishing LLC 2008.

This technique has been successfully implemented on many advanced TE compounds, such as Bi_2Te_3-based compounds (Poudel et al. 2008, Ma et al. 2008, Lan et al. 2009, Yu et al. 2009, Zhang et al. 2012), PbTe-based alloys (Yu et al. 2010), SiGe alloys (Wang et al. 2008, Joshi et al. 2008, Zhu et al. 2009), skutterudite (Yang et al. 2009, Bao et al. 2006, He et al. 2008, He et al. 2008), β-$FeSi_2$ (Ito et al. 2003, Cai et al. 2004, Ito et al. 2006), Half-Heuslers (Yan et al. 2011), MgSi alloys (Song et al. 2007a, Song et al. 2007b, Bux et al. 2011) and so on (Bottger et al. 2010, Shin et al. 2013). A summarized table of advanced TE materials prepared by hot pressing and ball milling method has been given by Lan et al. in Ref. (Lan et al. 2010). Under appropriate processing conditions, the produced TE nanocomposites have a reduced thermal conductivity and a higher ZT. For example, a peak ZT improvement from 1.0 to 1.4, a great breakthrough, is achieved in p-type $Bi_{2-x}Sb_xTe_3$ nanocomposites produced by a ball-milling and hot-pressing method, resulting from the decrease in thermal conductivity caused by the strong interface and point defect scattering of heat-carrying phonons (Poudel et al. 2008).

2.2 Chemical Synthesis and Spark Plasma Sintering Method

Chemical synthesis (also referred to as "bottom-up" approach) is using the chemical properties of single molecules to cause single-molecule components to self-organize or self-assemble into useful structures (Shi et al. 2013). Owing to the advantages in saving energy, convenient manipulation, excellent control over size and morphology, etc., chemical synthesis has become more popular in the fabrication of advanced TE nanostructures. The chemical synthesis attempts have so far included an acid pyrolysis process (Ritter and Maruthamuthu 1997), microemulsion process (Purkayastha et al. 2006), organometallic colloidal synthesis routes (Foos et al. 2001, Lu et al. 2005, Goldsmid and Douglas 1954, Zhao and Burda 2009, Scheele et al. 2009, Dirmyer et al. 2009, Zhao et al. 2010), soft template assisted hydrothermal solvothermal routes (Wang et al. 2005, Shi et al. 2006, Wang et al. 2008, Shi et al. 2008, Zhang et al. 2009, Shi et al. 2009, Datta et al. 2010), and microwave-assisted synthesis process (Zhou et al. 2008). Among all developed synthetic methods, the hydrothermal method based on a water system has attracted more and more attention because of its outstanding advantages, such as high yield, simple manipulation, easy control, uniform products, lower air pollution, low energy consumption and so on (Shi et al. 2013). Hydrothermal synthesis can be defined as a method of formation and growth of crystals by chemical reactions and solubility changes of substances in a sealed heated aqueous solution above ambient temperature and pressure (Feng and Xu 2001, Cundy and Cox 2003). Specifically, precursor materials (usually metal oxides or halides) in the specific end-product stoichiometry are combined with a reducing agent in the presence of an aqueous solution which is then loaded into an autoclave and sealed. The mixture is then pressurized and heated to a temperature above the critical point of the solvent, where it is held for a desired period of time, and then cooled to room temperature. The nanostructured product is then extracted and isolated from the residual salts (Bux et al. 2010). It can produce highly crystalline nanopowders, in most cases without further requirement for calcination, and it can also make doping of foreign ions relatively straightforward to realize, which is of great importance for the adjustment of carrier concentrations in the TE materials (Zhao et al. 2011). Precise control over hydrothermal synthetic conditions is a key to the success of the preparation of TE nanostructures. The controllable conditions can be classified as the interior reaction system conditions (such as concentration, pH value, time, pressure, organic additives or templates) and the exterior reaction environment conditions (such as the modes of input energy) (Deng et al. 2002, Deng et al. 2003a, Deng et al. 2003b, Zhao et al. 2004a, Zhao et al. 2004b, Zhao et al. 2005). Based on the adjustment of the two kinds of conditions, various hydrothermal synthetic strategies of TE nanostructures with different morphologies (e.g., nanoparticles, nanotubes, nanowires, nanobelts, nanopatelets, nanostring-cluster, etc.) have been developed (Shi et al. 2006, Zhang et al. 2009, Wang et al. 1999, Liu et al. 2002, Zhang et al. 2004, Zhao et al. 2005, Zhang et al. 2005, Mi et al. 2010, Gharleghi et al. 2014), which provides the possibility of achieving further improvement in TE efficiency because phonon scattering in nanostructured materials is dependent on the size and shape of the nanostructures. A good review of the reaction mechanism during a solvothermal reaction can be found in Ref. (Zhao et al. 2006).

Spark plasma sintering (SPS), commonly also defined as field assisted sintering (FAST) or pulsed electric current sintering (PECS) is a novel pressure assisted pulsed electric current sintering process utilizing ON-OFF DC pulse energizing, as is shown in Fig. 3 (Lakshmanan 2007). Some scholars believe that due to the repeated application of an ON-OFF DC pulse voltage and current in powder materials, the spark discharge point and the Joule heating point (local high temperature-state) are transferred and dispersed to the overall specimen (Lakshmanan 2007). The SPS process is based on the electrical spark discharge phenomenon: a high energetic, low voltage spark pulse current momentarily generates high localized temperatures, from several to ten thousand degrees between the particles resulting in

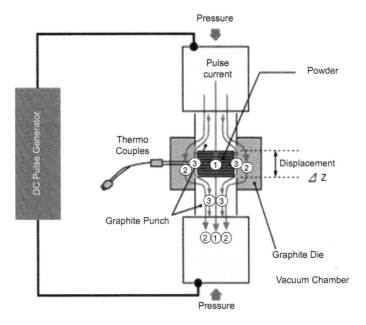

Figure 3. Schematic diagram of the sintering principle of spark plasma sintering. Reproduced with permission from ref. (Huang and Nayak 2012), an open access article under the terms of the Creative Commons Attribution License.

high thermal and electrolytic diffusion. During the SPS treatment, powders contained in a die can be processed for diverse novel bulk material applications, for example nanostructured materials, functional gradated materials, hard alloys, biomaterials, porous materials, etc., among which materials can be metals, inter-metallic, ceramics, composites or polymers. It is a novel technique used not only to sinter samples within a few minutes so as to minimize the post-synthetic crystal grain growth of nanoparticles, but to obtain densities as high as 99.9% (Tokita 1993).

Compacting chemistry-synthesized nanoparticles into dense samples by SPS is widely considered an efficient strategy to achieve low thermal conductivity due to the enhanced scattering of phonons at the numerous grain boundaries, and at the same time samples with high density after compaction could also show increased thermopower (Fan et al. 2011). Based on this combination, great achievements have been made in different bulk nanostructured TE system (Li et al. 2013). For example, Son et al. attempted on the large-scale synthesis of ultrathin Bi_2Te_3 nanoplates by the reaction between bismuth dodecanethiolate and tri-noctylphosphine telluride in the presence of oleylamine and subsequent spark plasma sintering to fabricate n-type nanostructured bulk TE materials (Son et al. 2012). They found that the grain size and density of Bi_2Te_3 nanostructured bulk materials were strongly dependent on the SPS temperature. Consequently, the highest ZT of 0.62 was achieved in sample sintered at 250°C. Park et al. described n-type nanostructured bulk TE La-doped $SrTiO_3$ materials produced by spark plasma sintering of chemically synthesized colloidal nanocrystals (Park et al. 2014). The results showed that nanostructured bulk La-doped $SrTiO_3$ exhibited a maximum ZT of 0.37 at 973 K at 9.0 at% La doping, making one of the highest values reported for doped $SrTiO_3$. Furthermore, the combination of chemical synthesis and spark plasma sintering has distinct advantages to achieve bulk nanostructured TE composites, where the nanometer-sized second-phase could disperse within the TE matrix homogeneously, leading not only to the reduced lattice thermal conductivity due to the newly built interfaces scattering of heat-carrying phonons, but simultaneously enhancing the Seebeck coefficient owing to electron energy filtering effect caused by the scattering of electrons on the band bending at the interfaces between nanoinclusions and the semiconductor host. Zhang et al. particularly, reported to introduce low-dimensional silver (silver nanowires or silver nanoparticles) into the Bi_2Te_3 matrix, where the Bi_2Te_3 nanopowders were prepared by the surfactant-mediated hydrothermal method and low-dimensional silver were obtained by using polyol reduction of silver nitrate, respectively, followed by spark plasma sintering (Zhang et al. 2014, Zhang et al. 2015). By regulating the content of low-dimensional silver and fine-tuning the architecture of nanostructured TE materials, they obtained more than three-fold improvement in the maximum ZT, as is shown in Fig. 4.

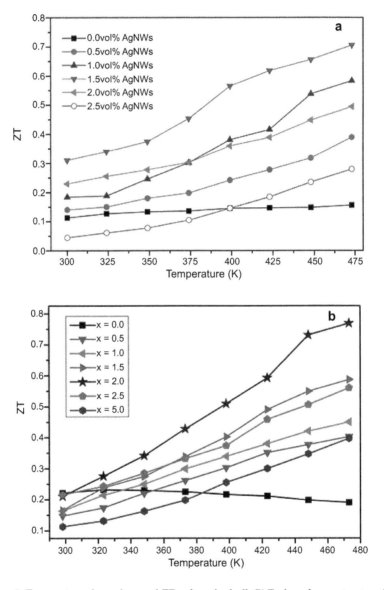

Figure 4. Temperature dependence of ZT values for bulk Bi_2Te_3-based nanostructured TE composites containing different content of **(a)** AgNWs, **(b)** AgNPs. Reproduced with permission from ref. (Zhang et al. 2014), copyright © Elsevier B.V. or its licensors or contributors. AIP Publishing LLC 2008 and (Zhang et al. 2015), copyright © WILEY-VCH Verlag GmbH & Co. KGaA, Weinheim 2014, respectively.

2.3 Melt Spinning and Spark Plasma Sintering Method

The melt spinning (MS), as a single roller rapid solidification technique, was developed by the metallurgical community in the 1960s to produce metastable solid solutions and amorphous phases. It is a powerful technique for rapid solidification which emphasizes the formation of alloys with a finer structure and less segregation during melt solidification. In the MS process, a jet of liquid metal is ejected from a nozzle and impinges on the outer surface of a moving substrate (rotating wheel), where a thin layer is formed from a melt puddle and rapidly solidifies as a continuous ribbon (Tkatch et al. 2002, Xie et al. 2013). The feasible adjustment of linear speed of copper roller (cooling rate) can realize the microstructural regulation and due to the ultra-high cooling rate in the melt spinning process, amorphous phases can be obtained (Tang et al. 2007).

The MS technique combined with a SPS process developed by the team in Wuhan University of Technology in China is an advanced method to fabricate bulk TE materials with a unique low-dimensional structure, as MS is an effective way to create a wide spectrum of multi-scale nanostructures and the SPS process can then control the as-formed nanostructures in the final product (Tang et al. 2007). The method has proved to be effective in reducing crystal size, enhancing phonon scattering, and improving TE performance, which has been used to produce high-performance Bi-Te compounds (Tang et al. 2007, Xie et al. 2009, Wang et al. 2011), filled skutterudites (Zhu et al. 2009, Li et al. 2008, Li et al. 2009, Tan et al. 2013), beta-Zn_4Sb_3 (Wang et al. 2011), Half-Heusler Alloys (Yu et al. 2010) and $AgSbTe_2$ (Du et al. 2011). Among them, Xie et al. reported that the p-type fine nanostructured $Bi_{0.52}Sb_{1.48}Te_3$ bulk material with unique microstructures obtained by MS combined with SPS presented the maximum ZT values of 1.56 at 300K, which was over 50% higher than that of the state-of-the-art commercial Bi_2Te_3-based materials (Xie et al. 2009).

2.4 Thermal Processing Techniques

Another method to produce advanced bulk nanostructured TE materials is to use thermal processing techniques to induce the formation of nanoscale precipitates, an approach which has been used successfully especially by the Kanatzidis group in several material systems such as $AgPb_mSbTe_{m+2}$ (Hsu et al. 2004, Quarez et al. 2005), $NaPb_mSbTe_{m+2}$ (Poudeu et al. 2006), $Ag_x(Pb,Sn)_mSb_yTe_{m+2}$ (Androulakis et al. 2006), and $Pb_{9.6}Sb_{0.2}Te_{10-x}Se_x$ (Poudeu et al. 2006). By choosing the appropriate compounds and subjecting the material to a thermal processing procedure, a metastable solid solution of different elements can be made to undergo spinodal decomposition or nucleation and growth mechanisms to create nanoscale features (Androulakis et al. 2007). A typical technique, known as matrix encapsulation, is using the fact that some materials are completely soluble in others in the liquid state but have very low or no solubility in the solid state (Sootsman et al. 2006). By rapidly cooling the liquid mixture from a molten homogeneous solution to the frozen state, the insoluble minority phase will precipitate since it reaches or exceeds the solid state solution limit, forming nanoparticles embedded in the host phase (Girard et al. 2010, Zhao et al. 2011, Girard et al. 2011, Androulakis et al. 2011). The technique used for the formation of precipitates was elaborated in Ref. (Zhao et al. 2014).

These techniques have been applied with considerable success to achieve high ZTs, such as ZT_{max} = ~ 1.8 for n-type $AgPbTe_2/PbTe$ (Hsu et al. 2004), 1.5 for n-type $Pb_{1-x}Sn_xTe/PbS$ (Androulakis et al. 2007), and 1.4 for p-type $Na_{0.8}Pb_{20}Sb_{0.6}Te_{22}$, ~ 1.5 for p-type $Na_{0.95}Pb_{20}SbTe_{22}$ (Poudeu et al. 2006). The advancements have primarily come from sizeable reductions in thermal conductivity (see Fig. 5), which is largely the result of the scattering of mid- and long-wavelength phonons at the interfaces of the nanoscale inclusions at high temperature (Sootsman et al. 2008). To give an idea of the magnitude of the reduction, bulk PbTe has a thermal conductivity of about 2.4 $W \cdot m^{-1}K^{-1}$ at 300 K, while PbTe with two percent Sb that is matrix encapsulated has a thermal conductivity of about 0.8 $W \cdot m^{-1}K^{-1}$ (Minnich et al. 2009). Similar results also appear in the PbSe (Lee et al. 2013) and PbS (Zhao et al. 2011, Zhao et al. 2012) systems.

2.5 Other Methods

Besides the above reviewed methods, advanced bulk nanostructured TE materials are also reported to be prepared by other methods or combinations. For example, the thermal conductivity of $Ag_{0.8}Pb_{18+x}SbTe_{20}$ nanocomposites decreases 40% after ball-milling and spark plasma sintering and a peak ZT of 1.3 has been reported at 300°C (Wang et al. 2006). $Ga_mSb_nTe_{1.5+m+n}$ nanocomposites with peak ZT = 0.98 can be prepared by the spark plasma sintering method from ball-milled nanoparticles (Cui et al. 2009). Zn_4Sb_3 nanocomposites can be produced by sintering cold-pressed samples (Ur et al. 2007). In $Bi_{100-x}Sb_x$ nanocomposites obtained by ball-milling and extrusion method, the thermal conductivity decreases 75% (Lopez et al. 1999). Some examples of *in situ* nanocomposites include Ag-Pb-Sb-Te (LAST) alloys, $PbTe-Sb_2Te_3$ eutectic alloys with self-assembled layer structures, GeTe-based amorphous/nanocrystal structures, and *in situ* coating via peritectic reactions in Mg_2SiSn alloys have also been reported (Lan et al. 2010). Actually, no one bulk processing method has shown a clear and consistent advantage over another in terms of achieving enhanced TE performance.

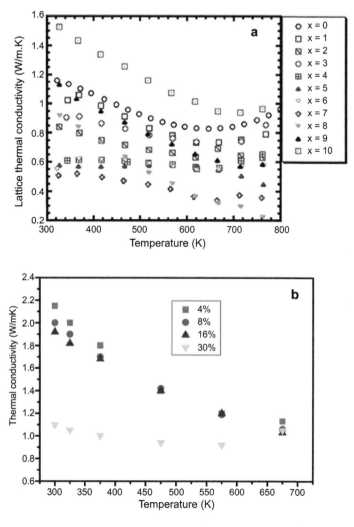

Figure 5. (a) Temperature dependence of the lattice thermal conductivity of $Pb_{9.6}Sb_{0.2}Te_{10-x}Se_x$ (x = 0 to 10) samples; **(b)** Thermal conductivity of $(Pb_{0.95}Sn_{0.05}Te)_{1-x}(PbS)_x$, x = 0.04, 0.08, 0.16 and 0.3. Adapted with permission from ref. (Poudeu et al. 2006), copyright © American Chemical Society 2006 and (Androulakis et al. 2007), copyright © American Chemical Society 2007, respectively.

3. Advanced Bulk Nanostructured Thermoelectric Materials

TE materials comprise a huge family, including different material systems from semimetal, semiconductor to ceramic, covering various crystalline forms from single crystal, polycrystal to nanocomposite, and containing varying dimensions from bulk, film, wire to cluster. Numerous investigations show that the TE figure of merit could be significantly improved in low dimensional materials, by enhancing power factor using quantum confinement effects and by decreasing thermal conductivity using phonon scattering effect (e.g., ZT = 2.4 for Bi_2Te_3/Sb_2Te_3 superlattice thin film and ZT = 3 for a PbSeTe/PbTe quantum dot superlattice) (Venkatasubramanian et al. 2001, Harman et al. 2005, Sun et al. 1999, Hicks et al. 1996, Yang et al. 2005, Broido et al. 2006, Kim et al. 2006), it has been considered more interesting to introduce the basic concepts of low dimensional systems into large bulk materials for the purpose of large scale commercial applications. Based on this strategy, many works have been done in recent years on advanced nanostructured bulk materials (Cao et al. 2008). Nanocomposite TE materials offer a promising approach for the preparation of bulk samples with nanostructured constituents. Such nanocomposite materials are handled easily from a properties-measurement/materials-characterization point of view; they can be assembled into a variety of desired shapes for device applications, and can be scaled up for commercial applications (Dresselhaus et al. 2007). Numerous TE materials systems have been developed and reviewed previously. Our goal herein is to update the new developments in some selected advanced bulk nanostructured TE materials including Bi_2Te_3, PbTe, Skutterudites, Half-Heusler and Si-Ge alloys.

3.1 Bismuth Telluride and Its Alloys

Bismuth telluride (Bi_2Te_3) and its alloys have been studied extensively since the 1960s (Goldsmid 1961). They are one of the most important TE materials found to date, which are used in the state-of-the-art devices for 200–400 K, such as TE refrigeration and thermopiles (Ioffe 1961, Rowe and Bhandari 1983). Bi_2Te_3 crystals belong to the rhombohedral crystal system with a layer structure (Chen et al. 2012). Each layer is composed of a Te-Bi-Te-Bi-Te unit, and each unit cell is composed of an ABC stacking of Te-Bi-Te-Bi-Te units along its c-axis direction. The lamellar structure and weak van der Waals bond between the two quintets make single-crystal p-type $Bi_{2-x}Sb_xTe_3$ alloys and n-type $Bi_2Te_{3-x}Se_x$ alloys susceptible to easy cleavage along their basal planes, perpendicular to the c-axis, and hence impart very poor mechanical properties. For this reason, their polycrystalline counterparts become a hot research topic, despite their worse TE performance. Bi_2Te_3 alloys possess high mean atomic weight, intrinsic low thermal conductivity, and the large Seebeck coefficient with relatively high carrier mobility achieved by multiple reasons such as its special band structure and suitable constituent atomic composition (Tritt 1999, Sales 2002). They were prepared traditionally by zone melting method, which only got the highest ZT close to 1, far from the ideal value for large-scale applications, especially in TE power generator (Tritt 1999). Since Hicks and Dresselhaus proved that low-dimensional materials could have higher ZT than their bulk analogs due both to their lower thermal conductivity and to quantum confinement effects (Hicks and Dresselhaus 1993), research interest in improving Bi_2Te_3 alloys has been rekindled, among which constructing bulk nanostructured Bi_2Te_3 material is considered a promising way. In the early 21st century, various morphologies of solvothermally or hydrothermally synthesized Bi_2Te_3 powders have been reported, including nanorods, polygonal nanosheets, polyhedral nanoparticles, and sheet-rods (Deng et al. 2002, Zhao et al. 2004a, Deng et al. 2003). Zhao et al. developed Bi_2Te_3 nanotubes by a hydrothermal method, and then embedded the nanotubes into the Bi_2Te_3 matrix grown by zone melting method. Since nanotubes have the structural features of holey and low dimensional materials, phonons was scattered strongly while power factor was influenced by the low dimensionality. It was found that the thermal conductivity was significantly reduced with the electrical conductivity much less affected and the greatest ZT value of 1.0 was obtained at 450 K for the n-type Bi_2Te_3 nanocomposite (Zhao et al. 2005). More strikingly, Poudel et al. prepared bulk nanostructured p-type $Bi_{2-x}Sb_xTe_3$ by ball milling and hot pressing the alloyed crystalline ingot as starting material, following the suggestion of using a random nanostructuring strategy (Poudel et al. 2008). In the study, they pushed the maximum ZT value up to 1.4 from 1.0 of their ingot counterpart, resulting from the decrease in thermal conductivity thanks to a strong interface and point defect scattering of heat-carrying phonons. Similarly, the team applied the same approach to synthesize p-type nanostructured BiSbTe bulk materials from elemental chunks Bi, Sb, and Te. A peak ZT of about 1.3 in the temperature range of 75 and 100°C was achieved, which was caused mostly by the reduced thermal conductivity (Ma et al. 2008). This simple BM-HP route was also applied to fabricate polycrystalline n-type $Bi_2Te_{3-x}Se_x$ nanocomposites. However, no obvious improvement in ZT value was obtained since the gain from the decreased lattice thermal conductivity was offset by the reduced power factor due to the decreased carrier mobility and the increased carrier thermal conductivity due to the increased carrier concentration (Liu et al. 2011). Grain growth is a major issue during hot pressing the nanopowders into dense bulk samples. To prevent grain agglomeration during ball milling and growth during hot pressing, organic agent (Oleic Acid, OA) as additive was added into the materials at the beginning of the ball milling process. The OA was reported to lower the surface energy and create uniformly smaller grains after hot pressing, leading to the decreased lattice thermal conductivity but a degraded power factor (Zhang et al. 2012).

Further enhancement in ZT requires a simultaneous increase in power factor along with a decrease in thermal conductivity. This might be achieved by the introduction of nanoinclusions, as was theoretically proposed by Faleev and Leonard (Faleev and Leonard 2008) and Zebarjadi et al. (Zebarjadi et al. 2009). They pointed out that band bending at the interface between nanoinclusions and the semiconductor host could cause energy-dependent scattering of electrons, namely, block low energy electrons and transmit high energy electrons, leading to an energy-filtering effect that increased the Seebeck coefficient for a given carrier concentration (Sumithra et al. 2011). By combining this theory with quantum confinement for phonon scattering on the nanostructure, attempts at introducing a second low-dimensional nanoscale phase (including metals, semimetals, semiconductors, ceramics and even carbon nanotubes and fullerenes) into the Bi_2Te_3 TE matrix have been reported. For example, Li et al. mixed SiC nanoparticles into the p-type BiSbTe matrix by ball milling the elemental powders of Bi, Sb, Te and nano-SiC with subsequent consolidation by SPS (Li et al. 2013). They demonstrated that SiC nanoinclusions possessing coherent interfaces with the $Bi_{0.3}Sb_{1.7}Te_3$ matrix increased both the Seebeck coefficient and the electrical conductivity, in addition to its effect of reducing lattice thermal conductivity by enhancing phonon scattering. The approach of mixing SiC nanoparticles into the BiSbTe matrix effectively enhanced its TE properties with a ZT value up to 1.33 at 373 K. Recently, Zhang et al. demonstrated a practical and feasible bottom-up chemistry approach to dramatically enhance TE properties of the Bi_2Te_3 matrix by means of exotically introducing silver nanoparticles (AgNPs) (Zhang et al. 2014). By regulating the content of AgNPs and fine-tuning the architecture of nanostructured TE materials, the lattice thermal conductivity was reduced significantly and the power factor was improved due to low-energy electron filtering and excellent electrical transport property of Ag itself. Table 1 summarizes some reported advanced bulk Bi_2Te_3-based nanostructured TE materials.

3.2 Lead Telluride and Its Alloys

For mid-temperature power generation (100–550°C), lead telluride (PbTe) has been found to have very good TE properties, which produces higher Seebeck coefficients compared to bismuth telluride. It crystallizes in the NaCl crystal structure with Pb atoms occupying the cationic sites and Te forming the anionic lattice. A band gap of 0.32 eV allows it to be optimized for power-generation applications and can be doped either n- or p-type with appropriate dopants (Sootsman et al. 2009). The TE properties of PbTe are reasonably well understood and have been previously reviewed

Table 1. A summary of some reported bulk Bi_2Te_3-based nanostructured TE materials.

Nanocomposites	Carries type	Preparation method	ZT_{max}	References
Micronsized Bi_2Te_3 + Bi_2Te_3 nanopowders	n	Zone melting/HS + HP	0.83 at 370 K	Ni et al. 2005
Micronsized Bi_2Te_3 + Bi_2Te_3 nanotubes	n	Zone melting/HS + HP	1.0 at 450 K	Zhao et al. 2005
Bi_2Te_3	p	MS + SPS	1.35 at 300 K	Tang et al. 2007
Bi_2Te_3 + Sb_2Te_3	p	HS + HP	1.47 at 450 K	Cao et al. 2008
BiSbTe	p	BM + HP	1.3 at 373 K	Ma et al. 2008
Micronsized $Bi_{0.4}Sb_{1.6}Te_3$ + nano-$Bi_{0.4}Sb_{1.6}Te_3$	p	Horizontal-BM + HP	1.8 at 316 K	Fan et al. 2010
$Bi_2Te_{2.7}Se_{0.3}$	n	BM + HP + Repress	1.04 at 398 K	Yan et al. 2010
Micronsized Bi_2Te_3 + nano-Bi_2Te_3	n	Horizontal-BM + HP	1.18 at 315 K	Fan et al. 2011
Bi_2Te_3 + Bi nanoinclusions	n	Planetary-BM + HP	0.45 at 373 K	Sumithra et al. 2011
$Cu_{0.01}Bi_2Te_{2.7}Se_{0.3}$	n	BM and dc-HP	0.99 at 448 K	Liu et al. 2011
$Bi_2(Te_{0.9}Se_{0.1})_3$ + multi-walled CNT	n	MA + HP	0.98 at 423 K	Park et al. 2011
$Bi_2Se_{0.3}Te_{2.7}$ + γ-Al_2O_3	n	BM + SPS	0.99 at 400 K	Li et al. 2011
$Bi_{0.5}Sb_{1.5}Te_3$ + Cu nanoparticles	p	Metal decoration + SPS	1.35 at 400 K	Lee et al. 2012
Bi_2Te_3 + single-walled CNT	n	Solid state reaction	1.25 at 280 K	Zhang et al. 2012
Bi_2Te_3	n	Solution-based chemical approach + SPS	~ 1.1 at 340 K	Saleemi et al. 2012
$Bi_{0.5}Sb_{1.5}Te_3$ + C60	p	Planetary mill + Sintering	1.17 at 450 K	Blank et al. 2012 Kulbachins-kii et al. 2012
Bi_2Te_3	n	Chemical synthesis + SPS	0.62 at 400 K	Son et al. 2012
$(Bi_{0.2}Sb_{0.8})_2Te_3$ + SbTe nanowire	p	MA + HP	1.33 at 348 K	Kim et al. 2012
Bi_2Te_3 + Bi_2Se_3 nanoparticles	n	Solid-state reactions	0.75 at 325 K	Kim et al. 2012
$Bi_{0.3}Sb_{1.7}Te_3$ + SiC nanoparticles	p	Planetary-BM + SPS	1.33 at 373 K	Li et al. 2013
Bi_2Te_3 + CNT	n	Chemical route + SPS	0.85 at 473 K	Kim et al. 2013

Table 1. contd....

Table 1. contd.

Nanocomposites	Carries type	Preparation method	ZT_{max}	References
$Bi_2Te_3 + Bi_2S_3$	n	Chemical synthesis + SPS	~ 0.48 at 500 K	Han et al. 2013
$Bi_2(Te_{0.9}Se_{0.1})_3$ + Zn nanoparticles	n	Zone melting	1.1 at 340 K	Wang et al. 2013
$Bi_2(Te_{0.9}Se_{0.1})_3$ + Cu nanoparticles	n	Zone melting	1.15 at 340 K	Wang et al. 2013
$Bi_{0.4}Sb_{1.6}Te_3 + Cu_7Te_4$ nanorods	p	Horizontal-BM + HP	1.14 at 444 K	Tan et al. 2013
$Bi_2Te_{2.7}Se_{0.3}$ + Al-doped ZnO	n	BM + HP	~ 0.85 at 323 K	Song et al. 2014
Bi_2Te_3 + Ag nanowires	n	HS + SPS	0.71 at 475 K	Zhang et al. 2014
Bi_2Te_3+ Si	n	MA + HP	0.48 at 425 K	Satyala et al. 2014
$(Bi_{0.2}Sb_{0.8})_2Te_3 + MoS_2$	p	MA + Hot extrusion	~ 0.9 at 370 K	Keshavarz et al. 2014
$Bi_{0.5}Sb_{1.5}Te_3$	p	BM + SPS + Hot forging	1.56 at 333 K	Jiang et al. 2014
$Sb_{1.6}Bi_{0.4}Te_3$ + Au/$AuTe_2$	p	γ-ray irradiation	1.01 at 423 K	Jung et al. 2014
Bi_2Te_3 + Ag nanoparticles	n	HS + SPS	0.77 at 475 K	Zhang et al. 2015

MA: mechanical alloying, SPS: spark plasma sintering, HP: hot-pressing, MS: melt spinning, BM: ball-milling, HS: hydrothermal synthesis.

(He et al. 2013, Zhao et al. 2014, Kanatzidis 2006). Our goal herein is to give an overview of most recent progress in improving the TE properties of PbTe-based materials by bulk nanostructuring.

The phenomenon of bulk nanostructuring in a high performance TE system was first observed in the so-called LAST-m ($AgPb_mSbTe_{2+m}$) system using the technique of transmission electron microscopy (Hsu et al. 2004). The LAST-m system is an interesting bulk-grown material that spontaneously forms nanostructures during cooling from the melt. The nanocomposite has many nanoscale in homogeneities embedded in the PbTe matrix, where the nanoinclusions of minor phases in $AgPb_mSbTe_{2+m}$ exhibit coherent or semicoherent interfaces with the matrix (Quarez et al. 2005). The spontaneous and ubiquitous nanoscale precipitation of second phases led to the very low lattice thermal conductivity, giving ZT ≈ 1.7 at approximately 700 K for $Ag_{1-x}Pb_{18}SbTe_{20}$ (Hsu et al. 2004). After this first observation, nanostructures were also observed in the systems LASTT ($Ag(Pb_{1-x}Sn_x)_mSbTe_{2+m}$) (Androulakis et al. 2006), SALT ($NaPb_mSbTe_{2+m}$) (Poudeu et al. 2006) and KPb_mSbTe_{2+m} (PLAT-m) (Poudeu et al. 2010). Among them, the ZT value of $Na_{0.95}Pb_{20}SbTe_{22}$ rises strongly with temperature and reaches 1 near 475 K and approximately 1.7 at 650 K, as is shown in Fig. 6a. This is one of the widest temperature ranges in which a single material exhibits a ZT value above 1. Subsequent studies showed that the inclusions in the LAST-m materials are formed via thermodynamic spinodal decomposition or nucleation and growth events during cooling (Cook et al. 2009). As a consequence of nucleation and growth of a second phase, coherent nanoscale inclusions form throughout the material, which are believed to result in scattering of acoustic phonons while causing only minimal scattering of charge carriers. Furthermore, good TE performance for LAST-m materials was also obtained by using MA and SPS method (Zhou et al. 2008). For example, Zhou et al. fabricated polycrystalline $Ag_{0.8}Pb_{18+x}SbTe_{20}$ materials by mechanical alloying of elemental powders followed by densifying through SPS and annealing for 30 days, achieving a high ZT value of 1.5 at 700 K, as is shown in Fig. 6b. Another important advance was the use of "incoherent" precipitates which show a clear boundary with the host to reduce the thermal conductivity of PbTe. Since the incoherent ones possess a large mismatch with the matrix, they work as nanoparticles and selectively scatter mid to long wavelength phonons, which work by different mechanisms from the coherent nanoinclusions that act as point defects to scatter short wavelength phonons. By incoherent nanoinclusions of Sb and Ag_2Te, a ZT of 1.4–1.5 has been achieved in n-type PbTe (Sootsman et al. 2008, Pei et al. 2011, Liu et al. 2012). But this is slightly lower

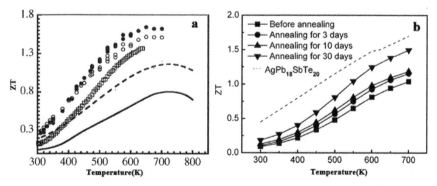

Figure 6. (a) Temperature dependence of the ZT for $Na_{0.95}Pb_{20}SbTe_{22}$ (●), $Na_{0.95}Pb_{19}SbTe_{21}$ (○), and $Na_{0.8}Pb_{20}Sb_{0.6}Te_{22}$ (□), compared to those of the state-of-the-art p-type PbTe (solid line) and TAGS (dashed line); **(b)** ZT value for $Ag_{0.8}Pb_{22.5}SbTe_{20}$ bulks before and after annealing for different times. Reproduced with permission from ref. (Poudeu et al. 2006), copyright © WILEY-VCH Verlag GmbH & Co. KGaA, Weinheim 2006 and (Zhou et al. 2008), copyright © American Chemical Society 2008, respectively.

than that obtained by using coherent nanoinclusions due to their adverse effects on charge carrier mobility.

Nanostructuring itself, however, scatters phonons with short and medium mean free paths ($\sim$ 3–100 nm), thus rendering only phonons with longer mean free paths unaffected. An additional and significant reduction in the lattice thermal conductivity may be achieved by further scattering of the phonons with longer mean free paths ($\sim$ 0.1–1 µm, that is, mesoscale), on which scale additional mechanisms of grain-boundary phonon scattering and impedance can be exploited (Biswas et al. 2012). Based on this idea, noticeable improvement was triumphantly made by a panoscopic approach: a hierarchical architecture across all relevant length-scales. In 2012, Biswas et al. reported that by harnessing integrated phonon scattering across multiple length scales-atomic-scale alloy scattering, scattering from nanoscale endotaxial precipitation and scattering from mesoscale grain boundaries-they achieved a record-high ZT value of $\sim$ 2.2 at 915 K in powder-processed and SPS-ed samples of PbTe-SrTe (4 mol%) doped with 2 mol% Na (Biswas et al. 2012). To date, the all-scale hierarchical architectures have been successfully applied to many lead chalcogenides PbQ (Q = Te, Se, and S) (Ahn et al. 2010, Biswas et al. 2011, Zhao et al. 2011, Zhao et al. 2012, Lee et al. 2013, Zhao et al. 2013).

Besides manipulating the various nanostructures, the band structure engineering could also help a lot to improve the power factor. The strategies to increase Seebeck coefficients without overshadowing electrical conductivity include: modifying band structure by degeneration of multiple valleys (Pei et al. 2011), electronic resonance states (Heremans et al. 2008), depressing bipolar effect at high temperature (Biswas et al. 2012), etc. (Chen et al. 2012). In 2008, an interesting enhancement of power factor was reported in bulk p-type PbTe doped with Tl by Heremans et al. (Heremans et al. 2008). In this case, instead of quantum confinement effects, resonant states formed due to the interaction of the Tl with the valence band of PbTe lead to an increase in the density of states (DOS) near the band edge. Optical measurements on 1.5% Tl-PbTe sample indicated that the resonant states were located $\sim$ 0.06 eV below the band edge, and increase the DOS by a factor of 2.6 above the bulk value. Electrical and thermal measurements showed a higher power factor for the 2% Tl-PbTe sample, which led to the doubling of ZT near 800 K. In 2011, Pei et al. found that a convergence of electronic band in $PbTe_{1-x}Se_x$ could achieve increased carrier mobility without negative impact to carrier mobility (Pei et al. 2011). Increasing the number of degenerate valleys increases the entropy per charge that can pass through the materials, thus resulting in a higher Seebeck coefficient and electrical conductivity. In this case, the convergence of the Σ band with 12 valleys and the L band with 4 valleys were achieved in heavily doped $Na_{0.02}Pb_{0.98}Te_{0.85}Se_{0.15}$, leading to an extraordinary ZT value of 1.8 at about

850 K. For other mechanisms known to increase the power factor see excellent reviews elsewhere (He et al. 2013, Zhao et al. 2014, Heremans et al. 2012, Pei et al. 2012).

In addition to solid-state reaction, there are more and more reports on PbTe-based nanocomposites made from wet-chemical approaches (Lin et al. 2012, Dong et al. 2013, Ibanez et al. 2013). For instance, Ibanez et al. detailed a rapid, high-yield and scalable colloidal synthetic route to prepare grams of PbTe@PbS coreshell nanocomposites (Ibanez et al. 2013). In the PbTe@PbS nanocomposites, synergistic nanocrystal doping effects resulted in up to 10-fold higher electrical conductivities with strongly reduced thermal conductivities. As a result, a record ZT of ~ 1.1 was obtained at 710 K, indicating the potential of the bottom-up assembly processes to achieve high TE performance.

3.3 Skutterudites

Although PbTe and its alloys have been playing a dominant role in the TE power generation application for more than 50 years, especially in the deep space exploring program (Liu et al. 2012), the severe environment contamination from lead and scarcity of telluride in the Earth's crust (0.001 ppm) have limited their large-scale applications. Skutterudites, hence, gradually become the most prospective candidates for intermediate temperature TE applications due to their merits such as the relatively high ZT value with good tunability by controlling their compositions and structures, relatively cheap and non-toxic components and wide application temperature up to 600°C (Han et al. 2014). Nowadays, skutterudites have been widely studied for their promising TE properties, and are regarded as potential candidates for next-generation TE materials for power generation using either solar energy or waste heat (Lan et al. 2010).

A binary skutterudite has a general chemical formula of MX_3, where M is a transition metal such as cobalt, rhodium, etc. and X stands for a metalloid such as phosphor, arsenic, or antimony. It crystallizes with the body-centered cubic space group (Im3). As is depicted in Fig. 7, a typical skutterudite unit cell contains 32 atoms arranged in eight groups of MX_3 blocks, where the transition metals M occupy the c sites and the Group V atoms X occupy the g sites. Each M metal atom lies in the center, which is octahedrally coordinated by six pnicogen atoms to form a MX_6 octahedron. The linked octahedra produce two large lattice voids (icosahedral voids), and each of the voids is surrounded by 12 X atoms in the structure (Shi et al. 2011). The voids are large enough to accommodate relatively large metal atoms, resulting in the formation of filled skutterudites (Nolas et al. 2006). Many different elements have been introduced into the voids, including lanthanide, actinide, alkaline-earth, alkali, thallium, and Group IV elements

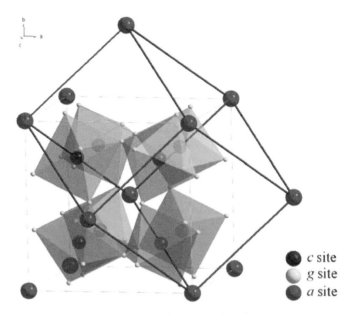

Figure 7. Crystal structure of a typical skutterudite material CoSb$_3$. Transition metal atoms, Co, occupy 8c sites and pnicogen atoms, Sb, occupy 24g sites. Guest atom I could be filled into the voids at 2a sites. Reproduced with permission from ref. (Shi et al. 2011).

(Nolas et al. 2006). Therefore, instead of the general formula of binary skutterudites (MX$_3$) mentioned above, one can also write the chemical formula as □M$_4$X$_{12}$, with the □ symbol illustrating the presence of the void (Sootsman et al. 2009). While the binary skutterudites have reasonably large Seebeck coefficients of ~ 200 µVK^{-1}, they still exhibit very high thermal conductivities. When a rare earth element is mixed with the binary skutterudite, the heavy rare earth atom occupies the empty space of the crystal, not only leading to large impurity scattering of phonons, but enhancing scattering of phonons due to the loosely bound heavy atoms rattling in their cages, and thus reducing thermal conductivity by an order of magnitude at room temperature. Such an idea initiated a wide investigation of filling elements to the cage-site, accelerating the development of skutterudite systems from single filling to double, triple and multiple filling.

Figure 8 briefly summarizes ZT values of skutterudite materials as a function of years. In 1994, Dr. Slack firstly proposed his ideal concept that the best TE materials should behave as "phonon glass electron crystal" (PGEC), that is, it would have the thermal properties of glass-like materials and the electrical properties of crystalline materials, where high mobility electrons are able to transport charge and heat freely, while the phonons are disrupted at the atomic scale from transporting heat (Slack 1995). Filled

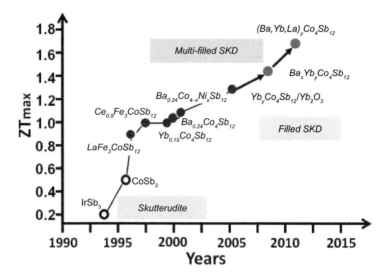

Figure 8. ZT$_{max}$ values of skutterudite materials as a function of years. Adapted with permission from ref. (Chen 2014), 33rd ICT Nashville, USA.

skutterudites, with guest filler ions in the void sites, were assumed to be potential candidates as ideal PGEC materials. In 1996, Dr. Sales significantly improved ZT value in p-type Fe-substituted La-filled skutterudites to be around 1.0, comparable to the state-of-the-art TE materials, which helped motivate much work in this system (Sales et al. 1996). Lots of theoretical and experimental efforts emerged in succession, making the filled skutterudites one of the most interesting and promising novel TE materials. By physically extrapolating transport data measured at low-temperature to the high-temperature range, the ZT value (1.0 at 800 K) was estimated to be able to reach around 1.4 at 1000 K in the p-type Fe-substituted Ce-filled skutterudites by Sales et al. in 1997 (Sales et al. 1997). Since TE technology requires both excellent p- and n-type materials with high ZTs used at identical working temperature, searching for matching n-type skutterudites became critical in this field. Rare-earth filled and Fe-substituted skutterudites are p-type, superior to other state-of-the-art TE materials. However, when it comes to preparing n-type skutterudites, it was found that rare earth fillers always occupied a small percentage of the intrinsic lattice voids (low filling fraction), and electron-deficient elements were often co-doped at Co or Sb sites to balance the charges, which usually drove the filled materials to p-type (Shi et al. 2011). Therefore, the search for excellent n-type skutterudites turned to other research directions: one is preparing doped skutterudite alloys through transition metal (such as Ni, Pd, Pt) at the Co sites or VIIA element at the Sb sites to provide extra electrons; the other approach is to partially fill guest atoms into the voids without charge-deficient element substitution in the framework to form filled skutterudites $I_yCo_4Sb_{12}$, where I represents guest atoms (Shi et al. 2011). Consequently, a high figure of merit (ZT = 1.0 at 600 K) was demonstrated by Nolas in the partially filled ytterbium skutterudite materials ($Yb_{0.19}Co_4Sb_{12}$), which showed an ideal combination for TE applications exemplifying the PGEC concept (Nolas et al. 2000). Soon afterwards, Chen et al. firstly published the alkaline earth metal Ba filled skutterudites with a measured ZT as high as 1.1 at 850 K prepared by ball-milling, followed by cold-pressing and sintering, expanding the scope of n-type skutterudites study (Chen et al. 2001). Although the ZT values of n-type partially single element-filled skutterudites have been significantly improved compared with that of binary skutterudites, they are still lower than the reported ZT values of the p-type materials mentioned above, and thus a further optimization to improve ZT in n-type skutterudites is required for real applications. With the rise of nanocomposites, it is expected that composites containing the filled $CoSb_3$ with higher Yb filling fractions and distributed second phase defects should have higher TE performance. In 2006, Zhao et al. synthesized composites containing Yb-filled $CoSb_3$ and well-distributed Yb_2O_3 particles by *in situ* reaction method (Zhao et al. 2006). The combination of the "rattling" of Yb ions inside the voids of $CoSb_3$ and the phonon scattering of the oxide defects resulted in a remarkable reduction in the lattice thermal conductivity. The maximum figures of merit reached 1.3 for the $Yb_{0.25}Co_4Sb_{12}/Yb_2O_3$ and 1.2 for the $Yb_{0.21}Co_4Sb_{12}/Yb_2O_3$ composites at 850 K. Since 2005, n-type partially filled skutterudites have been studied systematically in theory and lots of band structure information has been obtained (Shi et al. 2011). In-depth understanding of the correlation between TE transport properties and guest filling in the n-type partially filled skutterudites was achieved, which eventually provided a direct guidance for the optimal filler selection and innovative materials design. Recently, $CoSb_3$ with multiple fillers of Ba, La, and Yb were synthesized, and a ZT of 1.7 at 850 K was realized by Shi et al. (Shi et al. 2011), which is the highest reported value so far for skutterudites and is among the highest for all known bulk TE materials.

3.4 Half-Heusler Compounds

Another class of compounds of considerable interest as potential TE materials for the temperature range around 700 K and above are the half-Heusler (HH) intermetallic compounds formulated as MNiSn and MCoSb (M = Ti, Hf, or Zr, the most studied ones) (Sootsman et al. 2006). It has been reported that the MNiSn phases are promising n-type TE materials (Shen et al. 2001), and MCoSb phases are promising p-type materials (Ponnambalam et al. 2008). HH alloys have a MgAgAs type crystal structure, forming three interpenetrating face-centered-cubic sub-lattices with one Ni sub-lattice vacant (Hohl et al. 1999). They are small band gap semiconductors with a gap of 0.1–0.5 eV, exhibiting high Seebeck coefficient (S ≈ −300 μV·K^{-1}), low electrical resistivity values (0.1–8 mΩ·cm) and the relatively high lattice thermal conductivity (≈ 10 W·m^{-1}K^{-1}) at room temperature (Bhattacharya et al. 2000). One of the current challenges in HH alloys is to reduce the high thermal conductivity greatly or/and to further increase the power factor.

Initial attempts to improve ZT of HH alloys are focusing on the optimization of the power factor through substitutional doping and reduction in the thermal conductivity using mass fluctuation phonon scattering and strain field effects via solid-solution alloying. For example, in the ZrNiSn system, Uher's team introduced phonon mass fluctuation scattering by alloying Hf (atomicmass = 179) on the Zr (atomic mass = 91) site, leading to an overall reduction of the total thermal conductivity (Uher et al. 1999). Researches show that iso-electronic alloying does not introduce charge disorder in the crystal lattice but creates point defect scattering for phonons due to mass differences (mass fluctuations) and size and inter-atomic coupling force differences (strain field fluctuations) between the host atoms and the impurity atoms (Bhattacharya et al. 2000). And the team also explored the effect of less than 1 at.% Sb doping on the Sn site, which resulted in several orders of magnitude lowering of the resistivity without significantly reducing Seebeck coefficient. Coupled with partial substitution of Ni by Pd, they achieved a maximum ZT value of 0.7 at about 800 K in the $Zr_{0.5}Hf_{0.5}Ni_{0.8}Pd_{0.2}Sn_{0.99}Sb_{0.01}$ compound (Shen et al. 2001). For other HH systems with good TE performance by substitutional doping see excellent reviews elsewhere (Willem et al. 2014). Herein, we will focus on the recent progress in improving the TE performance of several HH compounds by utilizing the strategy of forming bulk HH nanocomposites.

The bulk HH nanocomposites can be categorized to two kinds in terms of the type of microstructure: (1) micro-scale HH matrix with nanoscale inclusions (see Fig. 9a), and (2) nanoscale HH phase with nanoinclusions (see Fig. 9b). To date, nanoinclusions in a micro-scale HH matrix can be prepared by *ex situ* mixing as well as *in situ* formation, and the nanoscale HH phase can be obtained by either ball milling or melt spinning, both with subsequent densification process by HP or SPS (Xie et al. 2012). In the early 21st century, Chen's team investigated the effects of γ-Al$_2$O$_3$ or ZrO$_2$ nanoparticles on the TE properties of the undoped ZrNiSn HH compound (Huang et al. 2003). They found that the reduction of approximately 25% and 35% in thermal conductivity could be achieved by introducing 6 vol.% γ-Al$_2$O$_3$ and ZrO$_2$ nanoparticles in the ZrNiSn system, respectively. Although the figure of merit is improved, the ZT value for the ZrNiSn/nanoinclusion is still low because of the limit of TE performance of the

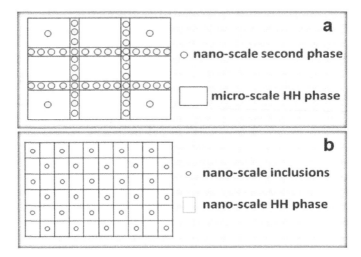

Figure 9. Two types of HH nanocomposites: **(a)** micro-scale HH matrix with nano-scale inclusions; **(b)** nano-scale HH phase with nanoinclusions. Reproduced with permission from ref. (Xie et al. 2012), an open access article under the terms of the Creative Commons Attribution License.

ZrNiSn matrix. Then they chose the doped alloy $Zr_{0.5}Hf_{0.5}Ni_{0.8}Pd_{0.2}Sn_{0.99}Sb_{0.01}$ as the matrix and the ZrO_2 nanoparticles as nanophase inclusions. By the solid state reaction and SPS process, the prepared-composites with 9 vol.% ZrO_2 nanoparticles had a ZT of about 0.75 at 800 K with an improvement of about 30%. To date, nanoscale secondary phases including C60 (Huang et al. 2005), WO_3 (Misra et al. 2011), and NiO (Yaqub et al. 2011) have also been chosen as nanoinclusions in the HH matrix, which are also added into the matrix through the *ex situ* mixing approach. In spite of the enhancement in ZT, there are still some disadvantages of the mechanical mixing, especially that it will usually induce extra defects that in turn results in the additional scattering of the charge carriers, leading to a noticeable increase in resistivity. Thereupon, Xie et al. attempted to form InSb nanoparticles *in situ* in a (TiZrHf) (CoNi)Sb matrix as nanoinclusions by a high-frequency induction melting method combined with a subsequent SPS process (Xie et al. 2010). The results showed that all three TE properties were improved simultaneously for certain concentrations of InSb nanoinclusions, resulting in a 160% improvement in ZT for the sample containing 1 at.% InSb nanoinclusions.

More recently, Poudeu et al. proposed an "Atomic-Scale Structural Engineering of Thermoelectric" (ASSET) approach to *in situ* form "half Huesler-full Heusler" (HH-FH) nanocomposites (Makongo et al. 2011). By embedding coherent well-dispersed sub-10 nm scale fragments of full-Heusler inclusions within the crystal lattice of a bulk n-type HH matrix, they achieved simultaneous large enhancements in both thermopower and electrical conductivity together with a moderate reduction in the total thermal conductivity, resulting in dramatic (up to 250%) improvements of the overall ZT. By extending the ASSET concept to the p-type $Ti_{0.5}Hf_{0.5}Co_{1+x}Sb_{0.9}Sn_{0.1}$ HH-FH nanocomposite systems, the team also found that nanometer scale inclusions with FH structure coherently embedded within the crystal lattice of the p-type HH matrix, greatly altered the electronic transport behavior of the existing ensemble of free charge carriers, boosting the thermopower and hole carrier mobility without increasing the total thermal conductivity (Sahoo et al. 2013). Their work has so far sparked a fast growing research interest on the application of similar concept to various combinations of HH-FH compositions as well as other TE material systems (Chai and Kimura 2012, Hazama et al. 2011, Douglas et al. 2012, Xie et al. 2013, Birkel et al. 2013).

Combining MS with SPS, Zhu et al. prepared (Zr,Hf)NiSn based HH nanocomposites (Yu et al. 2010, Xie et al. 2012, Yu et al. 2010). By adopting ball milling process with direct current hot pressing, Ren's team achieved more success in enhancing ZT of HH nanocomposites (Yan et al. 2011, Joshi et al. 2011). So far, the BM-HPed HH bulk nanocomposite have achieved the highest ZT ≈ 1.0 in the n-type $Hf_{0.75}Zr_{0.25}NiSn_{0.99}Sb_{0.01}$ nanocomposites, ZT = 0.8 in the p-type $Zr_{0.5}Hf_{0.5}CoSb_{0.8}Sn_{0.2}$ and ZT ≈ 1.0 in the p-type $Hf_{0.8}Ti_{0.2}CoSb_{0.8}Sn_{0.2}$ (Yan et al. 2012). More recently, Chen et al. successfully reduced Hf usage to 1/3 of the best reported composition and achieved peak ZT value of 1 for the $Hf_{0.25}Zr_{0.75}NiSn_{0.99}Sb_{0.01}$ nanocomposite at 600°C (Chen et al. 2013). Because among all elements in MNiSn based HH, Hf is the most expensive component, the reduction of Hf concentration is of great significance for such materials to be used in large-scale waste heat recovery.

3.5 Silicon Germanium Alloys

Silicon germanium (SiGe) alloys possesses high mechanical strength, high melting point, low vapor pressure, and resistance to atmospheric oxidation (Dismukes et al. 1964), and therefore, has been one of the main TE materials for power generation at high temperatures above 800°C. Improving the TE efficiency of SiGe alloys is especially interesting for industrial waste heat recovery, auto industry, and solar thermal power plants. Actually they have long been used in Radioisotope Thermoelectric Generators for deep-space missions to convert radioisotope heat into electricity since 1976 (Rowe 1995). From 1960s to 1990s, numerous efforts were made to improve the ZT of SiGe alloys, with the peak ZT of n-type SiGe reaching 1 at 900–950°C (Dismukes et al. 1964, Bhandari et al. 1980, Rowe et al. 1981, Slack and Hussain 1991, Vining 1991) and 0.65 for p-type SiGe alloys (Vining et al. 1991). New applications, especially heat conversion at high temperatures, demand higher ZTs for such material applications. As experimental and theoretical studies on heat conduction mechanisms suggested that an ordered superlattice structure is neither needed nor optimal for reducing the thermal conductivity (Chen 1998), extensive works were carried out on pursuing small grained SiGe alloy to reduce the thermal conductivity. However, it was found that when the grain size was reduced below microns, accompanying the thermal conductivity reduction was a similar reduction in electrical conductivity and a resulting degradation of ZT (Vining et al. 1991, Rowe et al. 1993). Until 2008, a pronounced enhancement in ZT values was demonstrated for n- and p-type SiGe alloys by using them in bulk nanostructured form. A ZT of 0.95 in p-type nanostructured bulk SiGe alloys was achieved by using a direct current-induced hot press of mechanically alloyed nanopowders that were initially synthesized by ball milling of commercial grade Si and Ge chunks with boron powder. The enhancement of ZT was due to a large reduction of thermal conductivity caused by the increased phonon scattering at the grain boundaries of the nanostructures combined with an increased power factor at high temperatures (Joshi et al. 2008). By using a similar nanostructure approach, a peak ZT of about 1.3 at 900°C in an n-type nanostructured SiGe bulk alloy was also achieved (Wang et al. 2008). However, these materials contain a fairly high concentration of Ge that is about a hundred times more expensive than Si. In 2009, Zhu et al. reported that a 5 at.% Ge replacing Si was sufficient to cause a further thermal conductivity reduction by a factor of 2 (Zhu et al. 2009). They obtained a ZT peak value of 0.94 in $Si_{95}Ge_5$ doped with GaP and P at ~ 900°C, which is significant, since a much smaller amount of expensive Ge was used. Recently, Basu et al. reported a further enhancement in the ZT of nanostructured n-type SiGe alloys prepared, using a ball mill and simple vacuum hot press technique (Bhattacharya et al. 2000). They demonstrated the highest ZT value of ~ 1.84 at 1073 K for n-type SiGe nanostructured bulk alloys, which has been the new record value for n-type SiGe alloys. The above reported studies on nanostructured $Si_{80}Ge_{20}$ alloys have employed hot-pressing to consolidate and densify the nanopowders, which however leads generally to some grain growth. Bathula et al. 2012 then synthesized phosphorous doped nanostructured n-type $Si_{80}Ge_{20}$ alloys employing high energy ball milling followed by rapid-heating using SPS (Basu et al. 2014). The rapid-heating rates, used in SPS, allowed the achievement of near-theoretical density in the sintered alloys, while retaining the nanostructural features introduced by ball-milling. Consequently, the nanostructured alloys displayed a low thermal conductivity (2.3 W/mK) and a high value of Seebeck cofficient (–290 μV/K) resulting in a significant enhancement in ZT to about 1.5 at 900°C.

In 2009, Mingo et al. proposed a "nanoparticle-in-alloy" approach to enhance the TE performance of SiGe (Mingo et al. 2009). With nanosized silicide and germanide fillers in the SiGe matrix, a 5-fold increase in the

ZT at room temperature and a 2.5 times increase at 900 K were achieved. Strong reductions in computed thermal conductivity were obtained for 17 different types of silicide nanoparticles. They also predicted that there must be an optimal nanoparticle size that could minimize the thermal conductivity of nanocomposite, and meanwhile, the nanoparticles would not impair the electrical conductivity because of size control. Experimentally, p-type nanostructured bulk $Si_{0.8}Ge_{0.2}$ composites with $CrSi_2$ nanocrystallite inclusions were synthesized via sintering approach (Zamanipour and Vashaee 2012). It was shown that the power factor was enhanced by 20% in the composite structure compared with $Si_{0.8}Ge_{0.2}$.

Another direction to enhance ZT has been improving the power factor that can be done by enhancing the electrical conductivity and/or the Seebeck coefficient. Zebarjadi et al. introduced a modulation-doping approach to nanocomposite SiGe-based bulk TE materials that uses an enhancement of the carrier mobility to increase the electrical conductivity (Zebarjadi et al. 2011). The proposed modulation-doped sample is a two-phase nanocomposite made out of two different types of nanograins. Rather than uniformly doping the sample, dopants are incorporated into only one type of nanograins. Charge carriers spill over from the doped nanograins to the undoped or lightly doped matrix phase, leaving behind ionized nanograins. Instead of the usually heavy uniform doping in TE materials, causing strong ionized impurity scattering of charges, ionized nanoparticles can be spatially placed much further apart in the modulation-doping scheme, leading to reduced electron scattering for higher mobility (Yu et al. 2012). Experimentally, the power factor of the p-type $Si_{86}Ge_{14}B_{1.5}$ uniform sample was improved by 40% using the modulation-doping approach, see Fig. 10a, which was achieved by using a thirty percent volume fraction of boron doped silicon nanoparticles in the intrinsic silicon germanium host matrix to make a $(Si_{80}Ge_{20})_{0.7}(Si_{100}B_5)_{0.3}$ sample (Zebarjadi et al. 2011). Meanwhile, a smaller improvement of about 20% was observed in the power factor of n-type samples $((Si_{80}Ge_{20})_{0.8}(Si_{100}P_3)_{0.2})$, as is shown in Fig. 10b. The enhancement of the power factor was attributed to the enhancement of the mobility by separating the carriers spatially from their parent impurity atoms. However, the ZTs were not increased due to the high thermal conductivities of the pure Si. In 2012, Yu et al. attempted to use $Si_{95}Ge_5$ as the matrix (instead of $Si_{80}Ge_{20}$) and $Si_{70}Ge_{30}P_3$ as the nanoparticles for the modulation doping (Yu et al. 2012). The resulted ZT was ~ 1.3 at 900°C for nanocomposites of $(Si_{95}Ge_5)_{0.65}(Si_{70}Ge_{30}P_3)_{0.35}$, about 30–40% higher than the equivalent uniform sample, resulting from the unaffected Seebeck coefficient combining with the enhanced electrical conductivity and the lower thermal conductivity.

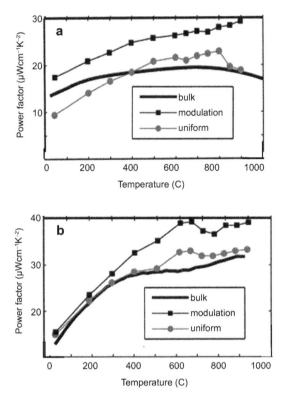

Figure 10. (a) Temperature dependence of the power factor of $(Si_{80}Ge_{20})_{70}(Si_{100}B_5)_{30}$ (black-filled squares), $Si_{86}Ge_{14}B_{1.5}$ (red-filled circles) and p-type SiGe (solid line); **(b)** Temperature dependence of the power factor of $(Si_{80}Ge_{20})_{80}(Si_{100}P_3)_{20}$ (black-filled squares), $Si_{84}Ge_{16}P_{0.6}$ (red-filled circles) and n-type SiGe (solid line). Adapted with permission from ref. (Zebarjadi et al. 2011), Copyright © American Chemical Society 2011.

4. Current Applications and Future Outlook

Since ZT = 1 (the benchmark for device applications) has been achieved in many advanced bulk nanostructured TE systems, module fabrication has been catching more and more attention, especially for power generation. For example, many research institutions including Jet Propulsion Laboratory (JPL) in the United States, National Institute of Advanced Industrial Science and Technology (AIST) in Japan, Shanghai Institute of Ceramics Chinese Academy of Sciences (SICCAS) in China, etc., have begun to conduct an intensive research on the $CoSb_3$-based TE power generation components. Among them, JPL produced efficient $Bi_2Te_3/CoSb_3$ segmented TE unicouples, reaching conversion efficiency of 12% at 550°C, and 15% at 670°C, which is almost double that of SiGe-based segmented TE devices at the same size (Zhao 2009). AIST reported that it developed Bi_2Te_3/$CoSb_3$ cascaded TE module with 10% efficiency of TE conversion under the temperature difference of 400°C, reaching the maximum output power for 4 W. Chen at al. fabricated Bi_2Te_3-$Yb_yCo_4Sb_{12}$ cascade module with maximum conversion efficiency of 10.1% at temperature difference of 500°C (Chen 2014). Furthermore, Poon et al. demonstrated a half-Heusler device made from $Hf_{0.3}Zr_{0.7}CoSn_{0.3}Sb_{0.7}$/nano-$ZrO_2$ and $Hf_{0.6}Zr_{0.4}NiSn_{0.995}Sb_{0.005}$ as n- and p-type materials, respectively (Poon et al. 2011). The maximum efficiency of this device can reach 8.7%, with a hot side temperature of 697°C and cold side temperature of 40°C. Such an efficiency is 10–15% higher than that of a PbTe-based device (Singh et al. 2009). A more recent work by Populoh et al. reported the performance of auni-leg device of $Ti_{0.33}Zr_{0.33}Hf_{0.33}NiSn$ (Populoh et al. 2013). With a hot side temperature of 595°C and a temperature difference of 565°C, the power density can reach 275 mW·cm^{-2}. This value is lower than a device with 16 segmented modules of p-type $YbNi_{0.4}Fe_{3.6}Sb_{12}$/$Bi_{0.25}Sb_{0.75}Te_3$ and n-type $Co_{0.9}(Pd,Pt)_{0.1}Sb_3$/$Bi_2Te_{2.7}Se_{0.3}$ that works at hot side temperature of 500°C and cold side temperature of 25°C (power density ~ 677 mW·cm^{-2}), which indicates that optimization of TE materials, contact materials and module design are essential to maximize the output power density (Chen et al. 2013).

To date, TE devices have been utilized or attempted to be used in areas like aerospace applications, transportation tools, industrial utilities, medical services, electronic devices and temperature detecting and measuring facilities (Zheng et al. 2014). Energy harvesting or scavenging systems is expected to replace the conventional batteries one day. For the waste heat recovery from the vehicle exhaust (at $\Delta T = 350$°C), the efficiency needs to be about 10% and the corresponding ZT is 1.25 to increase the mileage up to 10%. For the primary power generation, the net efficiency needs to be about 20% and have an average ZT of 1.5 or higher at $\Delta T = 800$°C. As ZT is 2, it appears to replace small internal combustion engines such as those used in lawn movers, blowers, and outboard motorboats. These engines would be very quiet and nearly vibration free. They could burn a wide spectrum of fuels like propane, butane, LNG and alcohols and would not necessarily depend on fossil oil as a fuel source. In another proposed co-generator concept, the solar spectrum is split into shorter wavelengths that yield high photovoltaic-conversion efficiency and longer wavelengths that heat thermo-electric generators (Alam and Ramakrishna 2013). Recently, Kraemer et al. reported a flat-panel solar TE generator (STEG) with a system efficiency of 4.6%–5.2% generated from a temperature difference of 180°C ($T_c = 20$°C and $T_h = 200$°C). This efficiency is 7–8 times higher than the previously reported best value for a flat-panel STEG, mainly due to the use of high performance nanostructured Bi_2Te_3-based TE materials (Kraemer et al. 2011). In addition, the sensors based on TE nanocomposites find huge applications in life sciences as well, such as fast DNA analysis on a nanoliter scale, continuous environmental or hazard assaying, real-time monitoring of complex biological processing, and to control the power supplies for remote-sensing systems (Alam and Ramakrishna 2013).

Overall, the field of thermoelectrics has long been recognized as a potentially transformative power generation technology and the field is now growing rapidly due to their ability to achieve the direct conversion between thermal energy and electric energy and to develop cost-effective, pollution-free forms of energy conversion. The preparation methods are becoming mature and stable, and developing towards the mass production. Recent achievements in advanced bulk nanostructrued TE materials and the corresponding TE devices have expedited the steps to practical application. The evaluation results indicate that the performance of TE devices for power generation improves gradually and presents well development tendency. However, when it comes to meeting the needs for industrial large-scale applications, challenges still lie ahead: (1) The ZT values still need further improvement because the real conversion efficiency of TE devices is much lower than the theoretical value and the values should keep as high as possible in a larger temperature range. (2) Many high ZT materials produced in the laboratory have no commercial potential due to the difficulties in large-scale preparation, as well as high fabrication cost and poor stability. (3) A thorough understanding of carrier and phonon transport mechanism in the complex nanocomposites is required for the development of the second generation bulk nanostructured materials with high TE performance. (4) The lack of accurate measurement protocol and mature instrument for evaluation of the TE elements and modules has created another barrier. (5) The service behavior under real and complex operation conditions (vibration, oxidation, thermal circling, etc.) should be further investigated in detail. Therefore, much more needs to be done in the future. All in all, continued efforts from cross-disciplinary research among various fields of science and engineering are required so as to gain a more thorough understanding of the rational design concepts, to explore the optimal synthetic route, to manipulate the structure at all length scales and to more accurately evaluate the optimized TE materials and modules and thus to accelerate the wide adoption of TE technologies in practical applications.

References

Ahn, K., M.K. Han, J.Q. He, J. Androulakis, S. Ballikaya, C. Uher, V.P. Dravid and M.G. Kanatzidis. 2010. Exploring resonance levels and nanostructuring in the PbTe-CdTe system and enhancement of the thermoelectric figure of merit. J. Am. Chem. Soc. 132: 5227–5235.

Alam, H. and S. Ramakrishna. 2013. A review on the enhancement of figure of merit from bulk to nano-thermoelectric materials. Nano Energy 2: 190–212.

Androulakis, J., C.H. Lin, H.J. Kong, C. Uher, C.I. Wu, T. Hogan, B.A. Cook, T. Caillat, K.M. Paraskevopoulos and M.G. Kanatzidis. 2007. Spinodal decomposition and nucleation and growth as a means to bulk nanostructured thermoelectrics: enhanced performance in $Pb_{1-x}Sn_xTe$-PbS. J. Am. Chem. Soc. 129: 9780–9788.

Androulakis, J., I. Todorov, J.Q. He, D.Y. Chung, V.P. Dravid and M.G. Kanatzidis. 2011. Thermoelectrics from abundant chemical elements: high-performance nanostructured PbSe-PbS. J. Am. Chem. Soc. 133: 10920–10927.

Androulakis, J., K.F. Hsu, R. Pcionek, H. Kong, C. Uher, J.J. D'Angelo, A. Downey, T. Hogan and M.G. Kanatzidis. 2006. Nanostructuring and high thermoelectric efficiency in p-type $Ag(Pb_{1-y}Sn_y)_{(m)}SbTe_{2+m}$. Adv. Mater. 18: 1170–1173.

Bao, S.Q., J.Y. Yang, J.Y. Peng, W. Zhu, X.A. Fan and X.L. Song. 2006. Preparation and thermoelectric properties of $La_xFeCo_3Sb_{12}$ skutterudites by mechanical alloying and hot pressing. J. Alloys Compd. 421: 105–108.

Basu, R., S. Bhattacharya, R. Bhatt, M. Roy, S. Ahmad, A. Singh, M. Navaneethan, Y. Hayakawa, D.K. Aswal and S.K. Gupta. 2014. Improved thermoelectric performance of hot pressed nanostructured n-type SiGe bulk alloys. J. Mater. Chem. A 2: 6922–6930.

Bathula, S., M. Jayasimhadri, N. Singh, A.K. Srivastava, J. Pulikkotil, A. Dhar and R.C. Budhani. 2012. Enhanced thermoelectric figure-of-merit in spark plasma sintered nanostructured n-type SiGe alloys. Appl. Phys. Lett. 101: 213902.

Bhandari, C.M. and D.M. Rowe. 1980. Silicon-germanium alloys as high-temperature thermoelectric materials. Contemp. Phys. 21: 219–242.

Bhattacharya, S., A.L. Pope, R.T. Littleton, T.M. Tritt, V. Ponnambalam, Y. Xia and S.J. Poon. 2000. Effect of Sb doping on the thermoelectric properties of Ti-based half-Heusler compounds. $TiNiSn_{1-x}Sb_x$. Appl. Phys. Lett. 77: 2476.

Birkel, C.S., J.E. Douglas, B.R. Lettiere, G. Seward, N. Verma, Y.C. Zhang, T.M. Pollock, R. Seshadri and G.D. Stucky. 2013. Improving the thermoelectric properties of half-Heusler TiNiSn through inclusion of a second full-Heusler phase: microwave preparation and spark plasma sintering of $TiNi_{1+x}Sn$. Phys. Chem. Chem. Phys. 15: 6990–6997.

Biswas, K., J.Q. He, G.Y. Wang, S.H. Lo, C. Uher, V.P. Dravid and M.G. Kanatzidis. 2011. High thermoelectric figure of merit in nanostructured p-type PbTe-MTe (M = Ca, Ba). Energy Environ. Sci. 4: 4675–4684.

Biswas, K., J.Q. He, I.D. Blum, C.I. Wu, T.P. Hogan, D.N. Seidman, V.P. Dravid and M.G. Kanatzidis. 2012. High-performance bulk thermoelectrics with all-scale hierarchical architectures. Nature 489: 414–418.

Blank, V.D., S.G. Buga, V.A. Kulbachinskii, V.G. Kytin, V.V. Medvedev, M.Y. Popov, P.B. Stepanov and V.F. Skok. 2012. Thermoelectric properties of $Bi_{0.5}Sb_{1.5}Te_3/C_{60}$ nanocomposites. Phys. Rev. B 86: 075426.

Bottger, P.H.M., K. Valset, S. Deledda and T.G. Finstad. 2010. Influence of ball-milling, nanostructuring, and Ag inclusions on thermoelectric properties of ZnSb. J. Electron. Mater. 39: 1583–1588.

Boukai, A.I., Y. Bunimovich, J.T. Kheli, J.K. Yu, W.A. Goddard and J.R. Heath. 2008. Silicon nanowires as efficient thermoelectric materials. Nature 451: 168–171.

Broido, D.A. and N. Mingo. 2006. Theory of the thermoelectric power factor in nanowire-composite matrix structures. Phys. Rev. B 74: 195325.

Bux, S.K., J.P. Fleurial and R.B. Kaner. 2010. Nanostructured materials for thermoelectric applications. Chem. Commun. 46: 8311–8324.

Bux, S.K., M.T. Yeung, E.S. Toberer, G.J. Snyder, R.B. Kaner and J.P. Fleurial. 2011. Mechanochemical synthesis and thermoelectric properties of high quality magnesium silicide. J. Mater. Chem. 21: 12259–12266.

Cai, K., E. Mueller, C. Drasar and C. Stiewe. 2004. The effect of titanium diboride addition on the thermoelectric properties of P-$FeSi_2$ semiconductors. Solid State Commun. 131: 325–329.

Cao, Y.Q., X.B. Zhao, T.J. Zhu, X.B. Zhang and J.P. Tu. 2008. Syntheses and thermoelectric properties of Bi_2Te_3/Sb_2Te_3 bulk nanocomposites with laminated nanostructure. Appl. Phys. Lett. 92: 143106.

Chai, Y.W. and Y. Kimura. 2012. Nanosized precipitates in half-Heusler TiNiSn alloy. Appl. Phys. Lett. 100: 033114.

Chen, G. 1998. Thermal conductivity and ballistic-phonon transport in the cross-plane direction of superlattices. Phys. Rev. B 57: 14958–14973.

Chen, L.D. 2014. Challenges in thermoelectric devices for power generation. ICT Nashville, TN, USA.

Chen, L.D., T. Kawahara, X.F. Tang, T. Goto, T. Hirai, J.S. Dyck, W. Chen and C. Uher. 2001. Anomalous barium filling fraction and n-type thermoelectric performance of $Ba_yCo_4Sb_{12}$. J. Appl. Phys. 90: 1864.

Chen, S. and Z.F. Ren. 2013. Recent progress of half-Heusler for moderate temperature thermoelectric applications. Mater. Today 16: 387–395.

Chen, S., K.C. Lukas, W.S. Liu, C.P. Opeil, G. Chen and Z.F. Ren. 2013. Effect of Hf concentration on thermoelectric properties of nanostructured n-type half-heusler materials $Hf_xZr_{1-x}NiSn_{0.99}Sb_{0.01}$. Adv. Energy Mater. 3: 1210–1214.

Chen, Z.G., G. Han, L. Yang, L. Cheng and J. Zou. 2012. Nanostructured thermoelectric materials: current research and future challenge. Prog. Nat. Sci. 22: 535–549.

Cook, B.A., M.J. Kramer, J.L. Harringa, M.K. Han, D.Y. Chung and M.G. Kanatzidis. 2009. Analysis of nanostructuring in high figure-of-merit $Ag_{1-x}Pb_mSbTe_{2+m}$ thermoelectric materials. Adv. Funct. Mater. 19: 1254–1259.

Cui, J.L., X.L. Liu, W. Yang, D.Y. Chen, H. Fu and P.Z. Ying. 2009. Thermoelectric properties in nanostructured homologous series alloys $Ga_mSb_nTe_{1.5(m+n)}$. J. Appl. Phys. 105: 063703.

Cundy, C.S. and P.A. Cox. 2003. The hydrothermal synthesis of zeolites: History and development from the earliest days to the present time. Chem. Rev. 103: 663–701.

Datta, A., J. Paul, A. Kar, A. Patra, Z.L. Sun, L.D. Chen, J. Martin and G.S. Nolas. 2010. Facile chemical synthesis of nanocrystalline thermoelectric alloys based on Bi-Sb-Te-Se. Crys. Growth Des. 10: 3983–3989.

Deng, Y., C.W. Nan, G.D. Wei, L. Guo and Y.H. Lin. 2003a. Organic-assisted growth of bismuth telluride nanocrystals. Chem. Phys. Lett. 374: 410–415.

Deng, Y., G.D. Wei and C.W. Nan. 2003b. Ligand-assisted control growth of chainlike nanocrystals. Chem. Phys. Lett. 368: 639–643.

Deng, Y., X. Zhou, G.D. Wei, J. Liu, C.W. Nan and S.J. Zhao. 2002. Solvothermal preparation and characterization of nanocrystalline Bi_2Te_3 powder with different morphology. J. Phys. Chem. Solids 63: 2119–2121.

Deng, Y., X.S. Zhou, G.D. Wei, J. Liu, J.B. Wu and C.W. Nan. 2003. Low temperature preparation and transport properties of ternary Pb-Bi-Te alloy. J. Alloys Compd. 350: 271–274.

Dirmyer, M.R., J. Martin, G.S. Nolas, A. Sen and J.V. Badding. 2009. Thermal and electrical conductivity of size-tuned bismuth telluride nanoparticles. Small 5: 933–937.

Dismukes, J.P., L. Ekstrom, E.F. Steigmeier, I. Kudman and D.S. Bears. 1964. Thermal and electrical properties of heavily doped Ge-Si alloys up to 1300 K. J. Appl. Phys. 35: 2899.

Dong, J.D., W. Liu, H. Li, X.L. Su, X.F. Tang and C. Uher. 2013. *In situ* synthesis and thermoelectric properties of PbTe-graphene nanocomposites by utilizing a facile and novel wet chemical method. J. Mater. Chem. A 1: 12503–12511.

Douglas, J.E., C.S. Birkel, M.S. Miao, C.J. Torbet, G.D. Stucky, T.M. Pollock and R. Seshadri. 2012. Enhanced thermoelectric properties of bulk TiNiSn via formation of a $TiNi_2Sn$ second phase. Appl. Phys. Lett. 101: 183902.

Dresselhaus, M.S., G. Chen, M.Y. Tang, R. Yang, H. Lee, D. Wang, Z.F. Ren, J.P. Fleurial and P. Gogna. 2007. New Directions for low-dimensional thermoelectric materials. Adv. Mater. 19: 1043–1053.

Du, B.L., H. Li, J.J. Xu, X.F. Tang and C. Uher. 2011. Enhanced thermoelectric performance and novel nanopores in $AgSbTe_2$ prepared by melt spinning. J. Solid State Chem. 184: 109–114.

Faleev, S.V. and F. Leonard. 2008. Theory of enhancement of thermoelectric properties of materials with nanoinclusions. Phys. Rev. B 77: 214304.

Fan, F.J., B. Yu, Y.X. Wang, Y.L. Zhu, X.J. Liu, S.H. Yu and Z.F. Ren. 2011. Colloidal synthesis of $Cu_2CdSnSe_4$ nanocrystals and hot-pressing to enhance the thermoelectric figure-of-merit. J. Am. Chem. Soc. 133: 15910–15913.

Fan, S.F., J.N. Zhao, J. Guo, Q.Y. Yan, J. Ma and H.H. Hng. 2010. P-type $Bi_{0.4}Sb_{1.6}Te_3$ nanocomposites with enhance figure of merit. Appl. Phys. Lett. 96: 182104.

Fan, S.F., J.N. Zhao, Q.Y. Yan, J. Ma and H.H. Hng. 2011. Influence of nanoinclusions on thermoelectric properties of n-type Bi_2Te_3 nanocomposites. J. Electron. Mater. 40: 1018–1023.

Feng, S.H. and R.R. Xu. 2001. New materials in hydrothermal synthesis. Acc. Chem. Res. 34: 239–247.

Foos, E.E., R.M. Stroud and A.D. Berry. 2001. Synthesis and characterization of nanocrystalline bismuth telluride. Nano Lett. 1: 693–695.

Gharleghi, A., Y.H. Pai, F.H. Lin and C.J. Liu. 2014. Low thermal conductivity and rapid synthesis of n-type cobalt skutterudite via a hydrothermal method. J. Mater. Chem. C 2: 4213–4220.

Girard, S.N., J.Q. He, C.P. Li, S. Moses, G.Y. Wang, C. Uher, V.P. Dravid and M.G. Kanatzidis. 2010. *In situ* nanostructure generation and evolution within a bulk thermoelectric material to reduce lattice thermal conductivity. Nano Lett. 10: 2825–2831.

Girard, S.N., J.Q. He, X.Y. Zhou, D. Shoemaker, C.M. Jaworski, C. Uher, V.P. Dravid, J.P. Heremans and M.G. Kanatzidis. 2011. High performance Na-doped PbTe-PbS thermoelectric materials: electronic density of states modification and shape-controlled nanostructures. J. Am. Chem. Soc. 133: 16588–16597.

Goldsmid, H.J. 1961. Recent studies of bismuth telluride and its alloys. J. Appl. Phys. 32: 2198.

Goldsmid, H.J. and R.W. Douglas. 1954. The use of semiconductors in thermoelectric refrigeration. J. Appl. Phys. 7: 386–390.

Han, C., Z. Li and S.X. Dou. 2014. Recent progress in thermoelectric materials. Chin. Sci. Bull. 59: 2073–2091.

Han, M.K., S. Kim, H.Y. Kim and S.J. Kim. 2013. An alternative strategy to construct interfaces in bulk thermoelectric material: nanostructured heterophase Bi_2Te_3/Bi_2S_3. RSC Adv. 3: 4673–4679.

Harman, T.C., M.P. Walsh, B.E. Laforge and G.W. Turner. 2005. Nanostructured thermoelectric materials. J. Electron. Mater. 34: L19–L22.

Harman, T.C., P.J. Taylor, M.P. Walsh and B.E. LaForge. 2002. Quantum dot superlattice thermoelectric materials and devices. Science 297: 2229–2232.

Hazama, H., M. Matsubara, R. Asahi and T. Takeuchi. 2011. Improvement of thermoelectric properties for half-Heusler TiNiSn by interstitial Ni defects. J. Appl. Phys. 110: 063710.

He, J.Q., M.G. Kanatzidis and V.P. Dravid. 2013. High performance bulk thermoelectric via a panoscopic approach. Mater. Today 16: 166–176.

He, Q.Y., Q. Hao, X.W. Wang, J. Yang, Y.C. Lan, X. Yan, B. Yu, Y. Ma, B. Poudel, G. Joshi, D.Z. Wang, G. Chen and Z.F. Ren. 2008. Nanostructured thermoelectric skutterudite $Co_{1-x}Ni_xSb_3$ alloys. J. Nano sci. Nanotechnol. 8: 4003–4006.

He, Q.Y., S.J. Hu, X.G. Tang, Y.C. Lan, J. Yang, X.W. Wang, Z.F. Ren, Q. Hao and G. Chen. 2008. The great improvement effect of pores on ZT in $Co_{(1-x)}Ni_{(x)}Sb_{(3)}$ system. Appl. Phys. Lett. 93: 042108.

He, Y., T. Day, T.S. Zhang, H.L. Liu, X. Shi, L.D. Chen and G.J. Snyder. 2014. High thermoelectric performance in non-toxic earth abundant copper sulfide. Adv. Mater. 26: 3974–3978.

Heremans, J.P., B. Wiendlocha and A.M. Chamoire. 2012. Resonant levels in bulk thermoelectric semiconductors. Energy Environ. Sci. 5: 5510–5530.

Heremans, J.P., V. Jovovic, E.S. Toberer, A. Saramat, K. Kurosaki, A. Charoenphakdee, S. Yamanaka and G.J. Snyder. 2008. Enhancement of thermoelectric efficiency in PbTe by distortion of the electronic density of states. Science 321: 554–557.

Hicks, L.D. and M.S. Dresselhaus. 1993. Effect of quantum-well structures on the thermoelectric figure of merit. Phys. Rev. B 47: 12727–12731.

Hicks, L.D., T.C. Harman, X. Sun and M.S. Dresselhaus. 1996. Experimental study of the effect of quantum-well structures on the thermoelectric figure of merit. Phys. Rev. B 53: 10493–10496.

Hochbaum, A.I., R. Chen, R.D. Delgado, W. Liang, E.C. Garnett, M. Najarian, A. Majumdar and P. Yang. 2008. Enhanced thermoelectric performance of rough silicon nanowires. Nature 451: 163–167.

Hohl, H., A.P. Ramirez, C. Goldmann, G. Ernst, B. Wolfing and E. Bucher. 1999. Efficient dopants for ZrNiSn-based thermoelectric materials. J. Phys.: Condens. Matter. 11: 1697–1709.

Hsu, K.F., S. Loo, F. Guo, W. Chen, J.S. Dyck, C. Uher, T. Hogan, E.K. Polychroniadis and M.G. Kanatzidis. 2004. Cubic $AgPb_mSbTe_{2+m}$: bulk thermoelectric materials with high figure of merit. Science 303: 818–821.

Huang, J.L. and P.K. Nayak. 2012. Effect of nano-TiN on mechanical behavior of Si_3N_4 based nanocomposites by Spark Plasma Sintering (SPS). http://cdn.intechopen.com.

Huang, X.Y., L.D. Chen, X. Shi, M. Zhou and Z. Xu. 2005. Thermoelectric performances of ZrNiSn/C-60 composite. Key Eng. Mater. 280–283, 385–388.

Huang, X.Y., Z. Xu, L.D. Chen and X.F. Tang. 2003. Effect of gamma-Al_2O_3 content on the thermoelectric performance of ZrNiSn/gamma-Al_2O_3 composites. Key Eng. Mater. 249: 79–82.

Hubert, M., E. Petracovschi, X.H. Zhang and L. Calvez. 2013. Synthesis of germanium-gallium-tellurium (Ge-Ga-Te) ceramics by ball-milling and sintering. J. Am. Ceram. Soc. 96: 1444–1449.

Ibanez, M., R. Zamani, S. Gorsse, J.D. Fan, S. Ortega, D. Cadavid, J.R. Morante, J. Arbiol and A. Cabot. 2013. Core-shell nanoparticles as building blocks for the bottom-up production of functional nanocomposites: PbTe-PbS thermoelectric properties. ACS Nano 7: 2573–2586.

Ioannou, M., E. Hatzikraniotis, C.B. Lioutas, T. Hassapis, T. Atlantis, K.M. Paraskevopoulos and T. Kyratsi. 2012. Fabrication of nanocrystalline Mg_2Si via ball milling process: structural studies. Powder Technol. 217: 523–532.

Ioffe, A.F. 1956. Thermoelements and thermoelectric cooling. London. UK: Infosearch.

Ioffe, F. 1961. Semiconductors thermoelements and thermoelectric cooling. Interscience, New York.

Ito, M., T. Tada and S. Hara. 2006. Thermoelectric properties of hot-pressed beta-$FeSi_2$ with yttria dispersion by mechanical alloying. J. Alloy. Compd. 2: 363–367.

Ito, M., T. Tada and S. Katsuyama. 2003. Thermoelectric properties of $Fe_{0.98}Co_{0.02}Si_2$ with ZrO_2 and rare-earth oxide dispersion by mechanical alloying. J. Alloy. Compd. 350: 296–302.

Jiang, Q.H., H.X. Yan, J. Khaliq, H.P. Ning, S. Grasso, K. Simpson and M.J. Reece. 2014. Large ZT enhancement in hot forged nanostructured p-type $Bi_{0.5}Sb_{1.5}Te_3$ bulk alloys. J. Mater. Chem. A 2: 5785–5790.

Joshi, G., H. Lee, Y. Lan, X. Wang, G. Zhu, D. Wang, R.W. Gould, D.C. Cuff, M.Y. Tang, M.S. Dresselhaus, G. Chen and Z.F. Ren. 2008. Enhanced thermoelectric figure-of-merit in nanostructured p-type silicon germanium bulk alloys. Nano Lett. 8: 4670–4674.

Joshi, G., X. Yan, H. Wang, W. Liu, G. Chen and Z.F. Ren. 2011. Enhancement in thermoelectric figure-of-merit of an n-type half-Heusler compound by the nanocomposite approach. Adv. Energy Mater. 1: 643–647.

Jung, D.Y., K. Kurosaki, S. Seino, M. Ishimaru, K. Sato, Y. Ohishi, H. Muta and S. Yamanaka. 2014. Thermoelectric properties of Au nanoparticle-supported $Sb_{1.6}Bi_{0.4}Te_3$ synthesized by a γ-ray irradiation method. Phys. Status Solidi B 251: 162–167.

Kanatzia, A., C. Papageorgiou, C. Lioutas and T. Kyratsi. 2013. Design of ball-milling experiments on Bi_2Te_3 thermoelectric material. J. Electron. Mater. 42: 1652–1660.

Kanatzidis, M.G. 2006. Nanostructured thermoelectrics: the new paradigm. Chem. Mater. 22: 648–659.

Keshavarz, M.K., D. Vasilevskiy, R.A. Masut and S. Turenne. 2014. Effect of suppression of grain growth of hot extruded $(Bi_{0.2}Sb_{0.8})_2Te_3$ thermoelectric alloys by MoS_2 nanoparticles. J. Electron. Mater. 43: 239–246.

Kim, H., M.K. Han, C.H. Yo, W. Lee and S.J. Kim. 2012. Effects of Bi_2Se_3 Nanoparticle inclusions on the microstructure and thermoelectric properties of Bi_2Te_3-based nanocomposites. J. Electron. Mater. 41: 3411–3416.

Kim, H.C., T.S. Oh and D.B. Hyun. 2000. Thermoelectric properties of the p-type Bi_2Te_3-Sb_2Te_3-Sb_2Se_3 alloys fabricated by mechanical alloying and hot pressing. J. Phys. Chem. Solids 61: 743–749.

Kim, K.T., S.Y. Choi, E.H. Shin, K.S. Moon, H.Y. Koo, G.G. Lee and G.H. Ha. 2013. The influence of CNTs on the thermoelectric properties of a CNT/Bi_2Te_3 composite. Carbon 52: 541–549.

Kim, M.Y., B.K. Yu and T.S. Oh. 2012. Thermoelectric characteristics of the p-type $(Bi_{0.2}Sb_{0.8})_2Te_3$ nanocomposites processed with SbTe nanowire dispersion. Electron. Mater. Lett. 8: 269–273.

Kim, W., J. Zide, A. Gossard, D. Klenov, S. Stemmer, A. Shakouri and A. Majumdar. 2006. Thermal conductivity reduction and thermoelectric figure of merit increase by embedding nanoparticles in crystalline semiconductors. Phys. Rev. Lett. 96: 045901.

Koch, C.C., R.O. Scattergood, K.M. Youssef, E. Chan and Y.T. Zhu. 2010. Nanostructured materials by mechanical alloying: new results on property enhancement. J. Mater. Sci. 45: 4725–4732.

Kraemer, D., B. Poudel, H.P. Feng, J.C. Caylor, B. Yu, X. Yan, Y. Ma, X.W. Wang, D.Z. Wang, A. Muto, K. McEnaney, M. Chiesa, Z.F. Ren and G. Chen. 2011. High-performance flat-panel solar thermoelectric generators with high thermal concentration. Nat. Mater. 10: 532–538.

Kulbachinskii, V.A., V.G. Kytin, M.Y. Popov, S.G. Buga, P.B. Stepanov and V.D. Blank. 2012. Composites of $Bi_{2-x}Sb_xTe_3$ nanocrystals and fullerene molecules for thermoelectricity. J. Solid State Chem. 193: 64–70.

Kuo, C.H., C.S. Hwang, M.S. Jeng, W.S. Su, Y.W. Chou and J.R. Ku. 2010. Thermoelectric transport properties of bismuth telluride bulk materials fabricated by ball milling and spark plasma sintering. J. Alloys Compd. 496: 687–690.

Lakshmanan, A. 2007. Sintering of ceramics-new emerging techniques. Intech, Janeza Trdine, Rijeka, Croatia.

Lan, Y.C., A.J. Minnich, G. Chen and Z.F. Ren. 2010. Enhancement of thermoelectric figure-of-merit by a bulk nanostructuring approach. Adv. Funct. Mater. 20: 357–376.

Lan, Y.C., B. Poudel, Y. Ma, D.Z. Wang, M.S. Dresselhaus, G. Chen and Z.F. Ren. 2009. Structure study of bulk nanograined thermoelectric bismuth antimony telluride. Nano Lett. 9: 1419–1422.

Lee, K.H., H.S. Kim, S.I. Kim, E.S. Lee, S.M. Lee, J.S. Rhyee, J.Y. Jung, I.H. Kim, Y. Wang and K. Koumoto. 2012. Enhancement of thermoelectric figure of merit for $Bi_{0.5}Sb_{1.5}Te_3$ by metal nanoparticle decoration. J. Electron. Mater. 41: 1165–1169.

Lee, Y., S.H. Lo, J. Androulakis, C.I. Wu, L.D. Zhao, D.Y. Chung, T.P. Hogan, V.P. Dravid and M.G. Kanatzidis. 2013. High-performance tellurium-free thermoelectrics: all-scale hierarchical structuring of p-type PbSe-MSe Systems (M = Ca, Sr, Ba). J. Am. Chem. Soc. 135: 5152–5160.

Li, D., X.Y. Qin, Y.F. Liu, N.N. Wang, C.J. Song and R.R. Sun. 2013. Improved thermoelectric properties for solution grown $Bi_2Te_{3-x}Se_x$ nanoplatelet composites. RSC Adv. 3: 2632–2638.

Li, F., X.Y. Huang, Z.L. Sun, J. Ding, J. Jiang, W. Jiang and L.D. Chen. 2011. Enhanced thermoelectric properties of n-type Bi_2Te_3-based nanocomposite fabricated by spark plasma sintering. J. Alloy Compd. 509: 4769–773.

Li, H., X.F. Tang, X.L. Su and Q.J. Zhang. 2008. Preparation and thermoelectric properties of high-performance Sb additional $Yb_{(0.2)}Co_{(4)}Sb_{(12+y)}$ bulk materials with nanostructure. Appl. Phys. Lett. 92: 202114.

Li, J.F., W.S. Liu, L.D. Zhao and M. Zhou. 2010. High-performance nanostructured thermoelectric materials. NPG Asia Mater. 2: 152–158.

Li, J.H., Q. Tan, J.F. Li, D.W. Liu, F. Li, Z.X. Li, M.M. Zou and K. Wang. 2013. BiSbTe-based nanocomposites with high ZT: the effect of SiC nanodispersion on thermoelectric properties. Adv. Funct. Mater. 23: 4317–4323.

Li, Q., Z.W. Lin and J. Zhou. 2009. Thermoelectric materials with potential high power factors for electricity generation. J. Electron. Mater. 38: 1268–1272.

Lin, R.C., G. Chen, J. Pei and C.S. Yan. 2012. Hydrothermal synthesis and thermoelectric transport property of PbS-PbTe core-shell heterostructures. New J. Chem. 36: 2574–2579.

Liu, H., X. Shi, F.F. Xu, L.L. Zhang, W.Q. Zhang, L.D. Chen, Q. Li, C. Uher, T. Day and G.J. Snyder. 2012. Copper ion liquid-like thermoelectric. Nat. Mater 11: 422–425.

Liu, H., J.Y. Wang, X.B. Hu, L.X. Li, F. Gu, S.S. Zhao, M.Y. Gu, R.I. Boughton and M.H. Jiang. 2002. Preparation of filled skutterudite nanowire by a hydrothermal method. J. Alloy Compd. 334: 313–316.

Liu, W.S., X. Yan, G. Chen and Z.F. Ren. 2012. Recent advances in thermoelectric nanocomposites. Nano Energy 1: 42–56.

Liu, W.S., Q.Y. Zhang, Y.C. Lan, S. Chen, X. Yan, Q. Zhang, H. Wang, D.Z. Wang, G. Chen and Z.F. Ren. 2011. Thermoelectric property studies on Cu-doped n-type $Cu_xBi_2Te_{2.7}Se_{0.3}$ nanocomposites. Adv. Energy Mater. 1: 577–587.

Lopez, R.M., A. Dauscher, H. Scherrer, J. Hejtmanek, H. Kenzari and B. Lenoir. 1999. Thermoelectric properties of mechanically alloyed Bi-Sb alloys. Appl. Phys. A: Mater. Sci. Process. 68: 597–602.

Lu, W., Y. Ding, Y. Chen, Z.L. Wang and J. Fang. 2005. Bismuth telluride hexagonal nanoplatelets and their two-step epitaxial growth. J. Am. Chem. Soc. 127: 10112–10116.

Ma, Y., Q. Hao, B. Poudel, Y. Lan, B. Yu, D.Z. Wang, G. Chen and Z.F. Ren. 2008. Enhanced thermoelectric figure-of-merit in p-type nanostructured bismuth antimony tellurium alloys made from elemental chunks. Nano Lett. 8: 2580–2584.

Makongo, J.P.A., D.K. Misra, X. Zhou, A. Pant, M.R. Shabetai, X. Su, C. Uher, K.L. Stokes and P.F.P. Poudeu. 2011. Simultaneous large enhancements in thermopower and electrical conductivity of bulk nanostructured half-Heusler alloys. J. Am. Chem. Soc. 133: 18843–18852.

Medlin, D.L. and G.J. Snyder. 2009. Interfaces in bulk thermoelectric materials a review for current opinion in colloid and interface science. Curr. Opin. Colloid In. 14: 226–235.

Mi, J.L., N. Lock, T. Sun, M. Christensen, M. Sondergaard, P. Hald, H.H. Hng, J. Ma and B.B. Iversen. 2010. Biomolecule-assisted hydrothermal synthesis and self-assembly of Bi_2Te_3 nanostring-cluster hierarchical structure. ACS Nano 4: 2523–2530.

Mingo, N., D. Hauser, N.P. Kobayashi, M. Plissonnier and A. Shakouri. 2009. "Nanoparticle-in-alloy" approach to efficient thermoelectrics: silicides in SiGe. Nano Lett. 9: 711–715.

Minnich, A.J., M.S. Dresselhaus, Z.F. Ren and G. Chen. 2009. Bulk nanostructured thermoelectric materials: current research and future prospects. Energy Environ. Sci. 2: 466–479.

Misra, D.K., J.P.A. Makongo, P. Sahoo, M.R. Shabetai, P. Paudel, K.L. Stokes and P.F.P. Poudeu. 2011. Microstructure and thermoelectric properties of mechanically alloyed $Zr_{0.5}Hf_{0.5}Ni_{0.8}Pd_{0.2}Sn_{0.99}Sb_{0.01}/WO_3$ half-heusler composites. Sci. Adv. Mater. 3: 607–614.

Ni, H.L., X.B. Zhao, T.J. Zhu, X.H. Ji and J.P. Tu. 2005. Synthesis and thermoelectric properties of Bi_2Te_3 based nanocomposites. J. Alloy Compd. 397: 317–321.

Nolas, G.S., J. Poon and M.G. Kanatzidis. 2006. Recent developments in bulk thermoelectric materials. MRS Bulletin 31: 199–205.

Nolas, G.S., M. Kaeser, R.T. Littleton and T.M. Tritt. 2000. High figure of merit in partially filled ytterbium skutterudite materials. Appl. Phys. Lett. 77: 1855.

Papageorgiou, C., E. Hatzikraniotis, C.B. Lioutas, N. Frangis, O. Valasiades, K.M. Paraskevopoulos and T. Kyratsi. 2010. Thermoelectric properties of nanocrystalline PbTe synthesized by mechanical alloying. J. Electron. Mater. 39: 1665–1668.

Park, D.H., M.Y. Kim and T.S. Oh. 2011. Thermoelectric energy-conversion characteristics of n-type $Bi_2(Te,Se)_3$ nanocomposites processed with carbon nanotube dispersion. Curr. Appl. Phys. 11: S41–S45.

Park, K., J.S. Son, S.I. Woo, K. Shin, M.W. Oh, S.D. Park and T. Hyeon. 2014. Colloidal synthesis and thermoelectric properties of La-doped SrTiO₃ nanoparticles. J. Mater. Chem. A 2: 4217–4224.

Pei, Y.Z., H. Wang and G.J. Snyder. 2012. Band engineering of thermoelectric materials. Adv. Mater. 24: 6125–6135.

Pei, Y.Z., J.L. Falk, E.S. Toberer, D.L. Medlin and G.J. Snyder. 2011. High thermoelectric performance in PbTe due to large nanoscale Ag_2Te precipitates and La doping. Adv. Funct. Mater. 21: 241–249.

Pei, Y.Z., X.Y. Shi, A. LaLonde, H. Wang, L.D. Chen and G.J. Snyder. 2011. Convergence of electronic bands for high performance bulk thermoelectrics. Nature 473: 66–69.

Pichanusakorn, P. and P. Bandaru. 2010. Nanostructured thermoelectrics. Mat. Sci. Eng. R 67: 19–63.

Pierrat, P., A. Dauscher, B. Lenoir, R.M. Lopez and H. Scherrer. 1997. Preparation of the $Bi_8Sb_{32}Te_{60}$ solid solution by mechanical alloying. J. Mater. Sci. 32: 3653–3657.

Ponnambalam, V., P.N. Alboni, J. Edwards, T.M. Tritt, S.R. Culp and S.J. Poon. 2008. Thermoelectric properties of p-type half-Heusler alloys $Zr_{(1-x)}Ti_{(x)}CoSn_{(y)}Sb_{(1-y)}$ (0.0 < x < 0.5; y = 0.15 and 0.3). J. Appl. Phys. 103: 063716.

Poon, S.J., D. Wu, S. Zhu, W.J. Xie, T.M. Tritt, P. Thomas and R. Venkatasubramanian. 2011. Half-Heusler phases and nanocomposites as emerging high-ZT thermoelectric materials. J. Mater. Res. 26: 2795–2802.

Populoh, S., O.C. Brunko, K. Gałązka, W.J. Xie and A. Weidenkaff. 2013. Half-heusler (TiZrHf) NiSn unileg module with high powder density. Materials 6: 1326–1332.

Poudel, B., Q. Hao, Y. Ma, Y. Lan, A. Minnich, B. Yu, X. Yan, D.Z. Wang, A. Muto, D. Vashaee, X. Chen, J. Liu, M.S. Dresselhaus, G. Chen and Z.F. Ren. 2008. High-thermoelectric performance of nanostructured bismuth antimony telluride bulk alloys. Science 320: 634–638.

Poudeu, P.F.P., A. Gueguen, C. Wu, T. Hogan and M.G. Kanatzidis. 2010. High figure of merit in nanostructured n-Type KPb_mSbTe_{m+2} thermoelectric materials. Chem. Mater. 22: 1046–1053.

Poudeu, P.F.P., J. Angelo, A.D. Downey, J.L. Short, T.P. Hogan and M.G. Kanatzidis. 2006. High thermoelectric figure of merit and nanostructuring in bulk p-type $Na_{1-x}Pb_mSb_yTe_{m+2}$. Angew. Chem. Int. Ed. 45: 3835–3839.

Poudeu, P.F.P., J. Angelo, H.J. Kong, A. Downey, J.L. Short, R. Pcionek, T.P. Hogan, C. Uher and M.G. Kanatzidis. 2006. Nanostructures versus solid solutions: Low lattice thermal conductivity and enhanced thermoelectric figure of merit in $Pb_{9.6}Sb_{0.2}Te_{10-x}Se_x$ bulk materials. J. Am. Chem. Soc. 128: 14347–14355.

Purkayastha, A., S. Kim, D.D. Gandhi, P.G. Ganesan, T.B. Tasciuc and G. Ramanath. 2006. Molecularly protected bismuth telluride nanoparticles: Microemulsion synthesis and thermoelectric transport properties. Adv. Mater. 18: 2958.

Quarez, E., K.F. Hsu, R. Pcionek, N. Frangis, E.K. Polychroniadis and M.G. Kanatzidis. 2005. Nanostructuring, compositional fluctuations, and atomic ordering in the thermoelectric materials $AgPb_mSbTe_{2+m}$. The myth of solid solutions. J. Am. Chem. Soc. 127: 9177–9190.

Ritter, J.J. and P. Maruthamuthu. 1997. Synthesis of fine-powder polycrystalline Bi-Se-Te, Bi-Sb-Te, and Bi-Sb-Se-Te alloys. Inorg. Chem. 36: 260–263.

Rowe, D.M. 1995. CRC handbook of thermoelectrics. CRC Press, Boca Raton, Florida.

Rowe, D.M. 2006. Thermoelectrics handbook: macro to nano. CRC Press, Boca Raton, Florida.

Rowe, D.M., L.W. Fu and S.G.K. Williams. 1993. Comments on the thermoelectric properties of pressure-sintered $Si_{0.8}Ge_{0.2}$ thermoelectric alloys. J. Appl. Phys. 73: 4683–4685.

Rowe, D.M., V.S. Shukla and N. Savvides. 1981. Phonon scattering at grain boundaries in heavily doped fine-grained silicon-germanium alloys. Nature 290: 765–766.

Rowe, D.W. and C.M. Bhandari. 1983. Modern thermoelectricity. Holt, Rinchalt and Wiston, London.

Sahoo, P., Y.F. Liu, J.P.A. Makongo, X.L. Su, S.J. Kim, N. Takas, H. Chi, C. Uher, X.Q. Pan and P.F.P. Poudeu. 2013. Enhancing thermopower and hole mobility in bulk p-type half-Heuslers using full-Heusler nanostructures. Nanoscale 5: 9419–9427.

Saleemi, M., M.S. Toprak, S.H. Li, M. Johnsson and M. Muhammed. 2012. Synthesis, processing, and thermoelectric properties of bulk nanostructured bismuth telluride (Bi_2Te_3). J. Mater. Chem. 22: 725–730.

Sales, B.C. 2002. Smaller is cooler. Science 295: 1248–1249.

Sales, B.C., D. Mandrus and R.K. Williams. 1996. Filled skutterudite antimonides: A new class of thermoelectric materials. Science 272: 1325–1328.

Sales, B.C., D. Mandrus, B.C. Chakoumakos, V. Keppens and J.R. Thompson. 1997. Filled skutterudite antimonides: electron crystals and phonon glasses. Phys. Rev. B 56: 15081.

Satyala, N., A.T. Rad, Z. Zamanipour, P. Norouzzadeh, J.S. Krasinski, L. Tayebi and D. Vashaee. 2014. Reduction of thermal conductivity of bulk nanostructured bismuth telluride composites embedded with silicon nano-inclusions. J. Appl. Phys. 115: 044304.

Scheele, M., N. Oeschler, K. Meier, A. Kornowski, C. Klinke and H. Weller. 2009. Synthesis and thermoelectric characterization of Bi_2Te_3 nanoparticles. Adv. Funct. Mater. 19: 3476–3483.

Shakouri, A. 2011. Recent developments in semiconductor thermoelectric physics and materials. Annu. Rev. Mater. Res. 41: 399–431.

Shen, Q., L. Chen, T. Goto, T. Hirai, J. Yang, G.P. Meisner and C. Uher. 2001. Effects of partial substitution of Ni by Pd on the thermoelectric properties of ZrNiSn-based half-Heusler compounds. Appl. Phys. Lett. 79: 4165.

Shi, S., M. Cao and C. Hu. 2009. Controlled solvothermal synthesis and structural characterization of antimony telluride nanoforks. Crys. Growth Des. 9: 2057–2060.

Shi, W., L. Zhou, S. Song, J. Yang and H. Zhang. 2008. Hydrothermal synthesis and thermoelectric transport properties of impurity-free antimony telluride hexagonal nanoplates. Adv. Mater. 20: 1892–1897.

Shi, W.D., J.B. Yu, H.S. Wang and H.J. Zhang. 2006. Hydrothermal synthesis of single-crystalline antimony telluride nanobelts. J. Am. Chem. Soc. 128: 16490–16491.

Shi, W.D., S.Y. Song and H.J. Zhang. 2013. Hydrothermal synthetic strategies of inorganic semiconducting nanostructures. Chem. Soc. Rev. 42: 5714–5743.

Shi, X., J. Yang, J.R. Salvador, M.F. Chi, J.Y. Cho, H. Wang, S.Q. Bai, J.H. Yang, W.Q. Zhang and L.D. Chen. 2011. Multiple-filled skutterudites: high thermoelectric figure of merit through separately optimizing electrical and thermal transports. J. Am. Chem. Soc. 133: 7837–7846.

Shi, X., S.Q. Bai, L.L. Xi, J. Yang, W.Q. Zhang, L.D. Chen and J.H. Yang. 2011. Realization of high thermoelectric performance in n-type partially filled skutterudites. J. Mater. Res. 26: 1745–1754.

Shin, D.K., K.W. Jang, S.C. Ur and I.H. Kim. 2013. Thermoelectric properties of higher manganese silicides prepared by mechanical alloying and hot pressing. J. Electron. Mater. 42: 1756–1761.

Singh, A., S. Bhattacharya, C. Thinaharan, D.K. Aswal, S.K. Gupta, J.V. Yakhmi and K. Bhanumurthy. 2009. Development of low resistance electrical contacts for thermoelectric devices based on n-type PbTe and p-type TAGS-85 (($AgSbTe_2)_{(0.15)}(GeTe)_{(0.85)}$). J. Phys. D.: Appl. Phys. 42: 015502.

Slack, G.A. 1995. CRC Handbook of thermoelectric. CRC Press, Boca Raton, Florida 407–440.

Slack, G.A. and M.A. Hussain. 1991. Phonon scattering at grain boundaries in heavily doped fine-grained silicon-germanium alloys. J. Appl. Phys. 70: 2694–2718.

Snyder, G.J. and E.S. Toberer. 2008. Complex thermoelectric materials. Nature materials 7: 105–114.

Son, J.S., M.K. Choi, M.K. Han, K. Park, J.Y. Kim, S.J. Lim, M. Oh, Y. Kuk, C. Park, S.J. Kim and T. Hyeon. 2012. N-type nanostructured thermoelectric materials prepared from chemically synthesized ultrathin Bi_2Te_3 nanoplates. Nano Lett. 12: 640–647.

Song, R., T. Aizawa and J. Sun. 2007a. Synthesis of $Mg_2Si_{1-x}Sn_x$ solid solutions as thermoelectric materials by bulk mechanical alloying and hot pressing. Mater. Sci. Eng. B 136: 111–117.

Song, R.B., Y.Z. Liu and T. Aizawa. 2007b. Thermoelectric properties of p-type $Mg_2Si_{0.6}Ge_{0.4}$ fabricated by bulk mechanical alloying and hot pressing. Phys. Status Solid RRL 1: 226–228.

Song, S.W., J.L. Wang, B. Xu, X.B. Lei, H.C. Jiang, Y.G. Jin, Q.Y. Zhang and Z.F. Ren. 2014. Thermoelectric properties of n-type $Bi_2Te_{2.7}Se_{0.3}$ with addition of nano-ZnO:Al particles. Materials Research Express 1: 035901.

Sootsman, J.R., D.Y. Chung and M.G. Kanatzidis. 2009. New and old concepts in thermoelectric materials. Angew. Chem. Int. Ed. 48: 8616–8639.

Sootsman, J.R., H. Kong, C. Uher, J.J. D'Angelo, C.I. Wu, T.P. Hogan, T. Caillat and M.G. Kanatzidis. 2008. Large enhancements in the thermoelectric power factor of bulk PbTe at high temperature by synergistic nanostructuring. Angew. Chem. Int. Ed. 47: 8618–8622.

Sootsman, J.R., R.J. Pcionek, H. Kong, C. Uher and M.G. Kanatzidis. 2006. Strong reduction of thermal conductivity in nanostructured PbTe prepared by matrix encapsulation. Chem. Mater. 18: 4993–4995.

Sumithra, S., N.J. Takas, D.K. Misra, W.M. Nolting, P.F.P. Poudeu and K.L. Stokes. 2011. Enhancement in thermoelectric figure of merit in nanostructured Bi_2Te_3 with semimetal nanoinclusions. Adv. Energy Mater. 1: 1141–1147.

Sun, X., Z. Zhang and M.S. Dresselhaus. 1999. Theoretical modeling of thermoelectricity in Bi nanowires. Appl. Phys. Lett. 74: 4005–4007.

Suryanarayana, C. 2001. Mechanical alloying and milling. Prog. Mater. Sci. 46: 1–184.

Tan, G.J., W. Liu, S.Y. Wang, Y.G. Yan, H. Li, X.F. Tang and C. Uher. 2013. Rapid preparation of $CeFe_4Sb_{12}$ skutterudite by melt spinning: rich nanostructures and high thermoelectric performance. J. Mater. Chem. A 1: 12657–12668.

Tan, L.P., T. Sun, S.F. Fan, L.Y. Ng, A. Suwardi, Q.Y. Yan and H.H. Hng. 2013. Facile synthesis of Cu_7Te_4 nanorods and the enhanced thermoelectric properties of Cu_7Te_4-$Bi_{0.4}Sb_{1.6}Te_3$ nanocomposites. Nano Energy 2: 4–11.

Tang, X.F., W.J. Xie, H. Li, W.Y. Zhao, Q.J. Zhang and M. Niino. 2007. Preparation and thermoelectric transport properties of high-performance p-type Bi_2Te_3 with layered nanostructure. Appl. Phys. Lett. 90: 012102.

Tkatch, V.I., I.L. Alexander, N.D. Sergey and G.R. Sergey. 2002. The effect of the melt-spinning processing parameters on the rate of cooling. Mater. Sci. Eng. A 323: 91–96.

Tokita, M. 1993. Trends in advanced SPS spark plasma sintering systems and technology. J. Soc. Powder Technol. 30: 790–804.

Tritt, T.M. 1999. Thermoelectric materials-holey and unholey semiconductors. Science 283: 804–805.

Uher, C., J. Yang, S. Hu, D.T. Morelli and G.P. Meisner. 1999. Transport properties of pure and doped MNiSn (M = Zr, Hf). Phys. Rev. B 59: 8615.

Ur, S.C., I.H. Kim and P. Nash. 2007. Thermoelectric properties of Zn_4Sb_3 processed by sintering of cold pressed compacts and hot pressing. J. Mater. Sci. 42: 2143–2149.

Venkatasubramanian, R., E. Siivola, T. Colpitts and B. O'Quinn. 2001. Thin-film thermoelectric devices with high room-temperature figures of merit. Nature 413: 597–602.

Vineis, C.J., A. Shakouri, A. Majumdar and M.G. Kanatzidis. 2010. Nanostructured thermoelectrics: big efficiency gains from small features. Adv. Mater. 22: 3970–3980.

Vining, C.B. 1991. A model for the high-temperature transport properties of heavily doped n-type silicon-germanium alloys. J. Appl. Phys. 69: 331–341.

Vining, C.B., W. Laskow, J.O. Hanson, V.D. Beck and P.D. Gorsuch. 1991. Thermoelectric properties of pressure-sintered $Si_{0.8}Ge_{0.2}$ thermoelectric alloys. J. Appl. Phys. 69: 4333–4340.

Wan, C.L., Y.F. Wang, N. Wang, W. Norimatsu, M. Kusunoki and K. Koumoto. 2010. Development of novel thermoelectric materials by reduction of lattice thermal conductivity. Sci. Technol. Adv. Mat. 11: 044306.

Wang, H., J.F. Li, C.W. Nan, M. Zhou, W. Liu, B.P. Zhang and T. Kita. 2006. High-performance $Ag_{0.8}Pb_{18+x}SbTe_{20}$ thermoelectric bulk materials fabricated by mechanical alloying and spark plasma sintering. Appl. Phys. Lett. 88: 092104.

Wang, L. and X.Y. Qin. 2003. The effect of mechanical milling on the formation of nanocrystalline Mg_2Si through solid-state reaction. Scr. Mater. 49: 243–248.

Wang, S.Y., H. Li, D.K. Qi, W.J. Xie and X.F. Tang. 2011. Enhancement of the thermoelectric performance of beta-Zn_4Sb_3 by *in situ* nanostructures and minute Cd-doping. Acta Mater. 59: 4805–4817.

Wang, S.Y., H. Li, R.M. Lu, G. Zheng and X.F. Tang. 2013. Metal nanoparticle decorated n-type Bi_2Te_3-based materials with enhanced thermoelectric performances. Nanotechnology 24: 285702.

Wang, S.Y., W.J. Xie, H. Li and X.F. Tang. 2011. Enhanced performances of melt spun $Bi_2(Te,Se)_3$ for n-type thermoelectric legs. Intermetallics 19: 1024–1031.

Wang, W., B. Poudel, J. Yang, D.Z. Wang and Z.F. Ren. 2005. High-yield synthesis of single-crystalline antimony telluride hexagonal nanoplates using a solvothermal approach. J. Am. Chem. Soc. 127: 13792–13793.

Wang, W., X. Yan, B. Poudel, Y. Ma, Q. Hao, J. Yang, G. Chen and Z.F. Ren. 2008. Chemical synthesis of anisotropic nanocrystalline Sb_2Te_3 and low thermal conductivity of the compacted dense bulk. J. Nanosci. Nanotechnol. 8: 452–456.

Wang, W.Z., Y. Geng, Y.T. Qian, Y. Xie and X.M. Liu. 1999. Synthesis and characterization of nanocrystalline Bi_2Se_3 by solvothermal method. Mater. Res. Bull. 34: 131–134.

Wang, X.W., H. Lee, Y.C. Lan, G.H. Zhu, G. Joshi, D.Z. Wang, J. Yang, A.J. Muto, M.Y. Tang, J. Klatsky, S. Song, M.S. Dresselhaus, G. Chen and Z.F. Ren. 2008. Enhanced thermoelectric figure of merit in nanostructured n-type silicon germanium bulk alloy. Appl. Phys. Lett. 93: 193121.

Willem, J., G. Bos and R.A. Downie. 2014. Half-Heusler thermoelectrics: a complex class of materials. J. Phys.: Condens. Matter 26: 433201.

Xie, W.J., S.Y. Wang, S. Zhu, J. He, X.F. Tang, Q.J. Zhang and T.M. Tritt. 2013. High performance Bi_2Te_3 nanocomposites prepared by single-element-melt-spinning spark-plasma sintering. J. Mater. Sci. 48: 2745–2760.

Xie, H.H., C. Yu, T.J. Zhu, C.G. Fu, G.J. Snyder and X.B. Zhao. 2012. Increased electrical conductivity in fine-grained (Zr,Hf)NiSn based thermoelectric materials with nanoscale precipitates. Appl. Phys. Lett. 100: 254104.

Xie, W.J., A. Weidenkaff, X.F. Tang, Q.J. Zhang, J. Poon and T.M. Trit. 2012. Recent advances in nanostructured thermoelectric half-heusler compounds. Nanomaterials 2: 379–412.

Xie, W.J., J. He, S. Zhu, X.L. Su, S.Y. Wang, T. Holgate, J.W. Graff, V. Ponnambalam, S.J. Poon, X.F. Tang, Q.J. Zhang and T.M. Tritt. 2010. Simultaneously optimizing the independent thermoelectric properties in (Ti,Zr,Hf)(Co,Ni)Sb alloy by *in situ* forming InSb nanoinclusions. Acta Mater. 58: 4705–4713.

Xie, W.J., X.F. Tang, Y.G. Yan, Q.J. Zhang and T.M. Tritt. 2009. Unique nanostructures and enhanced thermoelectric performance of melt-spun BiSbTe alloys. Appl. Phys. Lett. 94: 102111.

Xie, W.J., Y.G. Yan, S. Zhu, M. Zhou, S. Populoh, K. Gałazka, S.J. Poon, A. Weidenkaff, J. He, X.F. Tang and T.M. Tritt. 2013. Significant ZT enhancement in p-type Ti(Co, Fe)Sb-InSb nanocomposites via a synergistic high-mobility electron injection, energy-filtering and boundary-scattering approach. Acta Mater. 61: 2087–2094.

Yan, X., B. Poudel, Y. Ma, W.S. Liu, G. Joshi, H. Wang, Y.C. Lan, D.Z. Wang, G. Chen and Z.F. Ren. 2010. Experimental studies on anisotropic thermoelectric properties and structures of n-type $Bi_2Te_{2.7}Se_{0.3}$. Nano Lett. 10: 3373–3378.

Yan, X., G. Joshi, W.S. Liu, Y.C. Lan, H. Wang, S. Lee, J.W. Simonson, S.J. Poon, T.M. Tritt, G. Chen and Z.F. Ren. 2011. Enhanced thermoelectric figure of merit of p-type half-heuslers. Nano Lett. 11: 556–560.

Yan, X., W.S. Liu, H. Wang, S. Chen, J. Shiomi, K. Esfarjani, H.Z. Wang, D.Z. Wang, G. Chen and Z.F. Ren. 2012. Stronger phonon scattering by larger differences in atomic mass and size in p-type half-Heuslers $Hf_{1-x}Ti_xCoSb_{0.8}Sn_{0.2}$. Energ. Environ. Sci. 5: 7543–7548.

Yang, J., Q. Hao, H. Wang, Y.C. Lan, Q.Y. He, A. Minnich, D.Z. Wang, J.A. Harriman, V.M. Varki and M.S. Dresselhaus. 2009. Solubility study of Yb in n-type skutterudites $Yb_xCo_4Sb_{12}$ and their enhanced thermoelectric properties. Phys. Rev. B 80: 115329.

Yang, J.H. and T. Caillat. 2006. Thermoelectric materials for space and automotive power generation. MRS Bulletin 31: 224–229.

Yang, R.G., G. Chen and M.S. Dresselhaus. 2005. Thermal conductivity of simple and tubular nanowire composites in the longitudinal direction. Phys. Rev. B 72: 125418.

Yaqub, R., P. Sahoo, J.P.A. Makongo, N. Takas, P.F.P. Poudeu and K.L. Stokes. 2011. Investigation of the effect of NiO nanoparticles on the transport properties of $Zr_{0.5}Hf_{0.5}Ni_{1-x}Pd_xSn_{0.99}Sb_{0.01}$ (x = 0 and 0.2). Sci. Adv. Mater. 3: 633–638.

Yu, B., M. Zebarjadi, H. Wang, K. Lukas, H.Z. Wang, D.Z. Wang, C. Opeil, M.S. Dresselhaus, G. Chen and Z.F. Ren. 2012. Enhancement of thermoelectric properties by modulation-doping in silicon germanium alloy nanocomposites. Nano Lett. 12: 2077–2082.

Yu, B., Q.Y. Zhang, H. Wang, X.W. Wang, H.Z. Wang, D.Z. Wang, H. Wang, G.J. Snyder, G. Chen and Z.F. Ren. 2010. Thermoelectric property studies on thallium-doped lead telluride prepared by ball milling and hot pressing. J. Appl. Phys. 108: 016104.

Yu, C., T. Zhu, K. Xiao, J. Shen and X. Zhao. 2010. Microstructure and thermoelectric properties of (Zr,Hf)NiSn-based half-Heusler alloys by melt spinning and spark plasma sintering. Funct. Mater. Lett. 3: 227–231.

Yu, C., T.J. Zhu, K. Xiao, J.J. Shen, S.H. Yang and X.B. Zhao. 2010. Reduced grain size and improved thermoelectric properties of melt spun (Hf,Zr)NiSn half-Heusler alloys. J. Electron. Mater. 39: 2008–2012.

Yu, F.R., J.J. Zhang, D.L. Yu, J.L. He, Z.Y. Liu, B. Xu and Y.J. Tian. 2009. Enhanced thermoelectric figure of merit in nanocrystalline Bi_2Te_3 bulk. J. Appl. Phys. 105: 094303.

Zamanipour, Z. and D. Vashaee. 2012. Comparison of thermoelectric properties of p-type nanostructured bulk $Si_{0.8}Ge_{0.2}$ alloy with $S_{i0.8}Ge_{0.2}$ composites embedded with $CrSi_2$ nano-inclusisons. J. Appl. Phys. 112: 093714.

Zebarjadi, M., G. Joshi, G.H. Zhu, B. Yu, A. Minnich, Y.C. Lan, X.W. Wang, M.S. Dresselhaus, Z.F. Ren and G. Chen. 2011. Power factor enhancement by modulation doping in bulk nanocomposites. Nano Lett. 11: 2225–2230.

Zebarjadi, M., K. Esfarjani, A. Shakouri, J.H. Bahk, Z.X. Bian, G. Zeng, J. Bowers, H. Lu, J. Zide and A. Gossard. 2009. Effect of nanoparticle scattering on thermoelectric power factor. Appl. Phys. Lett. 94: 202105.

Zhang, G., W. Wang, X. Lu and X. Li. 2009. Solvothermal synthesis of V-VI binary and ternary hexagonal platelets: the oriented attachment mechanism. Gryst. Growth Des. 9: 145–150.

Zhang, H.T., X.G. Luo, C.H. Wang, Y.M. Xiong, S.Y. Li and X.H. Chen. 2004. Characterization of nanocrystalline bismuth telluride (Bi_2Te_3) synthesized by a hydrothermal method. J. Cryst. Growth 265: 558–562.

Zhang, L.Z., J.C. Yu, M.S. Mo, L. Wu, K.W. Kwong and Q. Li. 2005. A general *in situ* hydrothermal rolling-up formation of one-dimensional. single-crystalline lead telluride nanostructures. Small 1: 349–354.

Zhang, Q., Q.Y. Zhang, S. Chen, W.S. Liu, K. Lukas, X. Yan, H.Z. Wang, D.Z. Wang, C. Opeil, G. Chen and Z.F. Ren. 2012. Suppression of grain growth by additive in nanostructured p-type bismuth antimony tellurides. Nano Energy 1: 183–189.

Zhang, Q.H., X. Ai, L.J. Wang, Y.X. Chang, W. Luo, W. Jiang and L.D. Chen. 2015. Improved thermoelectric performance of silver nanoparticles-dispersed Bi_2Te_3 composites deriving from hierarchical two-phased heterostructure. Adv. Funct. Mater. 25: 966–976.

Zhang, Q.H., X. Ai, W.J. Wang, L.J. Wang and W. Jiang. 2014. Preparation of 1-D/3-D structured AgNWs/Bi_2Te_3 nanocomposites with enhanced thermoelectric properties. Acta Mater. 73: 37–47.

Zhang, Y., X.L. Wang, W.K. Yeoh, R.K. Zeng and C. Zhang. 2012. Electrical and thermoelectric properties of single-wall carbon nanotube doped Bi_2Te_3. Appl. Phys. Lett. 101: 031909.

Zhao, D.G. 2009. Doctoral dissertation. Shanghai Institute of Ceramics Chinese Academy of Sciences. China.

Zhao, L.D., J.Q. He, C.I. Wu, T.P. Hogan, X.Y. Zhou, C. Uher, V.P. Dravid and M.G. Kanatzidis. 2012. Thermoelectrics with earth abundant elements: high performance p-type PbS nanostructured with SrS and CaS. J. Am. Chem. Soc. 134: 7902–7912.

Zhao, L.D., J.Q. He, S.Q. Hao, C.I. Wu, T.P. Hogan, C. Wolverton, V.P. Dravid and M.G. Kanatzidis. 2012. Raising the thermoelectric performance of p-Type PbS with endotaxial nanostructuring and valence-band offset engineering using CdS and ZnS. J. Am. Chem. Soc. 134: 16327–16336.

Zhao, L.D., S. Hao, S.H. Lo, C.I. Wu, X. Zhou, Y. Lee, H. Li, K. Biswas, T.P. Hogan, C. Uher, C. Wolverton, V.P. Dravid and M.G. Kanatzidis. 2013. High thermoelectric performance via hierarchical compositionally alloyed nanostructures. J. Am. Chem. Soc. 135: 7364–7370.

Zhao, L.D., S.H. Lo, J.Q. He, H. Li, K. Biswas, J. Androulakis, C.I. Wu, T.P. Hogan, D.Y. Chung, V.P. Dravid and M.G. Kanatzidis. 2011. High performance thermoelectrics from earth-abundant materials: enhanced figure of merit in PbS by second phase nanostructures. J. Am. Chem. Soc. 133: 20476–20487. 236

Zhao, L.D., S.H. Lo, Y.S. Zhang, H. Sun, G.J. Tan, C. Uher, C. Wolverton, V.P. Dravid and M.G. Kanatzidis. 2014. Ultralow thermal conductivity and high thermoelectric figure of merit in SnSe crystals. Nature 508: 414–418.

Zhao, L.D., V.P. Dravid and M.G. Kanatzidis. 2014. The panoscopic approach to high performance thermoelectric. Energy Environ. Sci. 7: 251–268.

Zhao, X.B., T. Sun, T.J. Zhu and J.P. Tu. 2005. *In-situ* investigation and effect of additives on low temperature aqueous chemical synthesis of Bi_2Te_3 nanocapsules. J. Mater. Chem. 15: 1621–1625.

Zhao, X.B., T.J. Zhu and X.H. Ji. 2006. CRC Thermoelectrics handbook. CRC Press, Boca Raton.

Zhao, X.B., X.H. Ji, Y.H. Zhang and B.H. Lu. 2004a. Effect of solvent on the microstructures of nanostructured Bi_2Te_3 prepared by solvothermal synthesis. J. Alloys Compd. 368: 349–352.

Zhao, X.B., X.H. Ji, Y.H. Zhang, T.J. Zhu, J.P. Tu and X.B. Zhang. 2005. Bismuth telluride nanotubes and the effects on the thermoelectric properties of nanotube-containing nanocomposites. Appl. Phys. Lett. 86: 062111.

Zhao, X.B., Y.H. Zhang and X.H. Ji. 2004b. Solvothermal synthesis of nano-sized $La_xBi_{(2-x)}Te_3$ thermoelectric powders. Inorg. Chem. Commun. 7: 386–388.

Zhao, X.Y., X. Shi, L.D. Chen, W.Q. Zhang, S.Q. Bai, Y.Z. Pei, X.Y. Li and T. Goto. 2006. Synthesis of $Yb_yCo_4Sb_{12}/Yb_2O_3$ composites and their thermoelectric properties. Appl. Phys. Lett. 89(9): 092121.

Zhao, Y. and C. Burda. 2009. Chemical synthesis of $Bi_{0.5}Sb_{1.5}Te_3$ nanocrystals and their surface oxidation properties. ACS Appl. Mater. Inter. 1: 1259–1263.

Zhao, Y., J.S. Dyck, B.M. Hernandez and C. Burda. 2010. Enhancing thermoelectric performance of ternary nanocrystals through adjusting carrier concentration. J. Am. Chem. Soc. 132: 4982–4983.

Zhao, Y.X., J.S. Dyck and C. Burda. 2011. Toward high-performance nanostructured thermoelectric materials: the progress of bottom-up solution chemistry approaches. J. Mater. Chem. 21: 17049–17058.

Zheng, X.F., C.X. Liu, Y.Y. Yan and Q. Wang. 2014. A review of thermoelectrics research—recent developments and potentials for sustainable and renewable energy applications. Renew. Sust. Energ. Rev. 32: 486–503.

Zhou, B., Y. Ji, Y.F. Yang, X.H. Li and J.J. Zhu. 2008. Rapid microwave-assisted synthesis of single-crystalline $Sb_{(2)}Te_{(3)}$ hexagonal nanoplates. Crys. Growth Des. 8: 4394–4397.

Zhou, M., J.F. Li and T. Kita. 2008. Nanostructured $AgPb_mSbTe_{m+2}$ system bulk materials with enhanced thermoelectric performance. J. Am. Chem. Soc. 130: 4527–4532.

Zhu, T.J., Y.Q. Cao, Q. Zhang and X.B. Zhao. 2010. Bulk nanostructured thermoelectric materials: preparation, structure and properties. J. Electron. Mater. 39: 1990–1995.

Zhu, G.H., H. Lee, Y.C. Lan, X.W. Wang, G. Joshi, D.Z. Wang, J. Yang, D. Vashaee, H. Guilbert, A. Pillitteri, M.S. Dresselhaus, G. Chen and Z.F. Ren. 2009. Increased phonon scattering by nanograins and point defects in nanostructured silicon with a low concentration of germanium. Phys. Rev. Lett. 102: 196803.

Advanced Materials for Thermo-Responsive Applications

Yijun Zheng[1] and *Jiaxi Cui*[2,*]

ABSTRACT

Due to their sensitivities to environment temperature variety, thermo-responsive polymers constitute one of most promising advanced materials. These polymers can change their properties in response to a temperature stimulus and become candidates especially for fabricating soft actuators that can be applied externally in a non-invasive and non-contacting manner and thus display great potential application in engineering and medical fields. Two main classes of thermo-responsive polymeric actuators, liquid crystalline elastomers and thermo-responsive hydrogels, will be described respectively in this chapter. The content will involve organic chemistry, polymer chemistry, condensed matter physics, surface engineering, mechanical engineering, soft matter, devices, and medicine, etc. The chapter will be of interest to students, researchers, and engineers that are working on relative topics.

[1] 54 Oliver Street, Somerville, MA 02145, USA.
  Email: zhengyijun918@gmail.com
[2] School of Engineering and Applied Sciences, Harvard University, 58 Oxford Street, Cambridge, MA 02138, USA.
* Corresponding author: jiaxicui119@gmail.com

1. Introduction

Sensitivity and adaptability to the changing environment is one of most important features of advanced materials being developed in modern material science. Temperature change is one of the most common environmental phenomena. It relates to various aspects of our daily life. Various thermo-responsive materials have been developed to sense, detect, and utilize this change for improving our life; for example, fluids with thermo-induced volume expending/color shift are used to fabricate thermometer for recording temperature; bilayer structures made from metals with different expansion coefficients can bend with increasing temperature so provide a warming mechanism for fire accident, etc. Among these interesting thermo-responsive systems, thermo-responsive soft matters are attracting more and more attention due to their applications in wide range fields including drug delivery (Calejo et al. 2013, Matanovic et al. 2014, Schmaljohann 2006), seawater desalination (Zhao et al. 2013), passive cooling of building (Rotzetter et al. 2012), smart medical devices (Behl et al. 2010, Ratna and Karger-Kocsis 2008, Sokolowski et al. 2007), soft robots (Buguin et al. 2006, Yang et al. 2011), surgeries (Fujiwara 2012, Muramatsu et al. 2012, Strotmann et al. 2011), sensors (Islam et al. 2014, Pietsch et al. 2011, Zhang et al. 2015), water collection (Yang et al. 2013), etc. There are three main kinds of thermo-responsive soft matters: (1) shape-memory polymers; (2) liquid crystalline elastomers (or liquid crystal elastomers, LCEs); (3) thermo-responsive hydrogels. They have different working mechanisms that will be outlined briefly in this introductory section.

   Shape-memory polymers are able to change their shapes in response to external stimuli like temperature, pH, magnetic and electrical fields, and solvents, etc. (Behl et al. 2010, Hu et al. 2012, Lendlein and Kelch 2002, Xie 2011). When temperature is used as trigger approach, the change is called a thermo-induced shape-memory effect. Typical thermo-responsive shape-memory polymers are composed of a hard phase with high glass transition temperature (T_g) or a permanent network structure formed by covalent crosslinking mechanism and a switching phase with intermediate T_g or melting temperature that allows thermo-induced deformation. To obtain thermo-induced shape-memory effect, polymer materials are fabricated first to form their permanent shape. Afterwards, the materials can be deformed into temporary shape when heated above the T_g of switching phase (T_{trans}). The temporary shape is fixed by cooling to a temperature lower than the T_{trans}. In the temporary state the frozen polymer segments are entropically unfavorable (typical in a stretch state), which drives the materials to transform back to their permanent shape when the materials are heated above their T_{trans}. Such thermo-induced shape change is normally one-way and needs to be reprogrammed for other switching cycles.

LCEs are polymeric networks integrating mesogenic units. They combine polymeric elasticity and anisotropic liquid crystalline order (Broemmel et al. 2012, Ohm et al. 2010). In liquid crystalline phase the polymer chains are stretched to adapt an elongated conformation parallel or perpendicular with mesogenic director and then fixed by crosslinking reaction/interaction. When the elastomers are heated to induce phase transition from liquid crystalline to isotropic state, polymer chains turn to random coil state again, leading to macroscopic shape changes. Typically a contraction at the elongated direction of polymer chains as well as an expansion at perpendicular direction is observed during this heating process. It is a completely reversible transition and has been utilized as the main switching mechanism for developing artificial muscles (Mirfakhrai et al. 2007). There are two important things which need to be noted for getting large contractions: crosslinking density and mesogenic orientation. A low crosslinking density that leads to long polymer chains between net points, is favorable for deformation because the polymer chains can easily be deformed to high strains. In this case, typical elastic modulus for LCEs in isotropic phase (the phase that shows the actuating effect) is in the 10^5 Pa range or below, which allows large deformations. On the other hand, the mesogenic units have to be aligned uniformly over the whole sample (forming liquid crystalline monodomain) to produce the unique contraction of LCEs on a macroscopic scale. LCEs are well-known material systems and they have already been summarized in several reviews and books (Barclay and Ober 1993, Brand et al. 2006, Chambers et al. 2009, de Jeu 2012, Jiang et al. 2013, Ohm et al. 2010, Urayama 2007, Xie and Zhang 2005, Zentel 1994). In this chapter we will describe their typical preparing approaches briefly and then focus on the representative examples of recently developed systems as well as their novel applications.

Thermo-responsive hydrogel is a broad concept that includes various hydrogel materials showing responses to temperature change. In this chapter we narrow our definition to the hydrogel consisting of thermo-responsive polymers that undergo a liquid-liquid phase transition in response to temperature change, that is, phase separation that occurs from a well solvation state to precipitate state. The phase transition involves water swelling and release, accompanied by significantly change in the volume of the hydrogel. Thermo-responsive polymers are well-studied and show definite phase transition temperature, also known as the cloud point temperature (T_{cp}) since collapsed polymers lead to cloudy solution. When a phase separation occurs at elevated temperature, the temperature is named as lowest critical solution temperature (LCST) while reversed behavior is called as upper critical solution temperature (UCST) (Schild 1992, Seuring and Agarwal 2013). Hydrogels consisting of polymers with LCST shrink with increasing temperature, accompanied by various interesting behaviors such as volume and transparency changes, release of water-soluble species, etc. The aqueous properties of these thermo-responsive polymers have been recently reviewed elsewhere (Hocine and Li 2013, Seuring and Agarwal 2012b) and they are not strictly "materials", therefore will be introduced only briefly as the background in this chapter. We will focus instead on their hydrogel materials including the preparation, properties, and applications of these gels.

2. Liquid-Crystalline Elastomers

In 1975, de Gennes predicted that polymer networks incorporating mesogenic molecules as part of their architecture, would combine the key aspect of liquid crystals, orientational order and the essential aspect of solids, the ability to support sheer stress (de Gennes 1975). This concept was firstly realized by Finkelmann in 1981 in a silicone-based material system (Finkelmann et al. 1981). The combination of orientational order and mechanical stress in LCE materials leads to two complementary properties: mechanical deformations cause changes in orientational order as well as corresponding physical properties and inversely, changes in orientational order cause stress in the sample, then changes in shape. It is noted that various stimuli can be used to trigger liquid crystal phase transition such as temperature (Jiang et al. 2013), light (Ikeda et al. 2007), electrical field, etc. (Courty et al. 2003). The principle discussed here for LCEs is suitable for nearly all these systems but only thermo-induced examples will be introduced as they are considered within the scope of current chapter.

Liquid crystal (LC) is matter that can flow like a liquid but maintain its molecular orientation in a crystal-like way. Typical LC phases include nematic, smectic, cholesteric phase, etc. They have different order degrees. Molecules are in one-dimensional order at nematic phase but in two-dimensional order at smectic phase. Cholesteric phase is chiral nematic phase with helical packing. Order parameter Q is used to describe the average orientation tensor as (Palffy-Muhoray 2012):

$$Q = \; < P_2\cos\Theta > \; = \frac{1}{2} < 3\cos^2\Theta(t) - 1 >,$$

where Θ is the angle between the direction of a molecule and unit vector n. The n is also defined as the director that presents the direction of (domain) orientational order. When all molecules are parallel to each other like in crystal ($\Theta = 0$), Q has the maximal value of 1. In contrast, $Q = 0$ in isotropic state since molecules have the same probability directed in whichever direction in space in every moment. Typically Q value decreases with increase in temperature throughout the whole liquid crystal phase and falls discontinuously to zero when liquid crystal transfers to isotropic phase.

Without external stimuli bulk LC is composed of domains with different vector n, namely polydomain. Upon external stimuli all the domains can adopt the same vector, forming a state named as monodomain. Monodomain is a key parameter for LCE to get efficient macroscopic deformation.

Elastomer is crosslinking polymer networks. Crosslinking creates a topological relation between polymer segments, forming maximal entropy. Mechanical pressure can deform the conformation of polymer segments and then the entropy (S) fall. The free energy (F), which depends only on an entropy change itself driven by molecular shape change ($\Delta F = -T\Delta S > 0$), rises. It drives the elastomer to recover its original shape after mechanical force is released. If the material is incompressible, in the linear regime, the free energy can by described as

$$F_{el} = \frac{1}{2}Ye_{\alpha\beta}e_{\alpha\beta}$$

where Y is Young's modulus and e is strain tensor. Approximately, Young's modulus is given by

$$Y \approx kTp_c$$

where p_c is the crosslinking density of polymer networks. Therefore, the crosslinker used for preparing (liquid crystalline) elastomer is very important for final contraction effect. If an external stress $\sigma_{\alpha\beta}$ is applied, the elastic free energy becomes

$$F_{el} = \frac{1}{2}Ye_{\alpha\beta}e_{\alpha\beta} - \sigma_{\alpha\beta}e_{\alpha\beta}$$

Based on this equation, the strain e can be calculated by minimizing the elastic free energy in a equilibrium state.

LCE shows combined aspects of elasticity and liquid crystallinity and its free energy can be approximated simply as the sum of liquid crystalline and elastomeric contributions, as well as a coupling term. Take nematic LCE for example, the free energy takes a form:

$$F_{LCE} = F_{el} + F_{LC} - \gamma Q_{\alpha\beta}e_{\beta\alpha}$$

where the last term represents the simplest symmetry-allowed coupling between orientational order and strain. The coupling constant γ is proportional to the crosslink density and the step length anisotropy ($l_{\parallel} - l_{\perp}$, l: length tensor). This coupling suggests an interdependent affection between strain and order parameter. Several important relationships come out from these equations: (1) the effect of the order parameter on strain is the same as that of an external stress; and (2) the effect of strain on the orientational order is the same as that of an applied external field. It implies that the alignment of the director and orientational order (forming monodomain) can be controlled by applying uniaxial strain or external fields.

LCEs with monodomain are also named as single liquid crystal elastomers because of their anisotropic physical properties comparable to single crystals (Broemmel et al. 2012). In this chapter only orientated LCEs will be discussed since they show thermo-responsive deformations. The typical strategy to prepare orientated LCEs is shown in Fig. 1. In the presence of external field or surface effect, polydomain sample is converted to monodomain with macroscopically anisotropic chain conformations, followed by chemical/physical crosslinking reactions in the aligned state to yield orientated LCE actuator. In this section we will briefly discuss the synthetic routes for LCEs. LCEs prepared by these synthetic routes are polydomain and they show only thermo-induced optical properties as simple liquid crystal polymers. To obtain monodomain, orientation techniques are required and will be introduced as an independent section. In the last section representative examples developed recently will be discussed.

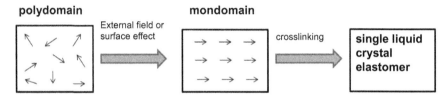

Figure 1. Schematic model for the principal route to single liquid crystal elastomer.

2.1 Synthetic Strategies for Liquid-Crystalline Elastomers

The mesogenic units in LCE can be either part of the polymer backbone (main-chain polymers) or attached in a side chain (side-chain polymers) or both. Side chain polymers are further classified into end-on and side-on model (main-chain polymers normally take an "end-on" form but there has also been a "side-on" example reported recently). These polymer networks are prepared by two synthetic routes: one-step or two-step method (Fig. 2).

In one-step method, liquid crystalline monomer is mixed with radical initiator and multifunctional crosslinker which may or may not be liquid crystalline. The resultant mixture can be cured under UV irradiation or at a certain temperature, in which polymerization and crosslinking reaction occur together, leading to a permanent shape. Typical examples are acrylate-functionalized mesogens mixing with diacrylates (crosslinker) and photoinitiators (Fig. 3) (Thomsen et al. 2001). A new approach based on in determinate photo-induced addition of thiols to olefin was recently developed (Fig. 4). The system contains a LC monomer with a vinyl and a mercapto group and a multifunctional crosslinker with two vinyl and

One-step

Two-step

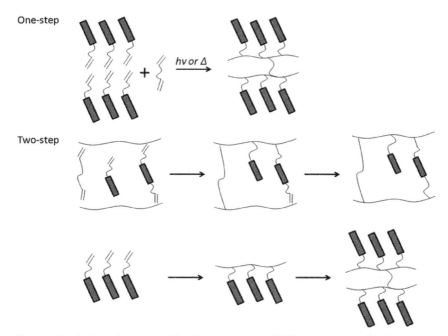

Figure 2. Synthetic pathways used for the preparation of LCEs. One-step method: a mixture of liquid crystalline monomer and crosslinker is polymerized, directly leading to an LCE. Two-step method: top, all components (polyhydrosiloxane chain, mesogens, and crosslinker) are mixed and reacted in a palladium-catalyzed reaction in two steps; down, liquid crystal polymer is prepared at first and then crosslinked to form LCE.

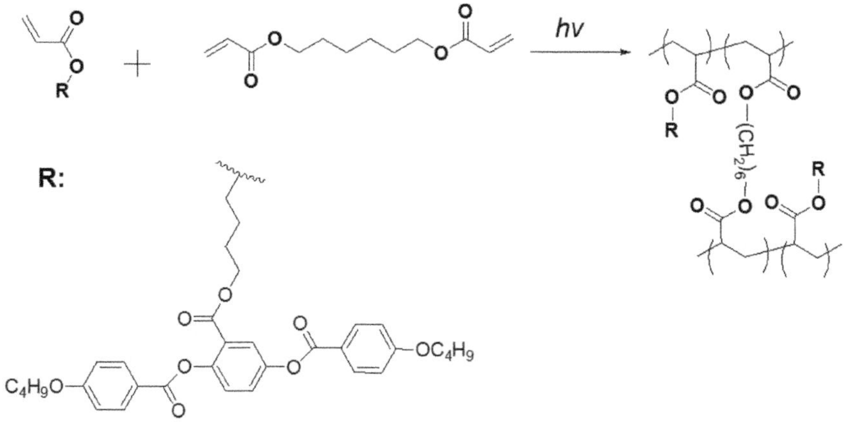

R:

Figure 3. LCE prepared from acrylate-functionalized mesogen and diacrylate.

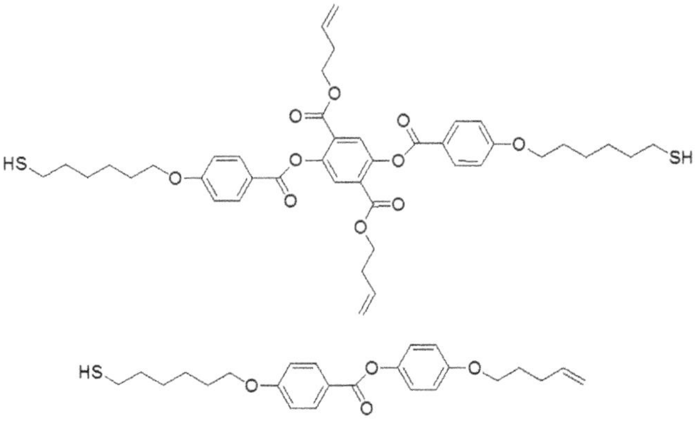

Figure 4. Structure of a mesogen and a mesogenic crosslinker used to prepare main-chain elastomer in a one-step process.

two mercapto groups (Yang et al. 2009). With their vinyl and thiol groups, the monomer and the crosslinker, acting as an AB monomers, undergo thiol-ene-polycondensation reaction to form polymer networks in one step. In one-step method, small molecules are used for starting materials. These compounds normally have low viscosities and phase transition temperatures (from crystalline to mesogenic phase), so can develop LC phase very fast at a relative low temperature. These properties are highly desirable for orientating mesogenic units and most of the orientation techniques for low-molecular-weight liquid crystals can be applied to induce the monodomain. Because of these advantages, this method is often performed in a complicated condition to create subtle structures like microfluidic setup or LCE micropattern (Buguin et al. 2006).

In a two-step method polymers are prepared at first and then crosslinked further to form elastomers. Based on polymer structures, they are further sub-classified into two groups: silicone-based polymers and liquid crystal polymers. The former are used most widely and first developed by Finkelmann and Rehage (Finkelmann and Rehage 1980). They are based on hydrosilylation reaction between terminal vinyl group of mesogen and Si–H bonds on silicone. Typically, a linear polyhydrosiloxane chain, mesogen molecules with vinyl and a methacryloyl-based crosslinking agent are mixed with catalyst as starting material (Fig. 5). Because of the significant higher reactivity of vinyl groups than methacryloyl groups, mesogen molecule is coupled at first to form a weakly crosslinked network. The mixture in this state, normally a very soft gel-like material, can be deformed or orientated before further crosslinking in second step. This method is very easy to perform and also takes advantage of elastic properties of silicone so that it can be used in various mesogenic groups without extensive changes in the whole system. Moreover, the soft gel-like intermediate is also convenient for

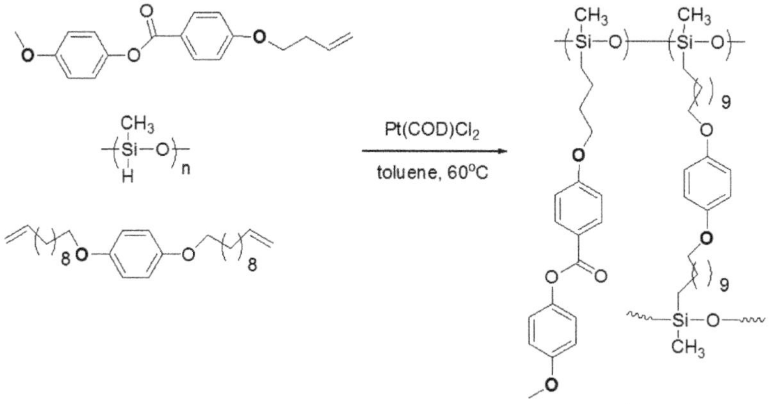

Figure 5. Siloxane-based LCE prepared by Finkelmann and Rehage (Finkelmann and Rehage 1980).

orientating. The main drawback is the impure final elastomers due to the incomplete reaction which leads to the residue of unreacted mesogens or crosslinkers. These low molecular molecules may aggregate and then phase separate. In the other group liquid crystal polymers with reacting groups are prepared first. These polymers can be purified and their properties studied well before further curing. Depending on the reacting groups, the liquid crystal polymers can self-crosslink or be connected by adding agents. Various well-developed reaction strategies can be designed for the crosslinking step, such as the coupling reactions between isocynanates and alcohols, azides and acetylenes (Xia et al. 2008), active esters and amines, photo-induced crosslinkings, etc. For example, Beyer et al. synthesized a kind of liquid crystal polyesters containing photo-crosslinkable groups which are able to crosslink in bulk film under UV irradiation (Fig. 6) (Beyer et al. 2007a, Beyer et al. 2007b). One of the advantages of using this method is that the mesogen orientating procedure can be conducted separately with crosslinking procedure. Different from low-molecular-weight mesogen, liquid crystal polymers have relative mechanical stability and may maintain their conformations formed in liquid crystalline phase when cooled down to glassy states.

Physical crosslinking is the other strategy to create LCEs. For example, triblock copolymers with hard terminal blocks and liquid crystalline middle block can self-assemble into a 3D network with aggregated hard phase as crosslinking domain (Ahir et al. 2006, Kempe et al. 2004). This type of crosslinking forms during the developing of the liquid crystalline phase and allows reversible de-fabrication of the LCE (thermoplastic when heated above the T_g of the hard-block). Thanks to the well-developed controlled polymerization, the synthesis of defined block copolymer is not a challenge any more (Cui et al. 2004, Xia et al. 2008).

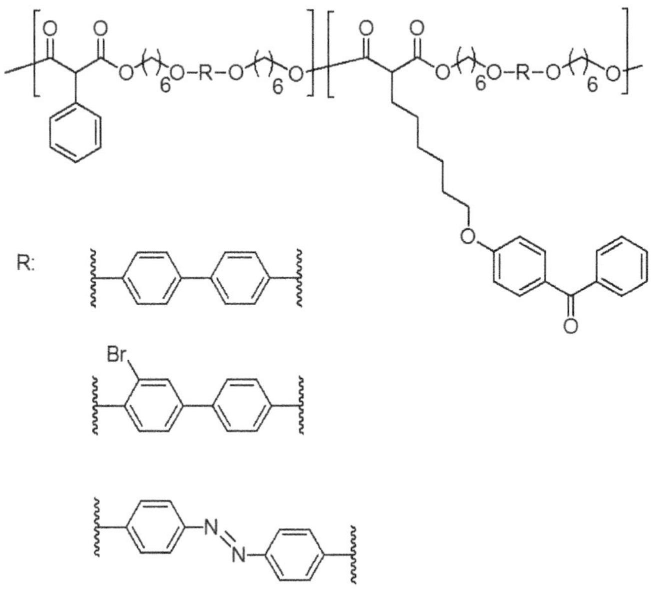

Figure 6. Smectic main-chain copolymers containing benzophenone moieties.

2.2 Orientation Techniques

As mentioned above, one of most important procedure in preparing thermo-deformable LCEs is the orientation of the mesogens to form a liquid crystalline monodomain. Orientation and crosslinking are often conducted together to fix the orientation effect. Typical orientation methods include mechanical shear, aligning polyimide layer, electric/magnetic field, surface tension, etc.

Mechanical shear is a widely used method to induce anisotropic environment. Under this force field (can be created by simple pressure, stretch, or spray effect like electrospinning) mesogens will adopt a regular alignment with respect to polymer chains they are bonding on, or will form, accompanied by a deformation of the material. Silicone-based LCEs are fabricated by this approach (Kaufhold et al. 1991). Weakly crosslinked precursors are prepared and stretched uniaxially to extend polymer chains. When conducted at a temperature of liquid crystalline phase, the mesogens align themselves in either parallel or perpendicular form to the polymer chains, forming a monodomain, which is locked further by crosslinking reaction. A similar approach is anisotropic deswelling developed by Kim and Finkelmann for orientating cholesteric liquid crystal: during slow drying of polymer solution, elastomer deswells only in one direction, accompanied by crosslinking reaction to fix the orientation effect (Kim and Finkelmann 2001).

Recently Ohm et al. developed a novel orientation method by using flow field to create mechanical shearing force (Ohm et al. 2009). A microfluidic setup is used, in which the mixture of liquid crystal monomer/crosslinker and immiscible fluid medium is pushed to flow through a thin capillary. A gradient velocity profile forms in the tube and then induces shear flow which will be transferred to disperse the liquid crystal monomer droplets. In the presence of the photoinitiator, the mixture of monomer and crosslinker can be cured by UV irradiation to fix the orientation when it passes through the tube. This stretching technique does not require pre-crosslinking since the material does not have to be stable mechanically to be stretched and can product a huge amount of microscopically sized actuators easily. Movement of the liquid crystal fluid can self-form from the shear effect. This effect becomes remarkable in microenvironments like nano hole. This strategy has been used by Ohm et al. to prepare nano LCE actuators by pouring liquid crystal fluid on a template with nano hole (Ohm et al. 2011a). A novel method to apply mechanical stretching is developed by Marshall et al. (Marshall et al. 2014). They embedded the LCE precursor in stretchable elastomer and then elongated the elastomer to induce orientation. Monodomain LCE is obtained after UV-curing at stretching state.

Aligning polyimide layer is able to induce mesogens to adopt a parallel packing by intermolecular interaction, forming a powerful technique to align liquid crystal molecules (Thomsen et al. 2001, Urayama et al. 2005, Yu and Ikeda 2006). This technique has been used to fabricate the liquid crystal cell used for display or research devices. Briefly, aligning polyimide layer is obtained by physical rubbing polyimide film coating on substrate. Liquid crystal monomers are melted on the substrate and cooled to their liquid crystalline phase. After polymerization, the resultant film is removed from the substrate with uniform orientation. This method is normally used to orientate low-molecular-weight compounds with low viscosity and only thin film can be prepared because the driving force for orientation is limited to the interface with the alignment layer. The huge advantage of this method is that it allows different orientations of the mesogens throughout the sample by using two substrates in a sandwich-like structure (Harris et al. 2005). For example, the top and the bottom of the substrates can have a planar orientation (parallel or perpendicular) or one planar orientation and one homeotropic alignment. It is very useful to create actuators with different deformation like coiling or bending, etc.

The electric field is an efficient method to induce the orientation of ferroelectric liquid crystalline (smectic C*) compounds due to their strong dipole features (Gebhard and Zentel 2000). For UV–induced crosslinking reaction, transparent indium tin oxide (ITO) is used as electrodes which are coated on two parallel surfaces of a reaction chamber to create an electric field. When liquid crystalline compound (e.g., polymer containing crosslinking groups) placed in the chamber is heated to its liquid crystal phase, an alternating rectangular or triangular electric field is applied at first to induce the formation of monodomain. Thus voltage is applied and the polymer is crosslinked by UV irradiation. In case of nematic liquid crystal materials, the magnetic field is used instead for orientation with an analogous method. In magnetic field mesogens prefer to be parallel to the field because of their strong diamagnetism. Generally nematic liquid crystal monomers have low viscosity at the liquid crystal phase so the permanent magnet is strong enough to orientate them, which simplifies the setup very much and allows a combination with commercial device for specific application (Buguin et al. 2006).

For minimizing surface energy, mesogens in thin films prefer to adopt a homeotropic packing on surfaces. This happens spontaneously for smectic-A and (chiral) nematic films and can be promoted by annealing the films. This method has been applied to prepare planar and curved smectic liquid crystalline films in the shape of inflated balloons and homogeneous planar oriented nematic elastomers (Schuering et al. 2001).

All the methods mentioned above involve a crosslinking process after alignment. Recently Pei et al. reported an example in which orientation can be conducted after crosslinking (Pei et al. 2014). Instead of normal permanent crosslinking mechanism, they introduced exchangeable links to get dynamic LCE materials (Fig. 7). At room temperature these materials maintain mechanical integrity and behave as traditional LCE networks. When the materials are heated to their transition temperature, exchanging reaction occurs, which allows reconfiguration of mesogens. Applying mechanical stretching under this condition leads to strong alignment of mesogens (biphenyl with smectic phase in their demonstration samples). This strategy allows for easy processing and alignment, as well as a possibility to re-mould. The authors claim that it might open a way to practical LCE actuators.

2.3 Soft Actuators of Liquid-Crystalline Elastomers

As discussed in the introductory part of this section, LCEs are able to change their shape reversibly when heated to induce phase transition from liquid crystal to isotropic phase. The deformation and strength produced in this process depends on various parameters including liquid crystal phase (nematic, cholesteric, smectic A, smectic C), attaching model of mesogens (main chain, end-on, side-on), chemical structure of mesogens and polymers, crosslinking density (Young's modulus), order parameter (Q), etc. The actuating effect can be estimated by the maximum strain and the maximum force. Maximum strain rates are easy to obtain by measuring the change in length. Thanks to Ohm's effort on comparing different

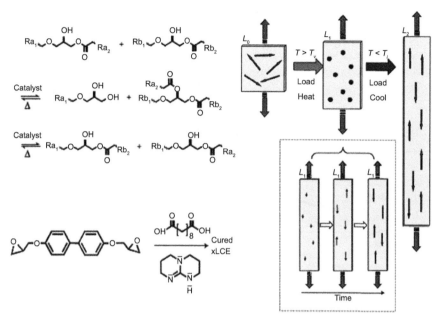

Figure 7. Reversible trans-esterification and the structure used for the dynamic LCE (left). Processing leading to an equilibrium monodomain alignment (right). Reprinted with permission from Ref. (Pei et al. 2014). Copyright 2014 Nature Publishing Group.

systems we can have a general feeling on this parameter (Ohm et al. 2012). The relative length can change more than 400% in main-chain LCEs, ~ 80% in side-on LCE, and ~ 40% in end-on LCE. It is noted that it is not a quantitative comparison because of the significant difference in chemical structures and orientation techniques. To measure the maximum force, the length of samples is kept constant, while the force exhibited by the sample is measured as a function of temperature. Thomsen et al. measure the force for the nematic side-on polyacrylate elastomer (Fig. 8) and the stress is in the range of 10^5 Pa (Thomsen et al. 2001). It is a fair force that can load weight. Anisotropy is the most important parameter for optimizing the actuating force in a given mesogen system. Deuterium nuclear magnetic resonance (^{2}H-NMR) and small angle neutron scattering (SANS) have been used to study the contribution of attaching the model on anisotropy and it is found that main-chain LC polymers normally have the strongest chain anisotropy, followed by the side-on systems, while end-on mesogens cause the lowest chain anisotropy. In addition, shorter spacer in the side-on systems increases the coupling interaction between mesogen and backbone and thereby the chain anisotropy.

Geometry is an important parameter of actuators (Burke and Mather 2013, Kim et al. 2014, Yang et al. 2011). The film is one of basic morphologies

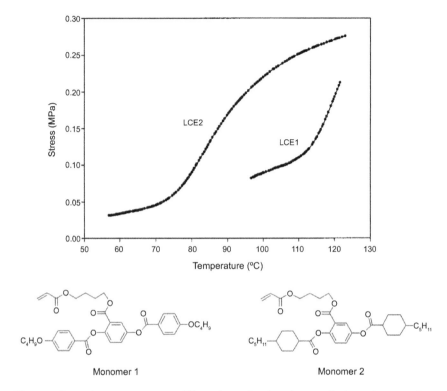

Figure 8. Isostrain measurement at 5% strain on heating the sample from the nematic to isotropic phase for LCE 1 and LCE 2. The sample's length is kept constant during the phase transition while the exerted force is measured as a function of temperature. LCE 1 consists of monomer 1 and a crosslinker; LCE 2 is a mixture of monomer 1 and 2 and the crosslinker. The preload stress of 62 and 30 kPa were applied respectively on LCE 1 and LCE 2. Reprinted with permission from Ref. (Thomsen et al. 2001). Copyright 2001 American Chemical Society.

for LCE actuators because of the convenience in its fabrication (nearly all orientation techniques can be used for this purpose) (Li et al. 2004). Minimizing the size of actuators is in favor of increasing responsive frequency because of the short thermo conducting time. Ohm et al. utilized microfluidic setup to prepare micrometer-sized LCE particles with different shapes directly from a liquid mixture of liquid crystal monomer, crosslinker, and photoinitiator (Fig. 9) (Ohm et al. 2011b). Taking advantage of the controlling ability of microfluidic setup on flow velocity in thin tubes, various shapes like spheres, disks, or rods have been made and their size, strength of the shape variation and the direction can be controlled well. These particles, suspended in a fluid medium, can undergo reversible thermo-induced deformation. The size can be further decreased to nano level by using template (Ohm et al. 2011a). In addition to microfluidic, electrospinning has also been used to create micrometer-sized LCE actuators

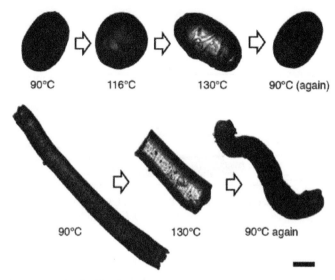

Figure 9. Micrometer-sized LCEs that shorten during actuation. Reprinted with permission of Ref. (Ohm et al. 2011b). Copyright 2011 American Chemical Society.

(Krause et al. 2007). Thin and highly oriented fibers (diameters of 0.1–5 mm) are fabricated by this method and these fibers display very high tensile strengths. When heated to induce phase transition, LCE fibers can contract to raise weight. These micro-size LCEs have been used as micropumps (Fleischmann et al. 2012) or photonic ink capsules (Lee et al. 2015).

One of the interesting applications of LCE is to fabricate micropattern on surfaces. Surfaces with micro structures have displayed attractive properties/application like wetting, adhesion, self-assembly, template, substrate for holding lubricant, etc. (del Campo et al. 2007, Drotlef et al. 2014, Greiner et al. 2009, Greiner et al. 2007, Reddy et al. 2007, Vogel et al. 2013, Wong et al. 2011, Yao et al. 2013). Most of these patterning surfaces are fabricated from available commercial precursors like PDMS, epoxy, polyurethane, etc. Although these materials are cheap and easy to operate, normally they do not respond to external stimuli (polyurethane shape memory material show irreversible deformation as heated). Introducing smart materials into patterning surfaces inspire various fascinating properties and then application. For this purpose, fabricating technique is the first challenge. Keller et al. figured out this issue by combining soft lithography and magnetic orientation technique to produce LCE patterning surfaces (Fig. 10) (Buguin et al. 2006). The setup is composed of a permanent magnet used to orientate mesogens, heating table for temperature control, polydimethylsiloxane (PDMS) mold to create microstructure, and UV lamp for photo-induced polymerization. Taking advantage of the transparency of PDMS, UV light can pass through the mold to trigger polymerization

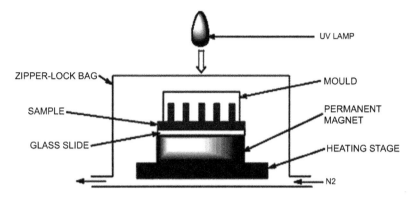

Figure 10. Experimental setup developed by Buguin et al. to prepare LCE microstructures. Reprinted with permission of Ref. (Buguin et al. 2006). Copyright 2006 American Chemical Society.

(and crosslinking reaction). In their case a mixture of nematic side-on monomer, crosslinker and photoinitiator is placed on a glass substrate and heated to melt, followed by coating negative PDMS mold. After cooling the mixture to liquid crystal phase in the magnetic field, mesogens form a monodomain which is stabilized by *in situ* photo-triggered polymerization and crosslinking. When the mold is peeled off, a surface consisting of regular microstructures like a cylinder or pillar, etc. is obtained from aligned LCEs. These micro pillars have a parallel anisotropy and therefore display a reduction in length when heated to isotropic phase. These thermo-responsive patterns can be used to mediate surface wetting properties.

Preparation of various micro-level LCE actuators seems to open one window to push these smart materials into real application. It is actually very difficult to play these accessories into devices due to a technique problem. Keller's great contribution of combining well-developed soft lithography already seems to figure out this problem. There is still one concern that the substrate holding LCE microstructures is also same LCE material. It raises several issues: (1) the substrate also shows thermo-induced deformation. It is difficult to distinguish the contraction effect of microstructure from the whole sample since the substrate might show significantly bigger deformation because it is normally much thicker. It becomes worse when the thickness of the substrate is difficult to control in current procedure. (2) The substrate will become soft when heated into isotropic phase and then lose mechanical supporting effect which is very important for real application. Cui et al. improved this by fabricating the microstructures on rigid poly(methyl methacrylate) (PMMA) substrate (Cui et al. 2012). They used similar setup suggested by Keller et al. In their method, liquid crystal mixture is first loaded into PDMS mold and then excess liquid crystal mixture is wiped, leading to dispersed pillars. This loaded mold is further combined with a glass substrate that is coated with fresh-prepared crosslinked PMMA. The final sample is located in a magnetic field placed on heating stage and then UV irradiated for polymerization. To obtain strong connection between PMMA substrate and LCE pillars, they optimized a lot of fabricating parameters including temperature, residual layer of the LC mixture on the PMMA backing, concentration of crosslinker, the thickness of PMMA layer, etc. On the sample obtained under optimized conditions, they found that only pillars display typical birefringence under polarized optical microscopy (Fig. 11). When heated from 70 to 90°C, these pillars display 17% contraction while PMMA substrate does not change at all. This LCE patterning surface allows reversible switching between adhesive and non-adhesive states upon application of a temperature change, representing the first example of a temperature-driven reversible gecko-like adhesive. In addition to reversible adhesive, LCE micropillars are potentially useful in haptic application (Torras et al. 2013).

Recently a direct laser writing technique is developed to fabricate high resolution 3D LCE microstructure by utilizing the photoalignment behavior of azobenzene-dye (Zeng et al. 2014). Azobenzene dyes can orient normal to the electric field vector of the linearly polarized light so that liquid crystal mixture containing azo-dye will undergo a shape change in response to light following the destruction of the liquid-crystalline structure order. Very recently this photoinduced alignment of azobenzene-based material is used to obtain voxelated LCE (Ware et al. 2015). An azobenzene-based polymer film is prepared as substrate. It is very convenient to control the located alignment of azo-dye in very small area by polarized light. When LC monomers are coated on these alignment substrates, their director will align to the local surface orientation of the photoalignment layer. The voxel of obtained LCE image depends on the resolution of azo-substrate and the size can be small to 0.01 mm^2. Based on this technique topological defects can be imprinted in LCEs. As a result of macroscopic azimuthal contraction (along the director) and radial expansion around each defect center, robust deformation from flat to cone with a change of 100 times in height is observed. This actuator can lift a load 147 times heavier than itself with a stroke of ~ 3000%.

Because of their shape change effect, LCEs are one kind of most important material for preparing artificial muscles. Typical deformation of LCEs themselves is simple uniaxial extensions or contractions. For better demonstration of their "muscle" function, LCEs are often fabricated into complex structures. The simplest model is bilayer structure (Agrawal et al. 2014, Greco et al. 2013). Recently Agrawal et al. described a simple LCE-polystyrene bilayer model (LCE-PS) that is able to show various thermo-induced shape changes like bending, twisting, and wrinkling, depending on system parameters of the thickness ratio of PS

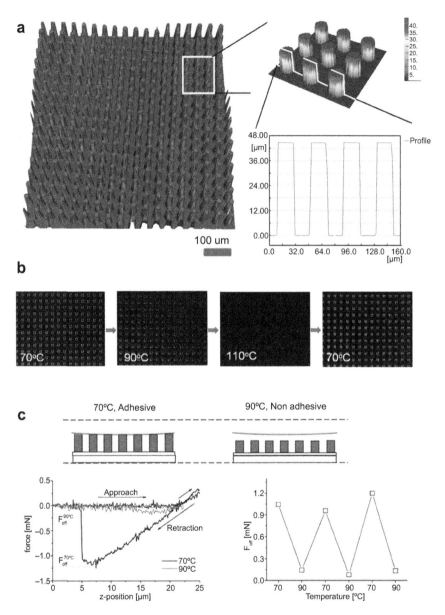

Figure 11. (a) LCE microstructure obtained by Cui et al.; **(b)** polarized optical microscopy images of the LCE microstructure with a transition from nematic to isotropic phase; **(c)** thermo-induced adhesion. Reproduced from Ref. (Cui et al. 2012) by permission of John Wiley & Sons Ltd.

to LCE, the overall aspect ratio of the bilayer, and patterning of LCE layer (Agrawal et al. 2014). When thin PS layer (18–101 nm) is used (LCE thickness: 0.480 mm), wrinkles are obtained because of the low mechanical support (the energy for deforming bilayer is significantly higher than that of the no-slip boundary). Increasing the thickness of PS leads to bending. An interesting thermo-responsive phenomenon is that the bending driven by LCE layer disappears when temperature is higher than T_g of PS. It is attributed to the thermo-induced creep of PS (thermoplastic) which releases the mechanical tension. In this case cooling will result in inverse bending due to the contraction of LCE layer.

The microvalve is the crucial component in microfluidics for controlling the flow (Auroux et al. 2002, Reyes et al. 2002). Sánchez-Ferrer et al. integrated an oriented nematic side-chain LCE into a silicon-based microstructured device for the use as a microvalve by two-step method (Sanchez-Ferrer et al. 2011). The actuation principle is based on the expansion of the LCE in the directions perpendicular to the director and the shrinkage in the direction parallel to the director. When the orientation is perpendicular to the flow channel, the thermo-induce expansion of the LCE microvalve in a limited room results in an abrupt buckling of the actuator forms in the middle and then closes the microchannel. In a prototype with the LCE microvalve of chip geometry of $10 \times 10 \times 1.04$ mm^3, a switching frequency of 0.01 Hz can be obtained and the authors claimed that using pulsed heating in an improved version can tune the volumetric flow rates and switching frequencies down to milliseconds.

3. Thermo-Responsive Hydrogels

As discussed in the introductory part, thermo-responsive hydrogels discussed in this chapter are composed of polymers that show LCST or UCST. For understanding the design and properties of thermo-responsive hydrogels from these polymers better, we will first discuss their basic solution properties. Polymers with LCST in water are soluble in aqueous solution at low temperature (favorable in free energy, $\Delta G < 0$) and phase separate as increase temperature (unfavorable in free energy, $\Delta G > 0$). It suggests both negative enthalpy (ΔH) and entropy (ΔS) for polymers dissolving in water. In other words, polymer chains should form favorable hydrogen bonding with water molecules and make them lose mobility (the loss of water molecular entropy is bigger that the increase of polymeric entropy, which leads to a negative entropic effect). The temperature response is contributed from the entropy-dominated dissolved behavior. Elevating temperature drives the hydrated water molecules back to the bulk water, leading to partially dehydrated polymer chains which will collapse and aggregate into a polymer rich phase. It is a completely reversible process with a sharp transition temperature. Typical polymers with LCST at ambient pressure are shown in Fig. 12. Poly(N-isopropylacrylamide) (PNIPAM) is the first and most widely studied temperature-responsive polymer in water (Fujishige et al. 1989, Schild 1992). It shows very robust phase behavior with a LCST of ~ 32°C, a temperature between body and room temperature. Its analogous poly(N-vinyl caprolactam) (PVCL) shows a LCST of ~ 31°C while poly(2-dimethylaminoethyl methacrylate) (PDMAEMA) has a LCST of ~ 52°C (Cui et al. 2013). Poly(oligo ethylene glycol (meth)

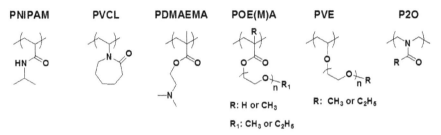

Figure 12. Typical polymers with LCST.

acrylate)s (POE(M)A) are a promising family because of their tunable LCST (depend on the terminal group and number of repeat ethylene glycol unit) (Ishizone et al. 2008), low T_g (in acrylate series), and convenient synthetic method of controlled radical polymerization (compared to PNIPAM and PVCL) (Lutz et al. 2006, Vancoillie et al. 2014). The preparation of poly(vinyl ether)s (PVEs) normally requires cationic polymerization so PVEs are not suitable to prepare hydrogel. Poly(2-oxazoline)s are a recently emerging class (Weber et al. 2012). They have LCSTs very closed to body temperature and the more important thing is that they do not show significant hysteresis in phase transition because they have low T_g and do not form intramolecular hydrogen bonding in collapsed state (which might be a favorable feature for fabricating sensitive hydrogel).

Compared to polymers with LCST in water, polymers with UCST in water are rare. From the thermodynamic point of view, to obtain UCST behavior, addition enthalpic tern represented association interaction between polymer chains has to be introduced in the Gibbs free energy equation. This interaction is dominant for polymer dissolving in water. With increasing temperature, the interaction strength decreases and then the hydration term becomes dominant. Therefore, polymers dissolve at high temperatures. Generally polymers with strong intermolecular interaction would show a UCST. Typical polymers with UCST are shown in Fig. 13. Poly(betaine)s like poly(2-dimethyl(methacryloxyethyl) ammonium propane sulfonate) (PDMMEAPS) and poly(3-(N-(3-methacrylamidopropyl)-N,N-dimethyl) ammonium propane sulfonate) (PNMAPDMAPS)

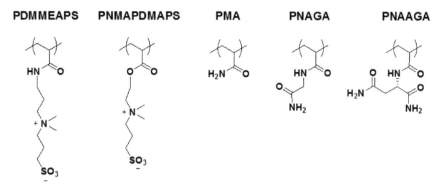

Figure 13. Typical polymers with UCST.

(Mary et al. 2007) are well-known examples. They are zwitterionic polymers and show strong electrostatic interactions so need additional enthalpy for dissolution. Poly(methacrylamide) (PMAm), poly(N-acryloylglycinamide) (PNAGA), and poly(N-acryloylasparaginamide) (PNAAGA) can form strong intermolecular hydrogen bonding and thereby also have UCST (Glatzel et al. 2011, Seuring and Agarwal 2010, 2012a,b).

A number of review articles have been published recently to discuss the behaviors and applications of these thermo-responsive polymers (Bunsow et al. 2010, Chen et al. 2010, Prabaharan and Mano 2006, Seuring and Agarwal 2012b, Wang et al. 2005a). In this section we will focus only on representative reports related to hydrogel materials. In this scope there are still so many reports on this topic that it is impossible to conclude them all in this review. So only representative examples reported recently about thermo-induced deformation, adhesion ability, some specific behaviors, etc., will be discussed.

3.1 Hydrogels for Thermo-Induced Actuation

Hydrogels made from polymers with LCST uptake water at a temperature lower than their LCST but release water when heated to a temperature higher than the LCST. Significant shape changes occur in this phase transition, which make them wonderful candidates for actuation under water (Benito-Lopez et al. 2014, Guan and Zhang 2011, Pelah et al. 2007, Piskin 2004, Wang et al. 2005b, Zhang et al. 2004, Zhang et al. 2011).

Breger et al. coated hydrogel consisting of poly(N-isopropylacrylamide-co-acrylic acid) (PNIPAM-AA) on an inactive substrate of polypropylene fumarate to fabricate a self-folding microgripper (Breger et al. 2015). Taking advantage of the large and reversible swelling response of the hydrogel in varying thermal environments, the microgripper can conduct pick-and-place tasks (Fig. 14). Based on finite element models they calculated and studied the influence of the thickness and the modulus of both the hydrogel and the stiff polymer layers on the self-folding characteristics of the microgrippers. In their demonstration system, PNIPAM-AA hydrogel prepared by them shows a critical temperature of 36°C, a temperature very close to the body one, at which hydrogel has a middle uptake of water and the microgripper is flat. Increasing temperature drives water release from the hydrogel, leading to fold to hydrogel side while decreasing temperature leads to inverse effect. These polymeric microgrippers are useful for soft robotic and surgical applications. Similar microstructures were also developed by Zheng et al. by using a tough temperature responsive hydrogel (Zheng et al. 2015).

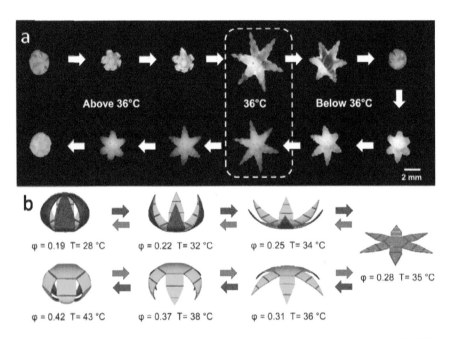

Figure 14. Thermal-responsive self-folding of a representative microgripper in H_2O: **(a)** experiment images and **(b)** computer simulation. Reprinted with permission from Ref. (Breger et al. 2015). Copyright 2015 American Chemical Society.

Actuation effect of thermo-responsive hydrogel was used by Sidorenko et al. to control nanostructures into complex micropatterns (Sidorenko et al. 2007). They integrated PNIPAM hydrogen with high-aspect-ratio silicon nano-columns, either free-standing or substrate-attached, to get a "soft-hard" hybrid structure. Under an environment of high humidity, hydrogel shows thermo-induced swelling or contraction, providing a force to orientate the nano-columns from tilted to perpendicular to the surface. Because the thickness of the hydrogel is very thin, a very fast trigger is possible in this system. Based on this hybrid structure, they demonstrated that a variety of elaborate reversibly actuated micropatterns can be obtained by controlling the stress field in the hydrogel. Such dynamic control over the movement and orientation of surface nanofeatures at micron and submicron scales may have applications in actuators, microfluidics, or responsive materials. The same research group further developed this system into more subtle structure and even to a homeostatic system (Fig. 15) (He et al. 2012, Zarzar and Aizenberg 2014). In the case of homeostatic system, chemical reaction releases heat to increase located temperature which drives the thermo-responsive hydrogel to collapse and then stretch the bending of microstructures caged with a catalyst. In the absence of a catalyst, the reaction is to stop on top layer and the temperature decreases back to room temperature at which hydrogel swell again to release microstructure back to the original environment, initiating another actuating cycle.

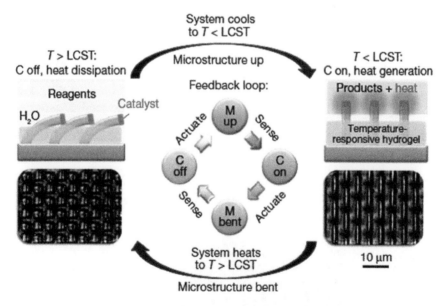

Figure 15. Schematic of the temperature-regulating mechano-chemical system showing a chemical-mechanical feedback loop. The mechanical action of the temperature-responsive gel is coupled with an exothermic reaction. The side view schematic and top-view microscope images depict on/off states of the reaction in the top layer. Reprinted with permission from Ref. (He et al. 2012). Copyright 2012 Nature Publishing Group.

3.2 Hydrogels with Thermo-Gating Recovery Ability

Self-healing is the ability to recover damage. Polymer materials with dynamic bondings display intrinsic self-healing ability (Blaiszik et al. 2010, Wu et al. 2008). To refill the damage area, polymers need to relocate, which required energy or association of solvent. When elevating temperature is normally used as a stimulus to trigger self-healing behavior, a hydrogel consisting of polymers with LCST can show an inverse behavior since polymer chains collapse at a high temperature to prevent polymer reconfiguration. Cui et al. developed a novel platform for designing self-healing hydrogel based on the crosslinking of complementary multiple hydrogen bonding (Cui and del Campo 2012, Cui et al. 2013). They integrated this reversible bonding into thermo-responsive polymers to get self-healing hydrogels and found that the self-healing ability of obtained hydrogels can be gated by temperature. With a PDMAEMA-based hydrogel they demonstrated that the hydrogel can recover the damage made by a knife in 5 min at room temperature (Fig. 16). At 50°C, a temperature higher than the LCST of the polymers in hydrogel (copolymerization decrease the LCST of PDMAEMA), self-healing

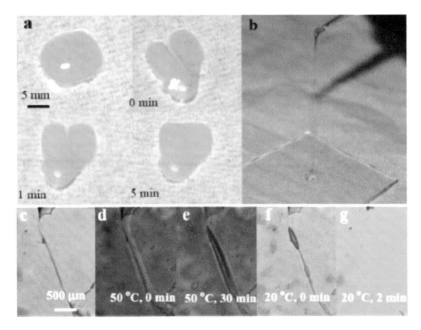

Figure 16. Demonstration of the self-healing **(a, c, d, e, f, g)** and the stretching **(b)** properties of the hydrogel consisting of PDMAEMA and hydrogen-bonding crosslinker. The gel in **(a)** and **(b)** was colored with methyl blue for better imaging. Optical microscopy images **(c–g)** were obtained from a hydrogel film with an incision **(c)** after annealing at 50°C **(d, e)** and subsequent cooling to 20°C **(f, g)**. Reproduced from Ref. (Cui and del Campo 2012) by permission of Royal Society of Chemistry.

is stopped because the collapse of polymer chains prevent the exchange reaction of hydrogen bonding. Self-healing occurs again when the sample is cooled down to 20°C.

3.3 Thermo-Responsive Hydrogels for Water Collection

When thermo-responsive polymers coating on surface, the surface properties can then be controlled by temperature. This concept has been used to prepare a smart surface with a switching ability between superhydrophilic and superhydrophobic (Sun et al. 2004). Recently Yang et al. applied this concept to collect water from fogs (Yang et al. 2013). They modified cotton with PNIPAM brush by surface-initiated controlled radical polymerization. Under a humidity of 96% at 23°C, these sponge-like cotton fibers absorb water up to 350 wt% of its weight without direct contact. Most of the uptake water can be released at 33°C. This uptake-and-release process is reversible and suggested to be useful for collecting water from the atmosphere converting it into recyclable water, for water flow-conduction and/or purification.

Similar control over water absorption and released in thermo-responsive hydrogel was used for passive cooling of buildings by Rotzetter et al. (Rotzetter et al. 2012). It is designed to decrease the great energy consumed used for building utilities, of which cooling and heating are the main contributors. The authors proposed that PNIPAM hydrogel coated on outside of building can absorb water during the rain which decreases environment temperature to a value lower than LCST of PNIPAM. When the temperature rises beyond the LCST of the hydrogel in the sun, water is expelled from the hydrogel, consuming a lot of energy for preventing temperature increase in the building. They demonstrated this concept in a model building coated with 3 mm thick sandwiched gel layer by comparing the controlling effect of thermo-responsive hydrogel and normal hydrogel. It was found that surface temperature reductions of up to 20°C under tropic solar irradiance were attained.

3.4 Thermo-Responsive Hydrogel Layers for Biomedicine

Various thermo-responsive polymers mentioned in the introductory part are biocompatible and it also a well known and well developed field for using PNIPAM as carrier in control release (Hu et al. 2013, Piskin 2004, Yoshida et al. 1993). There are a number of reviews which have discussed the application of thermo-responsive materials in control release (drug delivery) (Chiappetta and Sosnik 2007, Klouda and Mikos 2008, Liu et al. 2009, Lyon et al. 2009, Nakayama and Okano 2006, Schmaljohann

2006, Ward and Georgiou 2011, Xu and Liu 2008), and we will not present them here again. Another interesting topic on using thermo-responsive hydrogels in biomedicine is to prepare a cell sheet. It is developed by Okano's research group (Kobayashi and Okano 2013, Matsuzaka et al. 2013, Mizutani et al. 2008, Takahashi et al. 2012, Tang et al. 2014, Williams et al. 2009). They developed a thermo-responsive tissue culture surfaces by anchoring PNIPAM onto ordinary polystyrene surface (Fig. 17) (Takahashi et al. 2010). Obtained surfaces show temperature-responsive hydrophilic/hydrophobic changes. At typical culture conditions at 37°C, the surfaces are hydrophobic, allowing cells to attach, spread, and proliferate. Decreasing incubation temperature to 20°C at which the PNIPAM polymer becomes hydrophilic, lead to all the cultured cells to be harvested as intact sheets. This approach is useful for various cells and promoting to be an innovative technology for tissue and organ regeneration (Matsuura et al. 2014). Recently Guo et al. integrate phenylboronic acid (PBA) groups into the hydrogel layer and make the release of cell sheet saccharide-sensitive (Guo et al. 2015).

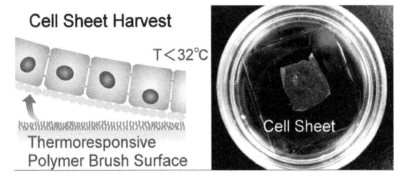

Figure 17. Schematic model of cell sheet delaminated from PNIPAM modified surface and obtained cell sheet. Reprinted with permission of Ref. (Takahashi et al. 2010). Copyright 2010 American Chemical Society.

4. Conclusion and Future Trends

In this chapter we have discussed two kinds of thermo-responsive soft matters: LCEs and hydrogels consisting of thermo-responsive polymers. Their properties are distinguishable. LCEs can be used in any kind of environment while hydrogels normally are used in aqueous or moist environment since swelling is required in the water resource. Monodomain LCEs display an anisotropic shape change without significant change in the whole volume while thermo-responsive hydrogels show an actuating ability dependent on volume changes as a consequence of swelling and shrinkage behaviors. Hydrogels are very easy to prepare because of their cheap commercial available precursors (monomer, crosslinker, and initiator) and simple fabrication procedures. However, hydrogels are weak materials and their mechanical force for actuating is relatively small even though volume changes in the swelling/shrinkage process could be very big. Recently the concept of tough gels (Gong 2010) has been introduced into thermo-responsive hydrogels (Zheng et al. 2015). It is a direction to figure out the issue of weakness. The driving force is still a challenge for their application for actuating purpose. Compared to this, LCEs display very impressive mechanical actuating behaviors in recent demonstrations. The main issues for LCEs are the expensive monomers and their complicated fabricating techniques. Recent attention in this field mainly focuses on developing new method to orientate the LCEs that are allowed to fabricate more complicate structures. Several important developments have been made and will break the bottleneck in this field. The state of no real application of LCE in our daily lives should be improved in the near future.

References

Agrawal, A., T.H. Yun, S.L. Pesek, W.G. Chapman and R. Verduzco. 2014. Shape-responsive liquid crystal elastomer bilayers. Soft Matter 10: 1411–1415.

Ahir, S.V., A.R. Tajbakhsh and E.M. Terentjev. 2006. Self-assembled shape-memory fibers of triblock liquid-crystal polymers. Adv. Funct. Mater. 16: 556–560.

Auroux, P.A., D. Iossifidis, D.R. Reyes and A. Manz. 2002. Micro total analysis systems. 2. Analytical standard operations and applications. Anal. Chem. 74: 2637–2652.

Barclay, G.G. and C.K. Ober. 1993. Liquid crystalline and rigid-rod networks. Prog. Polym. Sci. 18: 899–945.

Behl, M., M.Y. Razzaq and A. Lendlein. 2010. Multifunctional shape-memory polymers. Adv. Mater. 22: 3388–3410.

Benito-Lopez, F., M. Antonana-Diez, V.F. Curto, D. Diamond and V. Castro-Lopez. 2014. Modular microfluidic valve structures based on reversible thermoresponsive ionogel actuators. Lab Chip 14: 3530–3538.

Beyer, P., L. Braun and R. Zentel. 2007a. (Photo)crosslinkable smectic LC main-chain polymers. Macromol. Chem. Phys. 208: 2439–2448.

Beyer, P., E.M. Terentjev and R. Zentel. 2007b. Monodomain liquid crystal main chain elastomers by photocrosslinking. Macromol. Rapid Commun. 28: 1485–1490.

Blaiszik, B.J., S.L.B. Kramer, S.C. Olugebefola, J.S. Moore, N.R. Sottos and S.R. White. 2010. Self-healing polymers and composites. Annu. Rev. Mater. Res. 40: 179–211.

Brand, H.R., H. Pleiner and P. Martinoty. 2006. Selected macroscopic properties of liquid crystalline elastomers. Soft Matter 2: 182–189.

Breger, J.C., C. Yoon, R. Xiao, H.R. Kwag, M.O. Wang, J.P. Fisher, T.D. Nguyen and D.H. Gracias. 2015. Self-folding thermo-magnetically responsive soft microgrippers. ACS Appl. Mater. Interfaces 7: 3398–3405.

Broemmel, F., D. Kramer and H. Finkelmann. 2012. Preparation of liquid crystalline elastomers. Adv. Polym. Sci. 250: 1–48.

Buguin, A., M.H. Li, P. Silberzan, B. Ladoux and P. Keller. 2006. Micro-Actuators: when artificial muscles made of nematic liquid crystal elastomers meet soft lithography. J. Am. Chem. Soc. 128: 1088–1089.

Bunsow, J., T.S. Kelby and W.T.S. Huck. 2010. Polymer Brushes: routes toward mechanosensitive surfaces. Acc. Chem. Res. 43: 466–474.

Burke, K.A. and P.T. Mather. 2013. Evolution of microstructure during shape memory cycling of a main-chain liquid crystalline elastomer. Polymer 54: 2808–2820.

Calejo, M.T., S.A. Sande and B. Nystroem. 2013. Thermoresponsive polymers as gene and drug delivery vectors: architecture and mechanism of action. Expert Opin. Drug Delivery 10: 1669–1686.

Chambers, M., H. Finkelmann, M. Remskar, A. Sanchez-Ferrer, B. Zalar and S. Zumer. 2009. Liquid crystal elastomer-nanoparticle systems for actuation. J. Mater. Chem. 19: 1524–1531.

Chen, T., R. Ferris, J. Zhang, R. Ducker and S. Zauscher. 2010. Stimulus-responsive polymer brushes on surfaces: Transduction mechanisms and applications. Prog. Polym. Sci. 35: 94–112.

Chiappetta, D.A. and A. Sosnik. 2007. Poly(ethylene oxide)-poly(propylene oxide) block copolymer micelles as drug delivery agents: Improved hydrosolubility, stability and bioavailability of drugs. Eur. J. Pharm. Biopharm. 66: 303–317.

Courty, S., J. Mine, A.R. Tajbakhsh and E.M. Terentjev. 2003. Nematic elastomers with aligned carbon nanotubes: New electromechanical actuators. Europhys. Lett. 64: 654–660.

Cui, J. and A. del Campo. 2012. Multivalent H-bonds for self-healing hydrogels. Chem. Commun. 48: 9302–9304.

Cui, J., D.M. Drotlef, I. Larraza, J.P. Fernandez-Blazquez, L.F. Boesel, C. Ohm, M. Mezger, R. Zentel and A. del Campo. 2012. Bioinspired actuated adhesive patterns of liquid crystalline elastomers. Adv. Mater 24: 4601–4604.

Cui, J., D. Wang, K. Koynov and A. del Campo. 2013. 2-Ureido-4-Pyrimidone-based hydrogels with multiple responses. ChemPhysChem. 14: 2932–2938.

Cui, L., X. Tong, X. Yan, G. Liu and Y. Zhao. 2004. Photoactive thermoplastic elastomers of azobenzene-containing triblock copolymers prepared through atom transfer radical polymerization. Macromolecules 37: 7097–7104.

de Gennes, P.G. 1975. C. R. Acad. Sci. Paris 281: 101–103.

de Jeu, W.H. (ed.). 2012. Liquid Crystal Elastomers: Materials and Applications. Berlin Heidelberg: Springer-Verlag.

del Campo, A., C. Greiner, I. Alvarez and E. Arzt. 2007. Patterned surfaces with pillars with controlled 3D tip geometry mimicking bioattachment devices. Adv. Mater. 19: 1973–1977.

Drotlef, D.M., P. Bluemler and A. del Campo. 2014. Magnetically actuated patterns for bioinspired reversible adhesion (Dry and Wet). Adv. Mater. 26: 775–779.

Finkelmann, H., H. Kock and G. Rehage. 1981. Investigations on LC polysiloxans: 3. Liquid crystalline elastomers—a new type of liquid crystalline material. Makromol. Chem. Rapid Commun. 2: 317–322.

Finkelmann, H. and G. Rehage. 1980. Investigations on liquid crystalline polysiloxanes, 1. Synthesis and characterization of linear polymers. Makromol. Chem., Rapid Commun. 1: 31–34.

Fleischmann, E.K., H.L. Liang, N. Kapernaum, F. Giesselmann, J. Lagerwall and R. Zentel. 2012. One-piece micropumps from liquid crystalline core-shell particles. Nat. Commun. 3: 1178.

Fujishige, S., K. Kubota and I. Ando. 1989. Phase transition of aqueous solutions of poly(N-isopropylacrylamide) and poly(N-isopropylmethacrylamide). J. Phys. Chem. 93: 3311–3313.

Fujiwara, T. 2012. Thermo-responsive gels: biodegradable hydrogels from enantiomeric copolymers of poly(lactide) and poly(ethylene glycol). ACS Symp. Ser. 1114: 287–311.

Gebhard, E. and R. Zentel. 2000. Ferroelectric liquid crystalline elastomers, 1 variation of network topology and orientation. Macromol. Chem. Phys. 201: 902–910.

Glatzel, S., A. Laschewsky and J.F. Lutz. 2011. Well-defined uncharged polymers with a sharp UCST in water and in physiological milieu. Macromolecules 44: 413–415.

Gong, J.P. 2010. Why are double network hydrogels so tough? Soft Matter 6: 2583–2590.

Greco, F., V. Domenici, S. Romiti, T. Assaf, B. Zupancic, J. Milavec, B. Zalar, B. Mazzolai and V. Mattoli. 2013. Reversible heat-induced microwrinkling of PEDOT:PSS nanofilm surface over a monodomain liquid crystal elastomer. Mol. Cryst. Liq. Cryst. 572: 40–49.

Greiner, C., E. Arzt and A. del Campo. 2009. Hierarchical Gecko-like adhesives. Adv. Mater. 21: 479–482.

Greiner, C., A. del Campo and E. Arzt. 2007. Adhesion of Bioinspired Micropatterned Surfaces: Effects of Pillar Radius, Aspect Ratio, and Preload. Langmuir 23: 3495–3502.

Guan, Y. and Y. Zhang. 2011. PNIPAM microgels for biomedical applications: from dispersed particles to 3D assemblies. Soft Matter 7: 6375–6384.

Guo, B., G. Pan, Q. Guo, C. Zhu, W. Cui, B. Li and H. Yang. 2015. Saccharides and temperature dual-responsive hydrogel layers for harvesting cell sheets. Chem. Commun. 51: 644–647.

Harris, K.D., C.W.M. Bastiaansen, J. Lub and D.J. Broer. 2005. Self-assembled polymer films for controlled agent-driven motion. Nano Lett. 5: 1857–1860.

He, X., M. Aizenberg, O. Kuksenok, L.D. Zarzar, A. Shastri, A.C. Balazs and J. Aizenberg. 2012. Synthetic homeostatic materials with chemo-mechano-chemical self-regulation. Nature 487: 214–218.

Hocine, S. and M.H. Li. 2013. Thermoresponsive self-assembled polymer colloids in water. Soft Matter 9: 5839–5861.

Hu, J., Y. Zhu, H. Huang and J. Lu. 2012. Recent advances in shape-memory polymers: Structure, mechanism, functionality, modeling and applications. Prog. Polym. Sci. 37: 1720–1763.

Hu, L., A.K. Sarker, M.R. Islam, X. Li, Z. Lu and M.J. Serpe. 2013. Poly (N-isopropylacrylamide) microgel-based assemblies. J. Polym. Sci., Part A: Polym. Chem. 51: 3004–3020.

Ikeda, T., J.I. Mamiya and Y. Yu. 2007. Photomechanics of liquid-crystalline elastomers and other polymers. Angew. Chem., Int. Ed. 46: 506–528.

Ishizone, T., A. Seki, M. Hagiwara, S. Han, H. Yokoyama, A. Oyane, A. Deffieux and S. Carlotti. 2008. Anionic polymerizations of oligo(ethylene glycol) alkyl ether methacrylates: effect of side chain length and ω-alkyl group of side chain on cloud point in water. Macromolecules 41: 2963–2967.

Islam, M.R., A. Ahiabu, X. Li and M.J. Serpe. 2014. Poly (N-isopropylacrylamide) microgel-based optical devices for sensing and biosensing. Sensors 14: 8984–8995, 8912.

Jiang, H., C. Li and X. Huang. 2013. Actuators based on liquid crystalline elastomer materials. Nanoscale 5: 5225–5240.

Kaufhold, W., H. Finkelmann and H.R. Brand. 1991. Nematic elastomers. 1. Effect of the spacer length on the mechanical coupling between network anisotropy and nematic order. Makromol. Chem. 192: 2555–2579.

Kempe, M.D., N.R. Scruggs, R. Verduzco, J. Lal and J.A. Kornfield. 2004. Self-assembled liquid-crystalline gels designed from the bottom up. Nat. Mater. 3: 177–182.

Kim, C., S. Mukherjee, P. Luchette and P. Palffy-Muhoray. 2014. Director orientation in deformed liquid crystal elastomer microparticles. Soft Mater. 12: 159–165.

Kim, S.T. and H. Finkelmann. 2001. Cholesteric liquid single-crystal elastomers (LSCE) obtained by the anisotropic deswelling method. Macromol. Rapid Commun. 22: 429–433.

Klouda, L. and A.G. Mikos. 2008. Thermoresponsive hydrogels in biomedical applications. Eur. J. Pharm. Biopharm. 68: 34–45.

Kobayashi, J. and T. Okano. 2013. Thermoresponsive thin hydrogel-grafted surfaces for biomedical applications. React. Funct. Polym. 73: 939–944.

Krause, S., R. Dersch, J.H. Wendorff and H. Finkelmann. 2007. Photocrosslinkable liquid crystal main-chain polymers: thin films and electrospinning. Macromol. Rapid Commun. 28: 2062–2068.

Lee, S.S., B. Kim, S.K. Kim, J.C. Won, Y.H. Kim and S.H. Kim. 2015. Robust microfluidic encapsulation of cholesteric liquid crystals toward photonic ink capsules. Adv. Mater. 27: 627–633.

Lendlein, A. and S. Kelch. 2002. Shape-memory polymers. Angew. Chem., Int. Ed. 41: 2034–2057.

Li, M.H., P. Keller, J. Yang and P.A. Albouy. 2004. An artificial muscle with lamellar structure based on a nematic triblock copolymer. Adv. Mater. 16: 1922–1925.

Liu, R., M. Fraylich and B.R. Saunders. 2009. Thermoresponsive copolymers: from fundamental studies to applications. Colloid Polym. Sci. 287: 627–643.

Lutz, J.F., O. Akdemir and A. Hoth. 2006. Point by point comparison of two thermosensitive polymers exhibiting a similar LCST: is the age of Poly(NIPAM) over? J. Am. Chem. Soc. 128: 13046–13047.

Lyon, L.A., Z. Meng, N. Singh, C.D. Sorrell and A. St. John. 2009. Thermoresponsive microgel-based materials. Chem. Soc. Rev. 38: 865–874.

Marshall, J.E., S. Gallagher, E.M. Terentjev and S.K. Smoukov. 2014. Anisotropic colloidal micromuscles from liquid crystal elastomers. J. Am. Chem. Soc. 136: 474–479.

Mary, P., D.D. Bendejacq, M.P. Labeau and P. Dupuis. 2007. Reconciling low- and high-salt solution behavior of sulfobetaine polyzwitterions. J. Phys. Chem. B 111: 7767–7777.

Matanovic, M.R., J. Kristl and P.A. Grabnar. 2014. Thermoresponsive polymers: Insights into decisive hydrogel characteristics, mechanisms of gelation, and promising biomedical applications. Int. J. Pharm. 472: 262–275.

Matsuura, K., R. Utoh, K. Nagase and T. Okano. 2014. Cell sheet approach for tissue engineering and regenerative medicine. J. Controlled Release 190: 228–239.

Matsuzaka, N., M. Nakayama, H. Takahashi, M. Yamato, A. Kikuchi and T. Okano. 2013. Terminal-functionality effect of Poly(N-isopropylacrylamide) brush surfaces on temperature-controlled cell adhesion/detachment. Biomacromolecules 14: 3164–3171.

Mirfakhrai, T., J.D.W. Madden and R.H. Baughman. 2007. Polymer artificial muscles. Mater. Today 10: 30–38.

Mizutani, A., A. Kikuchi, M. Yamato, H. Kanazawa and T. Okano. 2008. Preparation of thermoresponsive polymer brush surfaces and their interaction with cells. Biomaterials 29: 2073–2081.

Muramatsu, K., M. Ide and F. Miyawaki. 2012. Biological evaluation of tissue-engineered cartilage using thermoresponsive poly(N-isopropylacrylamide)-grafted hyaluronan. J. Biomater. Nanobiotechnol. 3: 1–9.

Nakayama, M. and T. Okano. 2006. Intelligent thermoresponsive polymeric micelles for targeted drug delivery. J. Drug Delivery Sci. Technol. 16: 35–44.

Ohm, C., M. Brehmer and R. Zentel. 2010. Liquid crystalline elastomers as actuators and sensors. Adv. Mater. 22: 3366–3387.

Ohm, C., M. Brehmer and R. Zentel. 2012. Applications of liquid crystalline elastomers. Adv. Polym. Sci. 250: 49–93.

Ohm, C., N. Haberkorn, P. Theato and R. Zentel. 2011a. Template-based fabrication of nanometer-scaled actuators from liquid-crystalline elastomers. Small 7: 194–198.

Ohm, C., N. Kapernaum, D. Nonnenmacher, F. Giesselmann, C. Serra and R. Zentel. 2011b. Microfluidic synthesis of highly shape-anisotropic particles from liquid crystalline elastomers with defined director field configurations. J. Am. Chem. Soc. 133: 5305–5311.

Ohm, C., C. Serra and R. Zentel. 2009. A continuous flow synthesis of micrometer-sized actuators from liquid crystalline elastomers. Adv. Mater. 21: 4859–4862.

Palffy-Muhoray, P. 2012. Liquid crystal elastomers and light. Adv. Polym. Sci. 250: 95–118.

Pei, Z., Y. Yang, Q. Chen, E.M. Terentjev, Y. Wei and Y. Ji. 2014. Mouldable liquid-crystalline elastomer actuators with exchangeable covalent bonds. Nat. Mater. 13: 36–41.

Pelah, A., R. Seemann and T.M. Jovin. 2007. Reversible cell deformation by a polymeric actuator. J. Am. Chem. Soc. 129: 468–469.

Pietsch, C., U.S. Schubert and R. Hoogenboom. 2011. Aqueous polymeric sensors based on temperature-induced polymer phase transitions and solvatochromic dyes. Chem. Commun. 47: 8750–8765.

Piskin, E. 2004. Molecularly designed water soluble, intelligent, nanosize polymeric carriers. Int. J. Pharm. 277: 105–118.

Prabaharan, M. and J.F. Mano. 2006. Stimuli-responsive hydrogels based on polysaccharides incorporated with thermo-responsive polymers as novel biomaterials. Macromol. Biosci. 6: 991–1008.

Ratna, D. and J. Karger-Kocsis. 2008. Recent advances in shape memory polymers and composites: a review. J. Mater. Sci. 43: 254–269.

Reddy, S., E. Arzt and A. del Campo. 2007. Bioinspired surfaces with switchable adhesion. Adv. Mater. 19: 3833–3837.

Reyes, D.R., D. Iossifidis, P.A. Auroux and A. Manz. 2002. Micro total analysis systems. 1. Introduction, theory, and technology. Anal. Chem. 74: 2623–2636.

Rotzetter, A.C.C., C.M. Schumacher, S.B. Bubenhofer, R.N. Grass, L.C. Gerber, M. Zeltner and W.J. Stark. 2012. Thermoresponsive polymer induced sweating surfaces as an efficient way to passively cool buildings. Adv. Mater. 24: 5352–5356.

Sanchez-Ferrer, A., T. Fischl, M. Stubenrauch, A. Albrecht, H. Wurmus, M. Hoffmann and H. Finkelmann. 2011. Liquid-crystalline elastomer microvalve for microfluidics. Adv. Mater. 23: 4526–4530.

Schild, H.G. 1992. Poly(N-isopropylacrylamide): experiment, theory and application. Prog. Polym. Sci. 17: 163–249.

Schmaljohann, D. 2006. Thermo- and pH-responsive polymers in drug delivery. Adv. Drug Delivery Rev. 58: 1655–1670.

Schuering, H., R. Stannarius, C. Tolksdorf and R. Zentel. 2001. Liquid crystal elastomer balloons. Macromolecules 34: 3962–3972.

Seuring, J. and S. Agarwal. 2010. Non-ionic homo- and copolymers with H-donor and H-acceptor units with an UCST in water. Macromol. Chem. Phys. 211: 2109–2117.

Seuring, J. and S. Agarwal. 2012a. first example of a universal and cost-effective approach: polymers with tunable upper critical solution temperature in water and electrolyte solution. Macromolecules 45: 3910–3918.

Seuring, J. and S. Agarwal. 2012b. Polymers with upper critical solution temperature in aqueous solution. Macromol. Rapid Commun. 33: 1898–1920.

Seuring, J. and S. Agarwal. 2013. Polymers with upper critical solution temperature in aqueous solution: unexpected properties from known building blocks. ACS Macro Lett. 2: 597–600.

Sidorenko, A., T. Krupenkin, A. Taylor, P. Fratzl and J. Aizenberg. 2007. Reversible switching of hydrogel-actuated nanostructures into complex micropatterns. Science 315: 487–490.

Sokolowski, W., A. Metcalfe, S. Hayashi, L.H. Yahia and J. Raymond. 2007. Medical applications of shape memory polymers. Biomed. Mater. 2: S23–S27.

Strotmann, F., E. Bezdushna, H. Ritter and H.J. Galla. 2011. *In situ* forming hydrogels: a thermo-responsive polyelectrolyte as promising liquid artificial vitreous body replacement. Adv. Biomater.: B172–B180.

Sun, T., G. Wang, L. Feng, B. Liu, Y. Ma, L. Jiang and D. Zhu. 2004. Reversible switching between superhydrophilicity and superhydrophobicity. Angew. Chem., Int. Ed. 43: 357–360.

Takahashi, H., N. Matsuzaka, M. Nakayama, A. Kikuchi, M. Yamato and T. Okano. 2012. Terminally functionalized thermoresponsive polymer brushes for simultaneously promoting cell adhesion and cell sheet harvest. Biomacromolecules 13: 253–260.

Takahashi, H., M. Nakayama, M. Yamato and T. Okano. 2010. Controlled chain length and graft density of thermoresponsive polymer brushes for optimizing cell sheet harvest. Biomacromolecules 11: 1991–1999.

Tang, Z., Y. Akiyama and T. Okano. 2014. Recent development of temperature-responsive cell culture surface by using poly(N-isopropylacrylamide). J. Polym. Sci., Part B: Polym. Phys. 52: 917–926.

Thomsen III, D.L., P. Keller, J. Naciri, R. Pink, H. Jeon, D. Shenoy and B.R. Ratna. 2001. Liquid crystal elastomers with mechanical properties of a muscle. Macromolecules 34: 5868–5875.

Torras, N., K.E. Zinoviev, J. Esteve and A. Sanchez-Ferrer. 2013. Liquid-crystalline elastomer micropillar array for haptic actuation. J. Mater. Chem. C 1: 5183–5190.

Urayama, K. 2007. Selected issues in liquid crystal elastomers and gels. Macromolecules 40: 2277–2288.

Urayama, K., Y.O. Arai and T. Takigawa. 2005. Volume phase transition of monodomain nematic polymer networks in isotropic solvents accompanied by anisotropic shape variation. Macromolecules 38: 3469–3474.

Vancoillie, G., D. Frank and R. Hoogenboom. 2014. Thermoresponsive poly(oligo ethylene glycol acrylates). Prog. Polym. Sci. 39: 1074–1095.

Vogel, N., R.A. Belisle, B. Hatton, T.S. Wong and J. Aizenberg. 2013. Transparency and damage tolerance of patternable omniphobic lubricated surfaces based on inverse colloidal monolayers. Nat. Commun. 4: 2167.

Wang, J., Z. Chen, M. Mauk, K.S. Hong, M. Li, S. Yang and H.H. Bau. 2005a. Self-Actuated, thermo-responsive hydrogel valves for lab on a chip. Biomed. Microdevices 7: 313–322.

Wang, J., Z. Chen, M. Mauk, K.S. Hong, M. Li, S. Yang and H.H. Bau. 2005b. Self-Actuated, thermo-responsive hydrogel valves for lab on a chip. Biomed. Microdevices 7: 313–322.

Ward, M.A. and T.K. Georgiou. 2011. Thermoresponsive polymers for biomedical applications. Polymers 3: 1215–1242.

Ware, T.H., M.E. McConney, J.J. Wie, V.P. Tondiglia and T.J. White. 2015. Voxelated liquid crystal elastomers. Science 347: 982–984.

Weber, C., R. Hoogenboom and U.S. Schubert. 2012. Temperature responsive bio-compatible polymers based on poly(ethylene oxide) and poly(2-oxazoline)s. Prog. Polym. Sci. 37: 686–714.

Williams, C., Y. Tsuda, B.C. Isenberg, M. Yamato, T. Shimizu, T. Okano and J.Y. Wong. 2009. Aligned cell sheets grown on thermo-responsive substrates with microcontact printed protein patterns. Adv. Mater. 21: 2161–2164.

Wong, T.S., S.H. Kang, S.K.Y. Tang, E.J. Smythe, B.D. Hatton, A. Grinthal and J. Aizenberg. 2011. Bioinspired self-repairing slippery surfaces with pressure-stable omniphobicity. Nature 477: 443–447.

Wu, D.Y., S. Meure and D. Solomon. 2008. Self-healing polymeric materials: A review of recent developments. Prog. Polym. Sci. 33: 479–522.

Xia, Y., R. Verduzco, R.H. Grubbs and J.A. Kornfield. 2008. Well-defined liquid crystal gels from telechelic polymers. J. Am. Chem. Soc. 130: 1735–1740.

Xie, P. and R. Zhang. 2005. Liquid crystal elastomers, networks and gels: advanced smart materials. J. Mater. Chem. 15: 2529–2550.

Xie, T. 2011. Recent advances in shape memory polymer. Polymer 52: 4985–5000.

Xu, J. and S. Liu. 2008. Polymeric nanocarriers possessing thermoresponsive coronas. Soft Matter 4: 1745–1749.

Yang, H., A. Buguin, J.M. Taulemesse, K. Kaneko, S. Mery, A. Bergeret and P. Keller. 2009. Micron-sized main-chain liquid crystalline elastomer actuators with ultralarge amplitude contractions. J. Am. Chem. Soc. 131: 15000–15004.

Yang, H., G. Ye, X. Wang and P. Keller. 2011. Micron-sized liquid crystalline elastomer actuators. Soft Matter 7: 815–823.

Yang, H., H. Zhu, M.M.R.M. Hendrix, N.J.H.G.M. Lousberg, G. de With, A.C.C. Esteves and J.H. Xin. 2013. Temperature-triggered collection and release of water from fogs by a sponge-like cotton fabric. Adv. Mater. 25: 1150–1154.

Yao, X., Y. Hu, A. Grinthal, T.S. Wong, L. Mahadevan and J. Aizenberg. 2013. Adaptive fluid-infused porous films with tunable transparency and wettability. Nat. Mater. 12: 529–534.

Yoshida, R., K. Sakai, T. Okano and Y. Sakurai. 1993. Pulsatile drug delivery systems using hydrogels. Adv. Drug Delivery Rev. 11: 85–108.

Yu, Y. and T. Ikeda. 2006. Soft actuators based on liquid-crystalline elastomers. Angew. Chem., Int. Ed. 45: 5416–5418.

Zarzar, L.D. and J. Aizenberg. 2014. Stimuli-responsive chemomechanical actuation: a hybrid materials approach. Acc. Chem. Res. 47: 530–539.

Zeng, H., D. Martella, P. Wasylczyk, G. Cerretti, J.C.G. Lavocat, C.H., C. Parmeggiani and D.S. Wiersma. 2014. High-resolution 3d direct laser writing for liquid-crystalline elastomer microstructures. Adv. Mater. 26: 2319–2322.

Zentel, R. 1994. Liquid crystalline polymers. Top. Phys. Chem. 3: 103–141.

Zhang, Q., G. Vancoillie, M.A. Mees and R. Hoogenboom. 2015. Thermoresponsive polymeric temperature sensors with broad sensing regimes. Polym. Chem.: Ahead of Print.

Zhang, X.Z., D.Q. Wu and C.C. Chu. 2004. Synthesis, characterization and controlled drug release of thermosensitive IPN-PNIPAAm hydrogels. Biomaterials 25: 3793–3805.

Zhang, X., C.L. Pint, M.H. Lee, B.E. Schubert, A. Jamshidi, K. Takei, H. Ko, A. Gillies, R. Bardhan, J.J. Urban, M. Wu, R. Fearing and A. Javey. 2011. Optically- and thermally-responsive programmable materials based on carbon nanotube-hydrogel polymer composites. Nano Lett. 11: 3239–3244.

Zhao, Q., N. Chen, D. Zhao and X. Lu. 2013. Thermoresponsive magnetic nanoparticles for seawater desalination. ACS Appl. Mater. Interfaces 5: 11453–11461.

Zheng, W.J., N. An, J.H. Yang, J. Zhou and Y.M. Chen. 2015. Tough Al-alginate/Poly(N-isopropylacrylamide) hydrogel with tunable LCST for soft robotics. ACS Appl. Mater. Interfaces 7: 1758–1764.

10

Advanced Materials for Self-Healing Applications

Jinrong Wu,[1,*] *Li-Heng Cai*[2,*] and *Huanan Wang*[2,*]

ABSTRACT

Self-healing polymeric materials are inspired by biological systems where damage initiates an autonomous healing process without external intervention. Self-healing polymeric materials are defined as systems that after damage can revert to their original state with full or partial recovery of mechanical strength. According to the characteristics of self-healing materials, they could be categorized into stimuli-induced and autonomic classes. In this chapter we review the basic concepts that have been used for designing self-healing polymeric materials and their recent progress. We discuss the advantages and disadvantages of different types of self-healing polymeric materials and their potential applications. Finally we envision the future directions for the development of practically useful self-healing materials.

[1] College of Polymer Science and Engineering, Sichuan University, No. 24, Nanyi Section, Yihuan Raod, Chengdu 610065, China.

[2] School of Engineering and Applied Science, Harvard University, 9 Oxford Street, Cambridge, MA 02138, USA.

* Corresponding authors: wujinrong@scu.edu.cn; liheng.cai@gmail.com; hwang2@seas.harvard.edu

1. Introduction

Advanced materials are a significant driving force that boosts revolution in human society and still determine the pace of technology development. Self-healing materials are a category of advanced materials that can revert to their original state after damage with full or partial recovery of mechanical properties. As there is a strong demand in increasing the service life of materials and decreasing consumption of natural resources, self-healing materials have received enormous attention in recent years. Self-healing polymeric materials are inspired by biological systems in which damage initiates an autonomous healing process without external intervention (Wool 2008, Wu et al. 2008, Williams et al. 2008, Bergman and Wudl 2008). For example, an epidermal wound heals itself through three steps: formation of a blood clot, formation of a dense capillary network and a temporary granulation tissue and division and growth of epidermal cells (Epstein et al. 1999, Martin 1997). Learning from nature, material scientists have developed a variety of approaches to design self-healing materials, including polymers, metals and concretes.

The research on self-healing behavior of polymers started in 1980s with welding, which is based on inter-diffusion of polymers and consequent formation of entanglements at the interface between two fractured surfaces: the welding approach typically requires a temperature above the glass transition temperature (T_g) to promote polymer mobility. Another type of self-healing material is based on release of chemical healing agents upon crack formation. Pioneer works were performed by Dry and Sottos who used embedded hollow fibers containing encapsulated healing agents to repair cracks in composites in 1993 (Dry and Sottos 1993). However, this type of self-healing materials did not receive wide attention until White et al. first demonstrated autonomic healing without any external intervention in 2001 (White et al. 2001). Since then, various types of capsules have been developed, including hollow micro-particles and 3D vascular networks with different alignments. Meanwhile, covalent bonds that are reversible under external stimuli are introduced to design self-healing polymeric materials (Chen et al. 2002), which also boosts the development in this field greatly. Unlike self-healing materials based on embedded healing agents or stimuli induced reversible covalent bonds, an entirely different class of self-healing polymeric materials is based on weak reversible interactions such as hydrogen bonds. The first demonstration of a rubbery polymer network formed by hydrogen bonds was developed by Leibler group in 2008 (Cordier et al. 2008), which motivates extensive exploration of this type of self-healing polymeric materials afterwards.

According to the characteristics of self-healing materials, they could be categorized into stimuli-induced and autonomic classes. The stimuli-induced self-healing materials require external intervention, such as heat and light, to trigger the healing process; by contrast, autonomic self-healing materials recover their original mechanical properties by the crack initiated healing. In this chapter, we will discuss self-healing polymeric materials based on different healing mechanisms. We will identify the main features of different types of self-healing polymeric materials, discuss the potential applications and envision the future research directions of self-healing polymeric materials.

2. Stimuli Triggered Self-Healing Polymers

Ideally, self-healing materials need to "sense" a wound to initiate repair without external intervention. However, this is challenging in the synthetic polymeric systems as they are unlikely to possess such a smart response similar to active biological systems. Instead, many self-healing mechanisms of polymers require to be activated by external stimuli, such as heat and light.

2.1 Self-Healing Polymers Triggered by Heat

The self-healing polymers triggered by heat are stable under normal conditions. After cracking, the polymers can revert to their monomeric, oligomeric or un-crosslinked state upon heating. These monomeric, oligomeric or un-crosslinked units can reorganize to form new polymer networks during the subsequent cooling process to achieve self-healing (Bergman and Wudl 2008).

One important mechanism of heat-triggered self-healing is based on thermally activated Diels-Alder (DA) cycloaddition and retro-Diels-Alder (RDA) reaction, as shown in Fig. 1 (Chen et al. 2002, Chen et al. 2003, Murphy et al. 2008, Tian et al. 2010, Inglis et al. 2010, Liu and Chen 2007, Zeng et al. 2013, Bai et al. 2013). A conjugated diene and a dienophile easily undergo [4 + 2] cycloaddition to form unsaturated six-membered rings, which consist of more energetically stable σ-bonds. Nevertheless, the bond strength between diene and dienophile of the six-membered rings is weaker than typical carbon-carbon covalent bonds; this allows preferential breaking of the six-membered rings during damage. Heating of the damaged sample leads not only to further disconnection between diene and dienophile due to the RDA reaction, but also enhances the mobility of the disconnected groups. During the subsequent cooling process, the disconnected groups reform new adducts to heal the crack. According to the reactive functional groups, the DA/RDA systems can be classified into three main categories:

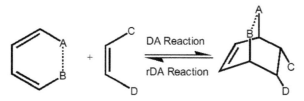

Figure 1. General schematic description of Diels-Alder reaction and retro-Diels-Alder reaction (Bergman and Wudl 2008).

furan-maleimide-based polymers, dicyclopentadiene-based polymers, and anthracene-based polymers (Bergman and Wudl 2008). So far only the furan-maleimide-based polymers are extensively investigated.

Since 1969, the DA/RDA reactions between furan and maleimide had been exploited to fabricate thermally reversible networks from both polymers and monomers carrying reactive furan and maleimide moieties (Craven 1969, Kennedy and Castner 1979, Chujo et al. 1990). But not until 2002 was the first example of using DA/RDA reactions for self-healing polymer reported by Wudl et al. (Chen et al. 2002). They synthesized a trivalent-maleimide monomer and a tetravalent-furan monomer; these star-shaped monomers polymerize through DA step growth, and generate a highly dense network with thermoset properties, as show in Fig. 2. This material has mechanical performances comparable to common structural epoxy resin; more importantly, the fractured sample can be rejoined with 57% recovery of the original tensile strength after a thermal treatment at 120–150°C. The incomplete healing is due to poor interfacial match-up between the fractured surfaces during the healing process.

In a rapid succession a number of works were reported to improve the self-healing efficiency of the similar systems based on furan and maleimide moieties. Wudl et al. (Chen et al. 2003) designed new divalent-maleimide monomers with a lower melting point, which react with the tetravalent-furan monomers to produce thermoset polymers with an improved healing efficiency of 83%. Broekhuis et al. (Zhang et al. 2009) prepared furan-functionalized polyketones (PK-furan) through the Paal-Knorr reaction of polyketones with furfurylamine. The PK-furan was crosslinked with bis-maleimide to produce highly crosslinked polymers. The thermal reversibility of the crosslinked PK-furan in dimethyl sulfoxide solvent was demonstrated by the reversible gelation upon heating and cooling, due to the DA/RDA reactions. Without the presence of solvent, the fracture bulk sample can be grounded into powders; these powders can be remolded to form a new specimen with the same mechanical properties as the original one. Thus, they claim a healing efficiency of 100%. Zhang et al. (Tian et al. 2010) designed a novel epoxy monomer, furfurylglycidyl ether,

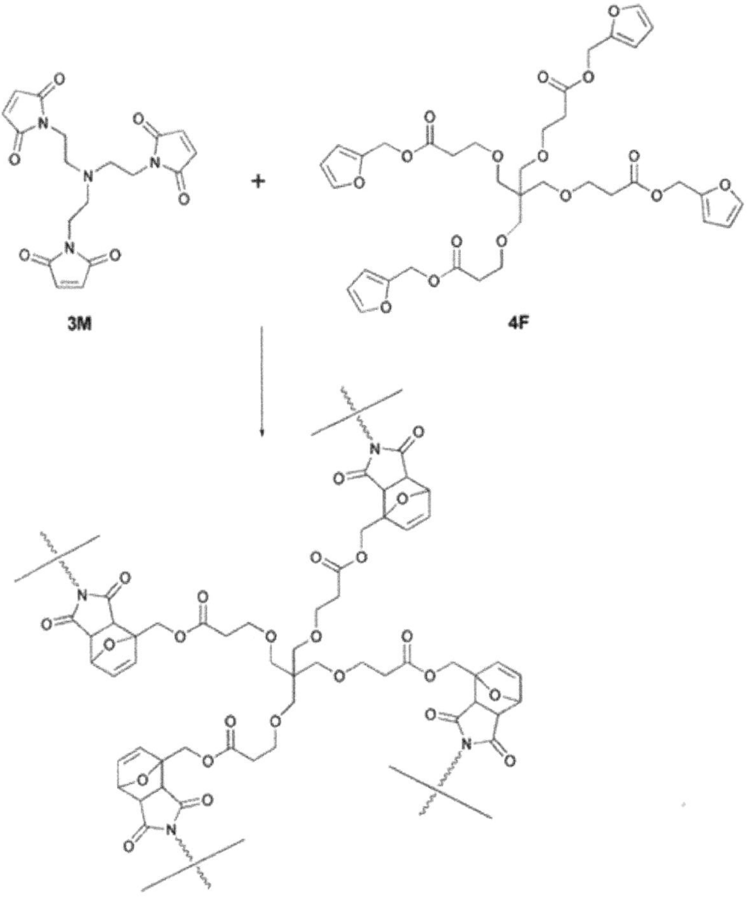

Figure 2. Preparation of a thermally re-mendable cross-linked polymer. Reprinted with permission from ref. (Bergman and Wudl 2008). Copyright 2008 The Royal Society of Chemistry.

which carries one furan and one epoxide on each of the ends. This monomer can be cured with bismaleimide and anhydride, resulting in an epoxy resin with healing efficiency of 96% after heat treatment at 110°C for 20 min and then at 80°C for 72 h.

In addition to improving the self-healing efficiency, efforts are also devoted to lower the healing temperature of the furan-maleimide based polymers. For example, Palmese et al. (Peterson et al. 2010) prepared an epoxy resin functionalized with many pendant furan moieties; after crack formation, a solution of bismaleimide was injected into the crack as a healing agent. The healing agent reacts with the pendant furan moieties, allowing crack repair at room temperature. The healing efficiency is even higher than 100% due to the solvent-mediated physical interlocking and chemical DA reaction. In a recent work from the same group, a solution of multimaleimide (MMI) is encapsulated and embedded into the furan functionalized epoxy resin; upon damage, the solution is released into the crack to enable the reaction between MMI and furan moieties, thus healing the thermoset (Pratama et al. 2013).

The chemical versatility of furan and maleimide moieties enables the preparation of a variety of self-healing polymers, including epoxy, polyketones, polyamide, polyurethane, polyacrylate, poly(ethylene adipate) and poly(N-acetylethyleneimine) (Liu and Chen 2007, Du et al. 2013, Kavitha and Singha 2009, 2010, Watanabe and Yoshie 2006, Chujo et al. 1990, Zhang et al. 2009, Tian et al. 2010, Peterson et al. 2010, Pratama et al. 2013). The richness in these polymer matrices is an advantage for potential applications of these materials. However, there are still some challenges for these materials: the working temperature of the furan-maleimide based polymers is generally lower than 120°C, which is too low for many applications; the high mending temperature is not easily accessible outdoor; the high mending temperature also leads to oxidative side-reactions; the synthesis of the DA monomers is very expensive and thus not realistic for industrial-level production.

2.2 Self-Healing Polymers Triggered by Light

Light stimulation is a convenient and powerful method, because the stimulation is readily accessible and easy to handle, and exposure can be limited to targeted areas which could avoid potential degradation of other parts of a sample. Moreover, light stimulation generally occurs at room temperature. Light of a certain wavelength convert some polymers into their monomeric, oligomeric or un-crosslinked state, which upon exposure to another light of different wavelength, can revert to the polymerized or crosslinked state. This process lays the foundation for self-healing polymers triggered by light stimulation.

Similar to the thermal-triggered DA/RDA reactions, an alternative type of [2 + 2] and [2 + 4] cycloaddition is initiated by a light; meanwhile, the resulting adducts with lower bond strength can readily be cleaved through the reverse cycloaddition under a mechanical stress or upon irradiation of a light with shorter wavelength. The molecular moieties used for the photoinitiated cycloaddition and reverse cycloaddition include cinnamate, coumarin, vinyl, maleimides and anthracene (Cardenas-Daw et al. 2012).

The first example of using photochemical cycloaddition for crack healing was demonstrated by Kim et al. (Chung et al. 2004). They synthesized a photo-crosslinkable monomer, which carries three optically active cinnamoyl groups. The cinnamoyl groups dimerize to form cyclobutane cross-links *via* photochemical [2 + 2] cycloaddition. The cyclobutane cross-links can reverse to the original cinnamoyl groups upon crack formation. After irradiation, re-photocycloaddition of cinnamoyl groups repairs the crack partially, as shown in Fig. 3a. However, the highest healing efficiency for this type of materials, 14%, is not sufficient for practical applications. Coumarin is another molecular moiety that easily forms reversible covalent bonds *via* photochemical [2 + 2] cycloaddition and reverse cycloaddition, as shown in Fig. 3b. Zhang and coworkers (Ling et al. 2012, Ling et al. 2011) introduced coumarin moieties either onto the backbones or as the side chains of polyurethane. When exposed to irradiation at 350 nm, coumarin moieties react with each other to form crosslinks between polyurethane chains; while mechanical damage or irradiation at 254 nm cleaves the dimers. The freshly fractured sample with coumarin on the backbone can fully restore its mechanical strength under UV irradiation, while the healing efficiency decreases successively during the subsequent repeated failure-repair tests. The reversible [4 + 4] cycloaddition of anthracene also provides a potential way for self-healing polymers, as shown in Fig. 3c. Landfester et al. (Froimowicz et al. 2011) designed a hyperbranched polyglycerol bearing many anthracene groups. Crosslinking *via* dimerization of the anthracene groups is induced when irradiated with 366 nm light; while the reversion of the dimerization takes place when exposed to 254 nm light. The self-healing ability of the hyperbranched polymer was tested by visual observation of closure of an artificial scratch.

Another class of photo-induced self-healing polymers is based on chain exchange reactions. Amamoto et al. (Amamoto et al. 2011) demonstrated that trithiocarbonates (TTC) undergo chain reshuffling reactions when exposed to UV irradiation at 330 nm, as shown in Fig. 4. Reversible addition–fragmentation chain-transfer (RAFT) copolymerization of *n*-butyl acrylate and a TTC crosslinker produces a self-healing polymer. Repeatable photo-induced self-healing can be accomplished both in the presence of a solvent and in the bulk state. Disulfide groups with low bond dissociation energy can also exchange under light irradiation. Thus, by incorporating thiuram disulfide (TDS) units into a low T_g polyurethane, the resulting crosslinked polymer shows successful attachment of surfaces of cut pieces proceeded under ambient conditions, when exposed to visible light from a commercial tabletop lamp (Amamoto et al. 2012).

Some cyclic ethers, such as oxetane (OXE), can form stable free radicals through photo-activated ring opening. Ghosh and Urban (Ghosh and Urban 2009) modified chitosan (CHI) with OXE; the resulting OXE-CHI react with tri-functional hexamethylene diisocyanate (HDI) and polyethylene glycol (PEG) to form a heterogeneously cross-linked polyurethane (PUR) network. Upon mechanical damage, the OXE rings open to form two reactive ends.

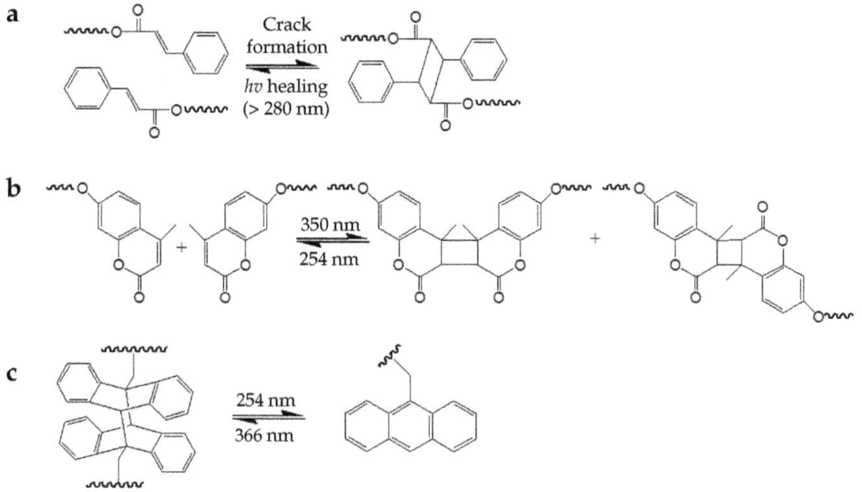

Figure 3. Self-healing via reversible photochemical cycloaddition reactions: **(a)** [2 + 2] cycloaddition of cinnamoyl groups; **(b)** [2 + 2] cycloaddition of coumarin groups; **(c)** [4 + 4] cycloaddition of anthracene groups.

When exposed to ultraviolet light, chitosan chain scission occurs; the resulting broken chitosan chains react with reactive oxetane ends to form new crosslinks and thus repair the network, as shown in Fig. 5. Optical observation demonstrates a healing of the cut on the film sample after 30 min of UV irradiation. The same group also demonstrates the similar self-healing characteristics of another PUR modified with five-membered cyclic ether (oxolane) (Ghosh et al. 2012).

2.3 Self-Healing Polymers Triggered by Other Stimuli

Catalyst stimulus: Transition-metal-catalyzed olefin metathesis reaction has the ability for shuffling strong C-C double bonds under ambient conditions, and thus offers an effective method for healing polymers. It has been shown that double bonds in the olefin-containing polymer networks rapidly exchange through a metallocyclobutane intermediate during metal-catalyzed olefin metathesis (Fig. 6) (Vougioukalakis and Grubbs 2009). This allows the re-organization of the topology of the network, while keeping the number of covalent bonds and cross-links unchanged, thus the mechanical integrity and strength of the network remain constant during the reaction process. Lu and Guan (Lu et al. 2012, Lu and Guan 2012) introduced the concept of olefin metathesis into polybutadiene loaded with Grubbs's second-generation Ru catalyst to prepare insoluble polymer networks with self-healing ability

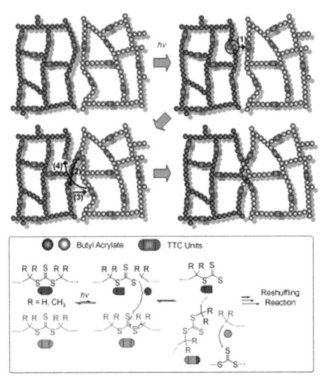

Figure 4. Chain reshuffling reactions of tithiocarbonates under UV irradiation. Reprinted with permission from ref. (Amamoto et al. 2011). Copyright 2011 Wiley-VCH Verlag GmbH & Co.

at sub-ambient temperature. They showed almost complete recovery of the tensile strength and strain after applying a pressure to the freshly cut samples for 6 h.

Redox stimulus: A disulfide bond can be cleaved to generate two thiol groups upon reduction. The thiol groups can reform new disulfide bridges under oxidizing conditions. This leads to exchange reactions between neighboring disulfide bonds, as shown in Fig. 7. Matyjaszewski and coworkers (Yoon et al. 2012) prepared poly(*n*-butyl acrylate) grafted star polymers by chain extension atom transfer radical polymerization (ATRP) from cross-linked cores that comprised of poly(ethylene glycol) diacrylate). By using the consecutive chain extension ATRP, bis(2-methacryloyloxyethyl disulfide) were further introduced to form S-S reversible cross-links at the branch peripheries. Due to the regeneration of S-S bonds *via* thiol–disulfide exchange reactions, a thin film of the polymer shows a rapid spontaneous self-healing behavior, but the efficiency of healing depends on the initial film thickness and the width of the cut.

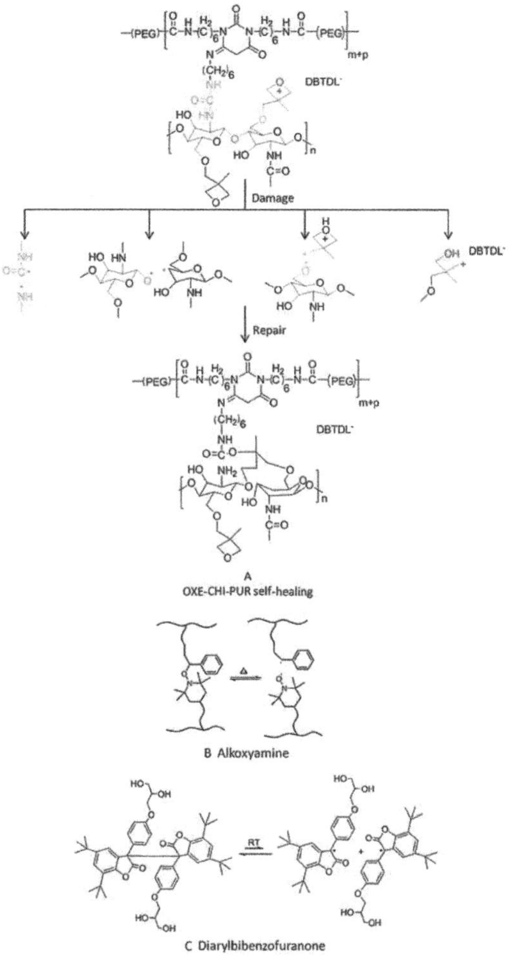

Figure 5. Self-healing *via* ring opening reactions of oxetane (OXE) in polyurethanes. Reprinted with permission from ref. (Yang and Urban 2013). Copyright 2013 The Royal Society of Chemistry.

Atmospheric gases stimulus: Plants assimilate carbon dioxide (CO_2) and water (H_2O) to produce organic compounds and heal their wounds. It will be advantageous if this process can be reproduced in synthetic materials. A recent work of Yang and Urban incorporates a monosaccharide, methyl-α-D-glucopyranoside (MGP) containing four reactive OH groups, into cross-linking reactions of HDI and PEG. The reaction was catalyzed by dibutyltin dilaurate to form MGP modified polyurethane (MGPPUR). The MGPPUR reacts with atmospheric CO_2 and H_2O, thus reforming covalent linkages capable of bridging cleaved network segments (Yang and Urban 2014).

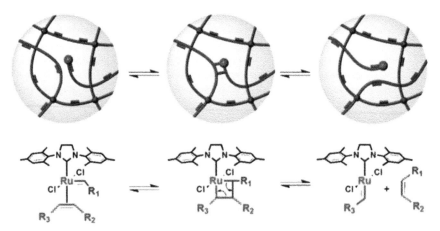

Figure 6. Concept of using olefin metathesis to make cross-linked polybutadiene malleable. Reprinted with permission from ref. (Lu et al. 2012). Copyright 2012 American Chemical Society.

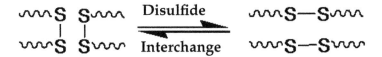

Figure 7. Exchange reactions between disulfide bonds.

Electrical stimulus: Conductive polymer composites (CPC) possess an alternative self-healing route. In conductive composites, the number of pathways for carrier transportation decreases upon material damage, reflected by an increase in the resistance at the site of damage. Applying an electrical field to the CPC generates local heating at the site of damage, which increases chain mobility and initiates self-healing mechanisms that can be activated by heat (Murphy and Wudl 2010). For example, Park et al. (Park et al. 2008) demonstrated a graphite/epoxy CPC which contains a mendomer with a dicyclopentadiene unit that can break into two cyclopentadiene groups through RDA reaction. Upon the application of an electrical field, the cyclopentadiene groups reunite due to heating. Huang et al. (Huang et al. 2013) fabricated a CPC based on few-layered graphene (FG) and thermoplastic polyurethane (TPU). The healing efficiency due to interfacial diffusion and formation of entanglements after applying a voltage that can be higher than 98%.

Electricomagnetic stimulus: This method bears the idea similar to the electrical-stimulus self-healing polymers. In general, when a polymer composite with distributed magnetic particles is subjected to a high-field magnetic field, the oscillation of the magnetic moment dissipates energy and generates heat, which can be employed to induce interfacial diffusion and re-entanglement and thus heal the polymer matrix. For example, Corten and Urban (Corten and Urban 2009) *in situ* synthesized *p*-methyl methacrylate/*n*-butylacrylate/ heptadecafluorodecyl methacrylate (*p*-MMA/*n*BA/HDFMA) copolymer colloidal particles in the presence of γ-Fe_2O_3 nanoparticles. It was found that the γ-Fe_2O_3 nanoparticles accumulated on the surface of p-MMA/ nBA/HDFMA; after coagulation, nanoparticles are uniformly dispersed throughout the composite, which shows superparamagnetic properties. When the cut film pieces are re-attached to form a physical contact with each other and exposed to an oscillating magnetic field, the cut can be healed and the re-joined film retains the original mechanical strength.

3. Autonomic Self-Healing Materials

The key feature of autonomic self-healing is that the material repairs by itself without external intervention. The self-repair following fracture is accomplished either through healing agents embedded into polymer matrix or through the formation of reversible bonds. In this section we discuss autonomic self-healing polymeric systems classified by their healing mechanisms.

3.1 Self-Healing Based on Release of Encapsulated Healing Agents

One of the most often used strategies to achieve autonomic self-healing is using embedded healing agents in polymeric matrix. The idea of self-healing system based on embedded healing reagents via micro-encapsulation approach was developed by Dry and collaborators (Dry 1994, 1996, Dry and Sottos 1993, Dry 1992) in the 1990s and later extended by White et al. (White et al. 2001, Toohey et al. 2007, Yang et al. 2008, Toohey et al. 2009, Patrick et al. 2014, Jackson et al. 2011, Hansen et al. 2009, Blaiszik et al. 2008, Blaiszik et al. 2009); an example of this self-healing system is illustrated in Fig. 8. In this system, chemicals are encapsulated into capsules, which are embedded into polymer matrix; whereas the catalyst is dispersed in the polymer matrix outside the capsules, as shown in Fig. 8. Upon damage, the capsules are ruptured by propagating cracks, resulting in the release of the chemicals into the cracks. Subsequent reaction between the chemicals with the aid of the dispersed catalyst heals the material and prevents further crack growth. One of the commonly used chemical reactions is ring-opening metathesis polymerization (ROMP) with the aid of Grubb's catalyst; this type of chemical reaction exhibits high metathesis activity and tolerance of a wide range of functional groups,

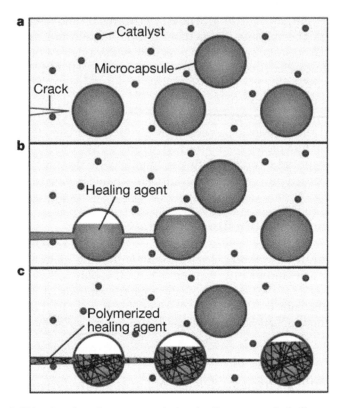

Figure 8. Self-healing based on encapsulation healing agents in hollow particles. A microencapsulated healing agent is embedded in a structural composite matrix containing a catalyst capable of polymerizing the healing agent. **(a)** Cracks form in the matrix wherever damage occurs; **(b)** the crack ruptures the microcapsules, releasing the healing agent into the crack plane through capillary action; **(c)** the healing agent contacts the catalyst, triggering polymerization that bonds the crack faces closed. Reprinted with permission from ref. (White et al. 2001). Copyright 2001 Nature Publishing Group.

oxygen and water, which favorably meet the diverse requirements of self-healing materials, including long shelf lifetime and rapid polymerization at ambient conditions.

Typical types of capsules include micro-hollow-particles (Brown et al. 2002, Brown et al. 2003, Kessler et al. 2003, Brown et al. 2005, Blaiszik et al. 2008, Jackson et al. 2011, Yang et al. 2008, Toohey et al. 2007, Blaiszik et al. 2009), hollow fibers (Pang and Bond 2005, Trask and Bond 2006, Trask et al. 2007, Williams et al. 2007, Yin, Rong et al. 2008, Yin et al. 2007, Yin, Zhou et al. 2008), and three dimensional vascular networks (Hoveyda and Zhugralin 2007, Toohey et al. 2009, Hansen et al. 2009, Patrick et al. 2014). Due to limited volume of micro-hollow-particles, the amount of chemicals carried by micro-hollow-particles often allows for single repair. To achieve multiple times of healing, vascular networks similar to that in skin are designed for synthetic materials, as shown in Fig. 9. Compared to micro particles, hollow fibers and 3D vascular networks draw chemicals far from the location where the crack occurs, enabling healing at the same crack multiple times. While using embedded capsules enables crack-triggered healing, together with its original version of using hollow particles, this method requires complicated fabrication of self-healing materials. Indeed, the fabrication typically involves preparation of capsules, mixing the capsules with polymeric matrix with the aid of solvents, and removing solvents afterwards. In addition, the mechanical properties of the capsules greatly influence the properties of the final polymeric composites. For example, tough capsules will not break easily on the crack and thus result in failure of releasing chemical agents and thus healing; by contrast, fragile capsules break easily, but likely to weaken

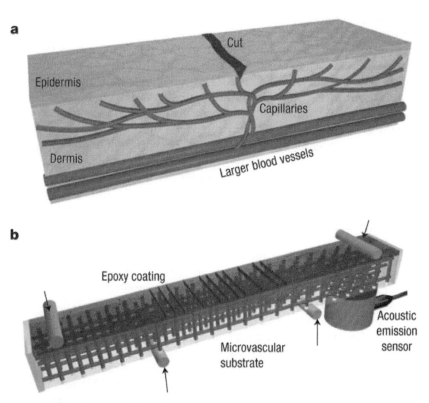

Figure 9. Self-healing materials with 3D microvascular networks. **(a)** Schematic diagram of a capillary network in the dermis layer of skin with a cut in the epidermis layer. **(b)** Schematic diagram of the self-healing structure composed of a microvascular substrate and a brittle epoxy coating containing an embedded catalyst. Reprinted with permission from ref. (Toohey et al. 2007). Copyright 2007 Nature Publishing Group.

the mechanical performance of the composites, as embedded capsules as well play an important role in determining the mechanical performance of the composites. In fact, the effect of embedded particles on properties of composites is an active area of research; the discussion of this topic is beyond the scope of this section (Balazs et al. 2006, Sarkar and Alexandridis 2015). Therefore, to balance efficiency of self-healing and mechanical performance of the polymer composites, the properties of capsules need to be tuned carefully. To make the capsules, various materials have been explored; these materials include urea-formaldehyde (UF) (Yuan et al. 2006, Cosco et al. 2007), melamine-formaldehyde (MF) (Yuan et al. 2008), melamine-urea-formaldehyde (MUF) (Liu et al. 2009), polyurethane (PU) (Cho et al. 2006), and acrylates (Xiao et al. 2009).

Unlike using chemical crosslinking agents, an alternative approach towards autonomic self-healing is to use solvents as healing agents (Caruso et al. 2007, Yuan et al. 2008, Neuser et al. 2012, Caruso et al. 2008, Blaiszik et al. 2009). The solvents are encapsulated in carriers; on a fracture, the solvents are released to facilitate the local mobility of polymer chains, resulting in formation of polymer entanglements across the crack interface. In addition to challenges due to introducing capsules, this type of self-healing materials has relatively poor mechanical performance due to the intrinsic properties of entanglements. Entanglements behave like permanent crosslinks at short time scales, and thus result in finite mechanical strength. However, the recovered mechanical strength due to entanglements is relatively weaker as the concentration of entanglements is intrinsically limited by material property; for example, typical elastic modulus for most polymers is on the order of MPa due to entanglements. Moreover, the entanglements relax on long time scales and no longer hold stress. Consequently, self-healing polymeric systems based on formation of entanglements at the crack interface have relatively weaker mechanical performance compared to covalently cross-linked systems.

In addition to taking the advantage of chemical crosslinking or formation of physical entanglements to achieve autonomic self-healing, another approach reported recently is to use nanoparticles as self-healing agents (Lee et al. 2004, Smith et al. 2005, Gupta et al. 2006, BaLazs 2007). This type of self-healing materials is based on entropic driving mechanisms; nanoparticles migrate through a composite material to the crack due to entropic immiscibility between particles and polymers. Polymers passing around nanoparticles are subjected to entropy unfavorable state, and tend to drive the particles toward the surface, or the crack. Meanwhile, these nanoparticles strengthen the interface at the crack and thus impede its growth.

In brief, self-healing polymeric materials based on the release of healing agents upon crack are one of the major classes of autonomic self-healing materials. In spite of their commonly complex fabrication process, these materials harness the power of chemical crosslinking to ensure good mechanical performance, though not always with high healing efficiency. Indeed, typical healing efficiency for the first time of healing is ~ 75%; however, the healing efficiency decreases dramatically for multiple self-healing on the same crack. Nevertheless, by adjusting the properties of capsules and exploring new type of healing agents may improve the mechanical performance of this type of self-healing materials and thus broaden their applications.

3.2 Self-Healing Based on Reversible Associations

An ideal self-healing polymeric material would fully recover its mechanical performance after healing and, even more, would be able to heal on unlimited times. In addition, healing of the material should not rely on any external intervention; the material should heal by itself once two fractured surfaces are brought into contact. To achieve this goal, new healing mechanisms rather than embedded healing reagents must be introduced. A possible self-healing mechanism is based on reversible associations that can break and reform. These associations include hydrogen bonds, associations due to metal-ligand coordination, and multiplets in ionomers. In this section we discuss the self-healing polymeric materials based on different types of reversible associations.

Hydrogen bonds. Hydrogen bonds are a type of most often seen reversible associations that exist in natural systems; examples include water, biomolecules, and polymeric systems with polarized motifs. An important feature of hydrogen bonds is that they can break and reform at room temperature, allowing relatively fast reaction whose results can be observed during experimental time scales. For example, a melt of polymers carrying such reversible associations has a significant higher viscosity compared to the polymer melt without such interactions (Sijbesma et al. 1997). However, due to the weakness in bonding strength, it is difficult to create hydrogen bond based polymeric systems with good mechanical properties. To overcome this challenge, molecules that can form stronger hydrogen bonds must be used and the concentration of hydrogen bonds must be increased. Indeed, by using motifs that can form multiple hydrogen bonds to achieve strong associations, Leibler group developed a novel and promising type of autonomic self-healing systems (Cordier et al. 2008). In this system, the multiple hydrogen bonds are amidoethyl imidazolidone, di(amido ethyl) urea and diamido tetraethyl triurea, as shown in Fig. 10. The di-functional groups form longer polymers, whereas the tri-functional groups serve as crosslinks, resulting in a polymeric system having rubbery behavior.

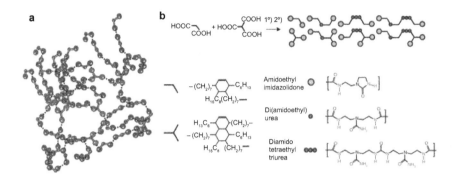

Figure 10. (a) Schematic view of a reversible network formed by mixtures of ditopic and tritopic molecules associated by directional interactions (represented by dotted lines). **(b)** A mixture of fatty diacid and triacid is condensed first with diethylene triamine and then reacted with urea giving a mixture of oligomers equipped with complementary hydrogen bonding groups: amidoethyl imidazolidone, di(amidoethyl) urea and diamido tetraethyl triurea (Cordier et al. 2008). Reprinted with permission from ref. (Cordier et al. 2008). Copyright 2008 Nature Publishing Group.

Inspired by the pioneering works from Leibler group, enormous efforts have been devoted to develop self-healing polymers using hydrogen bonds (Tournilhac et al. 2011, Montarnal et al. 2008, Montarnal et al. 2009, Maes et al. 2012). One notable example is polymeric networks based on ureido-pyrimidinone (UPy) moieties. UPys form quadruple hydrogen bonding, adding another level of tunability of hydrogen bonding strength (Burattini et al. 2010, Brunsveld et al. 2001). However, UPy molecules are highly immiscible with other polymers due to their anisotropic structure; in fact, they often form ordered structure, such as Pi-Pi stacks, and phase separate from other polymers in the matrix, resulting in an inhomogenous material and thus may impair the mechanical properties of the materials.

Ionic associations. One type of self-healing materials based on reversible ionic associations is poly (ethylene-co-methacrylic acid) copolymer, as illustrated in Fig. 11 (Dirama et al. 2008, Fall 2001, Kalista 2003, 2007, Kalista and Ward 2007, Varley and Zwaag 2008). To some extent, these ionomer films are not autonomic self-healing systems since their self-repair is achieved after projectile puncture. The penetration of these films by a projectile causes localized heating near the puncture and thus the ionomer material heals as a result of multiplet formation at a higher temperatures in the melt state (Fall 2001, Kalista 2003). Both experimental (Kalista 2007, Kalista and Ward 2007, Varley and Zwaag 2008, Varley and van der Zwaag 2008, Kalista et al. 2013) and simulation works (Dirama et al. 2008) are conducted to understand the self-healing mechanism of ionomers.

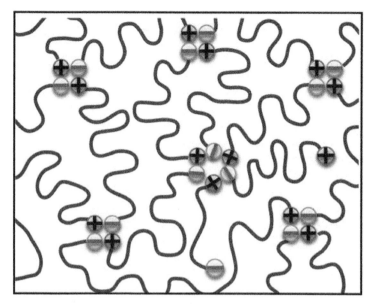

Figure 11. Self-healing polymer networks form based on ionic associations.

Hybrid self-healing systems. Self-healing polymeric materials based on reversible associations are often of poor mechanical performance due to the relatively weak reversible bonds compared to permanent covalent bonds. To improve the mechanical performance, stronger bonds are needed; a system that includes both weaker and stronger bonds is termed hybrid self-healing polymeric materials. One example is self-healing thermoplastic elastomers based on multiphase design, in which there are hard and soft phases, as illustrated in Fig. 12 (Chen et al. 2012). The soft phase contains polymers that carry reversible associations, enabling dynamic healing. In contrast, the hard phase behaves like permanent crosslinks below glass transition temperature. Similar to the concept of multiphase design, covalent crosslinks based on the dynamic olefin metathesis reaction can be incorporated into polymeric systems formed by reversible bonds (Neal et al. 2015). Another type of hybrid system is a self-healing polymer gel; the gel contains a covalently cross-linked permanent network onto which reversible metal-ligand coordinations are added (Kersey et al. 2007, Yount et al. 2005).

3.3 Theoretical Understanding

Autonomic self-healing materials have shown great promise as a new class of sustainable materials. Designing materials of better healing efficiency and mechanical performance requires better understanding of the healing

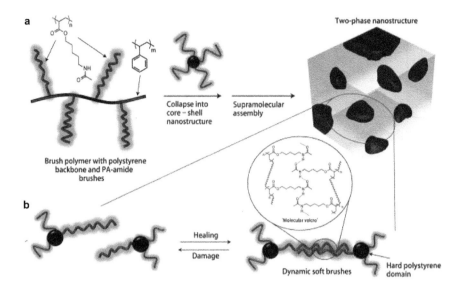

Figure 12. Design concept of self-healing polymers based on multiphases. **(a)** The hydrogen-bonding brush polymer self-assembles into a two-phase nanostructure morphology during processing. Polyvalent clusters of covalently linked associative hydrogen bonding interactions result in a mechanically stable connection that behaves like a permanent covalent linkage at short time scales. **(b)** The supramolecular connections between soft brushes can break and reform, enabling self-healing feature. Reprinted with permission from ref. (Chen et al. 2012). Copyright 2012 Nature Publishing Group.

mechanism; in particular, the dependence of healing efficiency on healing time, bond strength, and temperature and the dependence of mechanical performance on these materials. However, compared to enormous embodiment of experimental work, relatively less theoretical understanding (Hoy and Fredrickson 2009, Stukalin et al. 2013) has been developed for self-healing materials, in part due to the complexity in structure of the self-healing polymer networks.

To develop a predictive model for self-healing polymer networks, a good starting point is to consider model polymer networks formed by reversible associations. Polymers in reversible networks carry associative stickers. The dynamics of such polymers is largely determined by the number fraction and the average lifetime of associated groups (closed stickers) (Rubinstein and Semenov 1998, Semenov and Rubinstein 1998, Rubinstein and Semenov 2001). To understand the dynamics, various models have been developed; these include sticky Rouse model (Leibler et al. 1991, 1993, Rubinstein and Semenov 2001) that describes the dynamics of unentangled reversible networks and sticky reptation model for entangled case (Leibler et al. 1993, Rubinstein and Semenov 2001). Based on these works, the Rubinstein group developed a systematic theoretical description of the self-healing of unentangled reversible networks recently. In this work, a simple model of autonomic self-healing of unentangled polymer networks is considered. In this model one of the two end monomers of each polymer chain is fixed in space mimicking dangling chains attached to a polymer network, while the sticky monomer at the other end of each chain can form pairwise reversible bond with the sticky end of another chain, as shown in Fig. 13a-c.

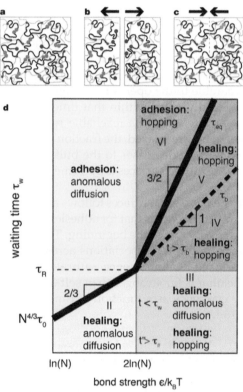

Figure 13. Polymer network capable of autonomous self-repair. **(a)** The ends of dangling chains carry groups that form reversible pairwise associations. **(b)** After the material is broken there are many reactive groups near the fractured surfaces. A significant fraction of these groups survives after extensive waiting time. **(c)** Many bonds are formed across the interface as the two fractured surfaces are brought into contact. Thin grey lines represent permanently cross-linked strands of the polymer network. **(d)** The diagram of kinetic regimes for healing processes depends on the waiting time, τ_w, after cut and bond strength, ϵ, which determines the equilibrium number density of open stickers. The thick solid line represents the bulk equilibration time, τ_{eq}, as function of bond strength (for $k_B T \ln N < \epsilon < 2k_B T \ln N$ and for $\epsilon > 2k_B T \ln N$, in which N is the length of polymers carrying associative groups), whereas the dashed line corresponds to the lifetime of the bond, τ_b, as a function of the strength of the bond. Thin solid lines correspond to the boundaries between different regimes. Vertical axis is logarithmic while the horizontal axis is linear. Reprinted with permission from ref. (Stukalin et al. 2013). Copyright 2013 America Chemical Society.

When two fractured surfaces are brought into contact immediately after break, the number of reformed reversible associations across the interface increases with healing time. For healing time shorter than the lifetime of an open sticker, a few associations are formed across the interface, as most open stickers are not able to find partners to link to within that short period. With the healing time increasing up to renormalized bond lifetime, the number of reformed bonds across the interface increases. At healing times comparable to chain relaxation time, the whole system relaxes and thus the number of reformed bonds across the interface saturates to the density of the un-cut sample.

If two fractured surfaces are not mended immediately but brought into contact after some waiting time, the number of available open stickers around fractured surfaces decreases (Cordier et al. 2008). At relatively short waiting time comparable to the open sticker lifetime, the fraction of open stickers decreases rapidly, since within that time most open stickers will find partners to associate with in a reachable region. At very long time scales when the polymers are relaxed, the fraction of available open stickers around a fractured surface decreases to the bulk value. In this situation, self-healing between fractured surfaces is the same as self-adhesion between unfractured surfaces.

Adhesion between two bare surfaces follows a similar process as that in self-healing. The only difference is that for adhesion the fraction of available open stickers is extremely low at the beginning. Thus for self-adhesion the maximum number of reformed associations across the interface is much smaller than that for self-healing. Hence, saturated recovered mechanical strength for self-healing is far stronger than self-adhesion. A full diagram of the self-healing efficiency as a function of time and bonding energy is presented in Fig. 13d.

4. Applications of Self-Healing Polymers

While the development of self-healing materials is still ongoing, their applications have been explored in various aspects. The ultimate target for self-healing polymers is probably to serve as smart protective coating for surface of devices and instruments, as well as structural materials in space, automobile, defense, and construction industries, in which materials capable of repairing themselves upon cracking are of particular importance in the safety and elongation of service time. We can even envisage that self-healing materials will be as smart as biological systems, which can respond to complex damage, renewing them over a period of long time. Due to the broad potential applications, the World Economic Forum in 2013 named self-healing materials as one of the top 10 emerging technologies. It was pointed out by the forum that a growing trend in biomimicry is the creation of non-living structural materials that also have the capacity to heal themselves when cut, torn or cracked. Self-healing materials which can repair damage without external intervention could give manufactured goods a longer lifetime and reduce the demand for raw materials, as well as improving the inherent safety of materials used in construction or to form the bodies of aircraft (World Economic Forum). In this section we discuss the current and potential applications of self-healing materials, particularly as protective surface coating and advanced bulk materials in biomedical applications.

4.1 Self-Healing Polymers for Surface Coating

Surface coating is probably the first step toward the application of self-healing materials. To date, various types of self-healing coatings have been developed, which possess the ability to heal structural damage and allow recovery of functions such as hydrophobicity, resistance to friction and corrosion (Binder 2013). For instance, the pioneering works of self-healing polymers from the group of Scott White at the University of Illinois and the groups of Bert Meijer and Rint Sijbesma at the Eindhoven University of Technology have led to start-up companies aiming at the applications of self-healing materials. The Illinois group founded Autonomic Materials in 2005. This company has developed three series of self-healing materials for coatings, including elastomeric, thermosetting and powder coatings. The elastomeric coatings incorporate a dual-capsule system based on the microencapsulation of vinyl-terminated PDMS resins and their corresponding curing agents. The size and concentration of these capsules are optimized to leverage the ability of elastomers to stretch for a more seamless repair of damage. The thermosetting coatings incorporate a dual-capsule system based on the microencapsulation of a proprietary blend of hydroxyl-terminated PDMS, silyl ethers, and various viscosity modifiers in one type of microcapsule and a corresponding catalyst in a second type of microcapsule. The powder coatings incorporate a dual-capsule system based on the microencapsulation of proprietary blends of PDMS resins and corresponding curing agents (AutonomicMaterials). The Eindhoven group co-founded SupraPolix; the key innovation of the company is based on the SupraB™ quadruple hydrogen bonding unit (also known as UPy or UreidoPyrimidone). Functionalization of polymers with UPy units leads to very strong interactions between the polymer chains. These interactions are reversible due to the sensitivity of hydrogen bonding to temperature, thus the materials self-heal after a shot heating cycle, or even more particularly, can repair themselves after damage by just reassembling the broken polymeric parts together. The self-healing materials have potential applications as adhesives, coatings, and elastomeric materials, which are beginning to reach the market (Suprapolix).

Many international companies also show interests in self-healing materials. For example, Nisan Motor Corporation designed a clear coating with a trade mark of Scratch Shield®. This coating is capable of repairing fine scratches, restoring painted surfaces close to their original state from one day to one week (NIssanMotorCorporation). Bayer Material Science developed polyurethane (PU) self-healing coatings utilize memory effect of polymers to repair scratches. In the PU coatings, fixed covalent bonds provide hardness and resistance, while reversible hydrogen bonds impart healing ability. When a scratch occurs, the polymer chains are physically displaced. The hydrogen bonds are disconnected and looking for new anchors, whereas the fixed chemical bonds are unaffected. Heat treatment, such as when a car sits out in the sun, will cause the material to return to their original shape due to the memory effect triggered by covalent bonds, and at the same time the disconnected hydrogen bonds will re-bond to form new hydrogen bonds. Consequently, a scratch on the material will disappear (BayerMaterialScience). NEI Corporation has developed a class of self-healing polymer nanocomposite coatings with a trade name of NANOMYTE®MEND. The coatings can heal both surface scratches and mesoscopic damage (e.g., micro-cracks and cavitation) by a two-step process: gap closure, followed by healing. Simultaneous crack closing and sealing is achieved by heating the surface of the coating, using a simple device such as a heat gun (NEICorporation).

4.2 Self-Healing Polymers for Biomedical Applications

One of the most amazing abilities of many human tissues, such as skin and bones, is their outstanding ability to self-heal, to regenerate and restore the structural and functional properties upon destruction caused by over-loading, trauma or diseases. Meanwhile, this considerable self-management of damage is the most important source of inspirations for the design of synthetic and artificial materials/devices capable of reversing the damage development, which in turn can serve as biomedical materials for repair and reconstruction of defective tissues/organs. In this section, we discuss the state-of-art of self-healing materials for biomaterials applications, and propose design principles for self-healing biomaterials with the potential applications as *in vivo* implantations.

4.2.1 Self-Healing Biopolymer Implants. Self-healing biopolymers, mainly hydrogels, are based on the direct association of molecular components through specific, noncovalent crosslinks within singular polymer (Zhang et al. 2012) or between binary or ternary polymers (Zhang and He 2015), including hydrogen bonds (Zhang et al. 2012), electrostatic or hydrophobic interactions, heterodimeric peptide/protein bindings (Lu, Charati et al. 2012, Foo et al. 2009), and guest-host interactions (Rodell, MacArthur et al. 2015). Among these physical interactions, hydrogen bonding is mostly used to create self-healing supramolecular materials. Self-healing hydrogels consist of single phase of biocompatible poly(vinyl alcohol) (PVA) was demonstrated by Zhang et al. (Zhang et al. 2012). The high concentration of free hydroxyl groups within PVA backbones facilitate physical crosslinked network that can heal without any external intervention. Other hydrogen bond-based hydrogels, such as poly(acryloyl-6-aminocaproic acid) (PA6ACA) (Phadke et al. 2012) and poly(DMAEMA-co-SCMHBMA) hydrogel (Cui et al. 2013) have been shown to possess self-healing ability, However, the applicability of these materials as *in vivo* implants are limited by cytotoxicity, compromised performance in physiological environment (Phadke et al. 2012), inert biological properties and non-biodegradability (Zhang et al. 2012).

More biocompatible supramolecular gels are developed by Rodell et al. using separate pendant modifications of biodegradable hyaluronic acid (HA) by the guest-host pair cyclodextrin and adamantine. This type of hydrogel displays shear-thinning and self-healing properties due to the reversible physical bonds. *In vivo* study has been conducted by applying such hydrogels as injectable biomaterials in porcine myocardium and rat epicardial infarct model, which shows desirable biocompatibility and capacity of delivering therapeutic components, thereby indicating their potential applications in soft tissue reconstruction (Rodell, MacArthur et al. 2015, Rodell, Wade et al. 2015).

4.2.2 Self-Healing Composite Implants. Although self-healing polymer-based biomaterials have exhibited promising properties for wide-spread applications in tissue regeneration, most of them possess relatively low mechanical strength, thereby limiting their further usage in load-bearing *in vivo* conditions such as regeneration of skeleton tissues. Therefore, composite materials with inherent ability of autonomic self-repair have recently been developed. Wang et al. invented a novel class of self-healing hydrogels formed by non-covalent interaction resulting from the specific design of a telechelic dendritic or poly (ethylene glycol) (PEG) molecules with multiple adhesive termini for binding to nanosized clay particles (Wang et al. 2010, Tamesue et al. 2013). By taking advantage of anionic-cationic interaction between inorganic and organic phase, a novel class of composite hydrogels with considerably strong mechanical strength, with elastic modulus on the order of 100 kPa, and rapid self-healing ability after destruction have been synthesized. Considering the desirable biocompatibility of PEG and osteogenicity of clay nanoparticles (Gaharwar et al. 2013), these composite biomaterials are of great potential for bone tissue regeneration.

To generate self-healing hydrogels for hard tissue regeneration, another strategy is to employ biominerals, such as calcium phosphate (CaP) and bio-glass, that can induce bone formation. Nejadnik et al. creatively developed a novel class of composite hydrogels composed of CaP nanoparticles and HA functionalized by bisphosphonate (BP), a small molecule drugs that have been widely used to treat osteoporosis due to its capability of forming strong coordination bonds with bone minerals CaP (Nejadnik et al. 2014). The reversible feature of CaP-BP interactions facilitated shear-thinning injection and rapid self-healing after network destruction. More importantly, *in vivo* animal experiments showed high target site retention after 4 week subcutaneous implantation, and improved bone regeneration after 4 weeks implantation in bone defect model.

Another self-healing biocomposite material is based on a bulk hydrogel system containing lipid vesicles loaded with Ca^{2+} ions as the healing agents for alginate matrices to be cross-linked. By introducing thermal sensitive liposomes encapsulated with healing agents, the hydrogels can be healed at physiological temperature. However, since the instability of the liposomes, such materials systems were incapable of multiple healing processes, which limited its applications for long-term *in vivo* implantation (Westhaus and Messersmith 2001).

4.3.3 Self-Healing Colloidal Gel Implants. Colloidal gels, a novel class of hydrogels that allow for a "bottom-up" approach for the design of biomaterials by employing biopolymer particles as building blocks, have recently emerged as an intriguing system in regenerative medicine. By employing non-covalent inter-particle associations including hydrophobic (Van Tomme 2007), electrostatic (Wang, Hansen et al. 2011, Wang et al. 2008) or magnetic (Xu et al. 2011) interactions, physically cross-linked network can be formed, which typically showed a shear-thinning and self-healing behavior (Wang, Leeuwenburgh et al. 2011). Colloidal hydrogels composed of oppositely charged gelatin nanoparticles have been developed recently and demonstrated to be shear-thinning and self-healing, which can be attributed to the fast and reversible re-establishment of the electrostatic interactions between anionic and cationic nanoparticles as well as re-arrangement of nanosphere packing upon gel equilibration (Wang, Hansen et al. 2011) (Fig. 14). Due to the use of nanoparticle which can also serve as delivery vehicles for controllable release of drugs for therapeutic purposes, the gels showed substantial capability of programmable release of multiple biomolecules by fine-tuning individual particle degradation rate

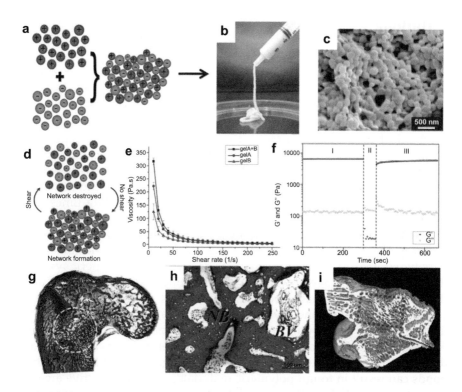

Figure 14. Schematic diagram **(a)** and resulting photographs **(b)** of injectable colloidal gels based on using oppositely charged gelatin nanospheres as building blocks. **(c)** Scanning electron microscopic images of the microstructure of the colloidal gels. **(d)** Schematic diagram showing the mechanism of shear-thinning and self-healing behavior of colloidal gels. **(e)** Shear-thinning of colloidal gels showing the decrease of gel viscosity as a function of shear rate. **(f)** Gel recovery after gel destruction assessed by monitoring G' and G" as a function of time: region I, II and III shows initial gel strength, gel destruction and gel recovery, respectively, of colloidal gels consisting oppositely charged gelatin nanosphere gels (15 w/v% solid content). Histological **(g, h)** and microCT **(i)** analyses of rat femoral condyles bone defect filled with growth factor-loaded colloidal gelatin gels 4 weeks after implantation. **(g, h)** Histological sections of the rat femoral condyle defects at low and high magnifications. The circular region of interest (ROI) at the defect site was schematically depicted in yellow (diameter of 3 mm). NB = new bone, BV = blood vessel-like structure. **(i)** 3D reconstructions of the explanted femoral condyles after microCT analysis. Cylinder-shaped defects are depicted in red. Each defect was sectioned in the middle of the defect along the longitudinal axis. Reprinted with permission from ref. (Wang, Hansen et al. 2011). Copyright 2011 Wiley-VCH Verlag GmbH & Co.

(Wang et al. 2012, Wang et al. 2013). Further *in vivo* studies confirmed their decent biological properties of the injectable gels as well as capacity to accelerate bone regrowth even in load-bearing physiological condition (van der Stok et al. 2013).

5. Outlook

Self-healing polymeric materials are a new class of materials that would significantly improve the safety, lifetime, energy efficiency and environmental impact of man-made materials. The research has greatly enriched the types of self-healing materials in the past 20 years. However, it remains a challenge to design self-healing polymeric materials with mechanical properties of practical usage, largely due to the intrinsic weakness of reversible bonds or the complexity in designing healing agents. For instance, most autonomic self-healing polymeric materials based on reversible bonds have fracture stress on the order of 1 MPa, and materials based on encapsulated healing agents are of limited healing times and expensive fabrication process.

A future direction of self-healing materials is to improve their mechanical performance such that they can be comparable to existing elastomers. A possible strategy to achieve this goal is to design hybrid polymeric systems that take the advantage of wisdom from existing polymeric systems and that of dynamic reversible polymeric materials. These systems would contain both covalent and reversible associations; covalent crosslinks are likely the major contributor to the mechanical properties of the materials, whereas reversible bonds, in addition to enable self-healing capability, can dissipate energy effectively. Energy dissipation introduced by reversible bonds can result in tough polymeric materials, which is an active area of current research. Indeed, the idea of introducing reversible bonds as energy dissipation mechanisms have been realized in hydrogels (Sun et al. 2012, Zhao 2014), but not yet for 'dry', solvent-free elastomers, largely due to the immiscibility of polar reversible bonds and non-polar covalent bonds.

The design of self-healing polymeric materials requires better understanding of the self-healing mechanisms at the molecular scale. The theory proposed by Rubinstein group represents an advance in understanding the healing mechanism of autonomic self-healing materials; however, for a proper understanding of the mechanical performance of self-healing materials, the effect of covalent bonds needs to be taken into account. In addition, existing theories apply only to elastomer rather than gels. More sophisticated, quantitative models are required to advance the design of self-healing materials of better properties.

Self-healing polymeric gels represent another future of self-healing polymeric materials. Gels are of particular importance as potential advanced materials in the biomedical applications. However, it remains a challenge to balance the biocompatibility and mechanical performance of self-healing gels, which significantly limits their clinical applications. Strategies that can circumvent these difficulties would open new avenue of the reach in self-healing polymeric materials.

References

Amamoto, Yoshifumi, Jun Kamada, Hideyuki Otsuka, Atsushi Takahara and Krzysztof Matyjaszewski. 2011. Repeatable photoinduced self-healing of covalently cross-linked polymers through reshuffling of trithiocarbonate units. Angewandte Chemie 123(7): 1698–1701.

Amamoto, Yoshifumi, Hideyuki Otsuka, Atsushi Takahara and Krzysztof Matyjaszewski. 2012. Self-healing of covalently cross-linked polymers by reshuffling thiuram disulfide moieties in air under visible light. Advanced Materials 24(29): 3975–3980.

AutonomicMaterials. http://www.autonomicmaterials.com/]. Available from http://www.autonomicmaterials.com/.

Bai, N., K. Saito and G.P. Simon. 2013. Synthesis of a diamine cross-linker containing Diels–Alder adducts to produce self-healing thermosetting epoxy polymer from a widely used epoxy monomer. Polym. Chem. 4(3): 724–730.

BaLazs, A.C. 2007. Modelling self-healing materials. Mater. Today 10(9): 18–23.

Balazs, A.C., T. Emrick and T.P. Russell. 2006. Nanoparticle polymer composites: where two small worlds meet. Science 314(5802): 1107–1110.

BayerMaterialScience. http://www.coatings.bayer.com/en/Technologies/Functional-Coatings/Self-Healing-Coatings.asp]. Available from http://www.coatings.bayer.com/en/Technologies/Functional-Coatings/Self-Healing-Coatings.asp.

Bergman, S.D. and F. Wudl. 2008. Mendable polymers. J. Mater. Chem. 18(1): 41–62.

Binder, W.H. 2013. Self-healing Polymers: From Principles to Applications: John Wiley & Sons.

Blaiszik, B.J., M.M. Caruso, D.A. McIlroy, J.S. Moore, S.R. White and N.R. Sottos. 2009. Microcapsules filled with reactive solutions for self-healing materials. Polymer 50(4): 990–997.

Blaiszik, B.J., N.R. Sottos and S.R. White. 2008. Nanocapsules for self-healing materials. Compos. Sci. Technol. 68(3-4): 978–986.

Brown, E.N., M.R. Kessler, N.R. Sottos and S.R. White. 2003. *In situ* poly(urea-formaldehyde) microencapsulation of dicyclopentadiene. J. Microencapsulation 20(6): 719–730.

Brown, E.N., N.R. Sottos and S.R. White. 2002. Fracture testing of a self-healing polymer composite. Exp. Mech. 42(4): 372–379.

Brown, E.N., S.R. White and N.R. Sottos. 2005. Retardation and repair of fatigue cracks in a microcapsule toughened epoxy composite—Part I: Manual infiltration. Compos. Sci. Technol. 65(15-16 SPEC. ISS.): 2466–2473.

Brown, E.N., S.R. White and N.R. Sottos. 2005. Retardation and repair of fatigue cracks in a microcapsule toughened epoxy composite—Part II: *In situ* self-healing. Compos. Sci. Technol. 65(15-16 SPEC. ISS.): 2474–2480.

Brunsveld, L., B.J.B. Folmer, E.W. Meijer and R.P. Sijbesma. 2001. Supramolecular polymers. Chemical Reviews 101(12): 4071–4097.

Burattini, S., B.W. Greenland, D.H. Merino, W. Weng, J. Seppala, H.M. Colquhoun, W. Hayes, M.E. Mackay, I.W. Hamley and S.J. Rowan. 2010. A healable supramolecular polymer blend based on aromatic pi-pi stacking and hydrogen-bonding interactions. J. Am. Chem. Soc. 132(34): 12051–12058.

Cardenas-Daw, C., A. Kroeger, W. Schaertl, P. Froimowicz and K. Landfester. 2012. Reversible photocycloadditions, a powerful tool for tailoring (Nano) materials. Macromol. Chem. Phys. 213(2): 144–156.

Caruso, M.M., B.J. Blaiszik, S.R. White, N.R. Sottos and J.S. Moore. 2008. Full recovery of fracture toughness using a nontoxic solvent-based self-healing system. Adv. Funct. Mater. 18(13): 1898–1904.

Caruso, M.M., D.A. Delafuente, V. Ho, N.R. Sottos, J.S. Moore and S.R. White. 2007. Solvent-promoted self-healing epoxy materials. Macromolecules 40(25): 8830–8832.

Chen, X., M.A. Dam, K. Ono, A. Mal, H. Shen, S.R. Nutt, K. Sheran and F. Wudl. 2002. A thermally re-mendable cross-linked polymeric material. Science 295(5560): 1698–1702.

Chen, X., F. Wudl, A.K. Mal, H. Shen and S.R. Nutt. 2003. New thermally remendable highly cross-linked polymeric materials. Macromolecules 36(6): 1802–1807.

Chen, Y.L., A.M. Kushner, G.A. Williams and Z.B. Guan. 2012. Multiphase design of autonomic self-healing thermoplastic elastomers. Nat. Chem. 4(6): 467–472.

Cho, S.H., H.M. Andersson, S.R. White, N.R. Sottos and P.V. Braun. 2006. Polydimethylsiloxane-based self-healing materials. Adv. Mater. 18(8): 997–1000.

Chujo, Y., K. Sada and T. Saegusa. 1990. Reversible gelation of polyoxazoline by means of Diels-Alder reaction. Macromolecules 23(10): 2636–2641.

Chung, C.-M., Y.-S. Roh, S.-Y. Cho and J.-G. Kim. 2004. Crack healing in polymeric materials via photochemical [2 + 2] cycloaddition. Chem. Mater. 16(21): 3982–3984.

Cordier, P., F. Tournilhac, C. Soulie-Ziakovic and L. Leibler. 2008. Self-healing and thermoreversible rubber from supramolecular assembly. Nature 451(7181): 977–80.

Corten, C.C. and M.W. Urban. 2009. Repairing Polymers using oscillating magnetic field. Adv. Mater. 21(48): 5011–5015.

Cosco, S., V. Ambrogi, P. Musto and C. Carfagna. 2007. Properties of poly(urea-formaldheyde) microcapsules containing an epoxy resin. J. Appl. Polym. Sci. 105(3): 1400–1411.

Craven, J.M. 1969. Cross-linked thermally reversible polymers produced from condensation polymers with pendant furan groups cross-linked with maleimides. Google Patents.

Cui, Jiaxi, Verónica San Miguel and Aranzazu del Campo. 2013. Light - triggered multifunctionality at surfaces mediated by photolabile protecting groups. Macromolecular Rapid Communications 34(4): 310–329.

Dirama, T.E., V. Varshney, K.L. Anderson, J.A. Shumaker and J.A. Johnson. 2008. Coarse-grained molecular dynamics simulations of ionic polymer networks. Mech. Time-Depend. Mater. 12(3): 205–220.

Dry, C. 1994. Smart multiphase composite-materials which repair themselves by a release of liquids which become solids. Smart Mater. 2189: 62–70.

Dry, C. 1996. Procedures developed for self-repair of polymer matrix composite materials. Compos. Struct. 35(3): 263–269.

Dry, C.M. 1992. Smart materials which sense, activate and repair damage; hollow porous fibers in composites release chemicals from fibers for self-healing, damage prevention, and/or dynamic control. In First European Conference on Smart Structures and Materials I First European Conference on Smart Structures and Materials.

Dry, C.M. and N.R. Sottos. 1993. Passive smart self-repair in polymer matrix composite materials. In 1993 North American Conference on Smart Structures and Materials: International Society for Optics and Photonics.

Dry, C. and N.R. Sottos. 1993. Passive Smart self-repair in polymer matrix composite-materials. Smart Mater. 1916: 438–444.

Du, P., X. Liu, Z. Zheng, X. Wang, T. Joncheray and Y. Zhang. 2013. Synthesis and characterization of linear self-healing polyurethane based on thermally reversible Diels–Alder reaction. RSC Adv. 3(35): 15475–15482.

Epstein, F.H., A.J. Singer and R.A.F. Clark. 1999. Cutaneous wound healing. N. Engl. J. Med. 341(10): 738–746.

Fall, Rebecca. 2001. Puncture Reversal of Ethylene Ionomers - Mechanis studies. Thesis, University Libraries Virginia Polytechnic Institute and State University, Blacksburg, Va.

Foo, Cheryl TS Wong Po, Ji Seok Lee, Widya Mulyasasmita, Andreina Parisi-Amon and Sarah C. Heilshorn. 2009. Two-component protein-engineered physical hydrogels for cell encapsulation. Proceedings of the National Academy of Sciences 106(52): 22067–22072.

Froimowicz, P., H. Frey and K. Landfester. 2011. Towards the generation of self-healing materials by means of a reversible photo-induced approach. Macromol. Rapid Commun. 32(5): 468–473.

Gaharwar, A.K., S.M. Mihaila, A. Swami, A. Patel, S. Sant, R.L. Reis, A.P. Marques, M.E. Gomes and A. Khademhosseini. 2013. Bioactive silicate nanoplatelets for osteogenic differentiation of human mesenchymal stem cells. Advanced Materials 25(24): 3329–3336.

Ghosh, B., K.V. Chellappan and M.W. Urban. 2012. UV-initiated self-healing of oxolane–chitosan–polyurethane (OXO–CHI–PUR) networks. J. Mater. Chem. 22(31): 16104–16113.

Ghosh, B. and M.W. Urban. 2009. Self-repairing oxetane-substituted chitosan polyurethane networks. Science 323(5920): 1458–1460.

Gupta, S., Q.L. Zhang, T. Emrick, A.C. Balazs and T.P. Russell. 2006. Entropy-driven segregation of nanoparticles to cracks in multilayered composite polymer structures. Nat. Mater. 5(3): 229–233.

Hansen, C.J., W. Wu, K.S. Toohey, N.R. Sottos, S.R. White and J.A. Lewis. 2009. Self-healing Materials with Interpenetrating Microvascular Networks. Adv. Mater. 21(41): 4143–+.

Hoveyda, A.H. and A.R. Zhugralin. 2007. The remarkable metal-catalysed olefin metathesis reaction. Nature 450(7167): 243–251.

Hoy, R.S. and G.H. Fredrickson. 2009. Thermoreversible associating polymer networks. I. Interplay of thermodynamics, chemical kinetics, and polymer physics. J. Chem. Phys. 131(22).

Huang, L., N. Yi, Y. Wu, Y. Zhang, Q. Zhang, Y. Huang, Y. Ma and Y. Chen. 2013. Multichannel and repeatable self-healing of mechanical enhanced graphene-thermoplastic polyurethane composites. Adv. Mater. 25(15): 2224–8.

Inglis, A.J., L. Nebhani, O. Altintas, F.G. Schmidt and C. Barner-Kowollik. 2010. Rapid bonding/debonding on demand: reversibly cross-linked functional polymers via Diels–Alder chemistry. Macromolecules 43(13): 5515–5520.

Jackson, A.C., J.A. Bartelt, K. Marczewski, N.R. Sottos and P.V. Braun. 2011. Silica-protected micron and sub-micron capsules and particles for self-healing at the microscale. Macromol. Rapid Commun. 32(1): 82–87.

Kalista, S.J. 2003. Self-healing of thermoplastic poly(ethylene-co-methacrylic acid) copolymers following projectile puncture. Master, University Libraries Virginia Polytechnic Institute and State University, Blacksburg, Va.

Kalista, S.J., Jr., T.C. Ward and Z. Oyetunji. 2007. Self-healing of poly(ethylene-co-methacrylic acid) copolymers following projectile puncture. Mech. Adv. Mater. Struct. 14(5): 391–397.

Kalista, S.J., J.R. Pflug and R.J. Varley. 2013. Effect of ionic content on ballistic self-healing in EMAA copolymers and ionomers. Polym. Chem. 4(18): 4910–4926.

Kalista, S.J. and T.C. Ward. 2007. Thermal characteristics of the self-healing response in poly (ethylene-co-methacrylic acid) copolymers. J. R. Soc. Interface 4(13): 405–411.

Kavitha, A.A. and N.K. Singha. 2009. "Click chemistry" in tailor-made polymethacrylates bearing reactive furfuryl functionality: a new class of self-healing polymeric material. ACS Appl. Mater. & Interface 1(7): 1427–1436.

Kavitha, A.A. and N.K. Singha. 2010. Smart "all acrylate" ABA triblock copolymer bearing reactive functionality via atom transfer radical polymerization (ATRP): demonstration of a "click reaction" in thermoreversible property. Macromolecules 43(7): 3193–3205.

Kennedy, J.P. and K.F. Castner. 1979. Thermally reversible polymer systems by cyclopentadienylation. I. A model for termination by cyclopentadienylation of olefin polymerization. J. Polym. Sci.: Polym. Chem. Ed. 17(7): 2039–2054.

Kersey, F.R., D.M. Loveless and S.L. Craig. 2007. A hybrid polymer gel with controlled rates of cross-link rupture and self-repair. J. R. Soc. Interface 4(13): 373–380.

Kessler, M.R., N.R. Sottos and S.R. White. 2003. Self-healing structural composite materials. Compos. Part A 34(8): 743–753.

Lee, J.Y., G.A. Buxton and A.C. Balazs. 2004. Using nanoparticles to create self-healing composites. J. Chem. Phys. 121(11): 5531–5540.

Leibler, L., M. Rubinstein and R.H. Colby. 1991. Dynamics of reversible networks. Macromolecules 24(16): 4701–4707.

Leibler, L., M. Rubinstein and R.H. Colby. 1993. Dynamics of telechelic ionomers—can polymers diffuse large distances without relaxing stress. J. Phys. II 3(10): 1581–1590.

Ling, J., M.Z. Rong and M.Q. Zhang. 2012. Photo-stimulated self-healing polyurethane containing dihydroxyl coumarin derivatives. Polymer 53(13): 2691–2698.

Ling, Jun, Min Zhi Rong and Ming Qiu Zhang. 2011. Coumarin imparts repeated photochemical remendability to polyurethane. J. Mater. Chem. 21(45): 18373–18380.

Liu, X., X. Sheng, J.K. Lee and M.R. Kessler. 2009. Synthesis and characterization of melamine-urea-formaldehyde microcapsules containing ENB-based self-healing agents. Macromol. Mater. Eng. 294(6-7): 389–395.

Liu, Y.L. and Y.W. Chen. 2007. Thermally reversible cross-linked polyamides with high toughness and self-repairing ability from maleimide- and furan-functionalized aromatic polyamides. Macromol. Chem. Phys. 208(2): 224–232.

Lu, Hoang D., Manoj B. Charati, Iris L. Kim and Jason A. Burdick. 2012. Injectable shear-thinning hydrogels engineered with a self-assembling Dock-and-Lock mechanism. Biomaterials 33(7): 2145–2153.

Lu, Y.-X., F. Tournilhac, L. Leibler and Z. Guan. 2012. Making insoluble polymer networks malleable via olefin metathesis. J. Am. Chem. Soc. 134(20): 8424–8427.

Lu, Y.X. and Z. Guan. 2012. Olefin metathesis for effective polymer healing via dynamic exchange of strong carbon-carbon double bonds. J. Am. Chem. Soc. 134(34): 14226–31.

Maes, F., D. Montarnal, S. Cantournet, F. Tournilhac, L. Corte and L. Leibler. 2012. Activation and deactivation of self-healing in supramolecular rubbers. Soft Matter 8(5): 1681–1687.

Martin, P. 1997. Wound healing—aiming for perfect skin regeneration. Science 276(5309): 75–81.

Montarnal, D., P. Cordier, C. Soulie-Ziakovic, F. Tournilhac and L. Leibler. 2008. Synthesis of self-healing supramolecular rubbers from fatty acid derivatives, diethylenetriamine, and urea. J. Polym. Sci. Part A-Polym. Chem. 46(24): 7925–7936.

Montarnal, D., F. Tournilhac, M. Hidalgo, J.-L. Couturier and L. Leibler. 2009. Versatile one-pot synthesis of supramolecular plastics and self-healing rubbers. J. Am. Chem. Soc. 131(23): 7966–+.

Murphy, E.B., E. Bolanos, C. Schaffner-Hamann, F. Wudl, S.R. Nutt and M.L. Auad. 2008. Synthesis and characterization of a single-component thermally remendable polymer network: Staudinger and Stille revisited. Macromolecules 41(14): 5203–5209.

Murphy, E.B. and F. Wudl. 2010. The world of smart healable materials. Prog. Polym. Sci. 35(1-2): 223–251.

Neal, J.A., D. Mozhdehi and Z. Guan. 2015. Enhancing mechanical performance of a covalent self-healing material by sacrificial noncovalent bonds. J. Am. Chem. Soc.

NEICorporation. http://neicorporation.com/products/coatings/self-healing-coatings/]. Available from http://neicorporation.com/products/coatings/self-healing-coatings/.

Nejadnik, M.R., X. Yang, M. Bongio, H.S. Alghamdi, J.J. Van den Beucken, M.C. Huysmans, J.A. Jansen, J. Hilborn, D. Ossipov and S.C. Leeuwenburgh. 2014. Self-healing hybrid nanocomposites consisting of bisphosphonated hyaluronan and calcium phosphate nanoparticles. Biomaterials 35(25): 6918–6929.

Neuser, S., V. Michaud and S.R. White. 2012. Improving solvent-based self-healing materials through shape memory alloys. Polymer 53(2): 370–378.

NIssanMotorCorporation. http://www.nissan-global.com/EN/TECHNOLOGY/OVERVIEW/scratch.html]. Available from http://www.nissan-global.com/EN/TECHNOLOGY/OVERVIEW/scratch.html.

Pang, J.W.C. and I.P. Bond. 2005. A hollow fibre reinforced polymer composite encompassing self-healing and enhanced damage visibility. Compos. Sci. Technol. 65(11-12): 1791–1799.

Park, J.S., K. Takahashi, Z. Guo, Y. Wang, E. Bolanos, C. Hamann-Schaffner, E. Murphy, F. Wudl and H.T. Hahn. 2008. Towards development of a self-healing composite using a mendable polymer and resistive heating. J. Compos. Mater. 42(26): 2869–2881.

Patrick, J.F., K.R. Hart, B.P. Krull, C.E. Diesendruck, J.S. Moore, S.R. White and N.R. Sottos. 2014. Continuous self-healing life cycle in vascularized structural composites. Adv. Mater. 26(25): 4302–4308.

Patrick, J.F., K.R. Hart, B.P. Krull, C.E. Diesendruck, J.S. Moore, S.R. White and N.R. Sottos. 2012. Rapid self-healing hydrogels. Proceedings of the National Academy of Sciences 109(12): 4383–4388.

Peterson, A.M., R.E. Jensen and G.R. Palmese. 2010. Room-temperature healing of a thermosetting polymer using the Diels–Alder reaction. ACS Appl. Mater. & Interface 2(4): 1141–1149.

Pratama, P.A., M. Sharifi, A.M. Peterson and G.R. Palmese. 2013. Room temperature self-healing thermoset based on the Diels–Alder reaction. ACS Appl. Mater. & Interface 5(23): 12425–12431.

Rodell, Christopher B., John W. MacArthur, Shauna M. Dorsey et al. 2015. Shear-thinning supramolecular hydrogels with secondary autonomous covalent crosslinking to modulate viscoelastic properties *in vivo*. Advanced Functional Materials 25(4): 636–644.

Rodell, Christopher B., Ryan J. Wade, Brendan P. Purcell, Neville N. Dusaj and Jason A. Burdick. 2015. Selective Proteolytic Degradation of Guest–Host Assembled, Injectable Hyaluronic Acid Hydrogels. ACS Biomaterials Science & Engineering.

Rubinstein, M. and A.N. Semenov. 1998. Thermoreversible gelation in solutions of associating polymers. 2. Linear dynamics. Macromolecules 31(4): 1386–1397.

Rubinstein, M. and A.N. Semenov. 2001. Dynamics of entangled solutions of associating polymers. Macromolecules 34(4): 1058–1068.

Sarkar, B. and P. Alexandridis. 2015. Block copolymer–nanoparticle composites: Structure, functional properties, and processing. Prog. Polym. Sci. 40(0): 33–62.

Semenov, A.N. and M. Rubinstein. 1998. Thermoreversible gelation in solutions of associative polymers. 1. Statics. Macromolecules 31(4): 1373–1385.

Sijbesma, R.P., F.H. Beijer, L. Brunsveld et al. 1997. Reversible polymers formed from self-complementary monomers using quadruple hydrogen bonding. Science 278(5343): 1601–1604.

Smith, K.A., S. Tyagi and A.C. Balazs. 2005. Healing surface defects with nanoparticle-filled polymer coatings: Effect of particle geometry. Macromolecules 38(24): 10138–10147.

Stukalin, E.B., L.H. Cai, N.A. Kumar, L. Leibler and M. Rubinstein. 2013. Self-healing of unentangled polymer networks with reversible bonds. Macromolecules 46(18): 7525–7541.

Sun, J.-Y., X.H. Zhao, W.R.K. Illeperuma et al. 2012. Highly stretchable and tough hydrogels. Nature 489(7414): 133–136.

Suprapolix. http://www.suprapolix.com/pages/polymers]. Available from http://www.suprapolix.com/pages/polymers.

Tamesue, Shingo, Masataka Ohtani, Kuniyo Yamada et al. 2013. Linear versus Dendritic molecular binders for hydrogel network formation with clay nanosheets: studies with aba triblock copolyethers carrying guanidinium ion pendants. Journal of the American Chemical Society 135(41): 15650–15655.

Tian, Q., M.Z. Rong, M.Q. Zhang and Y.C. Yuan. 2010. Synthesis and characterization of epoxy with improved thermal remendability based on Diels–Alder reaction. Polym. Int. 59(10): 1339–1345.

Toohey, K.S., C.J. Hansen, J.A. Lewis, S.R. White and N.R. Sottos. 2009. Delivery of two-part self-healing chemistry via microvascular networks. Adv. Funct. Mater. 19(9): 1399–1405.

Toohey, K.S., N.R. Sottos, J.A. Lewis, J.S. Moore and S.R. White. 2007. Self-healing materials with microvascular networks. Nat. Mater. 6(8): 581–585.

Tournilhac, F.G., D. Montarnal and L. Leibler. 2011. Self-assembly and self-healing of supramolecular polymers and networks. Paper read at Abstracts of Papers of the American Chemical Society, Aug 28.

Trask, R.S. and I.P. Bond. 2006. Biomimetic self-healing of advanced composite structures using hollow glass fibres. Smart Mater. Struct. 15(3): 704–710.

Trask, R.S., G.J. Williams and I.P. Bond. 2007. Bioinspired self-healing of advanced composite structures using hollow glass fibres. J. R. Soc. Interface 4(13): 363–371.

van der Stok, Johan, Huanan Wang, Saber Amin Yavari et al. 2013. Enhanced bone regeneration of cortical segmental bone defects using porous titanium scaffolds incorporated with colloidal gelatin gels for time- and dose-controlled delivery of dual growth factors. Tissue Engineering Part A 19(23-24): 2605–2614.

Van Tomme, Sophie R. 2007. Self-assembling microsphere-based dextran hydrogels for pharmaceutical applications, Utrecht University, Utrecht University.

Varley, R.J. and S. van der Zwaag. 2008. Development of a quasi-static test method to investigate the origin of self-healing in ionomers under ballistic conditions. Polym. Test. 27(1): 11–19.

Varley, R.J. and S. van der Zwaag. 2008. Towards an understanding of thermally activated self-healing of an ionomer system during ballistic penetration. Acta Mater. 56(19): 5737–5750.

Vougioukalakis, G.C. and R.H. Grubbs. 2009. Ruthenium-based heterocyclic carbene-coordinated olefin metathesis catalysts†. Chem. Rev. 110(3): 1746–1787.

Wang, H., O.C. Boerman, K. Sariibrahimoglu, Y. Li, J.A. Jansen and S.C. Leeuwenburgh. 2012. Comparison of micro- vs. nanostructured colloidal gelatin gels for sustained delivery of osteogenic proteins: Bone morphogenetic protein-2 and alkaline phosphatase. Biomaterials 33(33): 8695–703.

Wang, H., M.B. Hansen, D.W. Löwik, J. van Hest, Y. Li, J.A. Jansen and S.C. Leeuwenburgh. 2011. Oppositely charged gelatin nanospheres as building blocks for injectable and biodegradable gels. Adv. Mater. 23(12): H119–24.

Wang, H.N., Q. Zou, O.C. Boerman et al. 2013. Combined delivery of BMP-2 and bFGF from nanostructured colloidal gelatin gels and its effect on bone regeneration *in vivo*. J. Controlled Release 166(2): 172–181.

Wang, Huanan, Sander C.G. Leeuwenburgh, Yubao Li and John A. Jansen. 2011. The use of micro-and nanospheres as functional components for bone tissue regeneration. Tissue Engineering Part B: Reviews 18(1): 24–39.

Wang, Q., L. Wang, M.S. Detamore and C. Berkland. 2008. Biodegradable colloidal gels as moldable tissue engineering scaffolds. Advanced Materials 20(2): 236–239.

Wang, Qigang, Justin L. Mynar, Masaru Yoshida et al. 2010. High-water-content mouldable hydrogels by mixing clay and a dendritic molecular binder. Nature 463(7279): 339–343.

Watanabe, M. and Na. Yoshie. 2006. Synthesis and properties of readily recyclable polymers from bisfuranic terminated poly (ethylene adipate) and multi-maleimide linkers. Polymer 47(14): 4946–4952.

Westhaus, Eric and Phillip B. Messersmith. 2001. Triggered release of calcium from lipid vesicles: a bioinspired strategy for rapid gelation of polysaccharide and protein hydrogels. Biomaterials 22(5): 453–462.

White, S.R., N.R. Sottos, P.H. Geubelle et al. 2001. Autonomic healing of polymer composites. Nature 409(6822): 794–797.

Williams, H.R., R.S. Trask and I.P. Bond. 2007. Self-healing composite sandwich structures. Smart Mater. Struct. 16(4): 1198–1207.

Williams, K.A., D.R. Dreyer and C.W. Bielawski. 2008. The underlying chemistry of self-healing materials. MRS Bull. 33(8): 759–765.

Wool, R.P. 2008. Self-healing materials: a review. Soft Matter 4(3): 400–418.

WorldEconomicForum. https://agenda.weforum.org/2013/02/top-10-emerging-technologies-for-2013/]. Available from https://agenda.weforum.org/2013/02/top-10-emerging-technologies-for-2013/.

Wu, D.Y., S. Meure and D. Solomon. 2008. Self-healing polymeric materials: A review of recent developments. Prog. Polym. Sci. 33(5): 479–522.

Xiao, D.S., Y.C. Yuan, M.Z. Rong and M.Q. Zhang. 2009. Hollow polymeric microcapsules: Preparation, characterization and application in holding boron trifluoride diethyl etherate. Polymer 50(2): 560–568.

Xu, F., C.A.M. Wu, V. Rengarajan, T.D. Finley, H.O. Keles, Y. Sung, B. Li, U.A. Gurkan and U. Demirci. 2011. Three-dimensional magnetic assembly of microscale hydrogels. Advanced Materials 23(37): 4254–4260.

Yang, J., M.W. Keller, J.S. Moore, S.R. White and N.R. Sottos. 2008. Microencapsulation of isocyanates for self-healing polymers. Macromolecules 41(24): 9650–9655.

Yang, Y. and M.W. Urban. 2013. Self-healing polymeric materials. Chem. Soc. Rev. 42(17): 7446–67.

Yang, Y. and M.W. Urban. 2014. Self-repairable polyurethane networks by atmospheric carbon dioxide and water. Angew. Chem. Int. Ed. 53(45): 12142–12147.

Yin, T., M.Z. Rong, J. Wu, H. Chen and M.Q. Zhang. 2008. Healing of impact damage in woven glass fabric reinforced epoxy composites. Compos. Part A 39(9): 1479–1487.

Yin, T., M.Z. Rong, M.Q. Zhang and G.C. Yang. 2007. Self-healing epoxy composites— Preparation and effect of the healant consisting of microencapsulated epoxy and latent curing agent. Compos. Sci. Technol. 67(2): 201–212.

Yin, T., L. Zhou, M.Z. Rong and M.Q. Zhang. 2008. Self-healing woven glass fabric/epoxy composites with the healant consisting of micro-encapsulated epoxy and latent curing agent. Smart Mater. Struct. 17(1).

Yoon, J.A., J. Kamada, K. Koynov, J. Mohin, R. Nicolaÿ, Y. Zhang, A.C. Balazs, T. Kowalewski and K. Matyjaszewski. 2012. Self-healing polymer films based on thiol–disulfide exchange reactions and self-healing kinetics measured using atomic force microscopy. Macromolecules 45(1): 142–149.

Yount, W.C., D.M. Loveless and S.L. Craig. 2005. Small-molecule dynamics and mechanisms underlying the macroscopic mechanical properties of coordinatively cross-linked polymer networks. J. Am. Chem. Soc. 127(41): 14488–14496.

Yuan, L., G.Z. Liang, J.Q. Xie, L. Li and J. Guo. 2006. Preparation and characterization of poly(urea-formaldehyde) microcapsules filled with epoxy resins. Polymer 47(15): 5338–5349.

Yuan, Y.C., M.Z. Rong, M.Q. Zhang, J. Chen, G.C. Yang and X.M. Li. 2008. Self-healing polymeric materials using epoxy/mercaptan as the healant. Macromolecules 41(14): 5197–5202.

Zeng, C., H. Seino, J. Ren, K. Hatanaka and N. Yoshie. 2013. Bio-based furan polymers with self-healing ability. Macromolecules 46(5): 1794–1802.

Zhang, Hongji, Hesheng Xia a 311 nd Yue Zhao. 2012. Poly(vinyl alcohol) hydrogel can autonomously self-heal. ACS Macro Letters 1(11): 1233–1236.

Zhang, Xiaojie and Junhui He. 2015. Hydrogen-bonding-supported self-healing antifogging thin films. Sci. Rep. 5.

Zhang, Y., A.A. Broekhuis and F. Picchioni. 2009. Thermally self-healing polymeric materials: the next step to recycling thermoset polymers? Macromolecules 42(6): 1906–1912.

Zhao, X.H. 2014. Multi-scale multi-mechanism design of tough hydrogels: building dissipation into stretchy networks. Soft Matter 10(5): 672–687.

Zwaag, S. 2008. Self healing materials: an alternative approach to 20 centuries of materials science: Springer, Netherlands.

11

Advanced Materials for Soft Robotics

Li Wen,[1,]* *Daniel Vogt,*[2] *Zhenyun Shi,*[3] *Qi Shen*[4]
and *Ziyu Ren*[5]

ABSTRACT

Soft robotics, a novel member of the robotic family, may apply a
rich variety of flexible, advanced materials to generate movements
and sense the motion. This new robotic technology delivers a
variety of vital applications that can interact safely with human
beings and function effectively in complex natural environments,
including assistive medical and rehabilitation devices, search and
rescue, underwater exploration, space manipulation, industrial
operation, etc. Our understanding of the soft robotics is still in its

[1] Department of Mechanical Engineering and Automation, International Research Institute
for Multidisciplinary Science, Beihang University, Beijing, 100191, China.
[2] Wyss Institute for Biologically Inspired Engineering, Harvard University, Cambridge, MA
02138, USA.
Email: dvogt@seas.harvard.edu
[3] Department of Mechanical Engineering and Automation, Beihang University, Beijing, 100191,
China.
Email: shichong1983623@hotmail.com
[4] Department of Mechanical Engineering, University of Nevada, Las Vegas, NV 89154-4027,
USA.
Email: qi.shen@unlv.edu
[5] School of Mechanical Engineering and Automation, Beihang University, Beijing, 100191,
China.
Email: renziyu2013@gmail.com
* Corresponding author: liwen@buaa.edu.cn

infancy, the development of advanced materials that can be applied to soft robotics remain a key challenge. This chapter "Advanced materials for soft robotics" aims to meet the urgent requirements for research on advanced materials such as shape memory alloy, stretchable elastomeric sensor and Ionic Polymer-Metal composite that have very recently been used for soft robotics. Theoretical modeling, design, fabrication and experimental characterization included in this chapter could contribute to training of graduate students, robotic engineers and research scientists for understanding fundamental principles of advanced materials and how to apply advanced materials to robotics using a multidisciplinary science based approach.

1. New Trends in Robotics: Soft Robotics

Conventional robots have the capability of operating heavy weights and moving rapidly, but are poor at dealing with human beings, with tissue, with soft objects and objects of complex three-dimensional shapes. The state-of-art of soft robotics inspired a new wave of robotics, and is an innovative research field dedicated to the science and engineering of new advanced materials in machines. In order to interact with various natural environments, multidisciplinary in scope, *Soft Robotics* combines advances in biomedical engineering, biomechanics, material science, soft matter physics and artificial intelligence to provide comprehensive coverage of new approaches to construct devices that can undergo dramatic changes in shape, size, kinematics as well as other functions. In this chapter, we first provide a brief general overview of the advanced materials that can be applied to the soft robotics, then we introduce three representative advanced materials as new forms of actuators and sensors as well as some potential applications. Finally we end up with a discussion of the future perspectives of the soft robotics. This new technology delivers vital applications for a variety of purposes, including search and rescue field robotics, aeronautics space manipulation, industrial operation, assistive medical and rehabilitation devices, etc.

With growing interests in flexible origami of animals such as fish, octopus, jellyfish and caterpillar, etc., smart actuators and sensors have been developed to possess the several principal characteristics: high energy density, light weight and ease to fabricate. Definition of advanced materials, such as shape memory alloy, stretchable elastomeric material and Ionic Polymer-Metal composite and other new forms of both actuation and sensing would be critical to the development of Soft Robotics. In this chapter, we have introduced three representative flexible and advanced materials as new forms of actuators and sensors: Shape memory alloy, eGaIn elastomeric sensor and Ionic Polymer Metal Composites.

The shape memory alloy actuator is a representative actuating material due to its high mass-specific force, it can be widely used as the actuation element of a soft robot. Therefore it is chosen to be covered in this chapter. Further, sensors are essential for robots to interact with the world. Specific to soft robotics, which manipulating deformable and delicate conformal objects, the sensors embedded in soft robotics should be compliant in order to keep the entire system flexible and stretchable. Soft sensors containing micro-channels filled with eutectic Gallium Indium fluid (eGaIn) offer outstanding performance in terms of compliancy and electrical properties. This new type of soft sensor that can provide information of strain, pressure, curvature and shear, which is superb to previous flexible sensors, will be introduced in this chapter as well. Finally, we introduced Ionic Polymer Metal Composites (IPMC), which are made of an ionic polymer membrane with plated gold as electrodes chemically on both sides, which has the advantages of small activation voltage, low noise, and flexibility. These features promise IPMC potential applications on the miniature soft robots as function of both sensors & actuators.

2. Shape Memory Alloys for Soft Robotics

Shape memory alloys (SMAs) can generate force and actuation during the shape recovery process under heating, which makes the shape memory effect an important and innovative actuation principle in the micro robot system. The output force and driving pattern of the SMA actuator is based on the dimensions and shape of the actuator unit, which has shown excellent design flexibility on the joints of wider style robot systems. With growing interests in flexible structures such as fish fins, octopus tentacles, and creeping robots, SMA actuators have been developed to possess the following characteristics: high energy density, specific power and ease of fabrication. As a typical advanced material, a SMA unit can be applied both as an actuator and a sensor at the same time, making the unit a good substitute for muscles and nerves in soft robotics. In particular, the ductile deformation of SMA wires or slices can achieve flexible driving based on structural design. Its high power density supplies flexible design solutions, which can be implanted into tiny structures and maintain the adequate output. Various fabrication methodologies have been studied for combining the SMA unit with soft materials.

Shape memory alloys retain the memory of their original shape and can return to the pre-deformed shape after being heated to a critical temperature. A crystal phase transition appears during the temperature change and causes the materials to change between the high temperature phase (austenite) and the low temperature phase (martensite) and three different crystal structures (twinned martensite, detwinned martensite, and austenite) are included in the phase change (Fremond 1996). Below the martensite starting temperature, the SMA structure has low stiffness and inelastic straining. Above the austenite starting temperature, the shape memory effect (SME) appears by eliminating the detwinning strain; the maximum recoverable strain can be around 5%. Besides Young's modulus, these two phases also demonstrate obvious differences in thermal conductivity and resistivity.

Many different types of SMA materials were used for industry applications, including nickel-titanium-based and other iron-based and copper-based SMAs such as Fe-Mn-Si, Cu-Zn-Al, and Cu-Al-Ni. Other SMAs are more commercially available and are less expensive than the NiTi-based, but NiTi-based SMAs are most widely applied for better stability, feasibility, superior thermo mechanical performance, and biocompatibility. Most of the actuators and sensors discussed below are applications of NiTi, although NiTiCu is found more often in use in the medical field (Huang 2002). Aside from phase transformation as a response to temperature, the transformation can also be induced by mechanical stress, a phenomenon called superelasticity (SE). In the austenite phase, under certain stress, the alloy can be deformed up to 10% and recover its original shape after the stress is removed (Duerig and Pelton 1994).

The applications of SMA devices in various fields are developed around the SME and SE properties. SE properties are normally used to form specialized structures for the human skeleton, oral devices in the medical field, or couplings and fasteners in the aero field. Based on SME properties, SMAs are more widely used as actuators, as in arterial intervention devices and surgical instruments in medicine; as switches, plugs, circuit breakers, and transmission controllers for industrial applications; and as MEMS devices and aircraft wings in the military and civil fields (Sreekumar et al. 2007).

2.1 Structure Design of SMA

The inspiration for soft robot design often comes from nature, where the objects of imitation are normally derived from hydrostatic skeletons, muscular hydrostats, or other bio-inspired designs. Most soft animals (such as worms or actinia) have no rigid skeletons and are composed of epidermis, muscle, body fluid, and neural systems. The hydrostatic skeleton is one typical structure, in which the muscle forms a closed cavity full of internal fluid. The fluid maintains a constant volume and when the longitudinal or circumferential muscle contracts, the body expands in a perpendicular direction (Quillin 1999). Different from arthropod animals, worms change their body stiffness by fluid pressure, muscle tension, and the compression of tissue (Trimmer and Issberner 2007) and they control muscle contraction and expansion to achieve shape deformation and locomotion by creating traveling waves (Trueman 1975).

Many soft robots with hydrostatic skeletons have been developed based on SMA actuators. A constant volume design is a popular scheme that resembles the biological principle. Dario pioneered this kind of crawler robot (Menciassi et al. 2004), and the SoftBot that came after it is another typical case (Kate et al. 2008). It includes three main components: the silicone rubber body wall, the SMA spring actuator, and the control system. SMA coils are bonded inside the body wall and drive in a linear direction after heating, and the contoured body wall thickens and serves as the bias spring to supply the antagonistic force to the SMA actuator. Under the body wall is an inner compartment, and the space between the inner compartment and the body wall is sealed to allow transforming forces and regular stiffness with pressure. As shown in Fig. 1, by pulsed heating in each segment, the four SMA coils can adjust the contacts between the legs and the ground, creating a peristaltic posture that leads forward or crawls back. The force and contractions of the actuators are controlled by the heating frequency and duty cycle; a genetic algorithm was used to achieve more complex movement.

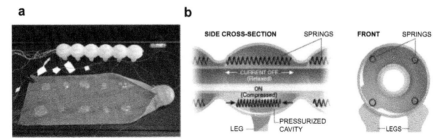

Figure 1. Soft-Bot. **(a)** Prototype of the Soft-Bot. **(b)** The model of the robotic caterpillar and its internal structure. Adapted permission from Dr. Barry Trimmer.

Different from the robot described above, the SMA actuator is set radially in the Meshworm robot (Kim et al. 2013), shown in Fig. 2. This is a constant length design that simulates *oligochaetes*. One segment of the robot body exhibits radial expansion, while the radial SMA coils contract in an adjacent segment, and the essential locomotion is derived from ground contact following the peristaltic wave. Linear potentiometers are set in each segment to detect the length change and provide close-looping control. Three control algorithms were analyzed to actualize lower energy consumption and a higher speed. Two longitudinal SMA coils were set on two sides of the body, and steering was incorporated into the motor ability. One important feature of a soft robot, low impedance to external pressure, is tested by hammer in this system.

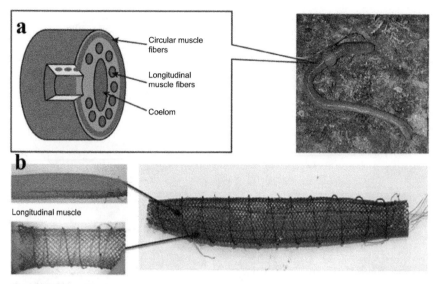

Figure 2. (a) Simulating the muscular structure of oligochaetes. **(b)** The Meshworm exhibits peristaltic locomotion by SMA actuators. Adapted with permission from ref. (Kim et al. 2013), copyright © Elsevier Publishing Group 2013.

The muscular hydrostatic structure is also without skeletal support, but unlike the case with a hydrostatic skeleton, it does not have a separate fluid cavity that is mainly composed of muscle tissue. The typical muscular hydrostatic structure is composed of transverse and longitudinal muscles, which always resist and work each other to apply vertical motion. The incompressible features of the muscle make it work similarly as fluid supplies pressure. Typical examples include the tongue, the elephant nose, and cephalopods.

Cephalopods are the most common objects whose motion is simulated, due to their intricate motion ability. Cephalopods can deform their shape to fill a narrow space, and the arms can apply complex manipulations. An octopus-inspired robot is always a hot topic for scientists (Zheng et al. 2014). An octopus arm has transverse, longitudinal, and oblique muscles that form a distinct structure that generates shortening, elongation, bending, or torsion (Kier and Stella 2007), as shown in Fig. 3a.

To develop octopus-inspired robots, one approach is to form the actuator system based on the real anatomy and mechanisms (Margheri et al. 2012), where the SMA actuator is always used to imitate the transverse, longitudinal muscles (Calisti et al. 2012, Mazzolai et al. 2012). As shown in Fig. 3b, a cylinder composed of plastic fiber braid and a silicone layer forms a robot arm with a highly deformable ability, and radial SMA

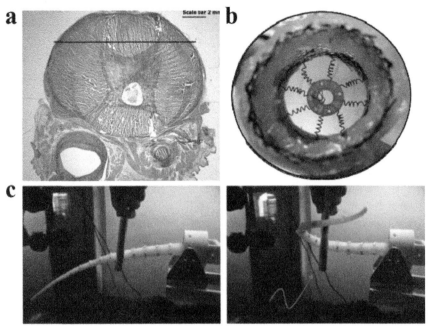

Figure 3. Octopus-inspired robot. **(a)** The cross-section morphology of the octopus arm. Adapted with permission from ref. (Margheri et al. 2012), copyright © IOP Publishing 2012. **(b)** The SMA springs inside the arm to achieve local actuation. **(c)** One soft robot arm wrapping around a bar. Adapted with permission from ref. (Mazzolai et al. 2012), copyright © IOP Publishing 2012.

coils are used to perform local deformations (Mazzolai et al. 2012). The arm was tested in the water for its wrapping ability and interaction with its surroundings, shown in Fig. 3c; it can bend, elongate, and shorten. The SMA actuator is also used to drive other bio-inspired soft robots, such as the Starfish-Like Soft Robot (Mao et al. 2014), the material characteristics make it a good choice to replace different muscles.

2.2 Controller of SMA

A SMA actuator is driven primarily by an electric heating current, and temperature control is the key to its implementation. These actuators are sensitive to changes in the environmental temperature and force load, both factors that can lead to the phase transformation of the materials and affect the driving performance. By studying the dynamics of the actuator, the phase transformation model and the convection heat transfer of SMA units, some open-loop control models have been established (Jala and Ashrafiuon 2006, Jayender et al. 2008). However, environmental changes result in varying control accuracy, which creates a major challenge. Experiments have shown that with external feedback, close-loop control provides SMA actuators with high control accuracy and position feedback (Hadi et al. 2010) and force feedback (Teh and Featherstone 2008) are commonly used.

In recent years, the resistance feedback control of SMA actuators has received more attention. One problem with developing soft robots is that we currently have no general theory of how to control such unconstrained structures. In the SMA-actuated soft robot, in order to achieve precise control with continuous deformation, the combination of actuation and sensing is a possible solution. During phase transformation, the resistance value of SMA materials can show a significant change, 8%–15%, which is clearly more than can be accounted for by differences occurring solely due to the shape change. By establishing a descriptive mathematical model of the strain-resistance relationship of SMA, a self-sensing feedback model can be achieved to apply accuracy control to a SMA-actuated soft robot (Wang et al. 2012b). When it comes to the limits in complex SMA unit shapes, the resistance value of coils or other shapes is hard to describe, which is why current research tends to be limited to the SMA wire actuator.

Both the shape changing and the force generation of SMAs derive from temperature alternation, and the phase transformation of the SMA and the temperature has nonlinear relationship, as shown in Fig. 4. To achieve robust control, methodologies have been used to describe the nonlinear function, such as neural networks, iterative learning, or fuzzy control. After establishing the control scheme of the whole soft robot, engineers build the models based on the locomotive patterns of soft-bodied animals

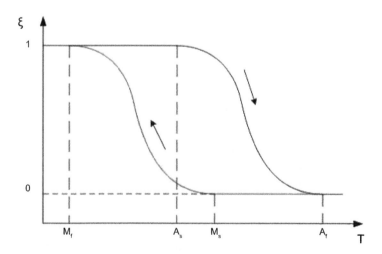

Figure 4. The hysteresis relationship of phase transformation and temperature for SMA actuators. Adapted with permission from Wikipedia.

and they consider the frequency, output force, and shape change mode of the corresponding SMA actuator to achieve motion schemes and to optimize efficiency. To control a SMA actuator for a soft robot, the technical combination of the material properties, the actuator design, and the feature extraction of the soft robot are all considered.

2.3 Fabrication of SMA

In most cases, when a SMA is used as the actuator for a soft robot driving system, the actuator is generally filamentous or in sheet form. SMA wires can deform along the axial direction or bend along the vertical direction, depending on the training process. Actuators with contraction along the axial direction are more widely used for significant larger outputs, and they can contract from heating and be extended under a bias force from another unit. To increase the deformation ratio, the SMA wires are sometimes wound into coiled springs, and in other cases, a linear shape is retained in order to supply a larger output.

For the SMA wire actuator, reciprocating motion can be achieved using different structures, for example, a passive bias force by spring (Lan and Yang 2009) or antagonistically paired wires (Teh and Featherstone 2008). In all these designs, the primary problems to be solved are slow response and cyclic degradation. The frequency of the SMA can be increased by a cooling system (Romano and Tannuri 2009) or repeat heating, with antagonistic actuation design being another option. Cyclical degradation is caused by the functional fatigue of SMA, where the material may lose its SME ability; this means that the driving displacement decreases with the cycle numbers. Experiments have shown that cyclical degradation can be improved with proper pre-straining and additional component design (e.g., spring and stopper).

In a soft robot actuator design, it is worth considering the use of soft materials with elasticity to work as additional components and to supply pre-straining to the SMA actuator to optimize the degradation property; in most cases, an antagonistic design demonstrates better function than a bias spring design. A SMA sheet actuator can provide more flexible motion based on different structure designs (Leester-Schadel et al. 2007), as shown in Fig. 5; the direction of motion can be vertical or can be along the tangent direction of the plane. The key properties such as deformation ratio and output force can also be varied by structural design.

A SMA sheet can be made by using several different processing methods, as in the shape deposition manufacturing (SDM) technology (Cho et al. 2009) shown in Fig. 6, as well as by laser cutting. One problem that needed to be solved for SMA actuators was how to embed the SMA into the soft materials; surface micromachining technology can be used

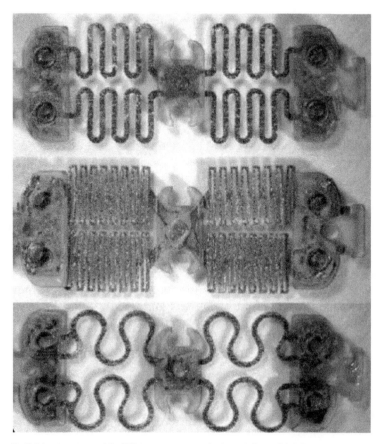

Figure 5. SMA actuators with different structure design. Adapted with permission from ref. (Leester-Schadel et al. 2007), copyright © SPIE 2007.

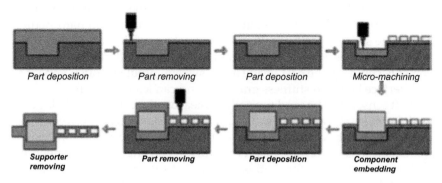

Figure 6. Schematic diagram of SDM process. Adapted with permission from ref. (Cho et al. 2009), copyright © Springer Publishing Company 2009.

to embed SMA wires into a 6 μm thick polyimide base (Kate et al. 2008). Previously, more processing was practiced to embed the SMA actuator into the robot body (Leester-Schadel et al. 2008), although a simple mechanical connection is still one popular option.

2.4 Brief Summary

Shape memory alloy actuators can be used widely as the driving element of a soft robot, due to their high mass-specific force. A general review of the material properties, structural design, and fabrication of SMA actuator is given in this chapter. The SME property of a SMA actuator makes it a suitable alternative to a muscle unit and by a multi-array of actuator settings, the system can achieve sophisticated motion. With the development of fabrication processes for SMA actuators and the use of embedding techniques, the bio structure design has become more flexible. Furthermore with the improvements of the self-sensing technology of the SMA actuator, more reliable and continuous control might be achieved in a soft deformation in the future. The design principle of the SMA actuator in soft robots will tend to show more complex morphological changes to simulate the muscle movement of soft animals by advanced fabrication. In addition the SMA might be used more frequently as a sensor in the bio-structure to achieve its self-sensing property.

3. Stretchable Elastomeric Sensor for Soft Robotics

Sensors are essential for robots to interact with the world. This is exacerbated in the field of soft robotics, where manipulating deformable and delicate objects is something that conformal and flexible soft robots are more adapted to do compared to traditional hard-bodied robots. Unlike in hard-bodied robots, the sensors used in soft robotics must be soft in order to keep the entire system completely flexible and stretchable. Although small rigid elements could still be cast inside a soft system, problems often appear at the interface between stiffness gradients and can lead to failure.

Soft sensors can be used typically to measure the curvature of a leg on a soft robot (Fig. 7a, Shepherd et al. 2011) or if there is a contact between a soft gripper and an object (Fig. 7b, Ilievski et al. 2011). In addition to soft-bodied robots, soft sensors can also be very useful for joint angle measurements on the human body (e.g., for motion tracking). As their high conformability reduced the shear on the skin, the sensors are very comfortable to wear and allow measurements without any impact on the user.

Several approaches can be used to fabricate sensors that are flexible and stretchable (Park et al. 2015). One strategy is to use regular conductive

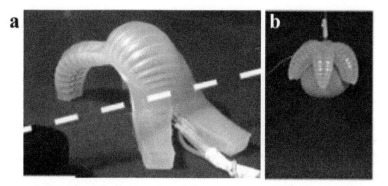

Figure 7. (a) A multi-gait soft robot. Adapted with permission from ref. (Shepherd et al. 2011), copyright © the National Academy of Sciences 2011. **(b)** Soft gripper lifting an egg. Adapted with permission from ref. (Ilievski et al. 2011), copyright © WILEY-VCH Verlag GmbH & Co. KGaA, Weinheim 2011.

materials such as metals or semiconductors, make them thin enough to allow them to bend and cut a meander pattern to allow them to stretch like a spring. An example of such stretchable electronics (Kim 2011) is commercialized by the company MC10 (Fig. 8), where ultra-thin, skin mounted sensors provide personalized biofeedback to the user. Although sensors fabricated using this approach can conform to most of the skin

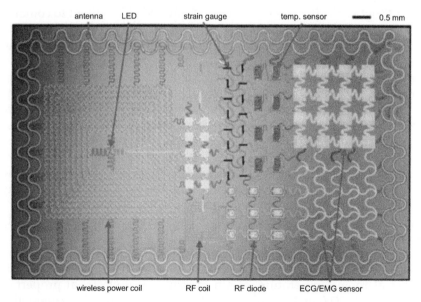

Figure 8. Epidermal electronics, commercialized by the company MC10. Adapted with permission from ref. (Kim 2011), copyright © The American Association for the Advancement of Science 2011.

on the human body, it is unable to conform to the high strains (> 100%, Menguc et al. 2014) that are experienced on joints such as on the elbow or the knee.

Another approach is to use only soft or liquid state materials. Indeed, many elastomers offer a high stretchability until failure (> 900%, Smooth-on) while maintaining a low shore hardness (lower than 1A). By using well known techniques from microfluidics, it is possible to pattern micro-channels inside the rubber and fill them with a conductive liquid. It is also possible to use other approaches to pattern liquid metals on surfaces (Joshipura et al. 2015). By reading the change in resistance as the sensor is stretched, one can calibrate a change of the electrical resistance as functions of the strain, normal pressure and other features that will be described in the following chapter. Other soft sensing techniques using capacitance (Fassler and Majidi 2013, Frutiger et al. 2015, Lipomi et al. 2011, Sun et al. 2014) or optic (Kadowaki et al. 2009) measurements are well studied, however, they are not covered in this chapter.

3.1 Design and Modeling of Stretchable Sensor

Soft sensors are designed primarily to provide information under strain as well as pressure. This design can nevertheless be modified to provide other kinds of information, such as curvature (Majidi et al. 2011) or shear (Vogt et al. 2013) on the surface. It can also be modified to combine modalities, such as sensing strain and normal pressure simultaneously (Park et al. 2012). The conductive liquid, with a resistivity ρ, plays an important role in the functionality of the soft sensor. One type of conductive liquid that is used is an ionic liquid (Cheung et al. 2008), which requires an alternating voltage to read a change of electrical resistance to avoid the electrolysis of the fluid. Another option is to use a mixture of conductive particles (such as carbon nanotubes (Yamada et al. 2011), carbon grease (Muth et al. 2014), etc.) and a carrying medium (e.g., grease or oil). However, the latter option presents the disadvantage that the percolation network between the particles is strongly altered under high deformation, resulting in strong peaks followed by decay in the electrical resistance reading (Muth et al. 2014), as shown on Fig. 9. This effect also alters the sensor's repeatability. Finally, the sensor's initial electrical resistance is relatively high (> kΩ).

In this chapter, the presented soft sensors use eutectic Gallium Indium (eGaIn, Fig. 10) as a conductive liquid. Because it is a metal which is liquid at room temperature, eGaIn presents ideal electrical properties ($\rho = 29.4 \times 10^{-6}\Omega$/cm, Chiechi et al. 2008). Compared to other conductive liquids, it nevertheless has a very high surface free energy (ca. 630 dynes/cm).

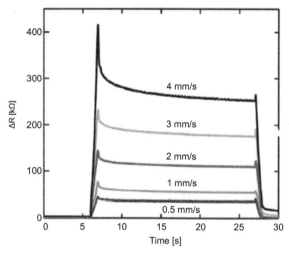

Figure 9. Step response for 3d-printed sensors for different printing speeds. One can notice peaks and decay due to the use of carbon grease as a conductive medium. Adapted with permission from ref. (Muth et al. 2014), copyright © John Wiley and Sons Company 2014.

Figure 10. Eutectic Gallium Indium (eGaIn). Adapted with permission from ref. (Chiechi et al. 2008), copyright © John Wiley and Sons Company 2007.

The simplest sensor design is a single microchannel of length L, width w and height h. The microchannel is filled with a conductive liquid of resistivity ρ. The overall resistance, R, measured at both ends is equivalent to:

$$R = \rho \frac{L}{wh} \tag{3-1}$$

When strain is applied to the soft sensor, the overall length of the micro-channel will increase and the overall cross-section will decrease as shown in Fig. 11a (left).

Equation (3-1) then becomes:

$$\Delta R = R - R_0 = \rho \frac{L + \Delta L}{(w + \Delta w)(h + \Delta h)} - \rho \frac{L}{wh} \tag{3-2}$$

Since $\epsilon = \frac{\Delta L}{L}$, one can replace Δw by $-v\epsilon w$ and Δh by $-v\epsilon h$. With a typical Poisson's ratio v of 0.5 for elastomers, equation (3-2) can then be simplified to:

$$\Delta R = \frac{\rho \varepsilon L(8 - \varepsilon)}{wh(2 - \varepsilon)^2} \tag{3-3}$$

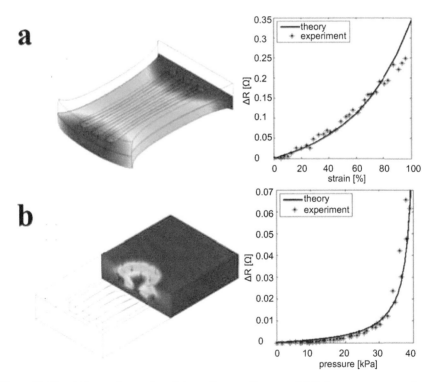

Figure 11. Simulation and experimental results on soft sensor. **(a)** Soft sensor response to strain. **(b)** soft sensor response to normal pressure. Adapted with permission from ref. (Park et al. 2012), copyright © IEEE 2012.

Figure 11a (right) shows the experimental and theoretical relationship between strain and electrical resistance change for a single microchannel.

Under normal contact pressure p of a microchannel, the overall length of the microchannel does not change, but its width and height will be decreased locally (Fig. 11a, left). With the assumption that the cross-section of the microchannel is square and E is the elastic modulus of the elastomer material, equation (3-1) becomes:

$$\Delta R = \frac{\rho L}{wh} \left\{ \frac{1}{1 - 2\left(1 - v^2\right)wp \, / \, Eh} - 1 \right\} \tag{3-4}$$

A comparison between the theoretical and experimental values is shown in Fig. 11b (right).

Without going into details, the micro-channel configuration can also be designed to sense other modalities such as curvature (Majidi et al. 2011) or shear (Vogt et al. 2013). Finally, it is also possible to measure several modalities simultaneously (Park et al. 2012) such as normal pressure and strain.

3.2 Fabrication and Characterization of Stretchable Sensor

In this section, the fabrication of a soft sensor is explained in detail. This soft sensor has been designed to be used on a wrist angle sensing band shown in Fig. 12 (Vogt and Wood 2014). It is primarily designed to measure strain, but will also be sensitive to normal pressure. Its overall dimensions are 45 × 15 × 0.7 mm. The main fabrication steps are shown in Fig. 13.

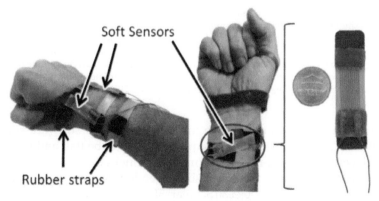

Figure 12. Wrist angle measurement using soft sensors. Adapted with permission from ref. (Vogt and Wood 2014), copyright © IEEE 2014.

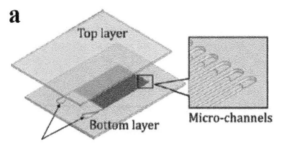

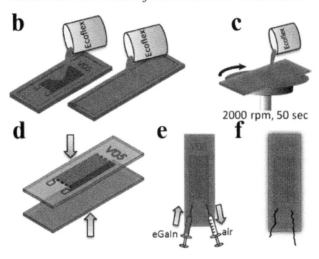

Figure 13. Main fabrication steps for soft sensors. **(a)** The schematic design of two molds. The top layer is flat and the bottom one contain the negative of the microchannel's design. **(b)** Schematic of pouring elastomers into the molds. The trapped air bubbles are evacuated by placing the molds in a vacuum for ten minutes. **(c)** Bonding two layers together by spin-coating a thin uncured wet layer of the same material on the surface of the flat layer. **(d)** Laying one layer over the other during bonding operation. It is important not to trap any air bubbles. **(e)** Filling Gallium Indium (eGaln) into the mold. One syringe is used to inject eGaln and the other one to remove air from the microchannels. **(f)** Connecting the soft sensor to an electronic readout circuit. Adapted with permission from ref. (Vogt and Wood 2014), copyright © IEEE 2014.

The first step is to design the molds with a CAD software such as Solidworks©. Two molds will be required. One will be flat, and the other one will contain the negative of the micro-channel's design.

The micro-channel cross-section size depends mostly on the technique to create the mold. Using an Objet 30 Scholar[1] 3D printer, a typical cross-section dimension is 300 × 300 mm. By using other mold fabrication techniques

---

[1] Objet 30, Stratasys, Eden Prairie, MN 55344, USA.

such as soft lithography, one can get micro-channel dimensions under 100 × 100 mm. By increasing the number of micro-channel loops one will increase the sensitivity of the soft sensor. In this example, the micro-channels are looped nine times to connect both extremities. At each extremity, reservoirs are added to ease the filling and wiring that will be described later. Depending on the mold fabrication technique, an additional step is required before pouring elastomer into the mold for the first time. 3D printed molds must be baked at 60°C for at least seven hours, to help dry the mold and prevent the elastomer from not curing properly when in contact of the mold. When using molds fabricated using soft lithography, it is necessary to silanize the wafer to prevent the elastomer from sticking to the surface of the molds once cured. This is done by placing the wafer in a vacuum chamber containing a recipient with a few drops of Trichloro (1H,1H,2H,2H-perfluoroctyl) silane[2] for more than three hours.

The next step is to pour elastomers into the molds (Fig. 13b). Elastomers are provided typically in a two part mixture and a planetary centrifugal mixer such as the Thinky © Mixer[3] offers the best mixing results. The elastomer used in this example will be Ecoflex® 00-304 as it offers a low shore hardness 00-30 and a high elongation at break (900% Smooth-on).[4] Another typical silicone rubber used for soft sensors is PDMS Sylgard 184.[5] This elastomer has a higher shore hardness than Ecoflex, but has the advantage of being able to bond layers together using oxygen plasma surface treatment. After pouring either of these elastomers, the next step is to place the molds in a vacuum for ten minutes to allow trapped air bubbles to evacuate. The molds can then be placed in an oven at 60°C to accelerate the curing (~ 20 min) or be left at room temperature for at least four hours.

Once both layers are cured, the next step is to bond them together. In the case of Ecoflex 00-30, this is done by spin-coating (2000 rpm, 50 sec) a thin uncured wet layer of the same material on the surface of the flat layer (Fig. 13c). The flat layer with the thin wet layer is then placed in an oven at 60°C for ~ 50 seconds to slightly cure the wet layer. If this step is omitted, the microchannels may be filled by capillary action when both layers are bonded together. On the other hand, if the flat layer is left too long in the oven, there is a risk that the wet layer will cure too much and the bonding will be very weak. When bonding both layers (Fig. 13d), it is important to not trap any air bubbles when laying one layer over the other. The soft sensor can then be left for curing for at least four hours at room temperature.

Once both layers are bonded together, the soft sensor is ready to be filled with eutectic Gallium Indium (eGaIn). Two syringes are required to

---

[2] Sigma Aldrich.
[3] Thinky Mixer ARE-310, Thinky USA, Laguna Hills, CA 23151, USA.
[4] Smooth-On Inc., Easton, PA, 18042, USA.
[5] Dow Corning, Midland, MI.

do this, one to inject the eGaIn, and the other to remove the air from the microchannels to ease the injection as shown in Fig. 13e. It is recommended to first inject air in one of the reservoirs to make sure that the microchannels have not been filled by capillary action. This will also facilitate the placement of the needle on the other reservoir for the syringe containing eGaIn. The air can then be removed and the eGaIn injection can start. It is recommended to tilt the syringe containing eGaIn slightly so as to not inject any air in the micro-channels which will result in an open circuit.

The last step is to insert wires (Fig. 13f) to be able to connect the soft sensor to an electronic readout circuit. It is recommended to use a small wire (e.g., 18 AWG) and to solder one extremity to the core of a 22 AWG solid core wire (~ 2 mm long). This will help anchor the wire in the reservoir. Once soldered to the core, the wire can be inserted into the soft sensor's reservoir following the hole created by the needle during the injection. To improve the sensor's robustness, a silicone adhesive such as Sil-Poxy®[6] can be used to glue the wires onto the soft sensor's surface.

To test the soft sensor, the easiest way is to connect it to an ohm-meter. Depending on the microchannel's design, the initial resistance should be a few ohms and a significant difference should occur when the soft sensor is stretched or compressed. Common issues can be the delamination of both layers, air injected in the microchannels, bad contact between the wire and the eGaIn in the reservoir, etc. Under strain, the higher stress concentration generally happens at the transition between soft and hard materials, in this case between the soft sensor's surface and the wire.

3.3 Brief Summary

In this chapter, a general overview is given about the theory and design of soft sensors which are essential for soft systems in order to interact with the world. In addition to do sensing on soft robots, soft sensors also have an important potential to be worn on the human body for motion tracking (Menguc et al. 2014, Vogt and Wood 2014). Compared to other soft sensing technologies, the main advantage of using a soft elastomer containing microchannels filled with a conductive liquid is to reach very high levels of elasticity and conformability.

4. Ionic Polymer-Metal Composite for Soft Robotics

Ionic Polymer Metal Composites (IPMCs) are innovative materials made of an ionic polymer membrane with plated gold as electrodes on both sides

---

[6] Smooth-On Inc., Easton, PA, 18042, USA.

chemically (Shahinpoor and Kim 2001, Shahinpoor and Kim 2005), which can serve as an alternative actuator as well as sensor for soft robotics. See Fig. 14a. The most interesting characteristics of IPMC are its softness and lightness; moreover, when a electric field is applied across the thickness of

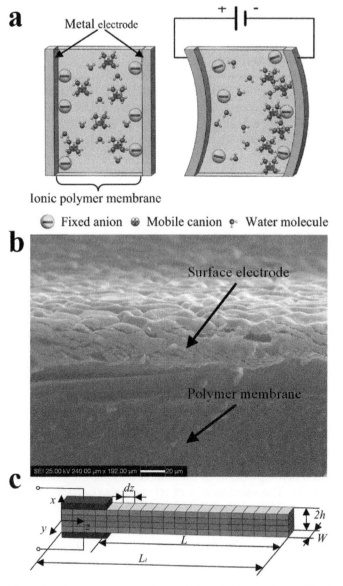

Figure 14. Physical and computation model for IPMC. **(a)** Illustration of IPMC operating principle. Adapted with permission from ref. (Shen et al. 2013), copyright © InTech 2013. **(b)** SEM image of IPMC cross-section. **(c)** Schematic of IPMC beam. Original figures.

direction, the IPMC would perform deformation. On the other hand, they generate a detectable voltage if subjected to a mechanical deformation. Briefly, it has the advantages of low activation voltage actuation (1-2V), limited power consumption, low noise, and high flexibility. Therefore the ionic polymer metal composite provides the possibility for developing a low-noise, micro-size, soft biomimetic robot.

4.1 Design and Modeling of IPMC

Several works concerning the modeling of IPMC have been reported. The influence of the electrode conductivity on the transduction behavior of IPMC was previously investigated by Shahinpoor and Kim (Shahinpoor and Kim 2000). Another model that describes electromechanical transduction in relation to the electrostatic interaction within the polymer was proposed by Nemat-Nasser and Li (Nemat-Nasser and Li 2000). The electrical dynamics of the IPMC based on Nemat-Nasser's work was elucidated by Chen and Tan (Chen and Tan 2008). Porfiri and his co-workers showed that the electrodes do affect the charge dynamics significantly and hence the actuation performance of IPMC (Porfiri 2009). An electrode model for IPMCs based upon a structural formation of electrode materials was proposed by Kim et al. (Kim et al. 2007).

The physical based model combines the effect of the electrode resistance change and the charge dynamics of the ionic polymer. Figure 14b shows the Scanning Electron Microscope (SEM) image of IPMC cross-section. The surface electrode and the inner polymer membrane composite of the IPMC. The length and width of the IPMC beam is L and W. The thickness of the polymer membrane and electrode are $2h$ and h_e. The unit length of the curved IPMC midline is assumed as dz. The unit cell length of the stretched electrode can be expressed as (Shen et al. 2014):

$$dL\left(\hat{h}, w(L,t)\right) = d\gamma\left(w(L,t)\right)\left(r\left(w(L,t)\right) + h + \hat{h}\right) \qquad (4\text{-}1)$$

where $\gamma(t)$ is the angle of the curved beam, $r(w(L, t))$ is the radius of the midline and $\hat{h}$ is the position of the unit cell along the thickness direction of the electrode. The expressions of $\gamma(t)$ and $r(w(L, t))$ are defined in ref. (Kim et al. 2007). The reciprocal of the electrode resistance is expressed as follows (Shen et al. 2014):

$$\frac{1}{R_e\left(w(L,t)\right)} = \int_0^{h_e} J_e \frac{\left(1-P_x\right)}{\left(r\left(w(L,t)\right)+h+\hat{h}\right)} \frac{d\hat{h}}{a} + \int_0^{h_e} \frac{P_x r(w(L,t))}{G_e\left(r\left(w(L,t)\right)+h+\hat{h}\right)L/W} \frac{d\hat{h}}{a}$$

$$= \left[J_e \frac{\left(1-P_x\right)}{a} + \frac{P_x W r(w(L,t))}{G_e aL}\right] \ln\left(\frac{r\left(w(L,t)\right)+h+h_e}{r\left(w(L,t)\right)+h}\right) \qquad (4\text{-}2)$$

with

$$J_e = \frac{(1-P_y)W}{E_e L} + \frac{P_y Wr(w(L,t))}{G_e L}, \quad E_e = F_e \frac{(1-P_z)}{r(w(L,t))} + G_e \frac{P_z}{r(w(L,t))},$$

$$F_e = \frac{\rho_p(a-2s_v)}{a^2} + \frac{2\rho_p \rho_v s_v}{2\rho_p s_v^2 + \rho_v(a^2 - 2s_v^2)}, \quad G_e = \frac{\rho_v(a-s_v)}{a^2} + \frac{\rho_p \rho_v s_v}{2\rho_p s_v^2 + \rho_v(a^2 - 2s_v^2)},$$

where ρ_p and ρ_v are the resistivity of the particle and void, a and s_v are the side length of the unit cell and void cube, P_x, P_z, P_z are the percentages of the defective cube in the x, y, z direction. By inversing (4-2), the expression of the resistance of the electrode can be obtained.

Analytical solutions are available for several special cases of geometric nonlinearity in a cantilever beam. Herein, the finite element approach is used to describe the dynamics of the IPMC strip. The IPMC actuator was assumed to be divided by a series of elements, where the voltage on each element is constant, as shown in Fig. 14c. The Nernst-Planck equation describes the cation migration and diffusion in the polymer element backbone as (Nemat-Nasser and Li 2000)

$$\frac{\partial C}{\partial t} + \nabla \cdot (-D\nabla C - \hat{z}\mu F C \nabla \phi) = 0 \tag{4-3}$$

where $C, t, \mu, D, F, \hat{z}$ and ϕ are the cation concentration, time, mobility of cations, diffusion constant, Faraday constant, charge number, and electric potential in the polymer element respectively. The ions move within the matrix of the polymer. The local charges will concentrate at the boundary between the polymer and surface electrodes. As a result, an electric field will increase in the opposite direction. The Poisson's equation describe the process as

$$\nabla \cdot \vec{E} = -\nabla^2 \phi = \frac{\rho}{\hat{\varepsilon}} \tag{4-4}$$

where $\hat{\varepsilon}$ and $\vec{E}$ are the absolute dielectric constant and strength of the electric field respectively. ρ is the charge density and is defined as

$$\rho = F(C - C_0) \tag{4-5}$$

where C_0 is the constant anion concentration. The time variable of $\rho(x, z, t)$ can be transformed in the Laplace form. The expression in the Laplace domain can be obtained as (Shen et al. 2014)

$$\rho(x,z,s)s - D\frac{\partial^2 \rho(x,z,s)}{\partial x^2} + \frac{\hat{z}\mu F^2 C_0}{\hat{\varepsilon}}\rho(x,z,s) = 0. \tag{4-6}$$

Nemat-Nasser and Li presented a theory that the charge density at the boundary of polymer is proportional to the induced stress σ4 and is expressed as

$$\sigma(\pm h, z, s) = \alpha \rho(\pm h, z, s).\qquad(4\text{-}7)$$

The boundary condition of (4-7) is $\rho(h,z,s) + \rho(-h,z,s) = 0$. The ionic flux in the x direction within the polymer is expressed as (Shen et al. 2014)

$$f(x,z,s) = -D[\frac{1}{F}\frac{\partial \rho(x,z,s)}{\partial x} - \frac{FC_0}{RT}E(x,z,s)]\qquad(4\text{-}8)$$

where R is the gas constant, and T is the absolute temperature. The boundary condition of (4-8) is expressed as

$$f(h,z,s) = f(-h,z,s) = 0.\qquad(4\text{-}9)$$

By solving (4-9) and with the condition of $E(\pm h,z,s) \neq 0$, the expression of ionic flux can be obtained. Based on Ramo-Shockley theorem, the local current density at an electrode boundary is expressed as

$$j(z,s) = \frac{1}{h}\int_{-h/2}^{h/2} f(x,z,s)dx.\qquad(4\text{-}10)$$

The electric potential on the surface of the IPMC element can be expressed as

$$\phi(\pm h, z, s) = \pm\frac{V(s)}{2} \mp \int_0^z j(z,s)W_e h_e R_e\left(w(L,s)\right)\frac{dz}{L}\qquad(4\text{-}11)$$

where $V(s)$ is the voltage applied to the clamp. With (4-10) and (4-11), the expression of $\rho(x,z,s)$ is derived. By relating the induced stress $\sigma(\pm h,z,s)$ to the bending moment, one can obtain

$$\sigma(\pm h, z, s) = \frac{\pm hM(z,s)}{I}\qquad(4\text{-}12)$$

where I is the moment of inertia of the IPMC element and $I = 2/3Wh^3$. With (4-7), (4-11) and (4-12), the bending moment can be obtained. With the linear beam theory, the tip displacement of the IPMC beam element relating to the z can be denoted as

$$\frac{d^2w(z,s)}{dz^2} = \frac{M(z,s)}{YI}\qquad(4\text{-}13)$$

where Y is the Young's modulus. The angle of the curved beam element is donated as

$$\tan d\varphi(z,s) = \frac{dw(z,s)}{dz}.\qquad(4\text{-}14)$$

By integrating the tip displacements of the elements, one can obtain the deformation of the IPMC as

$$w(L,s) = \int_0^L dz \left(d\varphi(z,s) + \frac{d\varphi(z,s)}{dz} \right). \qquad (4\text{-}15)$$

Solving (4-15) with boundary conditions $w(0,s) = 0$ and $\dfrac{dw(0,s)}{dz} = 0$, the deformation of the IPMC $w(L,s)$ under the voltage $V(s)$ can be obtained. The actuation bandwidth of an IPMC actuator is relatively low (under 10 Hz). To accommodate the vibration dynamics of the beam, we project $w(L,t)$ on the first mode of vibration expressed as

$$X_1(z) = ch\beta_1 z - \cos\beta_1 z - \frac{ch\beta_1 L + \cos\beta_1 L}{sh\beta_1 L + \sin\beta_1 L}(sh\beta_1 z - \sin\beta_1 z) \qquad (4\text{-}16)$$

where β_1 is the constant in relative with the natural frequency and the damping ratio. The actual tip displacement of the IPMC beam can be expressed as

$$w_l(L,t) = w(L,t)X_1(L). \qquad (4\text{-}17)$$

4.2 IPMC as Actuator for Fish-inspired Soft Underwater Robot

The actuation effect of IPMC is induced by the transport of hydrated cations and water molecules and the associated electrostatic interactions within the polymer membrane. Thus, the IPMC can work directly in water without waterproof processing. With the advantages of high flexibility, low drive voltage, and large bending deflection, IPMC is one of the promising advanced materials for micro biomimetic underwater propulsion (Shen et al. 2013). It should be noted that according to previous work, when compared to a rigid flapping foil, the thrust efficiency of a two-dimensional flapping foil with chord-wise flexibility is experimentally shown to increase significantly, i.e., up to 36% (Prempraneerach et al. 2003).

Several pioneering works concerning the IPMC actuated underwater robots have been reported. Tan and his group presented the speed model for an IPMC-propelled robotic fish (Chen et al. 2010, Mbemmo et al. 2008). Porfiri and co-workers developed a modeling framework for the surge, sway, and yaw motion predicting biomimetic underwater vehicles propelled by IPMC and investigated the IPMC beam's hydrodynamics using numerical computation and digital particle image velocimetry (DPIV) (Abdelnour et al. 2009, Aureli et al. 2010, Peterson et al. 2009). Kim proposed an analytical model of IPMC actuator dynamic characteristics, which can be used for modeling a single- or multi-segment IPMC actuator that operates in water (Yim et al. 2007). An efficiency model of a robotic fish was proposed, where the total power consumption of IPMC was considered in the model (Wang et al. 2012a). To study the robotic swimmer's thrust performance in micro-scale, an important dimensional parameter of fish, the Reynolds number Re, is introduced and is defined as

$$Re = UL_f / v \qquad\qquad (4\text{-}18)$$

where L_f is the fish length, U represents the steady swimming speed under the freely swimming condition, and v denotes the fluid kinematic viscosity. The Reynolds number of the robotic swimmer is low when its size is reduced. One could reduce the Reynolds number by increasing the kinematic viscosity of the fluid.

The robotic swimmer prototype consists of the rigid body shell, the soft IPMC stripe and the plastic tail fin, as shown in Fig. 15. It had a total length of 47.5 mm without the tail, 14.5 mm in height and 12 mm at its widest point. Its total weight approximated 5.05 g.

The hydrodynamic experiments were conducted in a horizontal low-velocity servo towing system with force detection apparatus on it. Figure 16 shows the mechanical components of the self-propelled experimental apparatus. The water tank, was filled with the mixture of water and glycerin. The force T_1, T_2 (see Fig. 16) was measured by the force transducers. The external force T_{ext} from the external apparatus acting on the swimmer can be measured by $T_{ext} = (T_1 - T_2)/100$.

Hydrodynamic experiments were performed to study the self-propelled speed and thrust force of IPMC swimmer. The robotic swimmer propelled

Figure 15. Prototype of the robotic swimmer. Adapted with permission from ref. (Shen et al. 2013), copyright © IOP Publishing 2013.

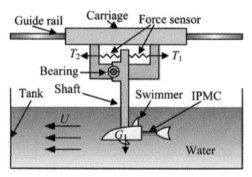

Figure 16. Illustration of experimental apparatus. Adapted with permission from ref. (Shen et al. 2013), copyright © IOP Publishing 2013.

by the IPMC tail swam freely in a tank, and its motion was recorded by a camera, which was set on top of the tank. A signal generator with a power amplifier provided the IPMC actuator sinusoidal signals. The captured images were analyzed. The velocities of the robotic swimmer were measured. Figure 17 shows the consecutive snapshot of the swimmer swimming in the tank.

The swimmer was towed under the servo towing system in the cruising speed measured above. For each cruising speed U_{exp}, the IPMC was under corresponding sinusoidal wave voltage input. According to the Newton's law, the robotic swimmer was considered to swim freely without the external force from the apparatus above acting on it, where the drag force F_D equals

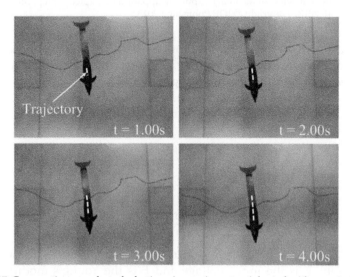

Figure 17. Consecutive snapshot of robotic swimmer in water. Adapted with permission from ref. (Shen et al. 2013), copyright © IOP Publishing 2013.

to the thrust force T_{exp}. When the robot swam under the speed U_{exp}, one could gain T_{exp} by measuring the drag force F_D. To obtain the drag force F_D, the input voltage of IPMC was set zero. Through towing the swimmer under the system at U_{exp}, the F_D was measured.

Figure 18 shows the experiment setup for the image capturing. The IPMC beam was fixed at one end and submerged in the fluids. A camera was set on top of the IPMC for the frame acquisition. Background lighting was turned off during testing while a spotlight was directed towards the reflective dots in order to increase the image contrast. The image analysis of the captured frames was performed.

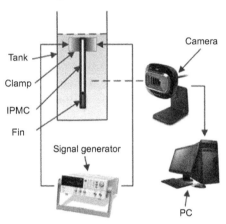

Figure 18. Schematic view of the image capturing apparatus. Adapted with permission from ref. (Shen et al. 2013), copyright © IOP Publishing 2013.

To measure the power consumption, the IPMC tail oscillated in the fluids under corresponding voltage input. The tip displacement s_f was measured through image processing. Then, the IPMC tail oscillated in air under the same voltage input and we measured the tip deflection s_a. The blocking force, F_b, of a cantilevered IPMC actuator can be written as (Lee et al. 2005)

$$F_b = \frac{3WHEd}{8L}V \tag{4-19}$$

with

$$d = \frac{2s_a H^2}{3L^2 V},$$

where W, H, L are the width, thickness and length of the IPMC beam respectively, V is the input voltage and E is the Young's modulus. The amplitude of the tail is far smaller than its length scale, so that the tail is supposed to bend periodically at the tip. The average power that the IPMC output in the fluid during one cycle at frequency f can be obtained as

$$\overline{P}_{exp} = \overline{P}_a - \overline{P}_f = \frac{\int_{t_s}^{1/f+t_s} F_b ds_a - \int_{t_s}^{1/f+t_s} F_b ds_f}{1/f}$$

$$= \frac{fWH^3 E \int_{t_s}^{1/f+t_s} s_a \left(ds_a - ds_f \right)}{4L^3} \tag{4-20}$$

where t_s is a random time point during IPMC tail performing, $\overline{P}_a$ and $\overline{P}_f$ are the power consumption of IPMC in air and fluid respectively.

Figure 19a-d shows the experimental results of swimmer velocities. At low viscosity, the optimal frequency for the speed approximated 1 Hz. When the viscosity increases, the optimal frequency for the speed decreases. The speed of the IPMC swimmer rises with the increasing of voltage amplitude, as shown in Fig. 19a and b. By comparing the experimental results of tail 1 and tail 2, one can find that the velocity of tail 2 is generally higher than that of tail 1. This means that the speed can be improved by increasing the voltage amplitude as well as the size of IPMC. It should be noted that as the viscosity rises, its velocity has a significant decrease, which is shown in Fig. 19c and d. The results also show that the robotic swimmer has relatively low speed at low Reynolds number.

The expression of the thrust efficiency can be presented as

$$\eta_{exp} = \frac{T_{exp} \cdot U_{exp}}{\overline{P}_{exp}}. \tag{4-21}$$

Based on the measured thrust force T_{exp}, speed U_{exp}, power consumption $\overline{P}_{exp}$, the experimental data of thrust efficiency η_{exp} was obtained. Figure 19e-h shows the experimental results of the thrust efficiency η_{exp}. Based on Fig. 19e and f, it was found that inceptively, the efficiency of robotic swimmer in water increases as the frequency increases. With the frequency being relatively high, the efficiency decreases. Its thrust efficiency reaches the maximum when the frequency is approximately 1 Hz. It was also noticed that the IPMC swimmer has relatively high thrust efficiency at high input voltage. As the viscosity increases, the optimal frequency for the thrust efficiency increases, shown in Fig. 19g and h. The thrust efficiency of IPMC swimmer varies with different tails, and each has its optimal frequency. Generally, the thrust efficiency of tail 2 is higher than that of tail 1. Thus it was demonstrated that by increasing the input voltage as well as the size of IPMC, the thrust efficiency of the robotic swimmer can be improved. It

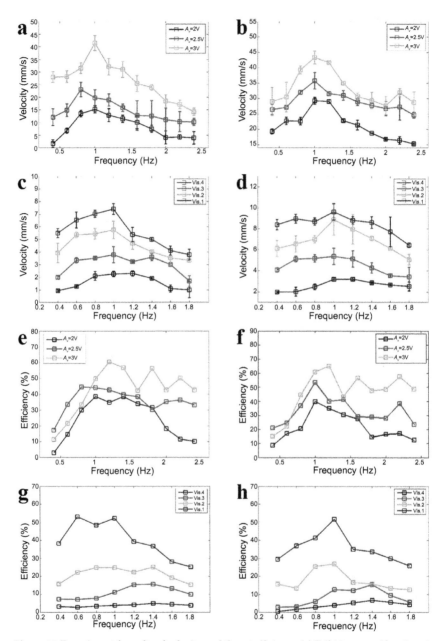

Figure 19. Experimental results of velocity and thrust efficiency. **(a)** Tail 1 in water (the viscosity is 1×10^{-6} m^2/s). **(b)** Tail 2 in water (the viscosity is 1×10^{-6} m^2/s). **(c)** Tail 1 in viscous fluid (A_v = 2.5 V). **(d)** Tail 2 in viscous fluid (A_v = 2.5 V). **(e)** Tail 1 in water (the viscosity is 1×10^{-6} m^2/s). **(f)** Tail 2 in water (the viscosity is 1×10^{-6} m^2/s). **(g)** Tail 1 in viscous fluid (A_v = 2.5 V). **(h)** Tail 2 in viscous fluid (A_v = 2.5 V). Original figures.

was also found that with the viscosity increasing, the Reynolds number and the thrust efficiency of the robotic swimmer decreases. One could indicate that the thrust efficiency of the robotic swimmer descends when its scale is reduced.

4.3 Brief Summary

In this section, two noteworthy efforts on the IPMC were proposed. Firstly, a physical based model of IPMC was presented. The model is based on the charge dynamics model of the polymer membrane and a microstructure model of the electrode. Secondly, an experimental approach was presented to test a robotic swimmer actuated by IPMC. The self-propelled speed, thrust force and power were measured and the thrust efficiency was calculated. With its characteristics of inherent softness, resilience and biocompatibility, IPMC has gained considerable research attention in the area of advanced materials, especially in actuation and sensing applications, such as artificial muscles, underwater actuators and advanced medical devices. The micro-fabrication of IPMCs is also enable scientists and engineers develop soft micro underwater actuators and sensors based on IPMC.

5. Conclusion

A general review of the material modeling, structural design, fabrication and experimental characterization of three representative advanced materials: SMA actuator, eGaln elastomeric sensor and Ionic Polymer Metal Composites are provided in this chapter. The design principle of the SMA actuator in soft robots will tend to show more complex morphological changes to simulate the muscle movement of soft animals by advanced fabrication. In addition the SMA might be used more frequently as a sensor in the bio-structure as well. Besides application on soft robotics, strtechable elastimeric sensors also have important potential to be worn on the human body for motion tracking (Menguc et al. 2014, Vogt and Wood 2014), which offers outstanding performance in terms of compliancy and electrical properties. We also showed characteristics of inherent softness, resilience and biocompatibility of IPMC. We consider new types of actuators, sensors and new approaches of combing different methods for integrating simple-to-control, robust, low cost soft robotics are quite promising and have not yet been explored.

Acknowledgements

The authors thank Prof. Robert Wood and Nick Bartlett for their help on improving the manuscript. Also thanks for supporting from the National Science Foundation support projects, China under contract number 61403012 (to Li Wen), Beijing Science Foundation support projects under contract number 4154077 (to Li Wen) and National Science Foundation support projects, China under contract number 61333016 (to Tan Min).

References

Abdelnour, K., E. Mancia, S.D. Peterson and M. Porfiri. 2009. Hydrodynamics of underwater propulsors based on ionic polymer-metal composites: a numerical study. Smart Mater. Struct. 18(8).

Aureli, M., V. Kopman and M. Porfiri. 2010. Free-locomotion of underwater vehicles actuated by ionic polymer metal composites. Ieee-Asme T Mech. 15(4): 603–614.

Calisti, M., A. Arienti, F. Renda, G. Levy, B. Hochner, B. Mazzolai, P. Dario and C. Laschi. 2012. Design and development of a soft robot with crawling and grasping capabilities. 2012 Ieee International Conference on Robotics and Automation (Icra) 4950–4955.

Chen, Z., S. Shatara and X.B. Tan. 2010. Modeling of biomimetic robotic fish propelled by an ionic polymer-metal composite caudal fin. Ieee-Asme T Mech. 15(3): 448–459.

Chen, Z. and X.B. Tan. 2008. A control-oriented and physics-based model for ionic polymer-metal composite actuators. Ieee-Asme T Mech. 13(5): 519–529.

Cheung, Y.-N., Y. Zhu, C.-H. Cheng, C. Chao and W.W.F. Leung. 2008. A novel fluidic strain sensor for large strain measurement. Sensors and Actuators A: Physical 147(2): 401–408.

Chiechi, R.C., E.A. Weiss, M.D. Dickey and G.M. Whitesides. 2008. Eutectic gallium-indium (EGaIn): A moldable liquid metal for electrical characterization of self-assembled monolayers. Angew Chem. Int. Edit. 47(1): 142–144.

Cho, K.J., J.S. Koh, S. Kim, W.S. Chu, Y. Hong and S.H. Ahn. 2009. Review of manufacturing processes for soft biomimetic robots. Int. J. Precis Eng. Man. 10(3): 171–181.

Duerig, T. and A. Pelton. 1994. Ti-Ni shape memory alloys. Materials Properties Handbook: Titanium Alloys 1035–1048.

Fassler, A. and C. Majidi. 2013. Soft-matter capacitors and inductors for hyperelastic strain sensing and stretchable electronics. Smart Materials and Structures 22(5).

Fremond, M. 1996. Shape memory alloy. Springer.

Frutiger, A., J.T. Muth, D.M. Vogt, Y. Mengüç, A. Campo, A.D. Valentine, C.J. Walsh and J.A. Lewis. 2015. Capacitive soft strain sensors via multicore–shell fiber printing. Advanced Materials 1521–4095.

Hadi, A., A. Yousefi-Koma, M.M. Moghaddam, M. Elahinia and A. Ghazavi. 2010. Developing a novel SMA-actuated robotic module. Sensor Actuat a-Phys 162(1): 72–81.

Huang, W. 2002. On the selection of shape memory alloys for actuators. Mater. Design 23(1): 11–19.

Ilievski, F., A.D. Mazzeo, R.F. Shepherd, X. Chen and G.M. Whitesides. 2011. Soft robotics for chemists. Angewandte Chemie 123(8): 1930–1935.

Jala, V.R. and H. Ashrafiuon. 2006. Robust control of a class of mechanical systems actuated by shape memory alloys. P. Amer. Contr. Conf. 1(12): 5923–5928.

Jayender, J., R.V. Patel, S. Nikumb and M. Ostojic. 2008. Modeling and control of shape memory alloy actuators. Ieee T Contr. Syst. T 16(2): 279–287.

Joshipura, I.D., H.R. Ayers, C. Majidi and M.D. Dickey. 2015. Methods to pattern liquid metals. Journal of Materials Chemistry C 3(16): 3834–3841.

Kadowaki, A., T. Yoshikai, M. Hayashi and M. Inaba. 2009. Development of Soft Sensor Exterior Embedded with Multi-axis Deformable Tactile Sensor System. Ro-Man 2009: The 18th Ieee International Symposium on Robot and Human Interactive Communication, Vols. 1 and 2: 1164–1169.

Kate, M., G. Bettencourt, J. Marquis, A. Gerratt, P. Fallon, B. Kierstead, R. White and B. Trimmer. 2008. SoftBot: A soft-material flexible robot based on caterpillar biomechanics. Tufts University, Medford, MA, 2155.

Kier, W.M. and M.P. Stella. 2007. The arrangement and function of octopus arm musculature and connective tissue. J. Morphol. 268(10): 831–843.

Kim, D.H. 2011. Epidermal electronics (vol. 333, pg. 838, 2011). Science 333(6050): 1703–1703.

Kim, S., C. Laschi and B. Trimmer. 2013. Soft robotics: a bioinspired evolution in robotics. Trends in Biotechnology 31(5): 287–294.

Kim, S.J., S.M. Kim, K.J. Kim and Y.H. Kim. 2007. An electrode model for ionic polymer-metal composites. Smart Mater. Struct. 16(6): 2286–2295.

Lan, C.C. and Y.N. Yang. 2009. A computational design method for a shape memory alloy wire actuated compliant finger. J. Mech. Design 131(2).

Lee, S., H.C. Park and K.J. Kim. 2005. Equivalent modeling for ionic polymer-metal composite actuators based on beam theories. Smart Mater. Struct. 14(6): 1363–1368.

Leester-Schadel, M., B. Hoxhold, S. Demming and S. Buttgenbach. 2007. SMA micro actuators for active shape control, handling technologies, and medical applications. Microtechnologies for the New Millennium. 2007: 65891D-65891D-10.

Leester-Schadel, M., B. Hoxhold, C. Lesche, S. Demming and S. Buttgenbach. 2008. Micro actuators on the basis of thin SMA foils. Microsyst. Technol. 14(4-5): 697–704.

Lipomi, D.J., M. Vosgueritchian, B.C.K. Tee, S.L. Hellstrom, J.A. Lee, C.H. Fox and Z.N. Bao. 2011. Skin-like pressure and strain sensors based on transparent elastic films of carbon nanotubes. Nat. Nanotechnol. 6(12): 788–792.

Majidi, C., R. Kramer and R.J. Wood. 2011. A non-differential elastomer curvature sensor for softer-than-skin electronics. Smart Materials and Structures 20(10).

Mao, S.X., E.B. Dong, H. Jin, M. Xu, S.W. Zhang, J. Yang and K.H. Low. 2014. Gait study and pattern generation of a starfish-like soft robot with flexible rays actuated by SMAs. J. Bionic. Eng. 11(3): 400–411.

Margheri, L., C. Laschi and B. Mazzolai. 2012. Soft robotic arm inspired by the octopus: I. From biological functions to artificial requirements. Bioinspir. Biomim. 7(2).

Mazzolai, B., L. Margheri, M. Cianchetti, P. Dario and C. Laschi. 2012. Soft-robotic arm inspired by the octopus: II. From artificial requirements to innovative technological solutions. Bioinspir. Biomim. 7(2).

Mbemmo, E., Z. Chen, S. Shatara and X.B. Tan. 2008. Modeling of biomimetic robotic fish propelled by an ionic polymer-metal composite actuator. 2008 Ieee International Conference on Robotics and Automation, Vols. 1-9, 1050-4729689-694.

Menciassi, A., S. Gorini, G. Pemorio and P. Dario. 2004. A SMA actuated artificial earthworm. 2004 Ieee International Conference on Robotics and Automation, Vols. 1–5, Proceedings, 1050-47293282-3287.

Menguc, Y., Y.L. Park, H. Pei, D. Vogt, P.M. Aubin, E. Winchell, L. Fluke, L. Stirling, R.J. Wood and C.J. Walsh. 2014. Wearable soft sensing suit for human gait measurement. Int. J. Robot Res. 33(14): 1748–1764.

Muth, J.T., D.M. Vogt, R.L. Truby, Y. Menguc, D.B. Kolesky, R.J. Wood and J.A. Lewis. 2014. Embedded 3D printing of strain sensors within highly stretchable elastomers. Advanced Materials 26(36): 6307–6312.

Nemat-Nasser, S. and J.Y. Li. 2000. Electromechanical response of ionic polymer-metal composites. J. Appl. Phys. 87(7): 3321–3331.

Park, J., I. You, S. Shin and U. Jeong. 2015. Material approaches to stretchable strain sensors. ChemPhysChem. 1439–7641.

Park, Y.L., B.R. Chen and R.J. Wood. 2012. Design and fabrication of soft artificial skin using embedded microchannels and liquid conductors. Ieee Sens. J. 12(8): 2711–2718.

Peterson, S.D., M. Porfiri and A. Rovardi. 2009. A particle image velocimetry study of vibrating ionic polymer metal composites in aqueous environments. Ieee-Asme T Mech. 14(4): 474–483.

Porfiri, M. 2009. Influence of electrode surface roughness and steric effects on the nonlinear electromechanical behavior of ionic polymer metal composites. Phys. Rev. E. 79(4).

Prempraneerach, P., F. Hover and M. Triantafyllou. 2003. The effect of chordwise flexibility on the thrust and efficiency of a flapping foil. In Proc. 13th Int. Symp. on Unmanned Untethered Submersible Technology: special session on bioengineering research related to autonomous underwater vehicles, New Hampshire.

Quillin, K.J. 1999. Kinematic scaling of locomotion by hydrostatic animals: ontogeny of peristaltic crawling by the earthworm lumbricus terrestris. Journal of Experimental Biology 202(6): 661–674.

Romano, R. and E.A. Tannuri. 2009. Modeling, control and experimental validation of a novel actuator based on shape memory alloys. Mechatronics 19(7): 1169–1177.

Shahinpoor, M. and K.J. Kim. 2000. The effect of surface-electrode resistance on the performance of ionic polymer-metal composite (IPMIC) artificial muscles. Smart Mater. Struct. 9(4): 543–551.

Shahinpoor, M. and K.J. Kim. 2001. Ionic polymer-metal composites: I. Fundamentals. Smart Mater. Struct. 10(4): 819–833.

Shahinpoor, M. and K.J. Kim. 2005. Ionic polymer-metal composites: IV. Industrial and medical applications. Smart Mater. Struct. 14(1): 197–214.

Shen, Q., K.J. Kim and T.M. Wang. 2014. Electrode of ionic polymer-metal composite sensors: Modeling and experimental investigation. J. Appl. Phys. 115(19).

Shen, Q., T.M. Wang, J.H. Liang and L. Wen. 2013. Hydrodynamic performance of a biomimetic robotic swimmer actuated by ionic polymer-metal composite. Smart Materials and Structures 22(7).

Shepherd, R.F., F. Ilievski, W. Choi, S.A. Morin, A.A. Stokes, A.D. Mazzeo, X. Chen, M. Wang and G.M. Whitesides. 2011. Multigait soft robot. Proceedings of the National Academy of Sciences 108(51): 20400–20403.

Smooth-on. Ecoflex (R) Series/ Super-Soft, Addition Cure Silicone Rubbers. In Smooth-on Inc.

Sreekumar, M., T. Nagarajan, M. Singaperumal, M. Zoppi and R. Molfino. 2007. Critical review of current trends in shape memory alloy actuators for intelligent robots. Ind. Robot 34(4): 285–294.

Sun, J.Y., C. Keplinger, G.M. Whitesides and Z.G. Suo. 2014. Ionic skin. Advanced Materials 26(45): 7608–7614.

Teh, Y.H. and R. Featherstone. 2008. An architecture for fast and accurate control of shape memory alloy actuators. Int. J. Robot Res. 27(5): 595–611.

Trimmer, B. and J. Issberner. 2007. Kinematics of soft-bodied, legged locomotion in Manduca sexta larvae. Biol. Bull-Us 212(2): 130–142.

Trueman, E.R. 1975. Locomotion of soft-bodied animals. Edward Arnold.

Vogt, D.M., Y.L. Park and R.J. Wood. 2013. Design and characterization of a soft multi-axis force sensor using embedded microfluidic channels. Ieee Sens. J. 13(10): 4056–4064.

Vogt, D.M. and R.J. Wood. 2014. Wrist angle measurements using soft sensors. In IEEE Sensors Conf.

Wang, T.M., Q. Shen, L. Wen and J.H. Liang. 2012a. On the thrust performance of an ionic polymer-metal composite actuated robotic fish: Modeling and experimental investigation. Sci. China Technol. Sc. 55(12): 3359–3369.

Wang, T.M., Z.Y. Shi, D. Liu, C. Ma and Z.H. Zhang. 2012b. An accurately controlled antagonistic shape memory alloy actuator with self-sensing. Sensors-Basel 12(6): 7682–7700.

Yamada, T., Y. Hayamizu, Y. Yamamoto, Y. Yomogida, A. Izadi-Najafabadi, D.N. Futaba and K. Hata. 2011. A stretchable carbon nanotube strain sensor for human-motion detection. Nat. Nanotechnol. 6(5): 296–301.

Yim, W., J. Lee and K.J. Kim. 2007. An artificial muscle actuator for biomimetic underwater propulsors. Bioinspir. Biomim. 2(2): S31–S41.

Zheng, T.J., Y.W. Yang, D.T. Branson, R.J. Kang, E. Guglielmino, M. Cianchetti, D.G. Caldwell and G.L. Yang. 2014. Control design of shape memory alloy based multi-arm continuum robot inspired by octopus. C Ind. Elect. Appl. 2156-23181108-1113.

12

Advanced Materials for Biomedical Engineering Applications

Guokui Qin[1,]* and *Xin Kai*[2]

ABSTRACT

This chapter describes mainly the smart design, structural formation, remarkable mechanical behavior and potential biomedical applications of selected natural protein-based advanced biomaterials including silk/silk-like polymers (SLPs), elastin/elastin-like polymers (ELPs), resilin/resilin-like polymers (RLPs) and other natural protein-based biopolymers. The reader will gain insight into the remarkable mechanical properties of the advanced biomaterials, the use of biotechnology to engineer the proteins and specific biomedical applications of these unique protein-based advanced biomaterials. The genetic manipulation and surface modification of these protein-based materials also reveal key relationships between structure and function in advanced biomaterials. The chapter will be involved in the interdisciplinary studies of protein-based advanced biomaterials for many potential

[1] Department of Chemical Engineering, Massachusetts Institute of Technology, 77 Massachusetts Avenue, Cambridge, MA 02139, USA.
[2] Ragon Institute of MGH, MIT and Harvard, Massachusetts General Hospital Cancer Center, Harvard Medical School, 400 Technology Square, Cambridge, MA 02139, USA.
Email: xinkai1980@gmail.com
* Corresponding author: Guokui.Qin@gmail.com; gkqin@mit.edu

applications and will be of interest to many—from graduate students getting started in their research to materials scientists and engineers. The interdisciplinary interchange is at the center of studies on protein-based natural advanced biomaterials. The information provided here including descriptions of advances in the biology, material properties, processing and biomedical applications of natural protein advanced biomaterials should be of interest to researchers in areas relevant to biomedical engineering, mechanical engineering, biology, physics, chemistry and clinical medicine.

1. Introduction

Biological materials or biomaterials can be defined as materials that are non-immunogenic, biocompatible and biodegradable, which can be functionalized with bioactive proteins and chemicals and serve the stated medical and surgical purposes (Cao and Wang 2009). Biopolymers, one group of polymeric biological materials are produced by living organisms for various functions such as information or energy storage, biocatalysts, stabilization and protection (Heim et al. 2010b). Polypeptides or proteins, polynucleic acids and polysaccharides are the examples of biopolymers with complex three-dimensional structures that are responsible for their highly specialized properties (Vendrely et al. 2008). For example, cellulose and chitin are the most abundant polysaccharides on earth and serve as important structural elements in plant cell walls and animal exoskeletons; collagen and elastin, the sequence-specific polypeptides, are synthesized by the DNA-directed templates as the main components of blood vessels, connective tissues and skins in animals and humans (Baier 1988, Eiras et al. 2010, Rusling et al. 2014). Biological synthesis and processing of biopolymers can provide important information on fundamental interactions involved in molecular recognition, self-assembly, and formation of biomaterials with well-defined architectures, features that are relevant for advanced biomaterial needs, such as for drug delivery and tissue engineering (Gagner et al. 2014, Kim 2013, Maskarinec and Tirrell 2005, Romano et al. 2011, Vendrely et al. 2008).

Proteins (known as polypeptides) are essential for all biological systems and contain the prominent secondary structures including α-helices, β-sheets, β-turns and random coils due to supramolecular interactions between side chains of amino acids, that functions in catalysis, binding, signal transduction, protection, and more (Heim et al. 2010b, Krishna and Kiick 2010, Maskarinec and Tirrell 2005, Vendrely et al. 2008). Natural structural proteins are the most versatile representatives of advanced biomaterials, such as silks (Omenetto and Kaplan 2010), elastins (Rodriguez-Cabello 2004), collagens (Chattopadhyay and Raines 2014) and keratins

(Mogosanu et al. 2014), are synthesized in higher organisms from combinations of up to 20 amino acid monomers and characterized by highly ordered domains in the materials formed from these protein polymers. Structural proteins are used for producing natural materials such as hair, connective tissue and silk, all of which show incredible and unique physical properties (Grove and Regan 2012). Because of their impressive mechanical properties, slow degradation *in vivo*, biocompatibility and versatile processing into many material formats, natural structural proteins are particularly suited for advanced biomaterials needs especially for biomedical engineering applications (Grove and Regan 2012).

This chapter describes the smart design, structural formation, remarkable mechanical behavior and potential biomedical applications for selected natural protein-based advanced biomaterials including silk/silk-like polymers (SLPs), elastin/elastin-like polymers (ELPs), resilin/resilin-like polymers (RLPs) and other natural protein-based biopolymers. The reader will gain insight into the remarkable mechanical properties of advanced biomaterials, the use of biotechnology to engineer the proteins, and specific biomedical applications of these unique protein-based advanced biomaterials. The genetic manipulation and surface modification of these protein-based materials also reveal key relationships between structure and function in advanced biomaterials. The chapter will look at the interdisciplinary studies on protein-based advanced biomaterials for many potential applications and will be of interest to many—from graduate students getting started in their research to materials scientists and engineers. The interdisciplinary interchange is at the center of studies on protein-based natural advanced biomaterials. The information provided here including descriptions of advances in the biology, material properties, processing and biomedical applications of natural advanced biomaterials should be of interest to researchers in areas relevant to biomedical engineering, mechanical engineering, biology, physics, chemistry and clinical medicine.

2. Natural Protein-Based Advanced Biomaterials

Structural proteins have been created in nature through billions of years of evolution for a wide variety of biological functions, and the translation of natural structural concepts into bio-inspired materials requires the combining of amino acid sequences and their associated folding patterns that can produce advanced biomaterials with elastic, rigid or tough behaviors (Annabi et al. 2013, Gagner et al. 2014, Main et al. 2013, Maskarinec and Tirrell 2005, Smeenk et al. 2005, van Hest and Tirrell 2001). Natural protein-based biomaterials exhibit desirable mechanical responses or behaviors, such as elasticity – undergo high deformation under stress without rupture, to recover the original state, once stress is removed. A growing number of advanced biomaterials based on silk, elastin and resilin biopolymers provide the challenging examples in materials design for material scientists and are considered as an alternative to conventional synthetic polymers, presenting a promising class of next generation advanced biomaterials for biomedical applications (Desai and Lee 2015).

In natural protein-based biopolymers discussed in this section, resilin and elastin have relatively high extensibility and resilience, but lack stiffness and strength when compared with the collagen and the silks (Su et al. 2014). Collagen and dragline silk are much stiffer materials, but lack the extensibility that is characteristic of the rubber-like proteins (Chattopadhyay and Raines 2014, Omenetto and Kaplan 2010, Su et al. 2014).The molecular origins of the remarkable physical/mechanical properties for protein biomaterials have not been completely understood. However, the primary amino acid sequences of these structural proteins have revealed the critical features that related to their unique structural and functional properties, that is, they are largely comprised of distinct tandem repeats of oligopeptide domains with a well-defined secondary structure, containing short amino acid sequences as protein polymer building blocks (of the order 5–20 residues) which tend to be rich in glycine residues and 'above average' fraction of proline residues (Kim 2013). Three-dimensional architectures are formed further through physically cross-linked networks via self-assembly or with chemical cross-linking to achieve desirable physical and mechanical characteristics (Lu et al. 2010, Lu et al. 2012, Murphy and Kaplan 2009).

2.1 Silk and Silk-Like Polymers (SLPs)

As one of the most ancient insect-derived advanced biomaterials due to its utility in the textile world, silk is originally produced by many insects for different purposes, such as cocoons for survival by silkworms, orb webs for prey capture by spiders, and nest construction by the Hymenoptera (bees, wasps, and ants) (Bellas et al. 2015, Fu et al. 2009, Sutherland et al. 2010, Veldtman et al. 2007). To date, the best-characterized silks include mainly the cocoon silk of the mulberry silkworm *Bombyx mori* and the dragline silk of spiders (*Nephila clavipes* and *Araneus diadematus*) and their astonishing properties have been studied heavily (Altman et al. 2003, Asakura et al. 2003, Fu et al. 2009, Jin and Kaplan 2003, Valluzzi et al. 2002). The high strength and elasticity of silks are the key to their potential utility in advanced biomaterials applications, influenced by temperature, state of hydration and extension rate (Madsen and Vollrath 2000, Shao and Vollrath 2002, Vollrath et al. 2001). In fact, silk is a remarkable advanced biomaterial as strong as aramid filaments such as Kevlar and superior to high-grade steel (Altman et al. 2003). Some spider silks can especially stretch to 140%

of their original length without breaking and hold their strength up to −140°C (Heim et al. 2009, Lewis 2006). However, spider silk is extreme light. A single strand of spider silk long enough to circle the Earth would weigh less than 500 g (Lewis 2006).

The primary amino acid sequences and thus the structure of silk are different for different species of silkworms and spiders with corresponding differences in molecular organization (Jin and Kaplan 2003, Omenetto and Kaplan 2010, Valluzzi et al. 2002). Generally, silks contain a high level of the amino acids glycine, alanine and serine, and have been characterized as natural block copolymers including hydrophobic blocks with short side-chain amino acids such as glycine and alanine, and hydrophilic blocks with larger side-chain amino acids, as well as charged amino acids. They are semicrystalline materials with either ordered molecular structures or β-sheets (crystallites), determining the mechanical properties of silks. Compared with globular proteins, the enhanced environmental stability of silk materials is attributed to the extensive hydrogen bonding and the hydrophobic nature of the protein, which leads to the formation of β-sheets or crystals (Fig. 1) (Keten et al. 2010). For example, an antiparallel β-sheet structure having extended polypeptide chains has been characterized for spider dragline silks and cocoon fibroin silks, with hydrogen bonds formed between the carbonyl oxygen atoms and amide hydrogen atoms from adjacent peptide chains, resulting in a pleated structure along the backbone of the peptide chain (Matsumoto et al. 2008, Rabotyagova et al. 2010, Valluzzi et al. 2002, Vepari and Kaplan 2007, Xu et al. 2014).

The *B. mori* cocoon silk is the most prominent silk production due to its use as a raw material in textiles and as medical sutures that are approved by the U.S. Food and Drug Administration (FDA) (Heim et al. 2010a, Heim et al. 2009, Heim et al. 2010b). They comprise of highly organized

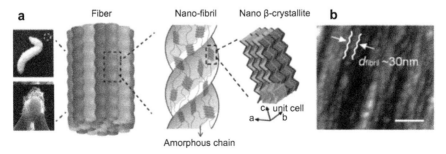

Figure 1. Hierarchical structure of *Bombyx mori* silkworm silk. **(a)** Silk fibers are composed of numerous interlocking nano-fibrils. **(b)** AFM image of the nano-fibrillar structure in silkworm silk with a sequence of linked segments (scale bar: 100 nm). Reproduced with permission from Ref. (Shao and Vollrath 2002), Copyright 2002 Nature Publishing Group 2002 (the cocoon image at lower left adapted), and Ref. (Xu et al. 2014), Copyright 2014 The Royal Society of Chemistry (the rest panels).

β-sheet regions (about 55% of the total structure), including two types of proteins, fibroin and sericin. Sericin is a family of antigenic glue like proteins, helping with the formation of the composite cocoon fibers in nature. Sericin must be removed for biomedical applications by degumming that is a typical process by boiling silk fibers in an aqueous solution of sodium carbonate (Gasperini et al. 2014). Separation is required for silk purification as the sericin proteins might cause an inflammatory response, and solubilizing the degummed silk fibers can be further fabricated and processed into various advanced material formats for a range of biomedical applications, including porous silk sponges, silk films, nano- or micro- scale coatings, hydrogels and nano- and micro- particles (Fig. 2) (Rockwood et al. 2011). The core filaments of cocoon or silkworm silk have at least two major fibroin proteins, a light chain (25 kDa) and a heavy chain (350 kDa) linked by disulfide bonds (Altman et al. 2003). The complete sequence of the fibroin heavy chain contains repetitive amino acids (-Gly-Ala-Gly-Ala-Gly-Ser-) along its sequence, forming a β-sheet secondary structure so that the methyl groups and hydrogen groups of opposing sheets interact. The hydrophobic domains play an important role in the final

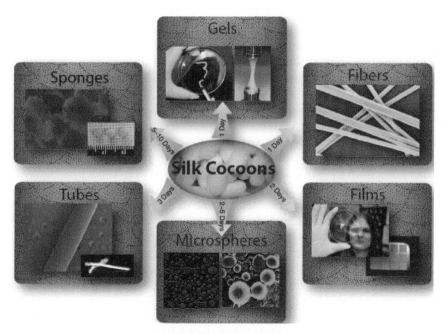

Figure 2. Schematic of material forms fabricated from silk fibroin using both organic solvent- and aqueous-based processing approaches. Overall, the silk fibroin extraction process takes 4 d and the time within the arrows indicates the time required to process the silk fibroin solution into the material of choice. Reprinted with permission from Ref. (Rockwood et al. 2011). Copyright 2011 Nature Publishing Group.

molecular assembly of the proteins into silk fibers, which are responsible for insolubility, leading to the high strength and thermal stability of the silk fibers (Fu et al. 2009). The materials properties of silk fibroins are determined by their special molecular structures which include mainly three different morphologies: (1) silk I, water soluble structure containing random coils and amorphous regions; (2) silk II, insoluble in several solvents (mild acid and alkaline conditions) with antiparallel β-sheets; (3) silk III, which consists of threefold polyglycine II-like helices (Valluzzi et al. 2002). In regenerated silk fibroins, the silk I structure easily converts to a β-sheet structure by chemical methods such as methanol treatment and silk II structure is more stable where the sheets are arranged back to back in alternation (Wilson et al. 2000). Silk II is insoluble and stabilized by strong hydrogen bonds and van der Waals forces that can be broken down by solvents with high ionic strength and high concentration of salts such as lithium bromide to obtain a water-soluble silk I random-coil conformation (Lu et al. 2010). Silk hydrogels can be further processed from silk I fibroin solutions in water with mild conditions, which can be influenced by mechanical stresses, protein concentration, temperature, pH and salt concentration in solution (Bellas et al. 2015, Wu et al. 2012, Yao et al. 2012, Yucel et al. 2009).

Spider silks are remarkable natural polymers and their molecular weights vary from 70–700 kDa with various protein sequences depending on the different spider species (Vepari and Kaplan 2007). The protein sequences with consensus repeat units have been identified (Fig. 3), including three main domains in natural sequences of spider silk: (1) a repetitive middle core domain where two basic sequences, crystalline (rigid) [poly(A) or poly(GA)] and less crystalline (highly elastic) (GGX or GPGXX) polypeptides alternate; (2) and (3) nonrepetitive N-terminal and C-terminal domains which are critical for pH-responsive fiber spinning in insect glands. Moreover, the polyalanine blocks can self-assemble into tightly packed β-sheets that are embedded in an amorphous matrix, leading to the extraordinary mechanical properties of the silk (Jin and Kaplan 2003, Rabotyagova et al. 2010, Valluzzi et al. 2002, Wilson et al. 2000). The dragline silks, major ampullate spidroin 1 and 2 (MaSp1 and MaSp2) have been investigated dramatically for recombinant expression because they can form the toughest fibers (Tokareva et al. 2014, Tokareva et al. 2013). Recent studies of chimeric silks with only one fifth of the native protein length, combining MaSp2 and flagelliform silk containing the elastic 'GPGGX' repeats, have demonstrated the ability to create a highly extensible yet strong silk-like polypeptide (SLP), providing a route for making light-weight advanced materials with high toughness and strength (Lewis 2006). In addition, minor ampullate silk possesses mechanical properties almost similar to major ampullate silks, but does not supercontract in water (Heim et al. 2009).

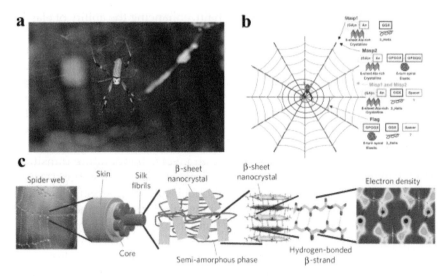

Figure 3. Hierarchical structure of spider silk. **(a)** An adult female orb weaver spider *Nephila clavipes* and her web. **(b)** Schematic overview of *N. clavipes* web composed of three different spider silk proteins and their structures. The coloured boxes indicate the structural motifs in silk proteins. An empty box marked '?' indicates that the secondary structure of the 'spacer' region is unknown. Note: MaSp1 or MaSp2: major ampullate spidroin 1 or 2; MiSp1 and 2: minor ampullate spidroin 1 and 2; Flag: flagelliform protein. **(c)** Schematic of the hierarchical spider silk structure that ranges from nano to macro, including the electron density at the Angstrom scale. Reproduced with permission from Ref. (Tokareva et al. 2013) (panel a & b adapted in an open access article under the terms of the Creative Commons Attribution License), and Ref. (Keten et al. 2010). Copyright 2010 Macmillan Publishers Limited (panel c adapted).

2.2 Elastin and Elastin-Like Polymers (ELPs)

Elastin is one of the main components of the extracellular matrix proteins present in blood vessels, lung epithelium, skin and other tissues where stretch and relax more than a billion times during life, providing structural integrity, high elasticity and resilience (Desai and Lee 2015). Elastin is a heavily cross-linked biopolymer with highly repetitive sequences that formed in the elastogenesis process. The cross-linked elastin is fibrous and hydrophobic, making it insoluble and difficult to isolate. Tropoelastin is a ~ 72 kDa soluble precursor of elastin and a highly repetitive protein with alternating elastic hydrophobic and lysine-rich hydrophilic peptide domains. The elastin protein sequence and genes have been identified for its biochemistry and structure, increasing our understanding of the role of elastin and its potential to biomedical applications (van Eldijk et al. 2012). The biodegradation sequence and specific cell adhesion motifs can be further added to the elastin polymer chain, enhancing the ability of elastin biopolymer for biomedical engineering applications.

In tropoelastin, lysine residues interspersed with alanine are mainly found in the hydrophilic domains, while the hydrophobic domains are composed of repetitive sequence units such as the tetra-, penta-, and hexa- peptides, containing 'VPGG', 'VPGVG', and 'VAPGVG', respectively (Annabi et al. 2013, Grove and Regan 2012, van Eldijk et al. 2012). The hydrophobic domains of tropoelastin are the source of elasticity and intrigue the unique thermal responsiveness, which is critical for mature elastin formation. This phenomenon occurs as an inverse temperature transition (ITT), which is also known as lower critical solution temperature (LCST) behavior, inducing the aggregation of tropoelastin. Tropoelastin is soluble with random-coil conformation in aqueous solutions under the transition temperature (T_t); upon increasing the temperature above its characteristic transition temperature (T_t), the tropoelastin molecule chain aggregates and folds, and its phase separates into a coacervated state, hydrophobically forming a regular, ordered beta-spiral structure stabilized by the interactions between their hydrophobic domains. This phenomenon is fully reversible by heating and cooling and thermodynamically controlled between room and body temperature, which can be influenced by amino acid composition/hydrophilicity, protein length/molecular weight (MW) and protein concentration, as well as ionic strength (salt concentration) and pH in the environment (Krishna and Kiick 2010, Li et al. 2014, van Eldijk et al. 2012) Most notably, it has been found the T_t of ELPs is inversely related to the ELP molecular weight and concentration, indicating the increase in ELP molecular weight or concentration results in a lower T_t of ELPs. In addition, the local pH can also influence the T_t of ELPs by influencing the amino acid sequences (Hassouneh et al. 2010, Rusling et al. 2014, Thapa et al. 2013).

Self-assembly is the transition process from the spontaneous organization of molecules under thermodynamical equilibrium conditions into structurally well-defined and rather stable arrangements through a number of non-covalent interactions (Daamen et al. 2007). Self-assembly is another important property of elastin and ELPs, leading to alignment of elastin molecules to intermolecular cross-linking under physiological conditions. The coacervation based on the LCST behavior of tropoelastin will induce the formation of ordered structure because raising the temperature and the release of water result in dehydration of the hydrophobic side chains, leading to the alignment of tropoelastin molecules or self-assembly (Li et al. 2014, Pinedo-Martin et al. 2014, van Eldijk et al. 2014). The self-assembly behavior of elastin-based biomaterials may be extremely valuable to obtain nano-scale advanced biomaterials with defined structure and mechanical properties, including nanotubes, nanofibres, nanoporous films and nanoparticles, providing the emerging and promising applications for cellular orientation and small-diameter blood vessels in soft tissue regeneration and for drug delivery or growth factor delivery devices (Daamen et al. 2007).

Previous studies have shown the possibility of creating self-assembled advanced systems with thermal responses by combining elastin-like polypeptides (ELPs) and globular proteins (Li and Kiick 2013a, Lv et al. 2010b, Xia et al. 2011b). For example, when introduced with ELPs, the new mCherry-ELPs protein fusion system can self-assemble into micelles or aggregated nanoparticles in solution, and further investigations demonstrated the behavior for order-disorder transition at high concentrations above T_t of ELPs (Qin et al. 2015). In addition, the feasibility of purification of specifically designed fusion proteins has been demonstrated on large scale. For example, proteins can be purified as fusions with ELPs by inverse transition cycling (ITC) (Fig. 4), where the thermo-sensitive solubility imparted by the ELP tag allows for large scale purification of fusion proteins at low cost (Bellucci et al. 2013, Meyer and

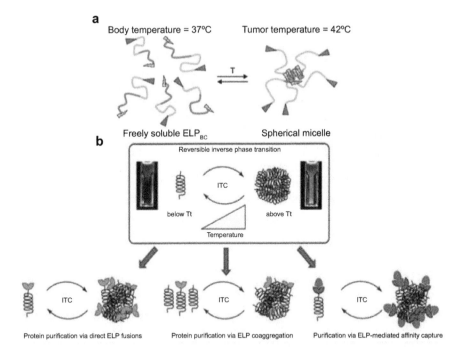

Figure 4. Thermal transition and self-assembly of ELPs. **(a)** Hyperthermia-triggered multivalency. Block copolymers consisting of two ELP blocks, a hydrophilic block and a hydrophobic block were designed. **(b)** Purification of ELPs by ITC is based on the reversible inverse phase transition. Reproduced with permission from Ref. (van Eldijk et al. 2012). Copyright 2011 Springer-Verlag Berlin Heidelberg.

Chilkoti 1999, Trabbic-Carlson et al. 2004a, Trabbic-Carlson et al. 2004b). The fusion protein containing an ELP fused with green fluorescent protein (GFP) was expressed successfully in *E. coli* with nutrient-rich medium without IPTG induction and purified at large scale, and the yield of resulting GFP/ELP fusion was extremely high up to 1.6 g/L of bacterial culture (Chow et al. 2006).

2.3 Resilin and Resilin-Like Polymers (RLPs)

Resilin was discovered in 1960 and it is a highly resilient protein that is a critical component within structures where energy storage and long-range elasticity are needed, such as the flight system of locusts, the jumping mechanism of fleas and the sound producing organ of cicadas (Qin et al. 2009). Resilin is a polymeric rubber-like protein with outstanding mechanical properties. For example, resilin could be stretched up to 3–4 times of its original length before breaking, demonstrating a remarkable capacity for stretching and immediately snap back to its resting length upon release of the tensile force, showing no deformation and great extensibility (Charati et al. 2009, Li and Kiick 2013b, Qin et al. 2012). Resilin is stable up to 140°C, and possesses high resilience (92% or more) and a very high fatigue lifetime, due to the covalent cross-linking between tyrosine residues, generating di- and tri- tyrosines, that is mediated through the action of peroxidases (Andersen 2010, Qin et al. 2011, Su et al. 2014). In the specialized cuticle regions of insects, resilin binds to the cuticle polysaccharide chitin via a chitin binding domain and is further polymerized through oxidation of the tyrosine residues resulting in the formation of di-tyrosine bridges and assembly of a high-performance protein – carbohydrate composite advanced materials (Qin et al. 2012, Qin et al. 2009, Qin et al. 2011).

The investigation of resilin advanced biopolymers have been increased dramatically since the gene CG15920 in *Drosophila melanogaster* was found to encode a resilin precursor due to its amino acid composition and an isoelectric point that resembled resilin closely, as well as the presence of an N-terminal signal peptide sequence for secretion (Andersen 2010, Li and Kiick 2013b, Su et al. 2014). Further sequence analysis showed that resilin protein was 620 amino acids long with highly conserved repeat sequences containing a greater percentage of acidic residues than collagen, elastin and silk fibroin and fewer non-polar residues than silk fibroin and elastin, which might be the reason of resilin's hydrophilicity as well as its low isoelectric point. Resilin also contains more tyrosine residues (~ 5% of the total weight) than the other structural proteins except silk fibroin. Chemical stability of the di- and tri- tyrosine cross-links indicates that resilin might be an ideal network and a variety of cross-linking strategies have been employed to introduce covalent cross-links in resilin (Li and Kiick 2013b,

Qin et al. 2012, Su et al. 2014). The soluble resilin protein from exon 1, rec1-resilin can create cross-linked hydrogels by reacting tyrosines using a peroxidase or through Ru (II) – mediated photo – cross-linking (Elvin et al. 2005). Other RLP (RLP_{12}) incorporating bioactive motif (GRGDSP) could be cross-linked through lysine residues and via THPP ((tris(hydroxymethyl)-phosphino)propionic acid), exhibiting the ability of stretching to average 180% before breaking and cell adhesion and NIH-3T3 proliferation (Desai and Lee 2015, Li and Kiick 2013b, Li and Kiick 2014, Su et al. 2014). The photo-cross-linked GB1-resilin biomaterials were also generated to mimic the unstructured and elastic features of the muscle behavior and served as a molecular spring where resilin was fused for unordered structures, demonstrating high resilience of > 99% and the stretching of 135% without breaking (Lv et al. 2010a).

Resilin protein has been identified for three exons (Fig. 5), including exon 1 (hydrophilic N-terminal segment being highly elastic), exon 2 (hydrophobic mid-segment containing chitin-binding domain (ChBD)) and exon 3 (hydrophilic C-terminal segment that can reversibly undergo conformational changes indicating energy storage). Exon 2 containing

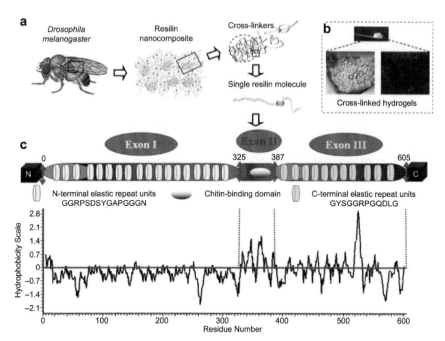

Figure 5. Primary sequence and structure model of resilin in *Drosophila melanogaster*. **(a)** Hierarchical structure of fruit fly resilin. The resilin fibrils with crosslinking consist of two major unstructured peptides derived from exon I and III of full-length resilin. **(b)** Cross-linked resilin hydrogels under ultraviolet. **(c)** Amino-acid sequence scheme of three exons in the full-length resilin protein and hydrophobicity index of the full-length resilin.

62 amino acids showed consensus to a Rebers – Riddiford sequence, and convinced the high affinity to chitin via the ChBD, providing evidences for its role in the formation of the resilin – chitin composites in the cuticle (Andersen 2010, Ardell and Andersen 2001, Qin et al. 2009). A N-terminal region for exon 1 is composed 18 pentadecapeptide repeats (GGRPSDSYGAPGGGN), while a C-terminal region comprising 11 tridecapeptide repeats (GYSGGRPGGQDLG) dominates exon 3, providing the basis for the development of recombinant resilin-like polypeptides (RLPs) that attempt to mimic and recreate the long-range elasticity of natural resilin (Li and Kiick 2013b, Qin et al. 2012, Qin et al. 2011, Su et al. 2014). Both these exons have a high content of glycine and proline, and lack sulfur-containing amino acids or tryptophan. Recent findings have shown the mainly unstructured and flexible chains of resilin and might form β-turns as well as more extended poly-proline II (PPII) secondary structures. Based on scanning probe microscopy (SPM) and tensile testing, the resilin exon 1 (the N-terminal segment) had up to 92% resilience and could be stretched to over 300% of its original length before breaking, exhibiting the near-ideal rubber elasticity, which is potentially useful in the design of advanced hydrogel structures with controlled morphology from resilin proteins that could be exploited as a reservoir for drugs, nanoparticles, enzymes, catalysts and sensor applications (Elvin et al. 2005). The resilin exon 3 (the C-terminal segment) has been studied for energy storage through a reversible conformation transition observed from random coil to β-turns by energy inputs including mechanical stretching and thermal treatments, explaining the molecular elasticity mechanisms for resilin in insects and enabling insects to jump and/or fly with great efficiency (Qin et al. 2012).

Recombinant resilin-like polypeptides (RLPs) have been shown to closely match native resilin in both physical and mechanical properties, composing of tandem repeats of consensus sequences from the N-terminal segment of resilin. For example, the nano-indentation studies by SPM or AFM for RLPs including rec1-resilin, An16 ((AQTPSSQYGAP)$_{16}$) and Dros 16 ((GGRPSDSYGAPGGGN)$_{16}$) confirmed the negligible hysteresis and resilience of 97%, 98% and 91%, respectively (Balu et al. 2014, Lyons et al. 2009, Nairn et al. 2008, Su et al. 2014, Truong et al. 2010). In addition, similar to elastin and ELP, resilin and RLP are also thermo-responsive and behavior as LCST with a sharp transition from hydrophilic to hydrophobic above a transition temperature, and the LCST of the protein occurs at a relatively high temperature of ~ 70°C due to the presence of many hydrophilic residues in RLP sequences (Desai and Lee 2015). Furthermore, the properties of the dual phase transition behavior and pH-responsiveness make resilin and resilin-mimetic advanced biomaterials the good candidates, allowing the control of cell adhesion and migration for biomedical engineering and the creation of a functional surface for biosensors (Liu et al. 2015, Su et al. 2014, Truong et al. 2010).

2.4 Other Natural Protein-Based Advanced Biomaterials

Collagens are the major proteins in the extracellular matrix (ECM) and characterized by their triple-helical molecular structure composed of the $(Gly\text{-}Xaa\text{-}Yaa)_n$ repeating amino acid sequence. They have high content of proline and require the post-translationally modified hydroxyproline (Hyp) to promote stabilization (An et al. 2013, Chattopadhyay and Raines 2014, Faraj et al. 2007). The most abundant collagens are used widely to form axially periodic fibrils in tendon, bone, cartilage and other tissues, playing an important role in cell signaling and development (Chattopadhyay and Raines 2014, Faraj et al. 2007, Nillesen et al. 2007). Collagen molecular structure in fibrils is characterized by the formation of right-handed triple helix, extending to 300 nm in length with 1.5 nm in diameter, and then self-assembled into higher-level supramolecular structures within the fibril. Collagen, one of the versatile structural proteins with triple helices, has mechanical properties and biological functions, providing building block for design of self-assembled advanced biomaterials and other applications. Currently available commercial collagen is derived mostly from animal sources and an alternative biosynthetic methods using genetic and protein engineering can be developed to overcome the potential problem such as immunological responses (Kim 2013). Collagen-like polypeptides (CLPs) can be designed and produced to enhance molecular organization and biological properties by recombinant DNA techniques, based on the most frequently used tripeptide sequences in natural collagen (Grove and Regan 2012, Kim 2013, Main et al. 2013). Recently a bacterial collagen domain was fused with a repetitive cocoon silk consensus sequence to generate the advanced collagen-silk chimeric proteins, allowing more rapid cell interactions with silk-based biomaterials and improving regulation of stem cell growth and differentiation and formation of artificial extracellular matrices useful for tissue engineering applications (An et al. 2013). In addition, collagen can be engineered to develop advanced functionalized collagens with new functional motifs such as repetitive cell binding domains, showing the ability to promote cell adhesion for drug delivery, tissue engineering and wound healing (Chattopadhyay and Raines 2014, Faraj et al. 2007, Nillesen et al. 2007).

Camouflage and signaling/communication are the natural optical features or coloration patterns widespread across the animal kingdom, from the most exotic iridescent patterns of butterfly species to the feathers of peacocks and other birds. Cephalopods can also rapidly alternate the color and reflectance of their skin in response to the environmental or external stimuli, such as specific light pulses, enzymatic reactions, or relative humidity (Grove and Regan 2012, Kim 2013, Krishna and Kiick 2010). Reflectins are a unique group of structural proteins involved in dynamic optical systems in cephalopods and function in camouflage by modulating incident light or bioluminescence. The specialized reflectin architectures have been found as the major component in flat, structural platelets in reflective tissues of the Hawaiian bobtail squid, *Euprymna scolopes* (Cephalopoda: Sepiolidae) (Crookes et al. 2004), producing structural color for camouflage that may have potential applications in the fields of advanced materials science and optical nanotechnology. Hawaiian bobtail squid reflectin proteins possess five repeating domains containing a highly conserved core subdomain, defined by the repeating motif $(M/FD(X)_5MD(X)_5MD(X)_{3/4})$, and could exhibit diverse morphologies, unusual solubility and self-organizing properties. The reflectin proteins can be further processed into thin films, diffraction grating structures, and fibers under various conditions (Crookes et al. 2004, Ghoshal et al. 2014, Izumi et al. 2010, Kramer et al. 2007). Previous studies on native incident light in the Loliginid squids have demonstrated that the dynamic, responsive and tunable optical function of iridophore cells was facilitated by the hierarchical supramolecular assembly of nanoscale reflectin protein particles that elicited large volume changes upon condensation (Tao et al. 2010). Furthermore, thin films created from the recombinant reflectin protein refCBA that reduced complexity compared to native reflectins display interesting optical features and diffraction patterns after self-assembly (Qin et al. 2013). Although little has been reported for the reflectin-mimetic biomaterials, biosynthesis of reflectin-like polypeptides based on repetitive sequences and multilayered thin films generated by bottom-up fabrication provide the opportunity in a range of camouflage and nanostructured advanced devices potentially for optical nanotechnology (Kim 2013).

3. Advanced Biomaterials for Biomedical Engineering Applications

Natural protein-based biopolymers like silk, elastin and collagen have promising advantages over synthetic polymers, providing an important set of advanced material options for biomaterials and scaffolds in biomedical and pharmaceutical applications. Diverse and unique biomechanical properties together with good biocompatibility and controllable biodegradability make natural protein-based biopolymers excellent candidates as advanced biomaterials for drug delivery, tissue engineering scaffolds and wound-healing matrices (House et al. 2010, Park et al. 2010, Wang et al. 2010, Wharram et al. 2010). The synthesized protein-based biopolymers can be further processed into various formats such as films, fibers, scaffolds and hydrogels to expand their applications as advanced biomaterials.

Among these natural polymers, silk-based biomaterials from silkworm cocoon silk have been used for sutures and the core fibroin fibers are comparable to most of the commonly used biomaterials in terms of biocompatibility after sericin is removed (Leal-Egana and Scheibel 2010). Natural spider silks have also demonstrated non-cytotoxicity, low antigenicity and non-inflammatory characteristics. Silk is classified as a non-biodegradable advanced biomaterial as a result of the wax coatings processed on silk fibers. However, recent studies of enzymatic degradation have shown that silk is susceptible to proteolytic degradation and slowly breaks down into smaller polypeptides and free amino acids over time with the adding of α-chymotrypsin that cleaves the less crystalline regions of the silk protein into peptides and protease XIV that degrades the antiparallel β-sheet structures of silks into nanofibrils and subsequently into nanofilaments (Horan et al. 2005, Numata and Kaplan 2010, Numata and Kaplan 2011). The degradation rate relies heavily on the beta-sheet content and is related to the preparation process of silk films or hydrogels. In addition, natural degradation products of silk crystals are not cytotoxicity, compared to cross-beta sheet crystals associated with cytotoxicity and amyloid-like deposits in Alzheimer's and related diseases (Horan et al. 2005, Numata and Kaplan 2010, Numata and Kaplan 2011). Several protein-degrading enzymes can also degrade elastin, including elastases, matrix metalloproteinases (MMPs) and cathepsins, and then interact with other ECM proteins to induce a broad range of biological activities (van Eldijk et al. 2012).

3.1 Sustained Drug Delivery and Controlled Drug Release

The goal of sustained drug delivery is to delivery the drug to the target therapeutic range, continuously maintain the constant drug concentrations within the therapeutically desirable range without peaks and valleys, and extend the inter-dose duration for chronic use medications, providing the potential clinical benefits such as reduction or elimination of unwanted side effects, low toxic thresholds, increased patient convenience and compliance, and enhanced efficacy and cost-efficiency (Pritchard et al. 2013, Pritchard and Kaplan 2011, Yucel et al. 2014b). Controlled degradation and release of drugs from the drug delivery system after accumulation at a specific site are the most important properties as required for regenerative medicine and drug delivery applications, and can be triggered by physiological stimuli such as pH, temperature and ionic strengths to release the encapsulated drugs. Particularly, the difference of extracellular pH of normal tissue (pH 7.2–7.4) and many solid tumors (pH 6.2–6.9) can be used to design a pH-sensitive advanced delivery system, improving the efficiency for the drug delivery application (Nitta and Numata 2013).

Protein-based natural polymers including silk, elastin and collagen have been explored as an advanced vehicle to deliver a wide range of bioactive molecules including genes, small molecules and biological drugs (Pritchard et al. 2013, Pritchard and Kaplan 2011, Yucel et al. 2014b). They are applied as drug carriers for cancer therapy (Yucel et al. 2014a), cartilage repair (Yodmuang et al. 2015) and vascular grafts (Liu et al. 2013) because of their biocompatibility, low toxicity, non-antigenicity, biodegradability, and tunable drug loading and release properties, as well as the abilities of emulsification, gelation, forming and water binding capacity (Numata and Kaplan 2010, Numata and Kaplan 2011, Yucel et al. 2014b). To match the needs of controlled drug delivery well, the combination of material synthesis, processing conditions, drug compounds used and finally drug release kinetics and mechanisms are needed to consider for any future advanced drug delivery. In protein biopolymer-based nanoparticle delivery systems, the design of specific sizes for drugs-loaded nanoparticles is one of the most important criteria to cross epithelial barriers, circulate in the blood vessels before reaching the target site and avoid the inflammatory or immunological responses (Nitta and Numata 2013). The sizes, shapes, solubility, biodegradability and surface properties of biopolymer-based nanoparticles need to be considered for cellular internalization via endocytosis to achieve the site-specific delivery and bioactive drug release at required rate and quantity (Nitta and Numata 2013). Furthermore, protein-based biomaterials can be engineered and incorporated directly with additional features, such as cell-specific targeting, to produce more efficient advanced drug delivery systems.

Silk protein-based materials have been considered for advanced drug delivery systems because of their unique mechanical properties, controlled biodegradation into non-inflammatory by-products, aqueous-based ambient purification and processing options, biocompatibility with sterilization methods and utility in drug stabilization (Pritchard et al. 2013, Yucel et al. 2014a, Yucel et al. 2014b). Various formats based on silk fibroins have been explored from the aqueous silk fibroin solution, processing into materials for advanced drug delivery such as hydrogels, films, tubes, nano/microspheres and transdermal micro-needles (Pritchard et al. 2013, Rabotyagova et al. 2010, Valluzzi et al. 2002). Silk micro- and nano- spheres with controllable sizes have been investigated for the studies of distribution and loading efficiency of drug molecules, resulting in different drug release behaviors in silks used with different hydrophobicity and charge (Wang et al. 2007). Silk film coating was also explored for small molecule drug delivery and the subsequent drug release was regulated by controlled drying and silk film treatment, with drug release profiles lasting (drug retention time) from a few hours to 10+ days, respectively. Further incorporation of protease inhibitors may enhance the ability to control local degradation rates of silk fibroin, improving the efficacy for controlled localization of drug release. Silk fibroins were blended with chitosan polymers to form < 100 nm nanoparticles for local and sustained therapeutic curcumin delivery to cancer cells (Kasoju and Bora 2012). The *in vitro* stability and half-life of insulin were also efficiently improved by conjugating with silk nanoparticles via covalent cross-linking (Humenik and Scheibel 2014, Klok et al. 2004, Lin et al. 2013).

Due to the most powerful property of its tunable ITT, elastin-like polypeptides have been used extensively for therapeutic drug delivery and targeting applications. These thermally associating advanced materials can also be applied for drug loading and release with desirable thermal response properties (Koria et al. 2011, Saxena and Nanjan 2015, Smits et al. 2015). The ITT of a desired ELP relies on hydrophobicity and the molecular weight of the ELP, showing different transition properties of ELPs. The transition temperature of an ELP designed by Chilkoti and coworkers can be tuned to about 41°C, allowing for the localization and remarkable accumulation of the ELP peptides in tumors through the induction of mild hyperthermia without any tags (Christensen et al. 2013, MacEwan and Chilkoti 2014, Rusling et al. 2014). The hydrophobic drug within ELP-drug conjugates might lower ELP ITT to 37°C to allow for physical gelation upon injection, improve residence time of the drug and enable efficient drug release over time to minimize side effects (Kimmerling et al. 2015, MacEwan and Chilkoti 2010, McDaniel et al. 2010, Wu et al. 2009). Recombinant ELPs can also be functionalized with specific targeting sequences or internalization peptides, enhancing the accumulation or intracellular delivery of drug carriers at the disease sites. The chimeric ELP developed with a tumor-homing AP1 peptide that targets cell surface interleukin-4 (IL-4) (an overexpressed cell surface marker in solid tumors) was shown to accumulate preferentially in tumors and significantly enhance the tissue localization (Sarangthem et al. 2013). The functionalization of ELPs fused with cell penetrating peptides (CPPs) has been evaluated to improve efficiency of cellular uptake and targets inside eukaryotic cells by non-specific, receptor-independent mechanisms (Bidwell and Raucher 2010, Massodi et al. 2005, Ryu et al. 2014). The delivery of kinase inhibitor peptide p21 drug using CPP-functionalized ELP cargo demonstrated the enhancement of the interaction of drug cargo with intracellular therapeutic targets and thereby increased drug efficacy (Bidwell and Raucher 2010, Massodi et al. 2005, Ryu et al. 2014). A different strategy to deliver drugs to a specific site rather than diffusing to all tissues and affecting normal cells is to use amphiphilic diblock ELP chains with a self-associating hydrophobic block fused with a hydrophilic block. The relatively hydrophobic domain in diblock ELPs may coacervate at a lower temperature, resulting in self-association while the other block still remains soluble, that is temperature-dependent amphiphilicity. Amphiphilic ELPs have demonstrated the high tumor vasculature retention once fused with tumor-targeting sequences NGR to target CD-13 receptors in tumor vasculature, providing the potential of these ELP-based nanoparticles for advanced targeted drug delivery (MacEwan and Chilkoti 2014, McDaniel et al. 2010, Wu et al. 2009).

The self-assembly investigation of silk-elastin-like polymers (SELPs) demonstrated the formation of the core of micelle-like nanoparticles by adding hydrophobic molecules with the size range from 20 to 150 nm in diameter, which is enough to cross the endothelial barrier, making them the good candidates as advanced drug delivery vehicles (Xia et al. 2014). Furthermore, protein-based biomaterials can be designed and synthesized by recombinant DNA techniques, expanding the versatility of protein-based advanced biomaterials with tightly controlled drug delivery capabilities (Numata et al. 2012). Silk copolymers have been engineered and incorporated with a specific peptide sequence for targeting and localization. For example, the combination of bioengineered silks with tumor homing peptides (THPs) would offer the opportunity to enable functionalization for targeted drug delivery, enhancing significantly the target specificity of the resulting nanoscale drug-loaded spheres to tumor cells with low toxicity (Numata and Kaplan 2010, Numata and Kaplan 2011, Numata et al. 2012).

3.2 Gene Delivery and Gene Therapy

It is believed the human disease can be cured by the transfer of genetic materials into specific patient cells to supply defective genes, and this strategy of gene therapy has been applied for many diseases including cancer, AIDS and cardiovascular diseases (Nitta and Numata 2013). The delivery of therapeutic genes into target cells in the patient is a promising approach for the treatment of various diseases, with either naked DNA or a viral vector used. To do so, the advanced gene delivery system with gene encapsulation must be small enough to internalize into cells and passage to the nucleus, escape endosome-lysosome processing and following endocytosis and finally protect the gene until it reaches its target site. To improve the safety and efficacy of gene delivery, current studies attempt to localize the gene delivery to particular tissue, protect the DNA from degradation, and control gene release profiles (Numata et al. 2010, Numata and Kaplan 2010, Numata et al. 2011, Numata et al. 2009).

Compared to other gene delivery vehicles such as liposomes and synthetic polymers, protein-based advanced biomaterials are used commonly to deliver plasmid DNA or adenoviral vectors due to their ability to be functionalized (Yucel et al. 2014b). An example from the amphiphilic diblock copolymer complexes composed of silk repetitive oligopeptide block and poly(L-lysine) block (pLL) shows the non-cytotoxicity ability to deliver plasmid DNA for non-viral gene therapy, where the anionic plasmid DNA (pDNA) can form ionic pairs with the cationic pLL block via electrostatic interactions. The resulting silk-pDNA complexes may then be further functionalized with cell-binding motif to enhance cell binding and modified with cell penetrating and cell membrane destabilizing peptides to improve transfection efficiency, allowing for cell-specific targeting and efficient gene transfer (Numata and Kaplan 2010, Numata et al. 2009, Yucel et al. 2014b). In addition, the transfection efficiency of silk-pDNA complexes modified with RGD motif might be determined by the number of RGD domain if applied the silk-pLL-RGD fusion block copolymers for gene delivery to several cell types (Kim et al. 2014, Kim 2013, Numata et al. 2010, Numata et al. 2009, Wu et al. 2012).

The stimuli-responsive ELPs and SELPs have potential to be served as advanced polymeric matrices for gene delivery, enabling the hydrogel formation once injected in the body while being liquid at room temperature, an attractive feature for any injectable system application. For example, recombinant SELP-47K hydrogels have been reported the controlled gene delivery with adenoviral vector delivery, showing the inverse DNA release and diffusivity related to the molecular weight of the plasmids used (Kim et al. 2012, Megeed et al. 2006, Swierczewska et al. 2008). The binding/releasing mechanism of DNA to SELPs was further explored by the influence of different factors such as ionic strength, DNA concentration, SELP concentration and molecular weight. The results have shown the increased release of DNA and adenovirus bound within polymeric matrices if increasing the ionic strength in buffer or lower concentration of the polymers (Kim et al. 2012, Megeed et al. 2006, Swierczewska et al. 2008). Particularly SELPs hydrogels loaded with adenovirus in a mouse model demonstrated greater reduction in tumor volume as compared to control injections of adenovirus in saline solution, providing the effective route for adenoviral gene therapy for cancer treatment (Numata and Kaplan 2010, Numata et al. 2009, Yucel et al. 2014b).

3.3 Tissue Regeneration and Tissue Engineering

The treatment of organ failure is heavily limited by donor supply and increasing morbidity, and the regeneration of functional tissue is still challenging to closely mimic the *in vivo* physiological microenvironment for desired cellular responses. Good communication between the host and implanted system is critical for substitution of a human body part with a material. The goal of tissue engineering is to regenerate tissue within suitable scaffold for implanting the constructed tissue at the target site. Using cells, scaffolds and appropriate growth factors in tissue engineering is a key approach in the treatments of tissue or organ failure. Advanced biomaterials 3D tissue engineering scaffold may provide a suitable microenvironment, acting as an architectural template (Kundu et al. 2013, Kundu et al. 2014, Kundu et al. 2010). Structural protein, being a component of natural tissues, is a rational choice to be used as porous 3D tissue scaffolds in tissue engineering, including silk, elastin and collagen. Functional requirements in tissue repair, regeneration and implantation for biomaterial scaffolds include providing support, surface topography and charge for cell attachment, mitogenesis and cell differentiation. Biomaterial scaffolds that mimic native extracellular matrix (ECM) have been studied to match the functional requirements for specific tissues, providing the potential to produce a functional tissue and organ (Annabi et al. 2013, Desai and Lee 2015, Gagner et al. 2014, Keatch et al. 2012, Khaing and Schmidt 2012, Kim 2013).

Silk protein fibroin is an attractive advanced biomaterial for tissue engineering because of the unique combination of elasticity and strength along with mammalian cell compatibility, which can be used effectively to produce a scaffolding material for development of advanced biomedical device (Desai and Lee 2015, Gagner et al. 2014, Keatch et al. 2012). Silk porous 3D sponges are ideal structures for tissue engineering scaffolds, which can be prepared by freeze drying, porogenic leaching and solid free form fabrication techniques with a good control over porosity and pore sizes. The resulting sponges possess the range of pore sizes from 60 to 250 µm, relying on the freezing temperature, pH and organic solvents. Due to the favorable tensile strength and their ability to be sterilized, silk-engineered scaffolds have been produced as substrates that mimic nanoscale properties of native ECM for cell attachment, cell proliferation and tissue regeneration (Bellas et al. 2015, Mandal et al. 2011, Park et al. 2012, Preda et al. 2013, Yodmuang et al. 2015) including tissue bone, ligaments, tendons, blood vessels and skin and cartilage (Dinis et al. 2013, Dinis et al. 2015, Elia et al. 2014, Hronik-Tupaj et al. 2013, Kimmerling et al. 2015, Liu et al. 2013, Lovett et al. 2015, Seib et al. 2013, Yodmuang et al. 2015). The mechanical and biological functions of protein-based biomaterials may be tailored by genetic engineering and surface chemical modification to produce advanced hybrid and composite systems and thus match tissue-specific needs (Vepari et al. 2010). For example, spider silks have been genetically engineered with various functionalities including the mineralizing domain R5 to perform bone like properties and dentin matrix protein 1 to mineralize calcium phosphate (CaP) (Wong Po Foo et al. 2006), and silk protein properties may be further enhanced through binding and delivery of cell signaling factors such as RGD-functionalized SLP to improve cell adhesion (Bini et al. 2006, Gil et al. 2010, Morgan et al. 2008). In addition, composite silk 3D scaffolds can be prepared to obtain good mechanical and biological outcomes by combining inorganic or organic fillers and by bio-mimicking approaches with other natural extracellular materials since the complex structure of native tissue requires a composite scaffolding material. The successful example is the using of composites of silk fibroin and human-like-collagen for the development of vascular constructs. Furthermore, silk layering with collagen-I could enhance the cell attachment and dispersion of keratinocytes cells, while silk coating with fibronectin might improve the cell adhesion and dispersion within the matrix for both keratinocytes and fibroblasts cells (An et al. 2013, Bhardwaj et al. 2015, Vasconcelos et al. 2008). Other silk composites successfully used include nano-fibrous silk-chitin, silk-collagen and silk fibroin-alginate blended scaffolds. All the findings suggest that the blending of silk fibroins with other natural materials may offer better prospects than pure silk fibroin for tissue regeneration.

Elastin is a main component of the extracellular matrix with non-immunogenic, biocompatible and biodegradable properties, providing the attractive advanced materials from elastin-derived biomaterials for tissue engineering. Easy purification by exploiting their ITT from bacteria with high yields of ELP protein production make ELPs the good candidate for cartilaginous, vascular, ocular and liver tissue regeneration. ELPs can be processed further to form various advanced material formats such as hydrogels, films and fibers to match the application needs. For example, ELP hydrogels were used to mimic ECM-like 3D environments for cell encapsulation in tissue engineering. The enzymatically cross-linked ELPs via lysine results in the encapsulation of chondrocytes and the formation of hyaline cartilage-like substrate rich in collagen II (McHale et al. 2005). By using cysteine-based disulfide bridge cross-links, the ELP hydrogels can be engineered further with gelation rate and gel stiffness controlled by H_2O_2 and protein concentration, enabling cell encapsulation and *in situ* formation of soft gels useful in tissue engineering (McHale et al. 2005, Trabbic-Carlson et al. 2003, Xu et al. 2012). ELP cross-linked hydrogels can also be cast into elastic films, performed the further cross-linking via lysine residues linking using chemicals, and fused with cell-adhesive motifs and related functional modules. For example, the simple surface coatings of ELPs fused with RGD and CS5 domain have been developed to investigate their biochemical effects on cells, generally improving cell-interactive properties of ELP scaffold surfaces (Heilshorn et al. 2003, Liu et al. 2004, Rodriguez-Cabello 2004). Additional studies on higher surface area of ELP fibers from concentrated protein solutions have demonstrated a better display of cell signaling modules to the interacting cells (Benitez et al. 2013).

Resilin and RLPs have also been proposed as the promising advanced biomaterials for tissue engineering due to their outstanding mechanical properties. For example, Resilin-based biomaterials (RZ10) has shown an unconfined compressive modulus similar to that of human cartilage, and the investigation of the chimeric material RZ10 fused with the RGD sequence (derived from fibronectin) has shown the faster cell spreading with well-organized actin structures of human mesenchymal stem cells (hMSCs), suggesting that RGD sequence can be recognized specifically and used for supporting of cell adhesion and spreading within resilin-based materials (Renner et al. 2012a, Renner et al. 2012b, Su et al. 2014). In addition, Resilin RLP12 hydrogels exhibited the comparable mechanical strength and extensibility similar to native vocal fold tissues at high frequencies that corresponded to the human voice (Li et al. 2013). Moreover, resilin RLP24 supported viability and spreading of encapsulated human aortic adventitial fibroblasts useful for cardiovascular applications (Li and Kiick 2013b, Li et al. 2011, Li et al. 2013).

A successful biomedical scaffold in tissue regeneration must allow homogeneous cell distribution within the whole cell culture matrix for inducing cellular activities such as cell attachment, proliferation and even differentiation, and then regenerate complex architectures of various tissues. Three-dimensional (3D) bioprinting technology is coupled with an accurate positioning system based on a deposition/encapsulation system. Recently it has been introduced in tissue engineering because of their efficiency and versatility in cell distribution within the 3D structures, enabling 3D printing of biocompatible advanced materials, cells and supporting components into complex 3D functional living tissues (Kolesky et al. 2014, Murphy and Atala 2014, Wang et al. 2005). With this particularly complex 3D encapsulating printer used, the protein-based biomaterials might be possible to translate directly into cell-laden scaffolds, and the sol-gel transition might also be designed and controlled by quickly heating (not too long to cause thermal shock to cells) and cooling cycle to obtain the advanced hydrogel from protein solutions (Gasperini et al. 2014). It has been believed that the combination of protein-based advanced biomaterials and directed 3D bioprinting technology will offer a promising strategy for future design and manufacture of soft and hard tissue regenerative substitutes and address the need for tissue and organs suitable for transplantation in regenerative medicine (Skardal and Atala 2014).

3.4 Wounds and Burns Dressing

Natural biopolymers including polysaccharides and fibrous proteins can be used widely for wounds and burns dressing materials that speed up the wound healing process because of their biocompatibility, biodegradability and similarity to the extracellular matrix (ECM), providing an optimal microenvironment for cell proliferation, migration and differentiation. Due to their three dimensional cross-linked polymeric networks, wounds and burns dressing materials made from natural biopolymers can maintain a suitable moisture and oxygen at the wound level, prevent and keep the wound mainly against microorganisms, and improve the wound healing process, which is useful for the regeneration and repairing of dermal and epidermal tissues (Mogosanu and Grumezescu 2014). Recent studies of several promising biopolymers will led to a substantial development of advanced wound dressings for regenerative medicine, such as silk fibroin and collagen (Ghezzi et al. 2011, Wlodarczyk-Biegun et al. 2014, Zhu et al. 2015).

As the natural biopolymer sutures for wound ligation with a long history of applications in the human body, silk fibers fulfill complex surgical needs for advanced wound dressing including good biocompatibility, slow and controllable biodegradability, flexibility (i.e., elasticity), and minimal inflammatory reaction (Altman et al. 2003, Fu et al. 2009, Heim et al. 2009, Lewis 2006, Wharram et al. 2010). Silk sutures have been applied for the treatments of skin wounds, lips, eyes and oral surgeries. Therefore, various systems with silk fibroins (SF) used have been explored for advanced wound dressing or healing, such as silk porous films (Gil et al. 2013), electrospun silk nanofibers with multiwalled carbon nanotubes (Jeong et al. 2014), silk-alginate-blended sponges/membranes (Mehta et al. 2015).

Compared to silk materials, other natural biomaterials including hyaluronic acid, gelatin and alginate may not provide sufficient mechanical strength, with the weakness of accelerated degradation (Altman et al. 2003, Wang et al. 2007, Wong Po Foo and Kaplan 2002). Collagen is the most abundant protein in the human body and the skin and is commonly used for wound dressings with minimal to moderate exudates (An et al. 2013, Muralidharan et al. 2013, Ruszczak 2003). Numerous studies of different collagen dressings have been reported for wounds and burns with various formulations, such as collagen sponges for deep skin wounds, collagen resorbable membranes for oral wounds, collagen electrospun nanofibrous scaffolds for wound repair, and collagen hydrogels for wound infections (Chang et al. 2010, Jorgensen 2003, Kim et al. 2015, Oryan 1995, Rudnick 2006). Collagen can also cross-link with other natural polymers to generate advanced composite wound dressing materials, such as collagen – alginic acid cross-linked biopolymers with thermostability and biodegradability (An et al. 2013, Sarithakumari and Kurup 2013, Sell et al. 2009).

4. Design and Exploration of Artificial Advanced Biomaterials

The fundamental understanding of the sequence – structure – function properties of naturally occurring structural proteins plays a key role in design and synthesis of novel advanced materials with the ability to self-assemble, respond to stimuli, and/or promote cell interactions. However, the design for advanced biomaterials from natural sources was limited by restricted material amounts, the heterogeneity of post-translational modifications, the inability to readily introduce point mutations into the sequences and other changes tailored for precise control over spatial and temporal release. Unlike the majority of synthetic chemical strategies used, the developments in genetic engineering and DNA manipulation techniques enable the optimization of structure and *de novo* design of protein-based advanced materials, allowing for the production of monodisperse polymers with interesting mechanical and biological properties. Such properties can be attributed to specifically defined structural modules, originating from a modular domain of natural proteins such as silk, elastin, collagen and resilin and unique secondary structures combined with great flexibility, as well as functional elements identified from other proteins such as cell binding sites or enzymatic domains (Grove and Regan 2012).

Recent advances in genetic engineering have provided a promising approach to design and synthesize artificial protein-based advanced biomaterials with similar behaviors to their native counterpart, enabling the self-assembly into fibrous structures with a regularly repeating and well-defined secondary structure. Various protein expression systems including bacteria, yeast, plant and mammalian cells have been investigated for cloning and expression of native and synthetic protein biopolymers to mimic the modular primary structure of proteins with unique physical and biological properties (Fahnestock et al. 2000, Tokareva et al. 2013, Winkler and Kaplan 2000). Among of them, *Escherichia coli* is widely used to express protein biopolymers such as silk, elastin, resilin and other biomimetic protein polymers, because of the easy genetic manipulation of the target genes and simple purification procedures afterwards, even if *E. coli* lacks post-translational modification for eukaryotic coding sequences. Moreover, besides making the large synthetic genes encoding the repetitive amino acid sequences and producing the repetitive protein biopolymers in various host cells, recombinant DNA techniques are ideally suited for the introduction of additional functionalities by re-engineering biologically functional peptide motifs such as cell-binding domains to the repetitive gene, showing their good biocompatibility and biodegradation when implanted and extending their biomedical applications as new advanced biomaterials including drug delivery and tissue engineering (Kim 2013).

Silk as a natural protein fiber plays crucial roles in the survival and reproduction of many silk-spinning insects (Sutherland et al. 2010), exhibiting different compositions, structures and properties based on special sources. Spider silks have not been commercialized for biomaterials production because of the predatory nature of spiders and the relatively low levels of silk production in an orb web. Recently the strategy of genetic engineering has provided new opportunities to overcome these limitations by cloning and expression of recombinant spider silk genes in bacteria, and multimeric and chimeric silk-like proteins (SLPs) fused in a single protein with multiple properties will be extremely valuable for advanced biomaterial investigations in various applications (Fig. 6). Genetically engineered SLPs may therefore be constructed using synthetic oligonucleotide version of consensus repeats based on highly repetitive amino acid sequences of silks (Altman et al. 2003, Tokareva et al. 2013, Wong Po Foo and Kaplan 2002). A variety of hosts were used for longer SLPs expression, including bacteria (*E. coli*), transgenic silkworms, transgenic plants and mammalian cells. Specially, various types of recombinant spider silks can be expressed in *E. coli* for structure characterization and assembly regulation through genetic manipulation (Partlow et al. 2014, Tokareva et al. 2013, Wong Po Foo and Kaplan 2002), allowing the precise control and efficient packing of silk proteins for the mechanical strength and stability of silk fibers, as well as the control of cell interactions and the rate of degradation (Altman et al. 2003, Horan et al. 2005, Omenetto and Kaplan 2010, Tokareva et al. 2013, Valluzzi et al. 2002). However, recombinant silk materials with limited SLP length (a critical factor defining silk mechanical properties) might not recapitulate the full mechanical potential of native silk fibers and the large-scale yield of spider silks with longer proteins remains challenging

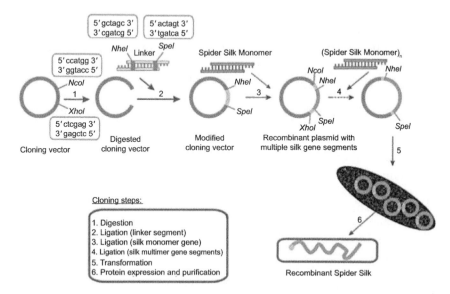

Figure 6. Recombinant DNA approach used to prepare silk-like proteins. Reprinted with permission from Ref. (Tokareva et al. 2013). An open access article under the terms of the Creative Commons Attribution License.

to express in bacteria, because of the insoluble expressed products and high glycine content within highly repetitive sequences. Previously, a SLP composed of spider silk MaSp1-derived sequence was successfully expressed in *E. coli* with a molecular weight of ~ 285 kDa, showing a comparable mechanical properties to native spider silks (Xia et al. 2010).

Elastins are extremely valuable for stimuli-responsive applications because of their high extensibility and stimuli-triggered self-assembly and molecule delivery. Natural elastin can be extracted from tissues by harsh alkaline treatments but with a poor yield (Gasperini et al. 2014). However, recombinant elastin-like polypeptides (ELPs) can recapture the original elasticity and thermo-responsiveness of elastins, which commonly use the hydrophobic domain-derived pentapeptide repeats 'VPGVG' (the most abundant sequence in natural human elastin) or more generally 'VPGXG' ('X' can be any amino acid except proline). Precise control over the ELP sequence can also be used to create the sequence architecture that enables ELP self-assembly. ELP diblock copolymers can be designed for self-assembled materials, consisting of a hydrophobic ELP segment fused to a hydrophilic ELP segment. An example from an elastin-derived material ELP_{BC} composed of a hydrophilic $[VPGEG(IPGAG)_4]_{14}$ block and hydrophobic $[VPGFG(IPGVG)_4]_{16}$ block connected via a VPGEG linker results in a self-assembly behavior of aggregation of the micelles into micron-scale particles (Le et al. 2013, Rodriguez-Cabello 2004, Wright and Conticello 2002). Recombinant ELP proteins are mainly expressed in *E. coli* and purified through relatively easy procedures due to the thermo-responsiveness of elastin with a reversible sol-gel transition upon heating. Recent studies on thioredoxin, tendamistat and virus capsid proteins have also demonstrated the potential of thermo-responsive ELP sequences as fusion tags for purification of other recombinant proteins (Meyer and Chilkoti 2004, Meyer et al. 2001, Rodriguez-Cabello 2004).

Recently, the strategies combining two or more structural proteins offer opportunities to create *de novo* non-natural chimeric advanced biomaterials tailored for specific applications. For example, the silk-elastin chimeric proteins have been generated to make the advanced silk-elastin-like polymers (SELPs) combining the high tensile strength of silk and high resilience of elastin in a single structure, where the silk blocks mimic the natural silkworm fibroin and tend to assemble into β-sheets while the elastin blocks are highly hydrated and disordered and provide thermo-response self-assembly. Different SELPs have been produced by varying the ratio of silk to elastin blocks, exhibiting large differences in solubility and different phase separation and self-assembly below and above the elastin transition temperature (Xia et al. 2011a). When engineered elastin blocks into chimeric fusion proteins, the overall secondary structure, stability and thermo-responsive self-assembly properties depend on the orientation and number of elastin blocks. Previous studies of chimeric system containing elastin blocks and coiled-coil matrix proteins have shown the decreased thermo transition temperatures from 27.8°C to 59.8°C if increasing elastin block number from one to four, as well as the importance of protein block directionality in fusion system for phase transition behavior, providing the additional tunable features for future advanced biomaterial design (Dai et al. 2011). The other chimeric protein based on resilin-elastin-collagen polypeptides (REC) was also designed and demonstrated the remarkable elasticity and self-assembly into fibrous structures with a Young's modulus between 0.1 and 3 MPa, much softer than collagen-like bundles (Bracalello et al. 2011). Specially the fully biosynthetic analogues to protein-polymer conjugates, mCherry-ELPs fusion proteins containing a thermo-responsive coil-like protein, ELP, and a globular protein, mCherry, have been developed, showing the self-assembly of biofunctional nanostructures such as hexagonal and lamellar phases in concentrated solutions. This new system provides a rich landscape to explore the capabilities of fusion architecture to control supramolecular assemblies for advanced heterogeneous biocatalysts (Qin et al. 2015).

5. Summary

Natural protein-based advanced biomaterials are reviewed in this chapter, looking at the recent advances in a broad range of natural polymeric biomaterials such as silk, elastin, resilin and others. Their protein composition and molecular structure, mechanical properties, and biomedical engineering applications were discussed, as well as the design, bioengineering, and processing of advanced biomaterials for biomedical engineering applications. Naturally occurring polypeptide-based biomaterials show incredibly outstanding properties when compared to synthetic polymer materials. Therefore, the likely roles for the compelling class of advanced biomaterials are likely to increase significantly, allowing mechanically robust, slowly degrading and versatile biomaterial designs with low inflammation and low immune response. Controllable processability and surface modifications also expand their utility in drug delivery and functional tissue engineering. To create novel and enough protein materials, the alternative way instead of growing natural sources including animals and insects, recombinant DNA technique has been introduced to generate these advanced biomaterials on a large scale. It is possible that the material structure and properties may be fine tuned and defined for protein-based natural biopolymers with the advances in protein engineering. A future challenge will be to scale-up the protein production and purification of recombinant protein biopolymers with all necessary modifications, and finally extend the application of advanced biomaterials in biomedical engineering.

References

Altman, G.H., F. Diaz, C. Jakuba, T. Calabro, R.L. Horan, J. Chen, H. Lu, J. Richmond and D.L. Kaplan. 2003. Silk-based biomaterials. Biomaterials 24: 401–416.

An, B., T.M. DesRochers, G. Qin, X. Xia, G. Thiagarajan, B. Brodsky and D.L. Kaplan. 2013. The influence of specific binding of collagen-silk chimeras to silk biomaterials on hMSC behavior. Biomaterials 34: 402–412.

Andersen, S.O. 2010. Studies on resilin-like gene products in insects. Insect Biochem. Mol. Biol. 40: 541–551.

Annabi, N., S.M. Mithieux, G. Camci-Unal, M.R. Dokmeci, A.S. Weiss and A. Khademhosseini. 2013. Elastomeric recombinant protein-based biomaterials. Biochem. Eng. J. 77: 110–118.

Ardell, D.H. and S.O. Andersen. 2001. Tentative identification of a resilin gene in Drosophila melanogaster. Insect Biochem. Mol. Biol. 31: 965–970.

Asakura, T., K. Nitta, M. Yang, J. Yao, Y. Nakazawa and D.L. Kaplan. 2003. Synthesis and characterization of chimeric silkworm silk. Biomacromolecules 4: 815–820.

Baier, R.E. 1988. Advanced biomaterials development from "natural products". J. Biomater. Appl. 2: 615–626.

Balu, R., N.K. Dutta, N.R. Choudhury, C.M. Elvin, R.E. Lyons, R. Knott and A.J. Hill. 2014. An16-resilin: an advanced multi-stimuli-responsive resilin-mimetic protein polymer. Acta Biomater. 10: 4768–4777.

Bellas, E., T.J. Lo, E.P. Fournier, J.E. Brown, R.D. Abbott, E.S. Gil, K.G. Marra, J.P. Rubin, G.G. Leisk and D.L. Kaplan. 2015. Injectable silk foams for soft tissue regeneration. Adv. Healthc Mater. 4: 452–459.

Bellucci, J.J., M. Amiram, J. Bhattacharyya, D. McCafferty and A. Chilkoti. 2013. Three-in-one chromatography-free purification, tag removal, and site-specific modification of recombinant fusion proteins using sortaseA and elastin-like polypeptides. Angewandte Chemie-International Edition 52: 3703–3708.

Benitez, P.L., J.A. Sweet, H. Fink, K.P. Chennazhi, S.V. Nair, A. Enejder and S.C. Heilshorn. 2013. Sequence-specific crosslinking of electrospun, elastin-like protein preserves bioactivity and native-like mechanics. Adv. Healthc Mater. 2: 114–118.

Bhardwaj, N., W.T. Sow, D. Devi, K.W. Ng, B.B. Mandal and N.J. Cho. 2015. Silk fibroin-keratin based 3D scaffolds as a dermal substitute for skin tissue engineering. Integr. Biol. (Camb.) 7: 53–63.

Bidwell, G.L. and D. Raucher. 2010. Cell penetrating elastin-like polypeptides for therapeutic peptide delivery. Adv. Drug Deliv. Rev. 62: 1486–1496.

Bini, E., C.W. Foo, J. Huang, V. Karageorgiou, B. Kitchel and D.L. Kaplan. 2006. RGD-functionalized bioengineered spider dragline silk biomaterial. Biomacromolecules 7: 3139–3145.

Bracalello, A., V. Santopietro, M. Vassalli, G. Marletta, R. Del Gaudio, B. Bochicchio and A. Pepe. 2011. Design and production of a chimeric resilin-, elastin-, and collagen-like engineered polypeptide. Biomacromolecules 12: 2957–2965.

Cao, Y. and B. Wang. 2009. Biodegradation of silk biomaterials. Int. J. Mol. Sci. 10: 1514–1524.

Chang, P.J., M.Y. Chen, Y.S. Huang, C.H. Lee, C.C. Huang, C.F. Lam and Y.C. Tsai. 2010. Morphine enhances tissue content of collagen and increases wound tensile strength. J. Anesth 24: 240–246.

Charati, M.B., J.L. Ifkovits, J.A. Burdick, J.G. Linhardt and K.L. Kiick. 2009. Hydrophilic elastomeric biomaterials based on resilin-like polypeptides. Soft Matter 5: 3412–3416.

Chattopadhyay, S. and R.T. Raines. 2014. Review collagen-based biomaterials for wound healing. Biopolymers 101: 821–833.

Chow, D.C., M.R. Dreher, K. Trabbic-Carlson and A. Chilkoti. 2006. Ultra-high expression of a thermally responsive recombinant fusion protein in *E. coli*. Biotechnol. Prog. 22: 638–646.

Christensen, T., W. Hassouneh, K. Trabbic-Carlson and A. Chilkoti. 2013. Predicting transition temperatures of elastin-like polypeptide fusion proteins. Biomacromolecules 14: 1514–1519.

Crookes, W.J., L.L. Ding, Q.L. Huang, J.R. Kimbell, J. Horwitz and M.J. McFall-Ngai. 2004. Reflectins: the unusual proteins of squid reflective tissues. Science 303: 235–238.

Daamen, W.F., J.H. Veerkamp, J.C. van Hest and T.H. van Kuppevelt. 2007. Elastin as a biomaterial for tissue engineering. Biomaterials 28: 4378–4398.

Dai, M., J. Haghpanah, N. Singh, E.W. Roth, A. Liang, R.S. Tu and J.K. Montclare. 2011. Artificial protein block polymer libraries bearing two SADs: effects of elastin domain repeats. Biomacromolecules 12: 4240–4246.

Desai, M.S. and S.W. Lee. 2015. Protein-based functional nanomaterial design for bioengineering applications. Wiley Interdiscip. Rev. Nanomed. Nanobiotechnol. 7: 69–97.

Dinis, T., G. Vidal, F. Marin, D. Kaplan and C. Egles. 2013. Silk nerve: bioactive implant for peripheral nerve regeneration. Comput. Methods Biomech. Biomed. Engin. 16 Suppl. 1: 253–254.

Dinis, T.M., R. Elia, G. Vidal, Q. Dermigny, C. Denoeud, D.L. Kaplan, C. Egles and F. Marin. 2015. 3D multi-channel bi-functionalized silk electrospun conduits for peripheral nerve regeneration. J. Mech. Behav. Biomed. Mater. 41: 43–55.

Eiras, C., A.C. Santos, M.F. Zampa, A.C. de Brito, C.J. Leopoldo Constantino, V. Zucolotto and J.R. dos Santos, Jr. 2010. Natural polysaccharides as active biomaterials in nanostructured films for sensing. J. Biomater. Sci. Polym. Ed. 21: 1533–1543.

Elia, R., C.D. Michelson, A.L. Perera, T.F. Brunner, M. Harsono, G.G. Leisk, G. Kugel and D.L. Kaplan. 2015. Electrodeposited silk coatings for bone implants. J. Biomed. Mater. Res. B Appl. Biomater. 103: 1602–1609.

Elvin, C.M., A.G. Carr, M.G. Huson, J.M. Maxwell, R.D. Pearson, T. Vuocolo, N.E. Liyou, D.C. Wong, D.J. Merritt and N.E. Dixon. 2005. Synthesis and properties of crosslinked recombinant pro-resilin. Nature 437: 999–1002.

Fahnestock, S.R., Z. Yao and L.A. Bedzyk. 2000. Microbial production of spider silk proteins. J. Biotechnol. 74: 105–119.

Faraj, K.A., T.H. van Kuppevelt and W.F. Daamen. 2007. Construction of collagen scaffolds that mimic the three-dimensional architecture of specific tissues. Tissue Eng. 13: 2387–2394.

Fu, C., Z. Shao and V. Fritz. 2009. Animal silks: their structures, properties and artificial production. Chem. Commun. (Camb) 6515–6529.

Gagner, J.E., W. Kim and E.L. Chaikof. 2014. Designing protein-based biomaterials for medical applications. Acta Biomater. 10: 1542–1557.

Gasperini, L., J.F. Mano and R.L. Reis. 2014. Natural polymers for the microencapsulation of cells. J. R. Soc. Interface. 11: 20140817.

Ghezzi, C.E., B. Marelli, N. Muja, N. Hirota, J.G. Martin, J.E. Barralet, A. Alessandrino, G. Freddi and S.N. Nazhat. 2011. Mesenchymal stem cell-seeded multilayered dense collagen-silk fibroin hybrid for tissue engineering applications. Biotechnol. J. 6: 1198–1207.

Ghoshal, A., D.G. DeMartini, E. Eck and D.E. Morse. 2014. Experimental determination of refractive index of condensed reflectin in squid iridocytes. J. R. Soc. Interface. 11: 20140106.

Gil, E.S., B.B. Mandal, S.H. Park, J.K. Marchant, F.G. Omenetto and D.L. Kaplan. 2010. Helicoidal multi-lamellar features of RGD-functionalized silk biomaterials for corneal tissue engineering. Biomaterials 31: 8953–8963.

Gil, E.S., B. Panilaitis, E. Bellas and D.L. Kaplan. 2013. Functionalized silk biomaterials for wound healing. Adv. Healthc Mater. 2: 206–217.

Grove, T.Z. and L. Regan. 2012. New materials from proteins and peptides. Curr. Opin. Struct. Biol. 22: 451–456.

Hassouneh, W., T. Christensen and A. Chilkoti. 2010. Elastin-like polypeptides as a purification tag for recombinant proteins. Curr. Protoc. Protein Sci. Chapter 6: Unit 6 11.

Heilshorn, S.C., K.A. DiZio, E.R. Welsh and D.A. Tirrell. 2003. Endothelial cell adhesion to the fibronectin CS5 domain in artificial extracellular matrix proteins. Biomaterials 24: 4245–4252.

Heim, M., C.B. Ackerschott and T. Scheibel. 2010a. Characterization of recombinantly produced spider flagelliform silk domains. J. Struct. Biol. 170: 420–425.

Heim, M., D. Keerl and T. Scheibel. 2009. Spider silk: from soluble protein to extraordinary fiber. Angew Chem. Int. Ed. Engl. 48: 3584–3596.

Heim, M., L. Romer and T. Scheibel. 2010b. Hierarchical structures made of proteins. The complex architecture of spider webs and their constituent silk proteins. Chem. Soc. Rev. 39: 156–164.

Horan, R.L., K. Antle, A.L. Collette, Y. Wang, J. Huang, J.E. Moreau, V. Volloch, D.L. Kaplan and G.H. Altman. 2005. *In vitro* degradation of silk fibroin. Biomaterials 26: 3385–3393.

House, M., C.C. Sanchez, W.L. Rice, S. Socrate and D.L. Kaplan. 2010. Cervical tissue engineering using silk scaffolds and human cervical cells. Tissue Eng. Part A 16: 2101–2112.

Hronik-Tupaj, M., W.K. Raja, M. Tang-Schomer, F.G. Omenetto and D.L. Kaplan. 2013. Neural responses to electrical stimulation on patterned silk films. J. Biomed. Mater. Res. A 101: 2559–2572.

Humenik, M. and T. Scheibel. 2014. Nanomaterial building blocks based on spider silk-oligonucleotide conjugates. ACS Nano 8: 1342–1349.

Izumi, M., A.M. Sweeney, D. Demartini, J.C. Weaver, M.L. Powers, A. Tao, T.V. Silvas, R.M. Kramer, W.J. Crookes-Goodson, L.M. Mathger, R.R. Naik, R.T. Hanlon and D.E. Morse. 2010. Changes in reflectin protein phosphorylation are associated with dynamic iridescence in squid. J. R. Soc. Interface 7: 549–560.

Jeong, L., M.H. Kim, J.Y. Jung, B.M. Min and W.H. Park. 2014. Effect of silk fibroin nanofibers containing silver sulfadiazine on wound healing. Int. J. Nanomedicine 9: 5277–5287.

Jin, H.J. and D.L. Kaplan. 2003. Mechanism of silk processing in insects and spiders. Nature 424: 1057–1061.

Jorgensen, L.N. 2003. Collagen deposition in the subcutaneous tissue during wound healing in humans: a model evaluation. APMIS Suppl. 1–56.

Kasoju, N. and U. Bora. 2012. Fabrication and characterization of curcumin-releasing silk fibroin scaffold. J. Biomed. Mater. Res. B Appl. Biomater. 100: 1854–1866.

Keatch, R.P., A.M. Schor, J.B. Vorstius and S.L. Schor. 2012. Biomaterials in regenerative medicine: engineering to recapitulate the natural. Curr. Opin. Biotechnol. 23: 579–582.

Keten, S., Z. Xu, B. Ihle and M.J. Buehler. 2010. Nanoconfinement controls stiffness, strength and mechanical toughness of beta-sheet crystals in silk. Nat. Mater. 9: 359–367.

Khaing, Z.Z. and C.E. Schmidt. 2012. Advances in natural biomaterials for nerve tissue repair. Neurosci. Lett. 519: 103–114.

Kim, D.G., E.Y. Kim, Y.R. Kim and I.S. Kong. 2015. Construction of chimeric human epidermal growth factor containing short collagen-binding domain moieties for use as a wound tissue healing agent. J. Microbiol. Biotechnol. 25: 119–126.

Kim, J.K., H.J. Kim, J.Y. Chung, J.H. Lee, S.B. Young and Y.H. Kim. 2014. Natural and synthetic biomaterials for controlled drug delivery. Arch. Pharm. Res. 37: 60–68.

Kim, J.S., H.S. Chu, K.I. Park, J.I. Won and J.H. Jang. 2012. Elastin-like polypeptide matrices for enhancing adeno-associated virus-mediated gene delivery to human neural stem cells. Gene Ther. 19: 329–337.

Kim, W. 2013. Recombinant protein polymers in biomaterials. Front Biosci. (Landmark Ed.) 18: 289–304.

Kimmerling, K.A., B.D. Furman, D.S. Mangiapani, M.A. Moverman, S.M. Sinclair, J.L. Huebner, A. Chilkoti, V.B. Kraus, L.A. Setton, F. Guilak and S.A. Olson. 2015. Sustained intra-articular delivery of IL-1RA from a thermally-responsive elastin-like polypeptide as a therapy for post-traumatic arthritis. Eur. Cell Mater. 29: 124–139; discussion 139–140.

Klok, H.A., A. Rosler, G. Gotz, E. Mena-Osteritz and P. Bauerle. 2004. Synthesis of a silk-inspired peptide-oligothiophene conjugate. Org. Biomol. Chem. 2: 3541–3544.

Kolesky, D.B., R.L. Truby, A.S. Gladman, T.A. Busbee, K.A. Homan and J.A. Lewis. 2014. 3D bioprinting of vascularized, heterogeneous cell-laden tissue constructs. Adv. Mater. 26: 3124–3130.

Koria, P., H. Yagi, Y. Kitagawa, Z. Megeed, Y. Nahmias, R. Sheridan and M.L. Yarmush. 2011. Self-assembling elastin-like peptides growth factor chimeric nanoparticles for the treatment of chronic wounds. Proc. Natl. Acad. Sci. USA 108: 1034–1039.

Kramer, R.M., W.J. Crookes-Goodson and R.R. Naik. 2007. The self-organizing properties of squid reflectin protein. Nat. Mater. 6: 533–538.

Krishna, O.D. and K.L. Kiick. 2010. Protein- and peptide-modified synthetic polymeric biomaterials. Biopolymers 94: 32–48.

Kundu, B., R. Rajkhowa, S.C. Kundu and X. Wang. 2013. Silk fibroin biomaterials for tissue regenerations. Adv. Drug Deliv. Rev. 65: 457–470.

Kundu, B., C.J. Schlimp, S. Nurnberger, H. Redl and S.C. Kundu. 2014. Thromboelastometric and platelet responses to silk biomaterials. Sci. Rep. 4: 4945.

Kundu, J., Y.I. Chung, Y.H. Kim, G. Tae and S.C. Kundu. 2010. Silk fibroin nanoparticles for cellular uptake and control release. Int. J. Pharm. 388: 242–250.

Le, D.H., R. Hanamura, D.H. Pham, M. Kato, D.A. Tirrell, T. Okubo and A. Sugawara-Narutaki. 2013. Self-assembly of elastin-mimetic double hydrophobic polypeptides. Biomacromolecules 14: 1028–1034.

Leal-Egana, A. and T. Scheibel. 2010. Silk-based materials for biomedical applications. Biotechnol. Appl. Biochem. 55: 155–167.

Lewis, R.V. 2006. Spider silk: ancient ideas for new biomaterials. Chem. Rev. 106: 3762–3774.

Li, L. and K.L. Kiick. 2013a. Resilin-based materials for biomedical applications. ACS Macro Letters 2: 635–640.

Li, L. and K.L. Kiick. 2013b. Resilin-based materials for biomedical applications. ACS Macro Lett. 2: 635–640.

Li, L. and K.L. Kiick. 2014. Transient dynamic mechanical properties of resilin-based elastomeric hydrogels. Front Chem. 2: 21.

Li, L., S. Teller, R.J. Clifton, X. Jia and K.L. Kiick. 2011. Tunable mechanical stability and deformation response of a resilin-based elastomer. Biomacromolecules 12: 2302–2310.

Li, L., Z. Tong, X. Jia and K.L. Kiick. 2013. Resilin-like polypeptide hydrogels engineered for versatile biological functions. Soft Matter 9: 665–673.

Li, N.K., F. Garcia Quiroz, C.K. Hall, A. Chilkoti and Y.G. Yingling. 2014. Molecular description of the LCST behavior of an elastin-like polypeptide. Biomacromolecules 15: 3522–3530.

Lin, Y., X. Xia, K. Shang, R. Elia, W. Huang, P. Cebe, G. Leisk, F. Omenetto and D.L. Kaplan. 2013. Tuning chemical and physical cross-links in silk electrogels for morphological analysis and mechanical reinforcement. Biomacromolecules 14: 2629–2635.

Liu, J.C., S.C. Heilshorn and D.A. Tirrell. 2004. Comparative cell response to artificial extracellular matrix proteins containing the RGD and CS5 cell-binding domains. Biomacromolecules 5: 497–504.

Liu, L., X. Zhang, X. Liu, J. Liu, G. Lu, D.L. Kaplan, H. Zhu and Q. Lu. 2015. Biomineralization of stable and monodisperse vaterite microspheres using silk nanoparticles. ACS Appl. Mater. Interfaces 7: 1735–1745.

Liu, S., C. Dong, G. Lu, Q. Lu, Z. Li, D.L. Kaplan and H. Zhu. 2013. Bilayered vascular grafts based on silk proteins. Acta Biomater. 9: 8991–9003.

Lovett, M.L., X. Wang, T. Yucel, L. York, M. Keirstead, L. Haggerty and D.L. Kaplan. 2015. Silk hydrogels for sustained ocular delivery of anti-vascular endothelial growth factor (anti-VEGF) therapeutics. Eur. J. Pharm. Biopharm.

Lu, Q., X. Hu, X. Wang, J.A. Kluge, S. Lu, P. Cebe and D.L. Kaplan. 2010. Water-insoluble silk films with silk I structure. Acta Biomater. 6: 1380–1387.

Lu, Q., H. Zhu, C. Zhang, F. Zhang, B. Zhang and D.L. Kaplan. 2012. Silk self-assembly mechanisms and control from thermodynamics to kinetics. Biomacromolecules 13: 826–832.

Lv, S., D.M. Dudek, Y. Cao, M.M. Balamurali, J. Gosline and H. Li. 2010a. Designed biomaterials to mimic the mechanical properties of muscles. Nature 465: 69–73.

Lv, S., D.M. Dudek, Y. Cao, M.M. Balamurali, J. Gosline and H.B. Li. 2010b. Designed biomaterials to mimic the mechanical properties of muscles. Nature 465: 69–73.

Lyons, R.E., K.M. Nairn, M.G. Huson, M. Kim, G. Dumsday and C.M. Elvin. 2009. Comparisons of recombinant resilin-like proteins: repetitive domains are sufficient to confer resilin-like properties. Biomacromolecules 10: 3009–3014.

MacEwan, S.R. and A. Chilkoti. 2010. Elastin-like polypeptides: biomedical applications of tunable biopolymers. Biopolymers 94: 60–77.

MacEwan, S.R. and A. Chilkoti. 2014. Applications of elastin-like polypeptides in drug delivery. J. Control. Release 190: 314–330.

Madsen, B. and F. Vollrath. 2000. Mechanics and morphology of silk drawn from anesthetized spiders. Naturwissenschaften 87: 148–153.

Main, E.R., J.J. Phillips and C. Millership. 2013. Repeat protein engineering: creating functional nanostructures/biomaterials from modular building blocks. Biochem. Soc. Trans. 41: 1152–1158.

Mandal, B.B., S.H. Park, E.S. Gil and D.L. Kaplan. 2011. Multilayered silk scaffolds for meniscus tissue engineering. Biomaterials 32: 639–651.

Maskarinec, S.A. and D.A. Tirrell. 2005. Protein engineering approaches to biomaterials design. Curr. Opin. Biotechnol. 16: 422–426.

Massodi, I., G.L. Bidwell, 3rd and D. Raucher. 2005. Evaluation of cell penetrating peptides fused to elastin-like polypeptide for drug delivery. J. Control Release 108: 396–408.

Matsumoto, A., A. Lindsay, B. Abedian and D.L. Kaplan. 2008. Silk fibroin solution properties related to assembly and structure. Macromol. Biosci. 8: 1006–1018.

McDaniel, J.R., D.J. Callahan and A. Chilkoti. 2010. Drug delivery to solid tumors by elastin-like polypeptides. Adv. Drug Deliv. Rev. 62: 1456–1467.

McHale, M.K., L.A. Setton and A. Chilkoti. 2005. Synthesis and *in vitro* evaluation of enzymatically cross-linked elastin-like polypeptide gels for cartilaginous tissue repair. Tissue Eng. 11: 1768–1779.

Megeed, Z., R.M. Winters and M.L. Yarmush. 2006. Modulation of single-chain antibody affinity with temperature-responsive elastin-like polypeptide linkers. Biomacromolecules 7: 999–1004.

Mehta, A.S., B.K. Singh, N. Singh, D. Archana, K. Snigdha, R. Harniman, S.S. Rahatekar, R.P. Tewari and P.K. Dutta. 2015. Chitosan sulfa-based three-dimensional scaffolds containing gentamicin-encapsulated calcium alginate beads for drug administration and blood compatibility. J. Biomater. Appl. 29: 1314–1325.

Meyer, D.E. and A. Chilkoti. 1999. Purification of recombinant proteins by fusion with thermally-responsive polypeptides. Nature Biotechnology 17: 1112–1115.

Meyer, D.E. and A. Chilkoti. 2004. Quantification of the effects of chain length and concentration on the thermal behavior of elastin-like polypeptides. Biomacromolecules 5: 846–851.

Meyer, D.E., K. Trabbic-Carlson and A. Chilkoti. 2001. Protein purification by fusion with an environmentally responsive elastin-like polypeptide: effect of polypeptide length on the purification of thioredoxin. Biotechnol. Prog. 17: 720–728.

Mogosanu, G.D. and A.M. Grumezescu. 2014. Natural and synthetic polymers for wounds and burns dressing. Int. J. Pharm. 463: 127–136.

Mogosanu, G.D., A.M. Grumezescu and M.C. Chifiriuc. 2014. Keratin-based biomaterials for biomedical applications. Curr. Drug Targets 15: 518–530.

Morgan, A.W., K.E. Roskov, S. Lin-Gibson, D.L. Kaplan, M.L. Becker and C.G. Simon, Jr. 2008. Characterization and optimization of RGD-containing silk blends to support osteoblastic differentiation. Biomaterials 29: 2556–2563.

Muralidharan, N., R. Jeya Shakila, D. Sukumar and G. Jeyasekaran. 2013. Skin, bone and muscle collagen extraction from the trash fish, leather jacket (Odonus niger) and their characterization. J. Food Sci. Technol. 50: 1106–1113.

Murphy, A.R. and D.L. Kaplan. 2009. Biomedical applications of chemically-modified silk fibroin. J. Mater. Chem. 19: 6443–6450.

Murphy, S.V. and A. Atala. 2014. 3D bioprinting of tissues and organs. Nat. Biotechnol. 32: 773–785.

Nairn, K.M., R.E. Lyons, R.J. Mulder, S.T. Mudie, D.J. Cookson, E. Lesieur, M. Kim, D. Lau, F.H. Scholes and C.M. Elvin. 2008. A synthetic resilin is largely unstructured. Biophys. J. 95: 3358–3365.

Nillesen, S.T., P.J. Geutjes, R. Wismans, J. Schalkwijk, W.F. Daamen and T.H. van Kuppevelt. 2007. Increased angiogenesis and blood vessel maturation in acellular collagen-heparin scaffolds containing both FGF2 and VEGF. Biomaterials 28: 1123–1131.

Nitta, S.K. and K. Numata. 2013. Biopolymer-based nanoparticles for drug/gene delivery and tissue engineering. Int. J. Mol. Sci. 14: 1629–1654.

Numata, K., J. Hamasaki, B. Subramanian and D.L. Kaplan. 2010. Gene delivery mediated by recombinant silk proteins containing cationic and cell binding motifs. J. Control Release 146: 136–143.

Numata, K. and D.L. Kaplan. 2010. Silk-based delivery systems of bioactive molecules. Adv. Drug Deliv. Rev. 62: 1497–1508.

Numata, K. and D.L. Kaplan. 2011. Differences in cytotoxicity of beta-sheet peptides originated from silk and amyloid beta. Macromol. Biosci. 11: 60–64.

Numata, K., A.J. Mieszawska-Czajkowska, L.A. Kvenvold and D.L. Kaplan. 2012. Silk-based nanocomplexes with tumor-homing peptides for tumor-specific gene delivery. Macromol. Biosci. 12: 75–82.

Numata, K., M.R. Reagan, R.H. Goldstein, M. Rosenblatt and D.L. Kaplan. 2011. Spider silk-based gene carriers for tumor cell-specific delivery. Bioconjug. Chem. 22: 1605–1610.

Numata, K., B. Subramanian, H.A. Currie and D.L. Kaplan. 2009. Bioengineered silk protein-based gene delivery systems. Biomaterials 30: 5775–5784.

Omenetto, F.G. and D.L. Kaplan. 2010. New opportunities for an ancient material. Science 329: 528–531.

Oryan, A. 1995. Role of collagen in soft connective tissue wound healing. Transplant Proc. 27: 2759–2761.

Park, S.H., E.S. Gil, H.J. Kim, K. Lee and D.L. Kaplan. 2010. Relationships between degradability of silk scaffolds and osteogenesis. Biomaterials 31: 6162–6172.

Park, S.H., E.S. Gil, B.B. Mandal, H. Cho, J.A. Kluge, B.H. Min and D.L. Kaplan. 2012. Annulus fibrosus tissue engineering using lamellar silk scaffolds. J. Tissue Eng. Regen. Med. 6 Suppl. 3: s24–33.

Partlow, B.P., C.W. Hanna, J. Rnjak-Kovacina, J.E. Moreau, M.B. Applegate, K.A. Burke, B. Marelli, A.N. Mitropoulos, F.G. Omenetto and D.L. Kaplan. 2014. Highly tunable elastomeric silk biomaterials. Adv. Funct. Mater. 24: 4615–4624.

Pinedo-Martin, G., E. Castro, L. Martin, M. Alonso and J.C. Rodriguez-Cabello. 2014. Effect of surfactants on the self-assembly of a model elastin-like block corecombinamer: from micelles to an aqueous two-phase system. Langmuir 30: 3432–3440.

Preda, R.C., G. Leisk, F. Omenetto and D.L. Kaplan. 2013. Bioengineered silk proteins to control cell and tissue functions. Methods Mol. Biol. 996: 19–41.

Pritchard, E.M., X. Hu, V. Finley, C.K. Kuo and D.L. Kaplan. 2013. Effect of silk protein processing on drug delivery from silk films. Macromol. Biosci. 13: 311–320.

Pritchard, E.M. and D.L. Kaplan. 2011. Silk fibroin biomaterials for controlled release drug delivery. Expert Opin. Drug Deliv. 8: 797–811.

Qin, G., P.B. Dennis, Y. Zhang, X. Hu, J.E. Bressner, Z. Sun, W.J. Crookes-Goodson, R.R. Naik, F.G. Omenetto and D.L. Kaplan. 2013. Recombinant reflectin-based optical materials. Journal of Polymer Science, Part B: Polymer. Physics 51: 254–264.

Qin, G., M.J. Glassman, C.N. Lam, D. Chang, E. Schaible, A. Hexemer and B.D. Olsen. 2015. Topological effects on globular protein-ELP fusion block. Adv. Funct. Mater. 25: 729–738.

Qin, G., X. Hu, P. Cebe and D.L. Kaplan. 2012. Mechanism of resilin elasticity. Nat. Commun. 3: 1003.

Qin, G., S. Lapidot, K. Numata, X. Hu, S. Meirovitch, M. Dekel, I. Podoler, O. Shoseyov and D.L. Kaplan. 2009. Expression, cross-linking, and characterization of recombinant chitin binding resilin. Biomacromolecules 10: 3227–3234.

Qin, G., A. Rivkin, S. Lapidot, X. Hu, I. Preis, S.B. Arinus, O. Dgany, O. Shoseyov and D.L. Kaplan. 2011. Recombinant exon-encoded resilins for elastomeric biomaterials. Biomaterials 32: 9231–9243.

Rabotyagova, O.S., P. Cebe and D.L. Kaplan. 2010. Role of polyalanine domains in beta-sheet formation in spider silk block copolymers. Macromol. Biosci. 10: 49–59.

Renner, J.N., K.M. Cherry, R.S. Su and J.C. Liu. 2012a. Characterization of resilin-based materials for tissue engineering applications. Biomacromolecules 13: 3678–3685.

Renner, J.N., Y. Kim, K.M. Cherry and J.C. Liu. 2012b. Modular cloning and protein expression of long, repetitive resilin-based proteins. Protein Expr. Purif 82: 90–96.

Rockwood, D.N., R.C. Preda, T. Yucel, X. Wang, M.L. Lovett and D.L. Kaplan. 2011. Materials fabrication from Bombyx mori silk fibroin. Nat. Protoc. 6: 1612–1631.

Rodriguez-Cabello, J.C. 2004. Smart elastin-like polymers. Adv. Exp. Med. Biol. 553: 45–57.

Romano, N.H., D. Sengupta, C. Chung and S.C. Heilshorn. 2011. Protein-engineered biomaterials: nanoscale mimics of the extracellular matrix. Biochim. Biophys. Acta 1810: 339–349.

Rudnick, A. 2006. Advances in tissue engineering and use of type 1 bovine collagen particles in wound bed preparation. J. Wound Care 15: 402–404.

Rusling, J.F., G.W. Bishop, N. Doan and F. Papadimitrakopoulos. 2014. Nanomaterials and biomaterials in electrochemical arrays for protein detection. J. Mater. Chem. B Mater. Biol. Med. 2.

Ruszczak, Z. 2003. Effect of collagen matrices on dermal wound healing. Adv. Drug Deliv. Rev. 55: 1595–1611.

Ryu, J.S., M. Kuna and D. Raucher. 2014. Penetrating the cell membrane, thermal targeting and novel anticancer drugs: the development of thermally targeted, elastin-like polypeptide cancer therapeutics. Ther. Deliv. 5: 429–445.

Sarangthem, V., E.A. Cho, S.M. Bae, T.D. Singh, S.J. Kim, S. Kim, W.B. Jeon, B.H. Lee and R.W. Park. 2013. Construction and application of elastin like polypeptide containing IL-4 receptor targeting peptide. PLoS One 8: e81891.

Sarithakumari, C.H. and G.M. Kurup. 2013. Alginic acid isolated from Sargassum wightii exhibits anti-inflammatory potential on type II collagen induced arthritis in experimental animals. Int. Immunopharmacol. 17: 1108–1115.

Saxena, R. and M.J. Nanjan. 2015. Elastin-like polypeptides and their applications in anticancer drug delivery systems: a review. Drug Deliv. 22: 156–167.

Seib, F.P., E.M. Pritchard and D.L. Kaplan. 2013. Self-assembling doxorubicin silk hydrogels for the focal treatment of primary breast cancer. Adv. Funct. Mater. 23: 58–65.

Sell, S.A., M.J. McClure, K. Garg, P.S. Wolfe and G.L. Bowlin. 2009. Electrospinning of collagen/biopolymers for regenerative medicine and cardiovascular tissue engineering. Adv. Drug Deliv. Rev. 61: 1007–1019.

Shao, Z. and F. Vollrath. 2002. Surprising strength of silkworm silk. Nature 418: 741.

Skardal, A. and A. Atala. 2014. Biomaterials for Integration with 3-D Bioprinting. Ann. Biomed. Eng.

Smeenk, J.M., M.B. Otten, J. Thies, D.A. Tirrell, H.G. Stunnenberg and J.C. van Hest. 2005. Controlled assembly of macromolecular beta-sheet fibrils. Angew. Chem. Int. Ed. Eng. l44: 1968–1971.

Smits, F.C., B.C. Buddingh, M.B. van Eldijk and J.C. van Hest. 2015. Elastin-like polypeptide based nanoparticles: design rationale toward nanomedicine. Macromol. Biosci. 15: 36–51.

Su, R.S., Y. Kim and J.C. Liu. 2014. Resilin: protein-based elastomeric biomaterials. Acta Biomater. 10: 1601–1611.

Sutherland, T.D., J.H. Young, S. Weisman, C.Y. Hayashi and D.J. Merritt. 2010. Insect silk: one name, many materials. Annu. Rev. Entomol. 55: 171–188.

Swierczewska, M., C.S. Hajicharalambous, A.V. Janorkar, Z. Megeed, M.L. Yarmush and P. Rajagopalan. 2008. Cellular response to nanoscale elastin-like polypeptide polyelectrolyte multilayers. Acta Biomater. 4: 827–837.

Tao, A.R., D.G. DeMartini, M. Izumi, A.M. Sweeney, A.L. Holt and D.E. Morse. 2010. The role of protein assembly in dynamically tunable bio-optical tissues. Biomaterials 31: 793–801.

Thapa, A., W. Han, R.H. Simons, A. Chilkoti, E.Y. Chi and G.P. Lopez. 2013. Effect of detergents on the thermal behavior of elastin-like polypeptides. Biopolymers 99: 55–62.

Tokareva, O., M. Jacobsen, M. Buehler, J. Wong and D.L. Kaplan. 2014. Structure-function-property-design interplay in biopolymers: spider silk. Acta Biomater. 10: 1612–1626.

Tokareva, O., V.A. Michalczechen-Lacerda, E.L. Rech and D.L. Kaplan. 2013. Recombinant DNA production of spider silk proteins. Microb. Biotechnol. 6: 651–663.

Trabbic-Carlson, K., L. Liu, B. Kim and A. Chilkoti. 2004a. Expression and purification of recombinant proteins from *Escherichia coli*: Comparison of an elastin-like polypeptide fusion with an oligohistidine fusion. Protein Science 13: 3274–3284.

Trabbic-Carlson, K., D.E. Meyer, L. Liu, R. Piervincenzi, N. Nath, T. LaBean and A. Chilkoti. 2004b. Effect of protein fusion on the transition temperature of an environmentally responsive elastin-like polypeptide: a role for surface hydrophobicity? Protein Engineering Design & Selection 17: 57–66.

Trabbic-Carlson, K., L.A. Setton and A. Chilkoti. 2003. Swelling and mechanical behaviors of chemically cross-linked hydrogels of elastin-like polypeptides. Biomacromolecules 4: 572–580.

Truong, M.Y., N.K. Dutta, N.R. Choudhury, M. Kim, C.M. Elvin, A.J. Hill, B. Thierry and K. Vasilev. 2010. A pH-responsive interface derived from resilin-mimetic protein Rec1-resilin. Biomaterials 31: 4434–4446.

Valluzzi, R., S. Winkler, D. Wilson and D.L. Kaplan. 2002. Silk: molecular organization and control of assembly. Philos. Trans. R Soc. Lond. B Biol. Sci. 357: 165–167.

van Eldijk, M.B., C.L. McGann, K.L. Kiick and J.C. van Hest. 2012. Elastomeric polypeptides. Top. Curr. Chem. 310: 71–116.

van Eldijk, M.B., F.C. Smits, N. Vermue, M.F. Debets, S. Schoffelen and J.C. van Hest. 2014. Synthesis and self-assembly of well-defined elastin-like polypeptide-poly(ethylene glycol) conjugates. Biomacromolecules 15: 2751–2759.

van Hest, J.C. and D.A. Tirrell. 2001. Protein-based materials, toward a new level of structural control. Chem. Commun. (Camb) 1897–1904.

Vasconcelos, A., G. Freddi and A. Cavaco-Paulo. 2008. Biodegradable materials based on silk fibroin and keratin. Biomacromolecules 9: 1299–1305.

Veldtman, R., M.A. McGeoch and C.H. Scholtz. 2007. Fine-scale abundance and distribution of wild silk moth pupae. Bull. Entomol. Res. 97: 15–27.

Vendrely, C., C. Ackerschott, L. Romer and T. Scheibel. 2008. Molecular design of performance proteins with repetitive sequences: recombinant flagelliform spider silk as basis for biomaterials. Methods Mol. Biol. 474: 3–14.

Vepari, C. and D.L. Kaplan. 2007. Silk as a Biomaterial. Prog. Polym. Sci. 32: 991–1007.

Vepari, C., D. Matheson, L. Drummy, R. Naik and D.L. Kaplan. 2010. Surface modification of silk fibroin with poly(ethylene glycol) for antiadhesion and antithrombotic applications. J. Biomed. Mater. Res. A 93: 595–606.

Vollrath, F., B. Madsen and Z. Shao. 2001. The effect of spinning conditions on the mechanics of a spider's dragline silk. Proc. Biol. Sci. 268: 2339–2346.

Wang, X., L. Sun, M.V. Maffini, A. Soto, C. Sonnenschein and D.L. Kaplan. 2010. A complex 3D human tissue culture system based on mammary stromal cells and silk scaffolds for modeling breast morphogenesis and function. Biomaterials 31: 3920–3929.

Wang, X., E. Wenk, A. Matsumoto, L. Meinel, C. Li and D.L. Kaplan. 2007. Silk microspheres for encapsulation and controlled release. J. Control. Release 117: 360–370.

Wang, Y., U.J. Kim, D.J. Blasioli, H.J. Kim and D.L. Kaplan. 2005. *In vitro* cartilage tissue engineering with 3D porous aqueous-derived silk scaffolds and mesenchymal stem cells. Biomaterials 26: 7082–7094.

Wharram, S.E., X. Zhang, D.L. Kaplan and S.P. McCarthy. 2010. Electrospun silk material systems for wound healing. Macromol. Biosci. 10: 246–257.

Wilson, D., R. Valluzzi and D. Kaplan. 2000. Conformational transitions in model silk peptides. Biophys. J. 78: 2690–2701.

Winkler, S. and D.L. Kaplan. 2000. Molecular biology of spider silk. J. Biotechnol. 74: 85–93.

Wlodarczyk-Biegun, M.K., M.W. Werten, F.A. de Wolf, J.J. van den Beucken, S.C. Leeuwenburgh, M. Kamperman and M.A. Cohen Stuart. 2014. Genetically engineered silk-collagen-like copolymer for biomedical applications: production, characterization and evaluation of cellular response. Acta Biomater. 10: 3620–3629.

Wong Po Foo, C. and D.L. Kaplan. 2002. Genetic engineering of fibrous proteins: spider dragline silk and collagen. Adv. Drug Deliv. Rev. 54: 1131–1143.

Wong Po Foo, C., S.V. Patwardhan, D.J. Belton, B. Kitchel, D. Anastasiades, J. Huang, R.R. Naik, C.C. Perry and D.L. Kaplan. 2006. Novel nanocomposites from spider silk-silica fusion (chimeric) proteins. Proc. Natl. Acad. Sci. USA 103: 9428–9433.

Wright, E.R. and V.P. Conticello. 2002. Self-assembly of block copolymers derived from elastin-mimetic polypeptide sequences. Adv. Drug Deliv. Rev. 54: 1057–1073.

Wu, X., J. Hou, M. Li, J. Wang, D.L. Kaplan and S. Lu. 2012. Sodium dodecyl sulfate-induced rapid gelation of silk fibroin. Acta Biomater. 8: 2185–2192.

Wu, Y., J.A. MacKay, J.R. McDaniel, A. Chilkoti and R.L. Clark. 2009. Fabrication of elastin-like polypeptide nanoparticles for drug delivery by electrospraying. Biomacromolecules 10: 19–24.

Xia, X.X., Z.G. Qian, C.S. Ki, Y.H. Park, D.L. Kaplan and S.Y. Lee. 2010. Native-sized recombinant spider silk protein produced in metabolically engineered *Escherichia coli* results in a strong fiber. Proc. Natl. Acad. Sci. USA 107: 14059–14063.

Xia, X.X., M. Wang, Y. Lin, Q. Xu and D.L. Kaplan. 2014. Hydrophobic drug-triggered self-assembly of nanoparticles from silk-elastin-like protein polymers for drug delivery. Biomacromolecules 15: 908–914.

Xia, X.X., Q. Xu, X. Hu, G. Qin and D.L. Kaplan. 2011a. Tunable self-assembly of genetically engineered silk—elastin-like protein polymers. Biomacromolecules 12: 3844–3850.

Xia, X.X., Q.B. Xu, X. Hu, G.K. Qin and D.L. Kaplan. 2011b. Tunable self-assembly of genetically engineered silk-elastin-like protein polymers. Biomacromolecules 12: 3844–3850.

Xu, D., D. Asai, A. Chilkoti and S.L. Craig. 2012. Rheological properties of cysteine-containing elastin-like polypeptide solutions and hydrogels. Biomacromolecules 13: 2315–2321.

Xu, G., L. Gong, Z. Yang and X.Y. Liu. 2014. What makes spider silk fibers so strong? From molecular-crystallite network to hierarchical network structures. Soft Matter 10: 2116–2123.

Yao, D., S. Dong, Q. Lu, X. Hu, D.L. Kaplan, B. Zhang and H. Zhu. 2012. Salt-leached silk scaffolds with tunable mechanical properties. Biomacromolecules 13: 3723–3729.

Yodmuang, S., S.L. McNamara, A.B. Nover, B.B. Mandal, M. Agarwal, T.A. Kelly, P.H. Chao, C. Hung, D.L. Kaplan and G. Vunjak-Novakovic. 2015. Silk microfiber-reinforced silk hydrogel composites for functional cartilage tissue repair. Acta Biomater. 11: 27–36.

Yucel, T., P. Cebe and D.L. Kaplan. 2009. Vortex-induced injectable silk fibroin hydrogels. Biophys. J. 97: 2044–2050.

Yucel, T., M.L. Lovett, R. Giangregorio, E. Coonahan and D.L. Kaplan. 2014a. Silk fibroin rods for sustained delivery of breast cancer therapeutics. Biomaterials 35: 8613–8620.

Yucel, T., M.L. Lovett and D.L. Kaplan. 2014b. Silk-based biomaterials for sustained drug delivery. J. Control. Release 190: 381–397.

Zhu, B., W. Li, R.V. Lewis, C.U. Segre and R. Wang. 2015. E-spun composite fibers of collagen and dragline silk protein: fiber mechanics, biocompatibility, and application in stem cell differentiation. Biomacromolecules 16: 202–213.

13

Advanced Materials for Solar Energy Harnessing and Conversion

Jian Liu

ABSTRACT

This chapter describes mainly the recent progress of the artificial photosynthesis—the way for mimicking natural counterpart to solve our current energy and environmental dilemma. It involves design and the architecting of advanced nanomaterials for light harvesting and light energy conversion purposes. The chapter incorporates interdisciplinary and diversified efforts from physicists, chemists and biologists for contributing to artificial photosynthesis research and development. Further development in the field will be of broad interests to related physicists, chemists, and biologists. Readers will gain insight into diversified routes to construct advanced light harvesting and conversion systems including conventional semiconductor engineering, morphology sculpture, reminiscent of bio-system, tandem system designs, etc. The proposed chapter targets college students, graduate and PhD students who are interested in recent advances on artificial photosynthesis. Postdoctoral students, some independent researchers and even some early-career research scientists are also listed as target readers of the proposed chapter.

2145 Sheridan Road, Department of Chemistry, Northwestern University, Evanston, IL 60208, US.
Email: Jianliu@northwestern.edu

1. Introduction

Innovation through scientific discovery is crucial for societal advancement. Energy and environmental issues are two major concerns in the 21st century for mankind to build a sustainable world (Faunce et al. 2013). To implement sustainable practices truly, energy must be harnessed more cleanly and efficiently and renewable energy source should be pursued to convert to storable form for utilization. This revolutionary change will require novel materials, new tools and new insights to ensure success. In addition to efforts in other fields to realize such a change, this chapter focuses mainly on the latest summary of advanced materials based solar energy harnessing and conversion from scientific community.

Nowadays, the diminishing reserve for fossil fuel and increasingly serious environmental pollution, especially an enormous green gas emission due to combustion of fossil fuel (and also some CH_4 from leakage) have forced mankind to reconsider our energy strategy. In order to power our modern life, enormous carbon dioxide and other emissions into the atmosphere trap heat, steadily drive up the planet's temperature, and produce significant and harmful impact on health, environment, and climate (Schneider 1989). We must seek a drastic change to maintain sustainability of the society of mankind (Lewis and Nocera 2006).

Hydrogen gas (H_2) has been identified as a promising energy carrier for driving vehicles in future because of its high-energy capacity and environmental friendliness (Züttel et al. 2011). However, hydrogen is not a naturally existing resource that could be collected directly from the ground. Hydrogen has to be generated on a large scale, and that can be done using a variety of resources. However, most of the current hydrogen production is not sustainable and environmental friendly—it is produced from steam methane reforming and therefore greenhouse gas carbon dioxide (CO_2) is produced as well. Water could be an ideal source of hydrogen by simply water splitting process. But, direct water splitting is impossible without external energy input. This necessitates an external applied bias for the desired water splitting reaction. If the driving force for water splitting is from a renewable energy source such as solar energy, hydrogen can be really considered as a green energy carrier for powering our future society. Solar energy is plentiful on the earth and utilizing hourly energy fully, reaching onto the earth's surface will meet world's annual energy needs. However, intermittent nature of this resource implies that energy storage will be a key component to a future renewable energy infrastructure. How to harness and convert solar energy fully to chemical energy (such as H_2) is key point in exploiting the potential of solar energy as an alternative energy source. Since water cannot absorb light, in order to accomplish water-splitting, photofunctional catalystic system has to be introduced.

Nature has its own solution for the energy issues and has also accomplished the task of water-splitting in the process of photosynthesis. Learning from nature can inspire us to pursue a more sustainable society driven by renewable energy (McEvoy and Brudvig 2006). Nature has evolved in photosynthesis a highly efficient and delicate means of doing this across a wide variety of scales from isolated bacterial colonies to large trees (Barber 2009). Natural photosynthesis is the most primary process on the earth to convert light to chemical energy (Kirk 1994, Round et al. 1990). Although photosynthesis is performed vastly by different species, the process always begins with light absorbance by chlorophyll pigments contained in reaction center proteins (Kay and Grätzel 1993). In plants, these proteins are held inside organelles called chloroplasts, which are most abundant in leaf cells, while in bacteria they are embedded in the plasma membrane. The heart of photosynthetic process is water splitting by sunlight into oxygen and biological "hydrogen" (Blankenship 2013). During the light reaction process, some energy is used to extract electrons from water, producing oxygen. Photosynthesis maintains oxygen levels in the atmosphere and supplies all the organic compounds and most of the energy necessary for life on Earth. However, biological "hydrogen" is not normally released into atmosphere but instead two further compounds are generated: reduced nicotinamide adenine dinucleotide phosphate (NADPH) and adenosine triphosphate (ATP), the "energy currency" of cells. Biological hydrogen is combined with carbon dioxide to make sugars and other organic molecules in dark reaction of photosynthesis. The simplified light reaction process is illustrated in Fig. 1.

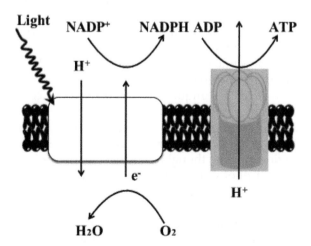

Figure 1. The simplified light reactions of photosynthesis. Light is absorbed and energy is used to drive electrons from water to generate NADPH and to drive protons across a membrane. These protons return through ATP synthase to make ATP.

Nature could shed great light for us to follow (Collings and Critchley 2007). A complete mimicry of natural photosynthesis is impossible at the current stage, but principle that nature does adopt for light harvesting and light energy conversion is worth following and mimicking (Steinberg-Yfrach et al. 1997). The fantastic natural photosynthesis process has always fascinated the scientific researchers. Substantial progress has been made in the exploration of photosynthesis, which could shed substantial light on how to construct artificial photosynthetic systems (Collings and Critchley 2007, Suga et al. 2014). Artificial photosynthesis, a term coined in 1980s, represents solar-to-chemical energy conversion process that is including solar-driven water splitting, CO_2 reduction and nitrogen (N_2) fixation and, etc. (Grätzel 1981). Artificial photosynthesis offers one possible solution to the pressing problems of global warming and diminishing fossil fuel resources (Kim et al. 2015). The basic concept is to mimic light (light harvesting) and dark (CO_2 reduction) processes of natural photosynthesis that converts light energy to chemical forms efficiently. This approach may include: (1) solar-to-electricity conversion; (2) solar-to-fuel production, including H_2, and methanol, etc.; (3) photoelectric cooperation for solar energy conversion (Grätzel 1981, O'regan and Grätzel 1991, Grätzel 2001, Tachibana et al. 2012, Bard 1980). These artificial photosynthetic technologies could be especially important because their implementation will reduce our greenhouse gas emissions and our dependence on fossil fuels (Collings and Critchley 2007). Hydrogen from water splitting provides the promising prospect for a clean society (Züttel et al. 2011). Hydrogen-driven future society also requests efforts in seeking high efficiency and abundant fuel cell (especially oxygen reduction reaction) catalyst, however, which is beyond the scope of the current chapter (Steele and Heinzel 2001).

How can we learn from natural photosynthesis to carry out artificial photosynthesis? Firstly, the way that we design the natural photosynthesis system is largely dependent on our knowledge and understanding of its natural counterpart. Fortunately, with great developments of modern biology and advanced technique, biological scientists could gain more understanding from nature, which could shed great light for chemists and physicists to develop a more advanced system to meet our long-term goals (Suga et al. 2014, McDermott et al. 1995). Secondly, materials design and engineering are thought to be the key points in realizing high efficiency artificial photosynthesis. Advances in materials science and technology offer great promise. The rapid development of nanomaterials in the last two decades specially laid the foundation for artificial photosynthesis investigation (Scholes et al. 2011). The investigation on artificial photosynthesis would also contribute to an overall development of materials science due to new phenomenon and new explanations appearing in related research. Designing efficient light harvesting, charge separation and transfer systems and catalytic nanostructured systems are the key in realizing artificial photosynthesis and contributing to highly efficient light energy conversion process (Wasielewski 1992, 2009, Savolainen et al. 2008). The construction of advanced light harvesting and converting system should be benefited by taking inspiration from natural photosynthesis. Biomimetic, emulating or duplicating bio-systems from molecular to nano- and macroscales, is a promising approach for constructing novel artificial systems and will be emphasized in the current chapter (Kalyanasundaram and Grätzel 2010).

In this chapter, the first part will illustrate the basic principle behind light harvesting and light energy conversion. Secondly, achieved progress on semiconductor nanomaterials based systems aiming for solar energy harnessing and conversion are highlighted. Special attention is paid in respect of engineering efficient visible-light-driven photocatalysts, morphology engineering for enhanced light harvesting, bio-inspired light harvesting systems, plasmonic photocatalysis, metal organic framework based artificial photosynthesis system and some tandem systems coupling the solar cell and photocatalytic system, etc. Solar energy conversion applications are based on photoelectrochemical water splitting and CO_2 reduction, N_2 fixation and some photo-enzymatic applications. Solar cell application represents a great scope of light energy conversion, but the chapter herein will focus on light-to-chemical energy conversion and solar cell will only be involved in the function for photoelectrolyzing water in hybrid/tandem systems (O'regan and Grätzel 1991, Khaselev and Turner 1998). The recent advances on advanced materials based tandem devices for artificial photosynthesis have been highlighted. The integrated module based device points out one practical way for further applications. Finally, summary and future outlook is made at the end of the chapter to share some personal prospects in the development of artificial photosynthesis.

2. Basic Principle

Materials summarized in the current chapter are mostly nanostructured. Usually, nanostructures can trap photos more efficiently by adopting appropriate geometric configuration comparing with bulky counterpart (Kamat 2007). Furthermore, under such reduced dimension, carrier diffusion lengths are comparable to the physical size and photogenerated carrier will be separated easily and readily for further applications. Therefore, light energy conversion performance will be enhanced greatly in nanostructured materials. This chapter will focus on fundamental properties of nanostructured materials and their applications in related light energy conversion fields. Special attention is paid to design and construction of advanced nanostructured semiconductor materials for light energy conversion purpose.

As illustrated in classical physics textbook, semiconductor is a solid substance that has conductivity between that of conductor and insulator (Kittel et al. 2005). Current conduction in a semiconductor occurs through the movement of free electrons and holes, collectively known as charge carriers. Adding impurity atoms to a semiconducting material, known as "doping" process externally, increases the number of charge carriers within semiconductor greatly. When a doped semiconductor contains mostly free holes, it is called "p-type" semiconductor; when it contains mostly free electrons, it is known as "n-type" semiconductor. The semiconductor materials used in electronic devices are doped under precise conditions to control the location and concentration of p- and n-type dopants. A single semiconductor crystal can have many p- and n-type regions; p-n junctions between these regions are responsible for the useful electronic behavior.

The unique electronic property of a semiconductor is characterized by its valence band (VB) and conduction band (CB). The VB of a semiconductor is formed by the interaction of the highest occupied molecular orbital (HOMO), while the CB is formed by the interaction of the lowest unoccupied molecular orbital (LUMO). There is no electron state between the top of the VB and the bottom of CB. The energy difference between CB and VB is called bandgap, which is usually denoted as Eg. The band structure, including bandgap and positions of VB and CB, is one of the important properties for a semiconductor photocatalyst, as it determines light absorption property as well as redox capability of a semiconductor (Fig. 2).

Photocatalytic reaction is initiated upon light irradiation excitation to generate electron-hole pairs (Fig. 3). When a semiconductor absorbs photons with energy equal to or greater than its E_g, electrons in VB will be excited to CB, while holes will be left in VB. These photogenerated electron-hole pairs might be further involved in following three possible processes: (i) migrate successfully to the surface of semiconductor; (ii) be captured by defect sites in bulk and/or on the surface region of semiconductor and (iii) recombine and release energy in the form of heat or photon. The last two processes are generally viewed as deactivation processes because in these two cases photogenerated charge carriers do not contribute to the desired photocatalytic reaction. Only photogenerated charges that could reach to the surface of the semiconductor could be available for photocatalytic reactions. The defect sites both in bulk and on the surface of the semiconductor, may serve as recombination centers for photogenerated electrons and holes, which will impair the efficiency of photocatalytic reaction.

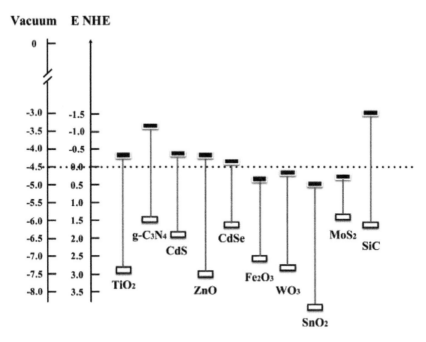

Figure 2. Absolute conduction band and valance band energy levels for some semiconducting photocatalysts with respect to normal hydrogen electrode (E vs. NHE) and Vacuum.

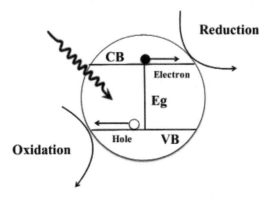

Figure 3. Schematic illustration of semiconductor excitation by suitable bandgap irradiation leading to the generation of electrons in the conduction band and holes in the valance band for reduction and oxidation reactions, respectively.

Since Fujishima and Honda reported PEC water splitting using a TiO_2 single crystal electrode in 1972, many researchers have been fascinated with the phenomenon and intensively studied water splitting using different semiconductor photoelectrodes and rationally extended to heterogeneous photocatalysis as a short-circuited PEC cell, as illustrated in Fig. 4a and b,

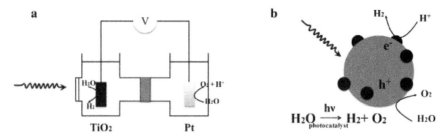

Figure 4. Schematic illustration of PEC water splitting **(a)** and heterogeneous process for water splitting **(b)**.

respectively. The principles of PEC cells can be extended to the design of particulate systems for carrying out the desired reactions including water splitting, CO_2 reduction and photocatalytic organic synthesis. For example, TiO_2 particles deposited with platinum nanoparticles can be considered a short-circuited PEC cell (Fig. 4b). The desired overall reaction occurs by electron and hole transfer and separation at two sites on the particle. These particulate systems have distinct advantage of being much simpler and less expensive to construct comparing with PEC cells. Moreover, efficiency of light harvesting and conversion in heterogeneous aqueous solution can be very high because of the strong light scattering and high surface area of nanoparticles, respectively.

Photocatalytic water splitting is an artificial photosynthesis process with photocatalysis in a PEC cell (or heterogeneous process) used for dissociation of water into its constituent parts, hydrogen (H_2) and oxygen (O_2). Theoretically and ideally, only solar energy, water, and a catalyst are required. In such systems, light energy is converted into chemical energy and the Gibbs free energy increases greatly by the following water splitting reaction in eqn (1):

$$H_2O + hv \rightarrow H_2 + \frac{1}{2}O_2 \quad \Delta G° = 237 \; kJ \; mol^{-1} \tag{1}$$

The positive change of the Gibbs free energy (237 kJ mol^{-1}) indicates the uphill nature of water splitting reaction. In order to accomplish water splitting, standard Gibbs free energy change $\Delta G° = 237$ kJ mol^{-1} or 1.23 eV is needed. The result of water-splitting process is that water molecules can be split into hydrogen and oxygen molecules under external driving force. The bandgap of the photocatalyst should be higher than 1.23 eV to accomplish the water-splitting task. The photogenerated holes and electrons are involved in water oxidation and water reduction processes, respectively. Under suitable illumination, electrons and holes could be generated in the CB and VB, respectively. For facilitating the water-splitting reaction, the prerequisite of the bandgap and potentials match of the CB and VB should be satisfied. Electrons populate in the CB and the energy level should be negative than H^+/H_2 (vs. NHE), while energy level of holes involved in the oxidation reaction should be more positive than O_2/H_2O (vs. NHE).

The reduction of CO_2 to useful chemicals such as methanol and formic acid has received a lot of attention as an alternative to depletion of fossil resources. As chemical reduction of CO_2 is energetically uphill due to its remarkable thermodynamic stability, this process requires a significant transfer of energy. The process can be driven by various forms of energy and thereby divided into different conversion mechanism such as direct or catalyzed chemical conversion, thermochemical conversion, electrochemical conversion and photochemical conversion. The high negative redox potential of CO_2/CO_2^- (−1.90 V vs. NHE, at pH 7.00) renders single electron reduction highly unfavorable due to the large reorganization energy between the linear molecule (CO_2) and bent radical anion (CO_2^-). In comparison, proton-coupled multi-electron transfer for CO_2 reduction is generally more favorable at the relatively low redox potential (vs. NHE, at pH 7.00) and the possible reduction products are thermodynamically more stable molecules, such as CO, HCHO, CH_3OH, or CH_4, depending on the number of transferred electrons ($2e^-$, $4e^-$, $6e^-$, and $8e^-$, respectively), as shown in following eqns (2)–(6):

$$CO_2 + 2e^- + 2H^+ \rightarrow HCOOH \qquad E^\circ = -0.61\ V \qquad (2)$$
$$CO_2 + 2e^- + 2H^+ \rightarrow CO + H_2O \qquad E^\circ = -0.53\ V \qquad (3)$$
$$CO_2 + 4e^- + 4H^+ \rightarrow HCOH + H_2O \qquad E^\circ = -0.48\ V \qquad (4)$$
$$CO_2 + 6e^- + 6H^+ \rightarrow CH_3OH + H_2O \qquad E^\circ = -0.38\ V \qquad (5)$$
$$CO_2 + 8e^- + 8H^+ \rightarrow CH_4 + 2H_2O \qquad E^\circ = -0.24\ V \qquad (6)$$

3. Recent Advances on Advanced Materials for Harnessing Solar Energy toward Artificial Photosynthesis

During the past four decades since 1972, considerable endeavors from both scientific and industrial communities have been devoted to the development of solar-harvesting devices that may become an important and reliable source of sustainable energy to mankind in the near future (Wrighton et al. 1975, Hardee and Bard 1977, Fan et al. 1980, Bard 1979, Frank and Bard 1977, Kraeutler and Bard 1978). Tremendous amounts of semiconductor materials have been developed and explored as photocatalysts under ultraviolet (abbreviated as UV) and visible light irradiation. The UV responsive photocatalysts are in large quantity and classified into four groups according to the literature including (a) d^0 metal oxide; (b) d^{10} metal oxide; (c) f^0 metal oxide; (d) non-oxide photocatalysts

(Chen et al. 2010). Some d^0 metal oxide based famous examples including TiO_2, titanates ($Na_2Ti_2O_5$ and $K_2Ti_4O_9$), $SrTiO_3$, and ZrO_2 could perform water splitting under UV irradiation in the presence of a sacrificial agent. Further coupling of these metal oxides with other semiconductors or noble metals could greatly accelerate the reaction (Duonghong et al. 1981, Jang et al. 2009, Shibata et al. 1987, Domen et al. 1980, Sayama and Arakawa 1994). Some metal oxides with d^{10} configuration or f^0 were also capable of performing water splitting under UV irradiation including Ga_2O_3, CeO_2 and so on (Yanagida et al. 2004, Kadowaki et al. 2007). Domen and coworkers did pioneering work on design and application of non-oxide GaN based materials for photoassisted water splitting (Maeda et al. 2006). UV light responsive materials have obvious drawbacks. Only 4% of the incoming solar energy lies in the UV range while visible light takes the majority of solar light spectrum. Therefore, developing visible-light-driven photocatalyst materials is desired greatly. A lot of strategies were developed to engineer the above-mentioned UV-responsive photocatalysts into visible-light-driven materials. There are also a lot of single-phase visible-light-driven photocatalysts reported every day both from scientific and industrial communities. The most famous examples are those including black TiO_2, $BiVO_4$, Bi_2WO_6, Ag_3PO_4, $g-C_3N_4$, Ta_3N_5, and so on (Kudo et al. 1999, Kudo et al. 1998, Bi et al. 2011, Wang et al. 2009, Hitoki et al. 2002, Feng et al. 2010, Chen et al. 2011, Zhang and Zhu 2005). Numerous materials have been developed at a rapid pace, however, the underlying mechanism remains the same both in PEC and photocatalytic processes.

Developing light harvesting materials is the key part in realizing real artificial photosynthesis devices (Scholes et al. 2011). Among variously investigated materials, metal oxides, especially TiO_2, ZnO, Fe_2O_3, WO_3, etc. have been mostly investigated (Chen and Mao 2007). Applications of metal oxides photocatalysts in PEC for water splitting and in heterogeneous photocatalysis process have been intensively investigated subjects (Grätzel 2001). Diverse efforts have been made to engineer nanostructured photocatalysts with desired dimensions (e.g., nanowires and nanotubes), porosity or hierarchical macro-/mesoporous structures, aiming to increase the effective surface area and reduce carrier diffusion lengths (Su et al. 2010, Park et al. 2006, Liu, Wang et al. 2015). In addition, improving the electronic structure is considered equally important as improving morphological factors for effective production, separation, transportation and collection of photoexcited charge carriers. Efficient charge separation is one of the most important factors that determine the eventual photocatalytic activities. Enormous strategies have been proposed for enhancing charge separation efficiency. Increasing crystallinity of semiconductor might diminish the formation of charge recombination defect sites, which would be beneficial for photocatalysis. Construction of various kinds of nanostructures such as nanowires, nanosheets and meso-/macroporous structures may also facilitate light harvesting, efficient charge transportation and separation efficiency (Chen et al. 2011).

The bio-inspired route suggests great possibilities for engineering efficient and highly selectively light harvesting and conversion systems. The rising of new materials also represents promising routes for artificial photosynthesis. Recently, the emergence of graphene presents new possibilities to engineer metal oxide and most importantly, the two dimensional graphene itself could be an excellent platform for realizing high-efficiency photocatalysis (Iwase et al. 2011, Yang et al. 2014). Metal-organic framework, abbreviated as MOF, is another rising material that revolutionizes the material field in relatively short time (Wang et al. 2012). Featuring with huge surface area, MOF suggests great possibility for acting as the scaffold for hosting photoactive component or providing active sites by itself (Zhang and Lin 2014). The related progress will be summarized in the following context. The combination of different processes and different reactions, named "tandem system", are also reviewed herein (Khaselev and Turner 1998, Luo et al. 2014).

3.1 Semiconductor Engineering for Light Energy Conversion

Solar-driven photocatalytic processes involving water splitting and CO_2 reduction is largely controlled by the semiconductor's capability of absorbing visible and infrared light, as well as its ability to suppress the rapid combination of photogenerated electrons and holes (Chen et al. 2010, Li et al. 2013, Zhang et al. 2009). Numerous methods have been developed to enhance light energy conversion efficiency, including semiconductors coupling, doping, etc. (Anderson and Bard 1995, Wang and Lewis 2006). Herein, the current section will focus only on routes to enhance light harvesting, while excellent representative review work on mediating charge separation and transfer can be found in the Reference (Wasielewski 1992).

Since nearly half of solar energy incident on the earth's surface lies in the visible region (400 nm $< \lambda <$ 800 nm), it is reasonable to envisage using mainly visible light to produce H_2 on a huge scale by PEC or photocatalytic water splitting processes. TiO_2 is investigated mostly in the photocatalytic field since n-type TiO_2 single crystal electrode was firstly discovered by Fujishima and Honda to be able to photoelectrochemically split water under UV light irradiation in 1972 and later platinum nanoparticles deposited TiO_2 powder was found to be capable of water splitting in the heterogeneous aqueous solution with methanol (Fujishima and Honda 1972). TiO_2 has many merits such as (i) low cost and abundant, (ii) chemically and biologically inert and photostable, (iii) nontoxic. However, the bandgap of TiO_2 is too wide and it is active only under UV light irradiation. Other low bandgap semiconductors such as CdS and Fe_2O_3 usually suffer from low photostability over a long time run (Meissner et al. 1988, Linsen Li et al. 2012). Graphitic carbon nitride (abbreviated as $g-C_3N_4$) was recently discovered to possess visible-light-driven photocatalytic property but still suffers from low efficiency (Wang et al. 2009). How to engineer the materials and extend light absorption spectrum to visible region and maintain long-term high photoactivity is the key issue in developing highly efficient visible-light-driven light energy conversion systems. This should facilitate and expedite the use of sunlight as an inexpensive and renewable energy source for photocatalytic fuel production.

Adding controlled metal or nonmetal impurities in chemical composition of TiO_2 will generate donor or acceptor states in the band gap. Almost from the very early stage of photocatalytic research in 1970s, the photoactivity of TiO_2 systems was substantially enhanced by the addition of noble metals such as Pt, Pd, Ir, and Ag onto the oxide surface (Kraeutler and Bard 1978, Baba et al. 1985, Subramanian et al. 2004). These metals act as electron trapping centers, facilitating efficient charge separation process (Subramanian et al. 2001).

The alteration of semiconductor catalytic properties by doping catalysts with ions was observed. Kiwi and Morrison found that doping anatase with lithium cation could enhance the photocatalytic hydrogen evolution performance greatly because of promoted conduction band electron transfer (Kiwi and Morrison 1984). However, lithium doped TiO_2 is active only under UV irradiation. Karakitsou and Verykios investigated the influence of altervalent cation doping of TiO_2 on its performance for photocatalytic hydrogen evolution (Karakitsou and Verykios 1993). They found that incorporation of cations of valence higher than that of Ti^{4+} (W^{6+}, Ta^{5+}, Nb^{5+}) into the crystal matrix of TiO_2 results in enhanced hydrogen evolution performance while the opposite is observed upon doping with cations of lower valence (In^{3+}, Zn^{2+}, Li^+). The doping didn't lead to the alternation of light absorption capacity of TiO_2 while the doping was thought to alter electron-hole generation and separation efficiency, which explained the activity difference for the altervalent cations doping of TiO_2. Anion doping, particularly nitrogen doping and carbon doping, has also been used widely to modify the electronic structures and consequently light absorption ranges of TiO_2 (Park et al. 2006, Asahi et al. 2001, Luo et al. 2004). It is recognized that this modification is affected sensitively both by the chemical state and spatial distribution of anion dopants. Only bulk doping of substitutional dopants for lattice oxygen can result in a desired bandgap narrowing by elevating the valence band edge of the photocatalyst (Varley et al. 2011).

Recently, Chen et al. developed a novel and inspiring hydrogenation process to treat TiO_2 to get surface disorder-engineered TiO_2 nanocrystals, displaying large absorption in the visible and infrared optical regime, with a black color. The hydrogenation treatment could enhance visible-light-driven photocatalytic capability substantially (Chen et al. 2011, Chen et al. 2013, Zhou et al. 2014). Liu et al. developed a red anatase by employing a pre-doped interstitial boron gradient to weaken nearby Ti-O bonds for the easy substitution of oxygen by nitrogen, and consequently it improves the nitrogen solubility substantially. Red anatase was also thought to avoid the formation of nitrogen-related Ti^{3+} due to the charge compensation effect by boron (Liu et al. 2012).

Overall water splitting to form hydrogen and oxygen over a heterogeneous photocatalyst using solar energy is a promising process for clean and recyclable hydrogen production. In recent years, numerous attempts have been made for the development of photocatalysts that work under visible light irradiation to utilize solar energy efficiently and a variety of photocatalytic systems have been developed. Domen and coworkers have reported that certain (oxy)nitrides, such as Ta_3N_5 and $LaTiO_2N$, are potential candidates as photocatalysts for overall water splitting under visible light. For example, GaN-ZnO and $ZnGeN_2$-ZnO solid solutions, known as $(Ga_{1-x}Zn_x)(N_{1-x}O_x)$ and $(Zn_{1+x}Ge)(N_2O_x)$, respectively, have been proved to be stable photocatalysts for visible-light-driven overall water splitting (Maeda et al. 2006).

Wang et al. reported α-Fe_2O_3 anchored $TiSi_2$ nanonets system that presented one of the highest external quantum efficiency (46% at $\lambda = 400$ nm) measured on α-Fe_2O_3 without intentional doping in a water-splitting environment (Lin et al. 2011). The introduction of $TiSi_2$ nanonet serves a dual role as a structural support and an efficient charge collector, aiming at maximizing photon-to-charge conversion due to the highly conductivity and suitably high surface areas of nanonets. Further studies indicated that $TiSi_2$ coupled WO_3 and TiO_2 could also lead to better performance due to enhanced light harvesting and better charge transport (Lin et al. 2009, Liu et al. 2011). Ye et al. found that Ag_3PO_4 semiconductor could harness visible light to oxidize water as well as decompose organic contaminants in aqueous solution (Ye et al. 2010). This suggests its great potential as a photofunctional material for both water splitting and waste-water cleaning.

3.2 Morphology Sculpture for Light Harvesting Purpose

Material and structure optimization are the key points determining the efficiency of light energy conversion in artificial systems. Nanostructures are powerful tools for the design and realization of integrated microscopic systems, and it is possible that their use will enable practical applications of efficient and affordable solar-to-fuel conversion. Biological materials are endless fountains of inspiration for the design and fabrication of such kinds of nanostructured materials (Zhou et al. 2011). A tremendous amount of nanomaterials have been designed and synthesized accordingly (Hulteen 1997). Some well-known examples are highlighted in the current section.

3.2.1 Nanowire. Fabrication of one-dimensional semiconductor nanostructures has received tremendous attention because of their rich physical and chemical properties that are dependent on crystal structure, size, morphology and shape (Alivisatos 1996). Nanowire morphology provides a large surface area for cocatalyst loading and electrochemical reaction sites, while at the same time resulting in improved charge collection efficiency, especially for indirect band gap semiconductors with short minority carrier diffusion lengths (Law et al. 2005, Wu et al. 2012, Jang et al. 2007). The introduction of nanowire morphology could contribute to improve the performance of corresponding photoanode materials. Prof. Peidong Yang made a lot of pioneering work on semiconductor nanowires based light energy conversion for artificial photosynthesis. They loaded platinum nanoclusters onto p-type silicon nanowire array photocathodes using atomic layer deposition technique for hydrogen evolution reaction. Furthermore, by using silicon and TiO_2 nanowires as building blocks, they constructed a tree-shaped nanowire based heterojunction device for solar water splitting by selectively loading hydrogen evolution reaction and oxygen evolution reaction electrocatalystst (Liu et al. 2013). CdS is an II-VI semiconductor, which has a direct band gap of 2.4 eV. Structuring CdS into nanowire has aroused a lot of interests and many routes were developed. Lee et al. 2009 synthesized CdS nanowire through a solvothermal route in ethylenediamine as single solvent and further application demonstrated that CdS nanowire was proved to possess high performance of PEC and photocatalytic activity from aqueous solution containing sulfide and sulfite as hole scavengers under visible light irradiation (Jang et al. 2007).

Liu and Antonietti found that certain diatom could direct the growth of the g-C_3N_4 nanowire by simply mixing the diatom and cyanamide under high temperature condensation (Fig. 5). By increasing condensation temperature and the cyanamide concentration further, the nanowire could develop further into nanoribbon and nanosheet by coalescence of the nearby nanowires (Liu and Antonietti 2013). Diatom substrate is not to be removed necessarily for photocatalytic characterization. The fact that attached nanocluster on the diatom could enhance light harvesting capability and diatom-based nanoclusters greatly, demonstrated excellent photocatalytic hydrogen production efficiency under visible light illumination compared to its bulky counterpart.

3.2.2 Nanotube. The nanotube investigated mostly is the TiO_2 based nanotube, which has demonstrated superior performances in a series of

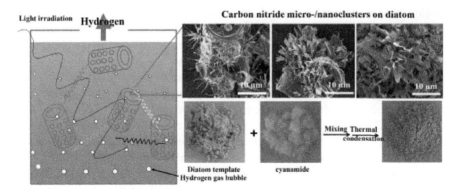

Figure 5. Diatom assisted synthesis of hierarchical micro-/nanoclusters for photocatalytic hydrogen evolution application. The simple fabrication route is also illustrated in the bottom right.

applications such as the dye-sensitized solar cells, PEC water splitting, hydrogen gas sensing and photocatalytical CO_2 reduction, etc. (Park et al. 2006, Sun et al. 2008, Kuang et al. 2008, Mor et al. 2004, Varghese et al. 2009, Macak et al. 2007). TiO_2 nanotubes were usually fabricated by anodization of titanium foil or titanium film on conductive glass substrate or templating them by anodic aluminum oxide. The tube wall thickness, tube length and tube diameter are all controllable and expected to affect physical and chemical properties (Grimes and Mor 2009, Prakasam et al. 2007). Detailed parameters could be modulated by electrolyte compositions, various ion species, and different voltage, respectively. Some review papers could be referred to for further detailed information (Kowalski et al. 2013). As for light harvesting and light energy conversion, nanotubular architecture allows for more efficient absorption of incident photons as well as decreased bulk recombination and vertical nanotube array alignment also facilitates good electrical transport. Due to intriguing properties of the nanotube, other materials are also sculptured to nanotubular architecture, such as $g\text{-}C_3N_4$, $CdIn_2S_4$, Ta_2O_5, WO_3, ZnO nanotubes (Kale et al. 2006, Liu et al. 2014, Zhao and Miyauchi 2008, Chu et al. 2009).

Some other conventional techniques like the electrospinning method could also be employed to fabricate tubular structures for light harvesting purpose (Li and Xia 2004). Zhao et al. designed biomimetic multichannel TiO_2 microtubes by a novel multifluidic compound-jet electrospinning technique. By removing inner fluids through calcinations, hollow multichannel structure was formed with controllable channel numbers by modulating the channel number of inner fluids. They further photocatalytic activity of the tubular TiO investigated and found out that introduction of a hollow structure can enhance photoactivity largely because of surface area enhancement. It was also believed that the hollow structure could offer a cooperative effect of trapping more gaseous molecules inside the channels and multiple reflection of incident light (Zhao et al. 2010).

Inspired by diatom frustule (Fig. 6a), Liu proposed a novel strategy for constructing nanotube based compartment nanoreactor (Fig. 6b and c). The SiO_2 nanotube was obtained by templating from nickel-hydrazine complex (Gao et al. 2011). The aspect ratios of the SiO_2 nanotube were tunable with adjusting surfactant amount and $NiCl_2$ concentration. The various aspect ratios (length to diameter) from 1 to 15 can be adjusted arbitrarily for the desired SiO_2 nanotube. By combining SiO_2 nanotube and cyanamide molecule and during *in situ* pyrolysis transformation of cyanamide into g-C_3N_4, melting cyanamide will diffuse into the interior of tube and condense and therefore separate nanotube into multicompartment reactor. By introducing the acid group on the outside of SiO_2 nanotube, combining with the intrinsic basic properties of g-C_3N_4, a tubular nanoreactor was obtained for Deacetalization-Henry cascade reactions. The g-C_3N_4 could also

Figure 6. Bio-inspired nanotube based nanoreactor for cascade reactions. **(a)** Optical microscopic observation of diatom frustule; **(b)** aggregates of tubular nanoreactor; **(c)** schematic illustration of cascade reaction on the functional compartment nanoreactor.

be processed to other nitride compound and further metal nanoparticles can also be introduced facilely to enhance multifunctionality.

3.2.3 Hierarchical Macro-/mesoporous Structures. Preparation of macroporous and mesoporous films and particles has been proposed as an effective strategy to increase the effective surface area of typical semiconductor photocatalysts, which would allow for increased light absorption and improved photocatalytic performance (Yu et al. 2007, Wang et al. 2005, Du et al. 2010, Liu et al. 2009, Rolison 2003, Li et al. 2012).

Yu et al. reported template-free construction of hierarchical macro-/mesoporous titania by simply adding tetrabutyl titanate to pure water and then calcined the materials at various temperatures (Yu et al. 2007). Wang et al. integrated light-harvesting macroporous channels into a mesoporous TiO_2 by simple hydrothermal reaction in the presence of nonionic surfactant as structure directing agent and characterized thermal

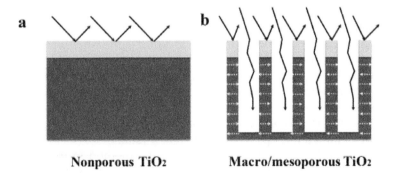

Figure 7. Light harvesting phenomenon on nonporous **(a)** and macro-/mesoporous TiO_2 **(b)**.

stability and photocatalytic activity of the resultant hierarchical macro-/mesoporous TiO_2 (Fig. 7) (Wang et al. 2005). The high photocatalytic performance of intact macro/mesoporous TiO_2 was claimed to be due to the existence of macrochannels that increases light harvesting efficiency and allows efficient diffusion of gaseous molecules. Iskandar et al. reported synthesis of macroporous brookite titania particles by using a spray-drying process with a precursor solution of brookite nanoparticles and colloidal particles (Iskandar et al. 2007). Macroporous particles outperformed dense particles in photocatalytic degradation of pollutant model as a result of their increased surface area and enhanced light harvesting capability.

3.2.4 Surface Controlled Growth of Micro- and Nanocrystallites. Fujishima's first water splitting experiment is based on rutile TiO_2 photoelectrode (Fujishima 1972). It's interesting to reconsider and reapply a single crystal in catalytic process. However, it is not reasonable to employ single crystals in real catalysis process directly because it would magnify the drawback of the single crystal systems and hide its benefits. The surface-controlled growth of micro- and nanocrystallites could serve the purpose of making most of the single crystal properties without sacrificing the benefits. Preparation of uniform, high-purity anatase single crystals with controllable crystallographic facets is promising for the catalytic and solar cell applications. However, the synthesis represents a great challenge. In 2008, Yang and co-workers reported greatly inspiring synthesis of micron-sized anatase TiO_2 crystallites with highly energetic (001) facets exposed and the percentage of exposed (001) facets was 47% (Yang et al. 2008). Xie and coworkers extended the synthesis strategy reported by Lu et al., and obtained rectangular nanosheet with exposed (001) facets (up to 89%) through simple hydrothermal route with the assistance of hydrofluoric acid solution (Han et al. 2009). The obtained TiO_2 nanosheets is demonstrated to possess excellent photocatalytic efficiency, superior to that of the commercially available Degussa P25, due to exposure of high percentage of highly reactive (001) facets. Crossland et al. developed a route for synthesizing micrometre-size mesoporous single-domain semiconductor particles based on crystal seeding in a sacrificial guiding template and successfully synthesized the mesoporous TiO_2 single crystal based on Lu's pioneering work on TiO_2 single crystal synthesis, which display higher conductivity and electron mobility than conventionally used nanocrystalline TiO_2 (Crossland et al. 2013). Furthermore, the authors fabricated a solid-state solar cell with a record of 7.2% efficiency under 150°C processing condition using the obtained mesoporous TiO_2 single crystal.

Yang et al. has developed a hydrothermal method in making facet-selective anatase TiO_2 microcrystals and nanocrystals by modifying the surface with fluorine (Yang et al. 2008). This work opens a new era in modulating TiO_2 photocatalytic performance by different crystal facet engineerings (Yang et al. 2009, Liu et al. 2010, Chen et al. 2010, Wu et al. 2011, Etgar et al. 2012). Recently, Snaith et al. moved the work further and the mesostructured TiO_2 single crystal was casted by spherical SiO_2 templating. The increased surface area renders TiO_2 microcrystal more promising and appealing for a wide variety of applications (Yang et al. 2008). Very recently, Yang further extended the synthesis strategy to TiO_2 single crystal with curved surface, which is enclosed by quasi-continuous high-index microfacets and thus has a unique truncated biconic morphology, which might open novel applications for the light energy conversion (Yang et al. 2014). Yu et al. found that varying the ratio between (001) and (101) facets of TiO_2 would affect photocatalytic CO_2 performance (Fig. 8). Furthermore, Yu et al. proposed a "surface heterojunction" concept to explain the difference in photocatalytic activity of TiO_2 with coexposed (001) and (101) facets by employing density functional theory calculations (Yu et al. 2014). The finding might point out a new way to engineer high efficiency photocatalytic CO_2 reduction catalyst. Ye et al. developed a high throughput route for fabrication single-crystalline Ag_3PO_4 rhombic dodecahedrons exposed with only {110} facets and cubes bounded entirely by {100} facets (Bi et al. 2011). The rhombic dodecahedrons turned out to

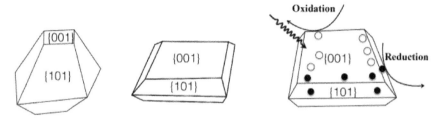

Figure 8. "Surface heterojunction" concept was employed to explain the difference in the photocatalytic activity of TiO_2 with differently coexposed (001) and (101) facets for CO_2 reductions.

possess a much higher photocatalytic performance for degradation of organic contaminants than that of the cubes. They ascribed the higher photocatalytic activity of rhombic dodecahedrons to the higher surface energy of {110} facets (1.31 J/m²) than that of {100} facets (1.12 J/m²). Morphological control of semiconductors allows preferential exposure of more active facets, which will greatly advance the development of highly efficient visible-light-driven photocatalysis.

3.2.5 Core-Shell Structures. Core-shell structures are well known in the materials field due to the spatial fixation and the easy modulation of the core/shell materials (Lauhon et al. 2002, Dabbousi et al. 1997). The core-shell structures could present great possibilities for increasing the light harvesting and facilitating light energy conversion process. The structures also impart materials some special properties not easily accessible by other structures, which would contribute greatly to the light energy conversion process. For example, by taking advantage of the magnetic core properties, coating magnetic core $Fe_3O_4@SiO_2$ with photoactive material (such as TiO_2), magnetically recoverable photocatalyst could be obtained (Ye et al. 2010). Ye and coworkers investigated the superior recoverable photocatalytic activity of the $Fe_3O_4@SiO_2/TiO_2$ composites by monitoring their constant photocatalytic activity during eighteen cycles of use (Ye et al. 2010).

In some cases, when multicomponent cocatalyst adopted the core/shell structures and attached to underlying semiconductor, significantly enhanced photocatalytic activity could be observed. Wang et al. prepared a core-shell-structured $Pt@Cu_2O$ co-catalyst on TiO_2, which can significantly promote photocatalytic reduction of CO_2 with H_2O to CH_4 and CO. It is proposed that the Cu_2O shell provides sites for preferential activation and conversion of CO_2 molecules in the presence of H_2O, while platinum core extracts the photogenerated electrons from TiO_2 (Zhai et al. 2013). Taking inspiration from the electron transfer system in natural photosynthesis, Tada and coworkers reported a bio-mimic $Au@CdS/TiO_2$ system, with the Au core mediating the electron transfer between the dually excited TiO_2 and CdS. The activity of ternary system was claimed to far exceed those of single- and two-component systems, as a result of vectorial electron transfer driven by two-step excitation of TiO_2 and CdS (Tada et al. 2006).

Dong et al. demonstrated controllable synthesis of several structures of multishelled ZnO hollow microspheres, by using carbonaceous microspheres as templates via a simple programmable heating process (Dong et al. 2012). Quadruple-shelled hollow microspheres with close double shells in the exterior and double-shelled hollow core have large surface areas and the ability to reflect and scatter light and thus show a high energy conversion efficiency of 5.6%. This work may present new opportunities for fabricating promising photocatalytic microreactor based on structural design and manipulation of hollow microspheres.

Chen et al. used multifluidic coaxial electrospinning approach creatively to fabricate core/shell ultrathin fibers with a novel nanowire-in-microtube structure (Chen et al. 2010). This method, developed by Chen et al., is an important extension of traditional electrospinning technique and it presents a diversified route to prepare core/shell fibers (Fig. 9). The hollow channel may introduce interesting and useful properties into conventional core/shell materials, which may find potential applications in many functional fields such as waveguides in optics, nanocables in microelectronics and other photocatalytic applications.

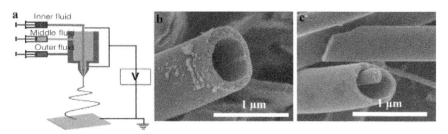

Figure 9. Schematic illustration of multifluidic coaxial electrospinning approach to fabricate core/shell ultrathin fibers with a novel nanowire-in-microtube structure and corresponding SEM image of the obtained materials. Courtesy, Dr. Hongyan Chen.

3.3 Natural Template Inspired Hierarchical Structures Construction

Besides chemical modification method to get different structured materials, physical engineering route also serve the same purpose. Because less than a few percent of solar photons can be absorbed by anatase limited by its large bandgap (3.2 eV), considerable effort has been devoted to extend the anatase absorption to visible light, boosting electron and hole lifetimes through doping and incorporating metals, or increasing surface area by making it mesoporous, as mentioned in previous sections. In addition to these chemical modifications, increasing path length of light has also been shown to improve efficiency of dye-sensitized TiO_2 solar cells, for example, by enhancing random light scattering through incorporation of large particles or spherical voids. Photonic crystals, termed as periodic dielectric structures that have a band gap that forbids propagation of a certain frequency range of light, serves the purpose of enhancing light and matter interaction.

One well-known property of photonic crystals, in which the group velocity of light becomes anomalously small near the wavelength of a stop band, can be understood by considering the dispersion curve (E versus k) of light in a periodic dielectric (Nishimura et al. 2003). Approaching the stop band from the long wavelength side, light can be described increasingly as a standing wave. On the red edge of stop band, peaks of this wave are primarily localized in the high dielectric part of the photonic crystal, and on the blue edge, they are localized in the low dielectric part. Slow photons found in photonic crystals have an immense potential for increasing the path length of light since the group velocity of light at these wavelengths is significantly reduced, thereby increasing the apparent thickness of the material. Slow photons can be observed in periodic photonic structures at energies just above and below the photonic stop band (Chen et al. 2008).

Liu et al. designed Pt-loaded TiO_2 hierarchical photonic crystal photocatalysts and pioneered in applying the system in the photocatalytic water splitting experiment (Fig. 10). The Pt-loaded TiO_2 hierarchical photonic crystal photocatalysts was demonstrated to be able to double hydrogen evolution relative to nanocrystalline TiO_2 due to slow photon enhancement at the stop band edge and multiple scatterings among photonic crystal segments (Liu et al. 2010). Further efforts including designing plasmonic photonic crystal photocatalyst and photonic crystal structured photoelectrode were made to further advance the application of the photonic crystal in light energy conversion field (Lu et al. 2012, Zhang et al. 2012).

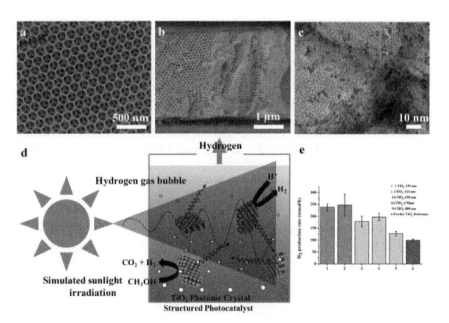

Figure 10. SEM and TEM characterizations of the Pt nanoparticles loaded TiO_2 photonic crystal: **(a)** and **(b)** SEM; **(c)** TEM; **(d)** Schematic illustration of Pt/TiO_2 photonic crystal involved photocatalytic hydrogen evolution in the presence of methanol as sacrificial agent; **(e)** The observed photocatalytic hydrogen evolution rate by different TiO_2 photonic crystal samples with different stop band positions.

By combining colloidal crystal template and amphiphilic triblock copolymer, Liu et al. also developed a novel hierarchically macro-/mesoporous Ti-Si oxide photonic crystal with highly efficient photocatalytic activity. It was found that thermal stability of mesoporous structures in the composite matrix were greatly improved due to the introduction of silica acting as glue, linking anatase nanoparticles together and photocatalytic activity of the Ti-Si oxide photonic crystals was affected by calcination condition (Liu et al. 2009). The hierarchically macro-/mesoporous Ti-Si oxide photonic crystals showed great performance for pollutant model degradation.

Two different research groups simultaneously proposed the photonic crystal concept in 1987 (Yablonovitch 1987, John 1987). But the prototype of photonic crystal always existed in nature, such as opal, diatom frustule, certain butterfly wing and, etc. (Parker and Townley 2007). Diatoms are unicellular photosynthetic organisms and can be found in nearly every aquatic habitat on earth, which are responsible for approximately one-fifth of the organic compounds yield (Gordon et al. 2009). As well known, diatom features the unique frustule architectures with hierarchical structures ranging from micrometric to nanometric scales, which are thought to contribute to their high photosynthesis efficiency. Therefore, imitating the structures of diatom frustule and reconstructing frustules with photoactive materials will be promising for the construction an artificial photosynthetic system for light energy conversion purpose. Liu and Antonietti employed diatom as the template to fabricate g-C_3N_4 coating onto the frustules to construct the highly efficient light harvesting system and applied the diatom-based light harvesting system for NADH regeneration for further enzymatic catalysis (Liu and Antonietti 2013). Further investigation shows that hierarchical carbon nitride nanorod array could also be fabricated from diatom frustule templating (Liu et al. 2014). The as-obtained hierarchical carbon nitride array was found to outperform bulky carbon nitride benchmark samples in NADH regeneration for driving three-dehydrogenase involved in CO_2 hydrogenation reaction.

Zhou et al. designed a route to engineer the semiconductor creatively (TiO_2, ZnO et al.) by templating with natural leaves (Fig. 11) (Zhou et al. 2010). The templating process could endow the material with the excellent light harvesting structures featured in the natural photosynthesis system and the possible dopant from the organic substance in the natural leaves. Liu and Yang et al. extended the fabrication strategy and smartly introduced another mesostructure directing agent to further increase surface area of photocatalyst and performance was greatly enhanced in degradation of pollutant model (Liu et al. 2013, Yang et al. 2015). The aquatic leaves were paid special attention due to the special aquatic environment featuring weak light intensity. Employing aquatic leaves as the template, new photocatalysts

Figure 11. Artificial leaves construction strategy by Zhou et al. Courtesy of Dr. Han Zhou.

could be obtained and applied for the photocatalytic degradation and water splitting. The strategy employed herein could provide a direct method to learn from nature.

Nam et al. demonstrated a biologically templated nanostructure for visible-light-driven water oxidation that involves a genetically engineered M13 virus scaffold to mediate the co-assembly of zinc porphyrin photosensitizer and iridium oxide hydrosol clusters as catalyst (Nam et al. 2010). Porous polymer microgels are employed as an immobilization matrix to improve the structural durability of the assembled nanostructures and to allow materials to be recycled. Their results suggested that bio-templated nanoscale assembly of functional components is a promising route to construct photocatalytic water splitting systems.

3.4 Advanced Two Dimensional Light Harvesting Systems

As a core organ in the natural photosynthesis, a granum is usually composed of 10–100 thylakoids in stacking structure with antenna pigment molecules and Integral membrane proteins in each thylakoid membrane (Fig. 12a). After light is harvested via the antenna chlorophyll pigment molecules, electrons are excited and transferred to reaction center chlorophyll, which allows the start of a flow of electrons down an electron transport chain that

Figure 12. (a) Schematic illustration of a granum with stacked thylakoids; **(b)** Thermal exfoliation method to construct g-C_3N_4 nanosheet; **(c)** and **(d)** Thylakoid inspired mesoporous carbon nitride sphere featuring with nanosheet as the building block.

leads to the ultimate reduction of NADP to NADPH. There are four key elements to carry out the primary reaction: (1) a stacked structure; (2) the antenna pigment molecules, which absorb the light; (3) the photosynthetic reaction center where photo-electric conversion takes place; and (4) a fast electron transport process, whereby photo-induced electrons are collected by the complex in each thylakoid and transported to the outside. There is a reason why nature has chosen the layered and stacked structure for thylakoids membranes, which is certainly beyond the current scope of the chapter. Taking inspiration from the stacked thylakoids membrane, the two dimensional light harvesting and conversion systems could be promising.

Graphene, with excellent electron conductivity, unique two-dimensional morphology and high transparency, is served as an ideal component to accommodate semiconductor components and wire photogenerated charge carriers from excited semiconductor components, thereby improving the efficiency of the photocatalytic processes. Tremendous effort is devoted to fabricating various graphene/semiconductor composites toward improved conversion of solar energy based on the observation and speculation that the photogenerated electrons from semiconductors (e.g., TiO_2, CdS) can be transferred readily into the two-dimensional graphene sheet.

Wang and Zhai et al. designed a TiO_2/graphene stacked structures by layer-by-layer assembly technique, aimed at imitating the function and structure of granum in green plants (Yang et al. 2012). With TiO_2 performing the role of the antenna pigment molecules and graphene replacing the b6f complex, the authors claimed that introduced graphene layer would promote the electron-hole separation efficiency.

g-C_3N_4 materials have gained intense interests from the scientific community because of their unique and excellent properties in thermal and chemical stability (Thomas et al. 2008). Specially, g-C_3N_4 is in the research focus due to the recently disclosed visible-light-driven photocatalytic properties (Wang et al. 2009). g-C_3N_4 materials are currently the mostly investigated polymeric semiconductors for light energy conversion, which is ascribed to the medium band gap, favorable band alignment and superior stability including chemical and thermal properties (Liu et al. 2013, Wang et al. 2012, Cao et al. 2015). g-C_3N_4 is also known as a layered material similar to graphite, where there were strong covalent C-N bonds instead of C-C bonds in each layer and weak van der Waals force across layers. In view of successful exfoliation of graphene from graphite, exfoliation of g-C_3N_4 to layered material is of great interests to the researchers. Ultrasonication and thermal exfoliation have been developed to obtain the large-scale g-C_3N_4 nanosheet and the excellent photocatalytic properties have been demonstrated, respectively (Fig. 12b) (Yang et al. 2013, Niu et al. 2012). In addition, Liu and Antonietti developed a confinement template method to obtain the thylakoid-mimic g-C_3N_4 mesosphere featuring nanosheet as the building block and high surface area (Huang et al. 2014). The obtained mesosphere demonstrated excellent efficiency for both photocatalytic NADH regeneration and hydrogen evolution (Fig. 12c and d).

Sun and coworkers reported a new but surprising strategy of using surface-functionalized small carbon nanoparticles to harvest visible photons for subsequent charge separation on the particle surface in order to drive the efficient photocatalytic CO_2 reduction process (Cao et al. 2015). The aqueous solubility of the catalysts enables photoreduction under more desirable homogeneous reaction conditions. In addition to CO_2 conversion, the nanoscale carbon-based photocatalysts are also useful for the photogeneration of H_2 from water under similar conditions.

Very recently, Kang and coworkers reported a metal-free carbon nanodot-carbon nitride nanocomposite and demonstrate its impressive performance for photocatalytic solar water splitting (Liu et al. 2015). They claimed that the quantum efficiencies of 16% at 420 nm wavelength, 6.29% at 580 nm wavelength, and 4.42% for at 600 nm wavelength could be obtained, respectively. The overall solar energy conversion efficiency of 2.0% could be acquired. The stable metal-free catalyst system consists of economic, abundant, environmentally benign materials and is very promising for the practical application in the near future.

3.5 Plasmonic Photocatalyst

Plasmonic photocatalysis, a term coined by Awazu et al. in 2008, has received a lot of attention recently (Awazu et al. 2008). Literally, plasmonic photocatalysis combines plasmonic properties of noble metals with conventional photocatalysis to achieve highly efficient solar energy conversion systems (Liu et al. 2011). Some excellent review on plasmonic photocatalysis can be referred for further details (Zhang et al. 2013). The disperse of noble metal nanoparticles (mostly Au, Ag and Cu) into semiconductor photocatalysts could lead to drastic enhancement of photoactivity under irradiation of UV and a broad range of visible light due to the formation of a Schottky junction and localized surface plasmon resonance (LSPR) deriving form the noble metal nanoparticles (Li and Antonietti 2013). The formation of Schottky junction would facilitate electrons and holes to separate in different directions once they are generated, which is well known as metal doping effect for semiconductor photocatalysis. The LSPR effect occurs when light interacts with noble nanoparticles that are comparable or smaller than the incident wavelength. The resonance wavelength for Au and Ag nanoparticles can be modulated to locate in the visible range or the near-UV range, depending on the size, shape and surrounding environment. Huang and coworkers reported one plasmonic photocatalyst Ag@AgCl, in which Ag nanoparticles are deposited on the surfaces of AgCl particles. The plasmonic photocatalyst was prepared by treating Ag_2MoO_4 with HCl to form AgCl powder with some Ag^+ ions in the surface region of AgCl particles reduced to Ag^0. The photocatalyst is claimed to be highly efficient for dyes degradation under visible light (Wang et al. 2008). Yu and Huang further combined the plasmonic photocatalyst Ag/AgCl with TiO_2 nanotube arrays by depositing AgCl nanoparticles into the self-organized TiO_2 NTs and then reducing partial Ag^+ ions in the surface region of the AgCl particles to Ag^0 species under *in situ* xenon lamp irradiation. The prepared metal-semiconductor nanocomposite plasmonic photocatalyst was reported to possess a highly visible-light photocatalytic activity for photocatalytic degradation of methyl orange in water (Yu et al. 2009). Zhu et al. reported that well-defined graphene oxide enwrapped Ag/AgX (X = Br, Cl) nanocomposites, which are composed of Ag/AgX nanoparticles and gauze-like graphene oxide nanosheets, could be facilely fabricated via a water/oil system (Fig. 13) (Zhu et al. 2011). The resultant graphene oxide-based hybrid nanocomposites could be used as a stable plasmonic photocatalyst for the photodegradation of

Figure 13. Fabrication procedure to Graphene oxide enwrapped Ag/AgX (X = Br, Cl) nanocomposites and the characterization results. Courtesy of Dr. Mingshan Zhu.

methyl orange pollutant under visible-light irradiation. The authors found that hybridization of Ag/AgX with GO nanosheets facilitates enhanced adsorptive capacity of Ag/AgX/GO to methyl orange molecules, decreasing the size of the Ag/AgX nanoparticles in Ag/AgX/GO, facilitating charge transfer, and suppressing recombination of electron-hole pairs in Ag/AgX/GO. All these factors might contribute to high photocatalytic degradation activity toward pollutant model. The LSPR could endow a visible light absorbance in addition to UV light absorbance of large bandgap semiconductor (TiO_2) and also enhanced visible light absorbance for low bandgap semiconductor (Fe_2O_3) (Thimsen et al. 2010, Ingram and Linic 2011). Lu et al. integrated plasmonic effect with the photonic crystal by depositing noble nanoparticles onto porous skeleton of the TiO_2 inverse opal photonic crystal (Lu et al. 2012). The researchers thought that excellent photocatalytic performance benefited from the cooperatively enhanced light harvesting owing both to the localized surface plasmon resonance of Au NPs, which extended the light response spectra and the photonic effect of the TiO_2 inverse opal, which intensified the plasmonic absorption by Au NPs (Liu et al. 2011, Shen et al. 2012).

3.6 Metal-Organic Frameworks Based Artificial Photosynthesis

Metal-organic frameworks (MOFs) are hybrid materials that consist of secondary building units and organic linkers (Li et al. 1999). The rational design of MOFs with tunable properties through a selective combination of metal ions and organic ligands has resulted in applications in gas adsorption, gas separation, drug delivery and catalysis, and some photocatalytic applications (Chen et al. 2005, Peng et al. 2014, Horcajada et al. 2010, Lee et al. 2009, Fu et al. 2012). Lin and coworkers developed a MOF photocatalyst doping with $Re(bpy)(CO_3)Cl$ complexes that reduced CO_2 to CO under UV light irradiation (Wang et al. 2011). However, the pioneering work involving MOF for photocatalytic CO_2 reduction ends up in very low efficiency. They further loaded ultrafine platinum nanoparticles into stable, porous, and phosphorescent metal-organic frameworks. The resulting Pt@MOF assemblies serve as effective photocatalysts for hydrogen evolution by synergistic photoexcitation of the MOF frameworks and electron injection into the platinum nanoparticles. The authors also claimed that the turnover number could be obtained up to 7000, highlighting the high efficiency in addition to the recyclable and reusable properties of the MOF based photocatalyst (Wang et al. 2012). Fu et al. further developed a visible-light-driven NH_2-MIL-125(Ti) with an amine-functionalized linker, which could reduce CO_2 to $HCOO^-$ in the presence of triethanolamine (Fu et al. 2012).

3.7 Reminiscent of Biosystem

The development of robust systems for conversion of solar energy into chemical fuels is an important subject in renewable energy research (Mershin et al. 2012, Lubner et al. 2011, Ciesielski et al. 2010, Yehezkeli et al. 2012, Badura et al. 2011, Yehezkeli et al. 2013). Key aspects are efficient and rapid catalysis of both fuel production (reduction of H_2O or CO_2), and water oxidation. Enzymes and some proteins often have extraordinary and unique capabilities as electrocatalysts (Hexter et al. 2012). Nature has created hydrogenase as a hydrogen evolution catalyst with a high turnover number at ambient conditions (Armstrong and Hirst 2011). Integration of enzyme into the artificial system can serve the purpose of solar fuel generation or the inspiration of designing the synthetic catalysts (Reisner et al. 2009, Kato et al. 2014).

Bruce and coworkers found that self-organized photosynthetic nanoparticle consisting of photosystem I from a thermophilic bacterium and cytochrome-c6 can and a platinum catalyst, could stably produce hydrogen *in vitro* upon illumination (Iwuchukwu et al. 2010). The self-organized photosynthetic nanoparticles allow electron transport from electron donor (sodium ascorbate) to photosystem I via cytochrome-c6 and finally to the platinum catalyst, where hydrogen gas is evolved. The authors also claimed that further replacement of platinum catalyst with renewable hydrogenase is feasible and might point to a more sustainable chemistry. Giraldo et al. reported augmenting plants' ability to capture light energy by 30 percent by embedding carbon nanotubes in the chloroplast, the plant organelle where photosynthesis takes place. In order to make it happen, the researchers employed a technique called vascular infusion to deliver nanoparticles into Arabidopsis thaliana, a small flowering plant. The nanotubes moved into the chloroplast and boosted photosynthetic electron flow by about 30 percent. But how that extra electron flow influences the plants' sugar production is still unclear. Using another type of carbon nanotube, they also modified plants to detect the gas nitric oxide (Giraldo et al. 2014). The work is very interesting as it firstly points out that the externally introduced nanostructures could contribute to the plant photosynthesis. Therefore, the term "plant nanobionics" coined by the authors, may contribute to the development of biomimetic materials for light energy conversion and biochemical detection in the near future.

Armstrong and Reisner developed hybrid and enzyme-modified nanoparticles that could produce H_2 using visible light as the energy source under ambient conditions. The [NiFeSe]-hydrogenase from Desulfomicrobium baculatum (Db [NiFeSe]-H) is attached to Ru dye-sensitized TiO_2, with triethanolamine as a sacrificial electron donor, producing H_2 at a turnover frequency of approximately 50 (mol H_2) s^{-1} (mol total hydrogenase)$^{-1}$ at pH 7 and 25°C (Reisner et al. 2009). The system shows high electrocatalytic stability under anaerobic conditions and also after prolonged exposure to air. The robust system is promising for small scale applications. They further found that TiO_2 nanoparticles modified by attachment of carbon monoxide dehydrogenase and a Ru photosensitizer could produce CO from CO_2 at a rate of 250 μmol of CO (g of TiO_2)$^{-1}$ h^{-1} when illuminated with visible light at pH 6 and 20°C (Woolerton et al. 2010). Reinser and coworkers also explored the integration of the hydrogenase and bio-inspired nickel catalyst with the visible-light-driven g-C_3N_4 (Caputo et al. 2014). As claimed by the authors, the semibiological and purely synthetic systems show catalytic activity during solar light irradiation with turnover numbers of more than 50000 mol H_2 (mol H_2ase)$^{-1}$

and approximately 155 mol H_2 (mol NiP)$^{-1}$ in redox-mediator-free aqueous solution at pH 6 and 4.5, respectively. The successful integration of hydrogenase and g-C_3N_4 suggests the great possibility to maximize the potential of hybrid system in the view of substantial progress on the g-C_3N_4 materials (Wang et al. 2012). However, the lifetime of device performance is usually determined by the gradual decreasing activity of dehydrogenase and natural enzyme based system generally does not meet high throughput demand as economically applicable catalysts. Exploring artificially dehydrogenase analogue further to replace the usage of the dehydrogenase is expected and doable.

Han et al. reported a robust and highly active system for solar hydrogen generation in water (pH = 4.5 with ascorbic acid as electron donor) including CdSe nanocrystals capped with dihydrolipoic acid as light absorber and a soluble Ni^{2+}-dihydrolipoic acid catalyst for proton reduction (Fig. 14) (Han et al. 2012). The system gives > 600000 turnovers, and has nearly constant activity for 15 days under 520 nm wavelength illumination. The quantum yields in water were estimated to be over 36%. This light-driven hydrogen evolution system consists of simple components containing only

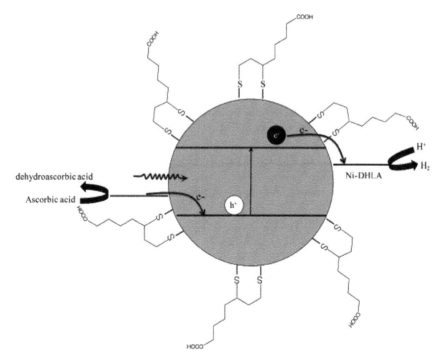

Figure 14. Schematic illustration of a robust H_2 photogeneration system from water employing CdS semiconductor nanocrystals and a nickel catalyst in the presence of Ascorbic acid.

Earth-abundant elements and could have a substantial impact on the sustainable production of chemical fuels in the future.

With efforts from biologists and chemists, crystallographic studies revealed that the active site of natural dehydrogenase system consists of two iron cations that are linked by a bridging dithiolate ligand. Rauchfuss and co-workers have shown that abiomimetic model of FeH can act as a catalyst for electrochemical hydrogen production (Gloaguen et al. 2001). Taking inspiration from this work, Sun and coworker further extended the biomimetic iron hydrogenase to the photochemical fuel production by covalently linking the hydrogenate mimic to the photosensitizer (Ott et al. 2003). Wu and coworkers have done a lot of work in the artificial photosynthesis by combining [FeFe]-hydrogenase mimic and variable quantum dots. By sensitization with CdTe quantum dots, the artificial water-soluble [FeFe]-hydrogenase mimic could produce hydrogen photocatalytically in the presence of formic acid as proton source and electron donor (Wang et al. 2011). Jian et al. further investigated that by the chitosan confinement for the H_2-evolving system, the durability and activity of the artificial [FeFe]-hydrogenase mimic are increased greatly by chitosan confinement for the H_2 evolving system, which could act as alternative for use in the future sustainable H_2 economy (Jian et al. 2013).

Using proteins and other biological structures as starting points, Kanatzidis and coworkers investigated porous chalcogenide framework that could contain immobilized redox-active centers ($[Fe_4S_4]^{2+}$, $[Mo_2Fe_6S_8(SPh)_3Cl_6]^{3-}$), linked by various thiostannate linking blocks ($[Sn_nS_{2n+2}; n = 1,2,4]$) covalently, and light-harvesting photo-redox dye molecules in a larger superstructure (Yuhas et al. 2011, Yuhas et al. 2011, Banerjee et al. 2015). These multifunctional chalcogels have been shown to reduce protons and carbon disulfide electrocatalytically and produce hydrogen photochemically and even reduce N_2 to NH_3 in the presence of sacrificial electron and hydrogen donor, respectively (Fig. 15). These gels are claimed to have a high degree of synthetic flexibility and might allow

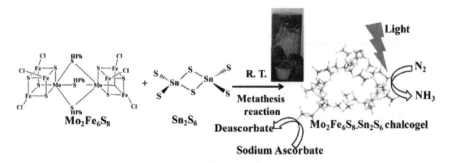

Figure 15. Schematic illustration of $Mo_2Fe_6S_8$-Sn_2S_6 chalcogel synthesis and its photocatalytic N_2 fixation property. The inset shows the digital image of as-formed chalcogel.

a wide range of light-driven processes relevant to the production of solar fuels, even though the efficiency is still very low, the next generation gel shows very promising prospect for photocatalytic N_2 fixation in ambient conditions.

Enzymatic catalysis features the mild reaction condition and highly selectivity merits. Dehydrogenase based enzymatic catalysis usually requests the continuous supply of NADH, a biological form of hydrogen, to act as cofactor for the reaction accomplishment. The high cost of enzyme-specific NADH limits the practical application of enzymatic catalysis. Inspired by light reaction of natural photosynthesis, NADH cofactor can also be regenerated by photocatalysis, which is more sustainable, green, and economic, comparing with the established but tedious enzymatic method and electrochemical method for NADH regeneration (Ryu et al. 2011, Nam et al. 2010, Kim et al. 2012, Kim et al. 2014). Using electron rich aromatic scaffolds with electronic properties similar to DNA or nucleobases might enable the direct electron transfer between photocatalyst and NAD^+ and consequently might serve the purpose of light induced substrate regeneration. g-C_3N_4, a "molecular fossil" material but with (photo)catalytic functional properties disclosed only in the recent years, could fulfill the conditions in principle (Wang et al. 2009). Liu and Antonietti made a lot of progress in this field. Inspired by the diatom photosynthesis, they developed diatom frustule template membrane reactor for such NADH regeneration in the presence of [Cp*Rh(bpy)Cl]Cl as the electron and proton mediator or even without the presence of the mediator (Fig. 16) (Liu and Antonietti 2013). With the involvement of mediator, the regenerated NADH is enzymatically active and could be used in the following enzymatic reactions. Without involvement of mediator, the pure g-C_3N_4 could also regenerate NADH. However, under such circumstance, all the resultant NADH is not 1,4-NADH. By mediating effect of a rhodium complex, NADH cofactor could be photocatalytically regenerated by the novel g-C_3N_4 array, which was obtained from the sacrificial diatom frustule templating (Liu et al. 2014). The rate of *in situ* NADH regeneration is high enough to reverse the biological pathway of the three consecutive dehydrogenase enzymes (formate dehydrogenase, formaldehyde dehydrogenase, and alcohol dehydrogenase), which then leads to the sustainable conversion of formaldehyde to methanol and also the reduction of carbon dioxide into methanol (Obert and Dave 1999, Cazelles et al. 2013). Further modification of g-C_3N_4 with various organic molecules could be a feasible route for realizing the metal-free photocatalytic regeneration of NADH (Pariente et al. 1994, Pauliukaite and Brett 2008, Long and Chen 1997). Liu further developed a sacrificial templating route for constructing mesoporous g-C_3N_4 sphere, which is excellent for NADH regeneration and *in situ* enzymatic conversion could be realized. Liu also developed a high-throughput method

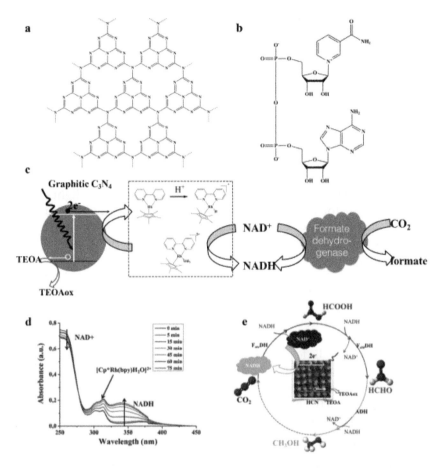

Figure 16. The interaction between g-C$_3$N$_4$ and β-NAD$^+$. **(a)** Illustration of a perfect g-C$_3$N$_4$ network constructed from heptazine building blocks. **(b)** Molecular structure of β-NAD$^+$. **(c)** Illustrated scheme of photocatalytic regeneration of NADH in the absence or presence of [Cp*Rh(bpy)H$_2$O]$^{2+}$ mediator and CO$_2$ reduction mediated by formate dehydrogenase in the presence of *in situ* photogenerated NADH. **(d)** Spectral measurements of NADH concentration in mediator [Cp*Rh(bpy)H$_2$O]$^{2+}$ involved photocatalytic regeneration process. **(e)** g-C$_3$N$_4$ involved photocatalytic cofactor regeneration for *in situ* three dehydrogenases involved enzymatic catalysis of CO$_2$ to methanol.

for synthesis of nanostructured g-C$_3$N$_4$ rod (Liu et al. 2014). By using such nanorods, NADH could be obtained efficiently and *in situ* NADH regeneration system was further coupled with L-glutamate dehydrogenase for sustainable generation of L-glutamate from α-ketoglutarate (Liu et al. 2014). Liu and Cazelles also extended the photocatalytic route to the electrochemical indirect regeneration of NADH by using the newly developed ultrathin g-C$_3$N$_4$ electrode and the [Cp*Rh(bpy)Cl]Cl (Liu et al.

2015, Cazelles et al. 2015). The inspiring works point promising and practical routes for NADH regeneration for enzymatic reactions involved in organic synthesis (Chenault and Whitesides 1987). There are some key issues to be solved prior to practical applications of the photocatalytic cofactor system, such as mediator immobilization issue and exploration of noble metal-free mediator system. The related research will greatly advance cofactor regeneration field and may also shed light on general photocatalysis.

3.8 Tandem Systems

The exciton (electron and hole pair) generation initiates when light shines onto semiconductor. Electron and hole should be efficiently separated for initiating further photocatalytic reactions. An ideal case should involve the usage of both charge carriers (hole and electron) for useful reduction and oxidation reactions simultaneously. But, unfortunately, either of the charge carriers has to be consumed purposely by the external sacrificial agent to liberate another charge carrier for the following photocatalytic reactions. To some extent, the introduction of the sacrificial agent diminishes the benefit of photocatalytic process. But, considering the low cost of the sacrificial agent and high value-added photocatalytic product, the exploration of photocatalysis is still desirable. However, exploring the efficient overall water splitting system should always be targeted. In the past, some photocatalysts were found to possess the capability for overall water splitting to hydrogen and oxygen simultaneously. But, efficiency is so low and stability of the photocatalytic system is also a major issue.

Taking inspiration from the Z-scheme in natural photosynthesis, a two-step photoexcitation mechanism by employing two different photocatalysts could be constructed, which could act as catalysts for H_2 evolution and O_2 evolution, respectively (Tada et al. 2006, Kato et al. 2004, Sekizawa et al. 2013). The shuttle redox couple is usually employed to combine the two half reactions (Sasaki et al. 2008). In this case, photoexcited electrons reduce water to hydrogen over the H_2 evolution catalyst while the holes left in the valence band of semiconductor catalyst oxidize the reductant to an oxidant. In the meanwhile, photoexcited holes on O_2 evolution photocatalyst could oxidize water to release oxygen and the electrons left in the conduction band of the O_2 evolution semiconductor photocatalyst reduce the oxidant to reductant. Therefore, the Z-scheme method requires two photocatalysts since water reduction and oxidation occur separately and electron transfer occurs from one photocatalyst to the other. The IO_3^-/I^- and Fe^{3+}/Fe^{2+} redox couple was usually employed as electron mediators between photoanode and photocathode (Miseki et al. 2013). Some semiconductors fulfilling the conditions of water oxidation could act as the photoanode (WO_3 and TiO_2), while some other semiconductors satisfying the requirement of water reduction could behave as photocathode (non-oxide photocatalysts such as CdS and g-C_3N_4) (Miseki et al. 2013, Martin et al. 2014). Arakawa and coworkers coupled two different semiconductor photocatalysts and a redox mediator for overall water splitting, mimicking the Z-scheme mechanism of photosynthesis. It was observed that H_2 was produced on a Pt-$SrTiO_3$ (Cr-Ta-codoped) photocatalyst by employing I^- electron donor under visible light irradiation. While Pt-WO_3 photocatalyst is capable for catalyzing O_2 evolution using IO_3^- electron acceptor under visible light irradiation, simultaneously (Sayama et al. 2002). Martin et al. reported that metal-free polymer semiconductor g-C_3N_4 could be integrated into a Z-scheme water splitting system coupling with $BiVO_4$ and WO_3 for overall water splitting, analogous to PSII and PSI in natural photosynthesis (Martin et al. 2014). Most importantly, hydrogen and oxygen can be evolved in an ideal ratio of 2:1, indicating the overall water splitting property.

Furthermore, the PEC process offers a promising route for avoiding the use of sacrificial agent for the water splitting or CO_2 reductions (Liu 2014). The direct photoelectrolysis of water at semiconductor electrolyte interface is not only of scientific interest but could also produce benefits such as cost and energy savings over coupled photovoltaic-electrolysis systems for H_2 production. Designing and engineering stable and visible-light-driven semiconductors that exhibit the optimal conduction and valence band edge alignment for PEC applications should be pursued and the efficiencies of PEC process could therefore be enhanced. In view of enormous solar cell advancements both from scientific and industrial communities, taking advantage of electricity from the solar cell could benefit the PEC process, which makes the PEC process more sustainable as it is without consumption of electricity from conventional fossil fuel combustion.

Mallouk and coworkers developed a photoassisted overall water splitting system by integration of heteroleptic ruthenium tris(bipyridyl) dye and $IrO_2 \cdot nH_2O$ nanoparticles into the PEC cell while the dye coordinated the $IrO_2 \cdot nH_2O$ nanoparticles and the porous TiO_2 electrode through chemical bonding linking, respectively (Youngblood et al. 2009). Under visible illumination in slightly acidic aqueous buffer, oxygen was generated at potentials of –325 mV vs Ag/AgCl and hydrogen was generated at the Pt cathode. However, the internal quantum yield for photocurrent generation was still very low (ca. 0.9%), which could be attributed to slow electron transfer from $IrO_2 \cdot nH_2O$ to the oxidized dye and the undesired high back electron transfer from TiO_2 to the dye. Turner and his coworker designed a monolithic photovoltaic-photoelectrochemical device ($GaInP_2$/GaAs p/n) for hydrogen evolution from water splitting. The hydrogen production efficiency of the systems is claimed to be 12.4%, based on the short circuit current and the lower heating value of hydrogen (Khaselev and Turner 1998). Luo et al. recently designed a highly efficient and low-cost water-splitting cell combining a solution-processed perovskite tandem solar cell and a bifunctional Earth-abundant catalyst (Luo et al. 2014). The NiFe layered double hydroxide acts as catalyst electrode and exhibits high activity toward both oxygen and hydrogen evolution reactions in alkaline electrolyte. The combination of the solar cell and water-splitting cell yields a water-splitting photocurrent density of around 10 mA/cm^2, corresponding to a solar-to-hydrogen efficiency of 12.3%. However, the poor stability of hybrid perovskite materials limits the cell lifetime and lead based perovskite solar cell might cause an environmental issue. Optimizing the stability of the perovskite further and the replacement of lead with other environmentally benign elements should contribute to the solar cell driven water splitting greatly.

4. Summary and Outlook

For decades, using solar energy to produce fuel may have appeared as a remote vision. Nature has evolved for millions of years to be a perfect prototype for mankind to follow. Artificial photosynthesis is about learning from nature, unfolding nature's secrets as it were and building artificial device and systems for storing solar energy in small molecule fuel, which is greatly interesting and important for both industrial and scientific communities. Rapid development of materials science field contributes greatly to the emergence and rising of artificial photosynthesis research.

Based on the development of semiconductor and nanomaterial fields, the artificial photosynthesis field established that diversified systems have been designed to mimic natural photosynthesis: sunlight is harvested and the light energy is used to convert water and carbon dioxide into fuels. In the past decades, substantial progress has been especially made considering the fact that a tremendous number of materials has been synthesized as photocatalysts. Nowadays, though the efficiency is still low, harnessing power of the sun has become a reality. A lot of excellent examples are highlighted in the current chapter. The highlighted scientific breakthroughs means that, at least in the laboratory, it is now possible to mimic photosynthesis partly and use sunlight, water and carbon dioxide to produce useful fuels. As also revealed in the current chapter, nanomaterials have been investigated intensively for the potential and promising solar fuel production. Nanostructures play important roles in directing and architecting efficient light harvesting systems. The basic properties of the nanomaterials coupling with special morphology architectures present endless possibilities for light energy conversion applications. Some excellent examples like hierarchically porous materials, nanotube, nanowire, bio-templated nanomaterials, bio-hybrid stand out as promising benchmark systems for constructing artificial photosynthesis systems.

There are still a lot of challenges ahead of us. Considering the increasingly serious environmental issues and diminishing fossil fuel resources, it is no exaggeration to say that survival of human civilization will depend on our ability to respond to this challenge. However, the overall progress in artificial photosynthesis is still limited though high efficiency photocatalyst materials are reported on a daily basis. In this field, a real breakthrough that could inspire immediate applications is desired. Furthermore, compared to the rapid progress in the solar cell field, especially the rising of perovskite solar cells with efficiency up to 20% in a short time, both challenges and chances are presented for researchers in the artificial photosynthesis field. Cheap and highly efficient solar cell driven water splitting could avoid the use of the sacrificial agent and might open a new era of artificial photosynthesis. The competition also exists, and rapid development of perovskite solar cell might drive people to reconsider the light harvesting components: solar cell or photocatalysts. The entire field deserves urgent action.

There are certainly great chances ahead of us in the pursuit of sustainable solar energy as the main energy source. Artificial photosynthesis represents a very promising prospect that solar fuel could drive the future human society. With great progress in further understanding natural photosynthesis and architecting artificial photosynthesis systems much more efficiently, the dream using solar energy to drive mankind society will become a reality in the future foreseen. This progress relies on the fact that chemists work closely with physicists, materials scientists and biologists, to develop high efficiency photocatalytic system. With concerted efforts from scientific and industrial communities, it should be possible to move the technologies of solar energy harnessing and conversion forward and the large-scale practical applications could be possible. The new materials, new phenomenon, and new explanations emerging in artificial photosynthesis investigation, will also advance the further development of materials science greatly.

Acknowledgements

J. Liu acknowledges the support of Alexander von Humboldt Foundation.

References

Alivisatos, A.P. 1996. Perspectives on the physical chemistry of semiconductor nanocrystals. J. Phys. Chem. 100: 13226–13239.

Anderson, C. and A.J. Bard. 1995. An improved photocatalyst of TiO_2/SiO_2 prepared by a sol-gel synthesis. J. Phys. Chem. 99: 9882–9885.

Armstrong, F.A. and J. Hirst. 2011. Reversibility and efficiency in electrocatalytic energy conversion and lessons from enzymes. Proc. Natl. Acad. Sci. USA 108: 14049–14054.

Asahi, R., T. Morikawa, T. Ohwaki, K. Aoki and Y. Taga. 2001. Visible-light photocatalysis in nitrogen-doped titanium oxides. Science 293: 269–271.

Awazu, K., M. Fujimaki, C. Rockstuhl, J. Tominaga, H. Murakami, Y. Ohki, N. Yoshida and T. Watanabe. 2008. A plasmonic photocatalyst consisting of silver nanoparticles embedded in titanium dioxide. J. Am. Chem. Soc. 130: 1676–1680.

Baba, R., S. Nakabayashi, A. Fujishima and K. Honda. 1985. Investigation of the mechanism of hydrogen evolution during photocatalytic water decomposition on metal-loaded semiconductor powders. J. Phys. Chem. 89: 1902–1905.

Badura, A., T. Kothe, W. Schuhmann and M. Rögner. 2011. Wiring photosynthetic enzymes to electrodes. Energy Environ. Sci. 4: 3263–3274.

Banerjee, A., B.D. Yuhas, E.A. Margulies, Y. Zhang, Y. Shim, M.R. Wasielewski and M.G. Kanatzidis. 2015. Photochemical nitrogen conversion to ammonia in ambient conditions with FeMoS-chalcogels. J. Am. Chem. Soc. 137: 2030–2034.

Barber, J. 2009. Photosynthetic energy conversion: natural and artificial. Chem. Soc. Rev. 38: 185–196.

Bard, A.J. 1979. Photoelectrochemistry and heterogeneous photo-catalysis at semiconductors. J. Photochem. 10: 59–75.

Bard, A.J. 1980. Photoelectrochemistry. Science 207: 139–144.

Bi, Y., S. Ouyang, N. Umezawa, J. Cao and J. Ye. 2011. Facet effect of single-crystalline Ag_3PO_4 sub-microcrystals on photocatalytic properties. J. Am. Chem. Soc. 133: 6490–6492.

Blankenship, R.E. 2013. Molecular Mechanisms of Photosynthesis. 2013. John Wiley & Sons, Hoboken, New Jersey.

Cao, S., J. Low, J. Yu and M. Jaroniec. 2015. Polymeric photocatalysts based on graphitic carbon nitride. Adv. Mater. 10.1002/adma.201500033.

Caputo, C.A., M.A. Gross, V.W. Lau, C. Cavazza, B.V. Lotsch and E. Reisner. 2014. Photocatalytic hydrogen production using polymeric carbon nitride with a hydrogenase and a bioinspired synthetic Ni catalyst. Angew. Chem. Int. Ed. 53: 11538–11542.

Cazelles, R., J. Drone, F. Fajula, O. Ersen, S. Moldovan and A. Galarneau. 2013. Reduction of CO_2 to methanol by a polyenzymatic system encapsulated in phospholipids-silica nanocapsules. New J. Chem. 37: 3721–3730.

Cazelles, R., J. Liu and M. Antonietti. 2015. Hybrid C_3N_4/fluorine-doped tin oxide electrode transfers hydride for 1, 4-NADH cofactor regeneration. ChemElectroChem 2: 333–337.

Chen, B., N.W. Ockwig, A.R. Millward, D.S. Contreras and O.M. Yaghi. 2005. High H_2 adsorption in a microporous metal-organic framework with open metal sites. Angew. Chem. 117: 4823–4827.

Chen, H., N. Wang, J. Di, Y. Zhao, Y. Song and L. Jiang. 2010. Nanowire-in-microtube structured core/shell fibers via multifluidic coaxial electrospinning. Langmuir 26: 11291–11296.

Chen, J.I., G. von Freymann, S.Y. Choi, V. Kitaev and G.A. Ozin. 2008. Slow photons in the fast lane in chemistry. J. Mater. Chem. 18: 369–373.

Chen, J.S., Y.L. Tan, C.M. Li, Y.L. Cheah, D. Luan, S. Madhavi, F.Y.C. Boey, L.A. Archer and X.W. Lou. 2010. Constructing hierarchical spheres from large ultrathin anatase TiO_2 nanosheets with nearly 100% exposed (001) facets for fast reversible lithium storage. J. Am. Chem. Soc. 132: 6124–6130.

Chen, X. and S.S. Mao. 2007. Titanium dioxide nanomaterials: synthesis, properties, modifications, and applications. Chem. Rev. 107: 2891–2959.

Chen, X., S. Shen, L. Guo and S.S. Mao. 2010. Semiconductor-based photocatalytic hydrogen generation. Chem. Rev. 110: 6503–6570.

Chen, X., L. Liu, Y.Y. Peter and S.S. Mao. 2011. Increasing solar absorption for photocatalysis with black hydrogenated titanium dioxide nanocrystals. Science 331: 746–750.

Chen, X., L. Liu, Z. Liu, M.A. Marcus, W.-C. Wang, N.A. Oyler, M.E. Grass, B. Mao, P.-A. Glans and Y.Y. Peter. 2013. Properties of disorder-engineered black titanium dioxide nanoparticles through hydrogenation. Sci. Rep. 3: 1510.

Chenault, H.K. and G.M. Whitesides. 1987. Regeneration of nicotinamide cofactors for use in organic synthesis. Applied Biochemistry and Biotechnology 14: 147–197.

Chu, D., Y. Masuda, T. Ohji and K. Kato. 2009. Formation and photocatalytic application of ZnO nanotubes using aqueous solution. Langmuir 26: 2811–2815.

Ciesielski, P.N., C.J. Faulkner, M.T. Irwin, J.M. Gregory, N.H. Tolk, D.E. Cliffel and G.K. Jennings. 2010. Enhanced photocurrent production by photosystem I multilayer assemblies. Adv. Funct. Mater. 20: 4048–4054.

Collings, A.F. and C. Critchley. 2007. Artificial Photosynthesis: from Basic Biology to Industrial Application. John Wiley & Sons, Hoboken, New Jersey.

Crossland, E.J., N. Noel, V. Sivaram, T. Leijtens, J.A. Alexander-Webber and H.J. Snaith. 2013. Mesoporous TiO_2 single crystals delivering enhanced mobility and optoelectronic device performance. Nature 495: 215–219.

Dabbousi, B., J. Rodriguez-Viejo, F.V. Mikulec, J. Heine, H. Mattoussi, R. Ober, K. Jensen and M. Bawendi. 1997. (CdSe)ZnS core-shell quantum dots: synthesis and characterization of a size series of highly luminescent nanocrystallites. J. Phys. Chem. B 101: 9463–9475.

Domen, K., S. Naito, M. Soma, T. Onishi and K. Tamaru. 1980. Photocatalytic decomposition of water vapour on an NiO-$SrTiO_3$ catalyst. J. Chem. Soc., Chem. Commun. 543–544.

Dong, Z., X. Lai, J.E. Halpert, N. Yang, L. Yi, J. Zhai, D. Wang, Z. Tang and L. Jiang. 2012. Accurate control of multishelled ZnO hollow microspheres for dye-sensitized solar cells with high efficiency. Adv. Mater. 24: 1046–1049.

Du, J., X. Lai, N. Yang, J. Zhai, D. Kisailus, F. Su, D. Wang and L. Jiang. 2010. Hierarchically ordered macro-mesoporous TiO_2-graphene composite films: Improved mass transfer, reduced charge recombination, and their enhanced photocatalytic activities. ACS Nano 5: 590–596.

Duonghong, D., E. Borgarello and Michael Grätzel. 1981. Dynamics of light-induced water cleavage in colloidal systems. J. Am. Chem. Soc. 103: 4685–4690.

Etgar, L., P. Gao, Z. Xue, Q. Peng, A.K. Chandiran, B. Liu, M.K. Nazeeruddin and M. Grätzel. 2012. Mesoscopic $CH_3NH_3PbI_3/TiO_2$ heterojunction solar cells. J. Am. Chem. Soc. 134: 17396–17399.

Fan, F.-R.F., H.S. White, B.L. Wheeler and A.J. Bard. 1980. Semiconductor electrodes. 31. Photoelectrochemistry and photovoltaic systems with n-and p-type tungsten selenide (WSe_2) in aqueous solution. J. Am. Chem. Soc. 102: 5142–5148.

Faunce, T., S. Styring, M.R. Wasielewski, G.W. Brudvig, A.W. Rutherford, J. Messinger, A.F. Lee, C.L. Hill, M. Fontecave and D.R. MacFarlane. 2013. Artificial photosynthesis as a frontier technology for energy sustainability. Energy Environ. Sci. 6: 1074–1076.

Feng, X., T.J. LaTempa, J.I. Basham, G.K. Mor, O.K. Varghese and C.A. Grimes. 2010. Ta_3N_5 nanotube arrays for visible light water photoelectrolysis. Nano Lett. 10: 948–952.

Frank, S.N. and A.J. Bard. 1977. Heterogeneous photocatalytic oxidation of cyanide ion in aqueous solutions at titanium dioxide powder. J. Am. Chem. Soc. 99: 303–304.

Fu, Y., D. Sun, Y. Chen, R. Huang, Z. Ding, X. Fu and Z. Li. 2012. An amine-functionalized titanium metal-organic framework photocatalyst with visible-light-induced activity for CO_2 reduction. Angew. Chem. Int. Ed. 51: 3364–3367.

Fujishima, A. and K. Honda. 1972. Electrochemical photolysis of water at a semiconductor electrode. Nature 238: 37–38.

Gao, C., Z. Lu and Y. Yin. 2011. Gram-scale synthesis of silica nanotubes with controlled aspect ratios by templating of nickel-hydrazine complex nanorods. Langmuir 27: 12201–12208.

Giraldo, J.P., M.P. Landry, S.M. Faltermeier, T.P. McNicholas, N.M. Iverson, A.A. Boghossian, N.F. Reuel, A.J. Hilmer, F. Sen and J.A. Brew. 2014. Plant nanobionics approach to augment photosynthesis and biochemical sensing. Nat. Mater. 13: 400–408.

Gloaguen, F., J.D. Lawrence and T.B. Rauchfuss. 2001. Biomimetic hydrogen evolution catalyzed by an iron carbonyl thiolate. J. Am. Chem. Soc. 123: 9476–9477.

Gordon, R., D. Losic, M.A. Tiffany, S.S. Nagy and F.A. Sterrenburg. 2009. The glass menagerie: diatoms for novel applications in nanotechnology. Trends Biotechnol. 27: 116–127.

Grätzel, Michael. 1981. Artificial photosynthesis: water cleavage into hydrogen and oxygen by visible light. Acc. Chem. Res. 14: 376–384.

Grätzel, M. 2001. Photoelectrochemical cells. Nature 414: 338–344.

Grimes, C.A. and G.K. Mor. 2009. TiO$_2$ Nanotube Arrays: Synthesis, Properties, and Applications. Springer-Verlag GmbH, Berlin, Germany.

Han, X., Q. Kuang, M. Jin, Z. Xie and L. Zheng. 2009. Synthesis of titania nanosheets with a high percentage of exposed (001) facets and related photocatalytic properties. J. Am. Chem. Soc. 131: 3152–3153.

Han, Z., F. Qiu, R. Eisenberg, P.L. Holland and T.D. Krauss. 2012. Robust photogeneration of H$_2$ in water using semiconductor nanocrystals and a nickel catalyst. Science 338: 1321–1324.

Hardee, K.L. and A.J. Bard. 1977. Semiconductor electrodes X. Photoelectrochemical behavior of several polycrystalline metal oxide electrodes in aqueous solutions. J. Electrochem. Soc. 124: 215–224.

Hexter, S.V., F. Grey, T. Happe, V. Climent and F.A. Armstrong. 2012. Electrocatalytic mechanism of reversible hydrogen cycling by enzymes and distinctions between the major classes of hydrogenases. Proc. Natl. Acad. Sci. USA 109: 11516–11521.

Hitoki, G., A. Ishikawa, T. Takata, J.N. Kondo, M. Hara and K. Domen. 2002. Ta$_3$N$_5$ as a novel visible light-driven photocatalyst. Chem. Lett. 736–737.

Horcajada, P., T. Chalati, C. Serre, B. Gillet, C. Sebrie, T. Baati, J.F. Eubank, D. Heurtaux, P. Clayette and C. Kreuz. 2010. Porous metal-organic-framework nanoscale carriers as a potential platform for drug delivery and imaging. Nat. Mater. 9: 172–178.

Huang, J., M. Antoniett and J. Liu. 2014. Bio-inspired carbon nitride mesoporous spheres for artificial photosynthesis: photocatalytic cofactor regeneration for sustainable enzymatic synthesis. J. Mater. Chem. A. 2: 7686–7693.

Hulteen, J. 1997. A general template-based method for the preparation of nanomaterials. J. Mater. Chem. 7: 1075–1087.

Ingram, D.B. and S. Linic. 2011. Water splitting on composite plasmonic-metal/semiconductor photoelectrodes: evidence for selective plasmon-induced formation of charge carriers near the semiconductor surface. J. Am. Chem. Soc. 133: 5202–5205.

Iskandar, F., A.B.D. Nandiyanto, K.M. Yun, C.J. Hogan, K. Okuyama and P. Biswas. 2007. Enhanced photocatalytic performance of brookite TiO$_2$ macroporous particles prepared by spray drying with colloidal templating. Adv. Mater. 19: 1408–1412.

Iwase, A., Y.H. Ng, Y. Ishiguro, A. Kudo and R. Amal. 2011. Reduced graphene oxide as a solid-state electron mediator in Z-scheme photocatalytic water splitting under visible light. J. Am. Chem. Soc. 133: 11054–11057.

Iwuchukwu, I.J., M. Vaughn, N. Myers, H. O'Neill, P. Frymier and B.D. Bruce. 2010. Self-organized photosynthetic nanoparticle for cell-free hydrogen production. Nat. Nanotechnol. 5: 73–79.

Jang, J.S., U.A. Joshi and J.S. Lee. 2007. Solvothermal synthesis of CdS nanowires for photocatalytic hydrogen and electricity production. J. Phys. Chem. C 111: 13280–13287.

Jang, J.S., S.H. Choi, D.H. Kim, J.W. Jang, K.S. Lee and J.S. Lee. 2009. Enhanced photocatalytic hydrogen production from water-methanol solution by nickel intercalated into titanate nanotube. J. Phys. Chem. C 113: 8990–8996.

Jian, J.-X., Q. Liu, Z.-J. Li, F. Wang, X.-B. Li, C.-B. Li, B. Liu, Q.-Y. Meng, B. Chen and K. Feng. 2013. Chitosan confinement enhances hydrogen photogeneration from a mimic of the diiron subsite of [FeFe]-hydrogenase. Nat. Commun. 4: 2695.

John, S. 1987. Strong localization of photons in certain disordered dielectric superlattices. Phys. Rev. Lett. 58: 2486.

Kadowaki, H., N. Saito, H. Nishiyama and Y. Inoue. 2007. RuO$_2$-loaded Sr^{2+}-doped CeO$_2$ with d0 electronic configuration as a new photocatalyst for overall water splitting. Chem. Lett. 36: 440–441.

Kale, B.B., J.O. Baeg, S.M. Lee, H. Chang, S.J. Moon and C.W. Lee. 2006. CdIn$_2$S$_4$ nanotubes and "Marigold" nanostructures: a visible-light photocatalyst. Adv. Funct. Mater. 16: 1349–1354.

Kalyanasundaram, K. and M. Grätzel. 2010. Artificial photosynthesis: biomimetic approaches to solar energy conversion and storage. Curr. Opin. Biotechnol. 21: 298–310.

Kamat, P.V. 2007. Meeting the clean energy demand: nanostructure architectures for solar energy conversion. J. Phys. Chem. C 111: 2834–2860.

Karakitsou, K.E. and X.E. Verykios. 1993. Effects of altervalent cation doping of titania on its performance as a photocatalyst for water cleavage. J. Phys. Chem. 97: 1184–1189.

Kato, H., M. Hori, R. Konta, Y. Shimodaira and A. Kudo. 2004. Construction of Z-scheme type heterogeneous photocatalysis systems for water splitting into H_2 and O_2 under visible light irradiation. Chem. Lett. 33: 1348–1349.

Kato, M., J.Z. Zhang, N. Paul and E. Reisner. 2014. Protein film photoelectrochemistry of the water oxidation enzyme photosystem II. Chem. Soc. Rev. 43: 6485–6497.

Kay, A. and Michael Grätzel. 1993. Artificial photosynthesis. 1. Photosensitization of titania solar cells with chlorophyll derivatives and related natural porphyrins. J. Phys. Chem. 97: 6272–6277.

Khaselev, O. and J.A. Turner. 1998. A monolithic photovoltaic-photoelectrochemical device for hydrogen production via water splitting. Science 280: 425–427.

Kim, D., K.K. Sakimoto, D. Hong and P. Yang. 2015. Artificial photosynthesis for sustainable fuel and chemical production. Angew. Chem. Int. Ed. 54: 3259–3266.

Kim, J.H., M. Lee, J.S. Lee, C.B. Park. 2012. Self-assembled light-harvesting peptide nanotubes for mimicking natural photosynthesis. Angew. Chem. Int. Ed. 51: 517–520.

Kim, J.H., D.H. Nam and C.B. Park. 2014. Nanobiocatalytic assemblies for artificial photosynthesis. Curr. Opin. Biotechnol. 28: 1–9.

Kirk, J.T.O. 1994. Light and Photosynthesis in Aquatic Ecosystems. Cambridge University Press, Cambridge, United Kingdom.

Kittel, C., P. McEuen and P. McEuen. 2005. Introduction to Solid State Physics, 8th Edition. John Wiley & Sons, Hoboken, New Jersey.

Kiwi, J. and C. Morrison. 1984. Heterogeneous photocatalysis. Dynamics of charge transfer in lithium-doped anatase-based catalyst powders with enhanced water photocleavage under ultraviolet irradiation. J. Phys. Chem. 88: 6146–6152.

Kowalski, D., D. Kim and P. Schmuki. 2013. TiO_2 nanotubes, nanochannels and mesosponge: Self-organized formation and applications. Nano Today 8: 235–264.

Kraeutler, B. and A.J. Bard. 1978. Heterogeneous photocatalytic preparation of supported catalysts. Photodeposition of platinum on titanium dioxide powder and other substrates. J. Am. Chem. Soc. 100: 4317–4318.

Kuang, D., J. Brillet, P. Chen, M. Takata, S. Uchida, H. Miura, K. Sumioka, S.M. Zakeeruddin and M. Grätzel. 2008. Application of highly ordered TiO_2 nanotube arrays in flexible dye-sensitized solar cells. ACS Nano 2: 1113–1116.

Kudo, A., K. Ueda, H. Kato and I. Mikami. 1998. Photocatalytic O_2 evolution under visible light irradiation on $BiVO_4$ in aqueous $AgNO_3$ solution. Catal. Lett. 53: 229–230.

Kudo, A., K. Omori and H. Kato. 1999. A novel aqueous process for preparation of crystal form-controlled and highly crystalline $BiVO_4$ powder from layered vanadates at room temperature and its photocatalytic and photophysical properties. J. Am. Chem. Soc. 121: 11459–11467.

Lauhon, L.J., M.S. Gudiksen, D. Wang and C.M. Lieber. 2002. Epitaxial core-shell and core-multishell nanowire heterostructures. Nature 420: 57–61.

Law, M., L.E. Greene, J.C. Johnson, R. Saykally and P. Yang. 2005. Nanowire dye-sensitized solar cells. Nat. Mater. 4: 455–459.

Lee, J., O.K. Farha, J. Roberts, K.A. Scheidt, S.T. Nguyen and J.T. Hupp. 2009. Metal-organic framework materials as catalysts. Chem. Soc. Rev. 38: 1450–1459.

Lewis, N.S. and D.G. Nocera. 2006. Powering the planet: Chemical challenges in solar energy utilization. Proc. Natl. Acad. Sci. USA 103: 15729–15735.

Li, D. and Y. Xia. 2004. Electrospinning of nanofibers: reinventing the wheel? Adv. Mater. 16: 1151–1170.

Li, H., M. Eddaoudi, M. O'Keeffe and O.M. Yaghi. 1999. Design and synthesis of an exceptionally stable and highly porous metal-organic framework. Nature 402: 276–279.

Li, L., Y. Yu, F. Meng, Y. Tan, R.J. Hamers and S. Jin. 2012. Facile solution synthesis of α-$FeF_3 \cdot 3H_2O$ nanowires and their conversion to α-Fe_2O_3 nanowires for photoelectrochemical application. Nano Lett. 12: 724–731.

Li, R., F. Zhang, D. Wang, J. Yang, M. Li, J. Zhu, X. Zhou, H. Han and C. Li. 2013. Spatial separation of photogenerated electrons and holes among {010} and {110} crystal facets of BiVO$_4$. Nat. Commun. 4: 1432.

Li, X.-H. and M. Antonietti. 2013. Metal nanoparticles at mesoporous N-doped carbons and carbon nitrides: functional Mott-Schottky heterojunctions for catalysis. Chem. Soc. Rev. 42: 6593–6604.

Li, Y., Z.Y. Fu and B.L. Su. 2012. Hierarchically structured porous materials for energy conversion and storage. Adv. Funct. Mater. 22: 4634–4667.

Lin, Y., S. Zhou, X. Liu, S. Sheehan and D. Wang. 2009. TiO$_2$/TiSi$_2$ heterostructures for high-efficiency photoelectrochemical H2O splitting. J. Am. Chem. Soc. 131: 2772–2773.

Lin, Y., S. Zhou, S.W. Sheehan and D. Wang. 2011. Nanonet-based hematite heteronanostructures for efficient solar water splitting. J. Am. Chem. Soc. 133: 2398–2401.

Liu, C., J. Tang, H.M. Chen, B. Liu and P. Yang. 2013. A fully integrated nanosystem of semiconductor nanowires for direct solar water splitting. Nano Lett. 13: 2989–2992.

Liu, G., L.-C. Yin, J. Wang, P. Niu, C. Zhen, Y. Xie and H.-M. Cheng. 2012. A red anatase TiO$_2$ photocatalyst for solar energy conversion. Energy Environ. Sci. 5: 9603–9610.

Liu, J., M. Li, J. Wang, Y. Song, L. Jiang, T. Murakami and A. Fujishima. 2009. Hierarchically macro-/mesoporous Ti-Si oxides photonic crystal with highly efficient photocatalytic capability. Environ. Sci. Technol. 43: 9425–9431.

Liu, J., G. Liu, M. Li, W. Shen, Z. Liu, J. Wang, J. Zhao, L. Jiang and Y. Song. 2010. Enhancement of photochemical hydrogen evolution over Pt-loaded hierarchical titania photonic crystal. Energy Environ. Sci. 3: 1503–1506.

Liu, J., M. Li, J. Zhou, C. Ye, J. Wang, L. Jiang and Y. Song. 2011. Reversibly phototunable TiO$_2$ photonic crystal modulated by Ag nanoparticles' oxidation/reduction. Appl. Phys. Lett. 98: 023110.

Liu, J. and M. Antonietti. 2013. Bio-inspired NADH regeneration by carbon nitride photocatalysis using diatom templates. Energy Environ. Sci. 6: 1486–1493.

Liu, J., J. Huang, D. Dontosova and M. Antonietti. 2013. Facile synthesis of carbon nitride micro-/nanoclusters with photocatalytic activity for hydrogen evolution. RSC Adv. 3: 22988–22993.

Liu, J., Q. Yang, W. Yang, M. Li and Y. Song. 2013. Aquatic plant inspired hierarchical artificial leaves for highly efficient photocatalysis. J. Mater. Chem. A 1: 7760–7766.

Liu, J. 2014. NextGen Speaks. Science 343: 26–26.

Liu, J., R. Cazelles, A. Galarneau and M. Antonietti. 2014. Bioinspired construction of ordered carbon nitride array for photocatalytic mediated enzymatic reductions. Phys. Chem. Chem. Phys.

Liu, J., J. Huang, H. Zhou and M. Antonietti. 2014. Uniform graphitic carbon nitride nanorod for efficient photocatalytic hydrogen evolution and sustained photoenzymatic catalysis. ACS Appl. Mater. Interfaces 6: 8434–8440.

Liu, J., H. Wang, Z.P. Chen, H. Moehwald, S. Fiechter, R. van de Krol, L. Wen, L. Jiang and M. Antonietti. 2015. Microcontact-printing-assisted access of graphitic carbon nitride films with favorable textures toward photoelectrochemical application. Adv. Mater. 27: 712–718.

Liu, J., Y. Liu, N. Liu, Y. Han, X. Zhang, H. Huang, Y. Lifshitz, S.-T. Lee, J. Zhong and Z. Kang. 2015. Metal-free efficient photocatalyst for stable visible water splitting via a two-electron pathway. Science 347: 970–974.

Liu, R., Y. Lin, L.Y. Chou, S.W. Sheehan, W. He, F. Zhang, H.J. Hou and D. Wang. 2011. Water splitting by tungsten oxide prepared by atomic layer deposition and decorated with an oxygen-evolving catalyst. Angew. Chem. Int. Ed. 50: 499–502.

Liu, S., J. Yu and M. Jaroniec. 2010. Tunable photocatalytic selectivity of hollow TiO$_2$ microspheres composed of anatase polyhedra with exposed {001} facets. J. Am. Chem. Soc. 132: 11914–11916.

Long, Y.-T. and H.-Y. Chen. 1997. Electrochemical regeneration of coenzyme NADH on a histidine modified silver electrode. J. Electroanal. Chem. 440: 239–242.

Lu, Y., H. Yu, S. Chen, X. Quan and H. Zhao. 2012. Integrating plasmonic nanoparticles with TiO_2 photonic crystal for enhancement of visible-light-driven photocatalysis. Environ. Sci. Technol. 46: 1724–1730.

Lubner, C.E., A.M. Applegate, P. Knörzer, A. Ganago, D.A. Bryant, T. Happe and J.H. Golbeck. 2011. Solar hydrogen-producing bionanodevice outperforms natural photosynthesis. Proc. Natl. Acad. Sci. USA 108: 20988–20991.

Luo, H., T. Takata, Y. Lee, J. Zhao, K. Domen and Y. Yan. 2004. Photocatalytic activity enhancing for titanium dioxide by co-doping with bromine and chlorine. Chem. Mater. 16: 846–849.

Luo, J., J.-H. Im, M.T. Mayer, M. Schreier, M.K. Nazeeruddin, N.-G. Park, S.D. Tilley, H.J. Fan and M. Grätzel. 2014. Water photolysis at 12.3% efficiency via perovskite photovoltaics and Earth-abundant catalysts. Science 345: 1593–1596.

Macak, J.M., M. Zlamal, J. Krysa and P. Schmuki. 2007. Self-organized TiO_2 nanotube layers as highly efficient photocatalysts. Small 3: 300–304.

Maeda, K., K. Teramura, D. Lu, T. Takata, N. Saito, Y. Inoue and K. Domen. 2006. Photocatalyst releasing hydrogen from water. Nature 440: 295–295.

Martin, D.J., P.J.T. Reardon, S.J. Moniz and J. Tang. 2014. Visible light-driven pure water splitting by a Nature-inspired organic semiconductor-based system. J. Am. Chem. Soc. 136: 12568–12571.

McDermott, G., S. Prince, A. Freer, A. Hawthornthwaite-Lawless, M. Papiz, R. Cogdell and N. Isaacs. 1995. Crystal structure of an integral membrane light-harvesting complex from photosynthetic bacteria. Nature 374: 517–521.

McEvoy, J.P. and G.W. Brudvig. 2006. Water-splitting chemistry of photosystem II. Chem. Rev. 106: 4455–4483.

Meissner, D., R. Memming and B. Kastening. 1988. Photoelectrochemistry of cadmium sulfide. 1. Reanalysis of photocorrosion and flat-band potential. J. Phys. Chem. 92: 3476–3483.

Mershin, A., K. Matsumoto, L. Kaiser, D. Yu, M. Vaughn, M.K. Nazeeruddin, B.D. Bruce, M. Grätzel and S. Zhang. 2012. Self-assembled photosystem-I biophotovoltaics on nanostructured TiO_2 and ZnO. Sci. Rep. 2: 234.

Miseki, Y., S. Fujiyoshi, T. Gunji and K. Sayama. 2013. Photocatalytic water splitting under visible light utilizing I_3^-/I^- and IO_3^-/I^- redox mediators by Z-scheme system using surface treated PtO_x/WO_3 as O_2 evolution photocatalyst. Catal. Sci. Technol. 3: 1750–1756.

Mor, G.K., M.A. Carvalho, O.K. Varghese, M.V. Pishko and C.A. Grimes. 2004. A room-temperature TiO_2-nanotube hydrogen sensor able to self-clean photoactively from environmental contamination. J. Mater. Res. 19: 628–634.

Nam, D.H., S.H. Lee and C.B. Park. 2010. CdTe, CdSe, and CdS nanocrystals for highly efficient regeneration of Nicotinamide cofactor under visible light. Small 6: 922–926.

Nam, Y.S., A.P. Magyar, D. Lee, J.-W. Kim, D.S. Yun, H. Park, T.S. Pollom, D.A. Weitz and A.M. Belcher. 2010. Biologically templated photocatalytic nanostructures for sustained light-driven water oxidation. Nat. Nanotechnol. 5: 340–344.

Nishimura, S., N. Abrams, B.A. Lewis, L.I. Halaoui, T.E. Mallouk, K.D. Benkstein, J. van de Lagemaat and A.J. Frank. 2003. Standing wave enhancement of red absorbance and photocurrent in dye-sensitized titanium dioxide photoelectrodes coupled to photonic crystals. J. Am. Chem. Soc. 125: 6306–6310.

Niu, P., L. Zhang, G. Liu and H.-M. Cheng. 2012. Graphene-like carbon nitride nanosheets for improved photocatalytic activities. Adv. Funct. Mater. 22: 4763–4770.

O'regan, B. and M. Grätzel. 1991. A low-cost, high-efficiency solar cell based on dye-sensitized colloidal TiO2 films. Nature 353: 737–740.

Obert, R. and B.C. Dave. 1999. Enzymatic conversion of carbon dioxide to methanol: enhanced methanol production in silica sol-gel matrices. J. Am. Chem. Soc. 121: 12192–12193.

Ott, S., M. Kritikos, B. Åkermark and L. Sun. 2003. Synthesis and structure of a biomimetic model of the iron hydrogenase active site covalently linked to a ruthenium photosensitizer. Angew. Chem. Int. Ed. 42: 3285–3288.

Pariente, F., E. Lorenzo and H. Abruna. 1994. Electrocatalysis of NADH oxidation with electropolymerized films of 3, 4-dihydroxybenzaldehyde. Anal. Chem. 66: 4337–4344.

Park, J.H., S. Kim and A.J. Bard. 2006. Novel carbon-doped TiO_2 nanotube arrays with high aspect ratios for efficient solar water splitting. Nano Lett. 6: 24–28.

Parker, A.R. and H.E. Townley. 2007. Biomimetics of photonic nanostructures. Nat. Nanotechnol. 2: 347–353.

Pauliukaite, R. and C. Brett. 2008. Poly (neutral red): Electrosynthesis, characterization, and application as a redox mediator. Electroanalysis 20: 1275–1285.

Peng, Y., Y. Li, Y. Ban, H. Jin, W. Jiao, X. Liu and W. Yang. 2014. Metal-organic framework nanosheets as building blocks for molecular sieving membranes. Science 346: 1356–1359.

Prakasam, H.E., K. Shankar, M. Paulose, O.K. Varghese and C.A. Grimes. 2007. A new benchmark for TiO_2 nanotube array growth by anodization. J. Phys. Chem. C 111: 7235–7241.

Reisner, E., D.J. Powell, C. Cavazza, J.C. Fontecilla-Camps and F.A. Armstrong. 2009. Visible light-driven H_2 production by hydrogenases attached to dye-sensitized TiO_2 nanoparticles. J. Am. Chem. Soc. 131: 18457–18466.

Rolison, D.R. 2003. Catalytic nanoarchitectures-the importance of nothing and the unimportance of periodicity. Science 299: 1698–1701.

Round, F.E., R.M. Crawford and D.G. Mann. 1990. The Diatoms: Biology and Morphology of the Genera. Cambridge University Press, Cambridge, United Kingdom.

Ryu, J., S.H. Lee, D.H. Nam and C.B. Park. 2011. Rational design and engineering of quantum-dot-sensitized TiO_2 nanotube arrays for artificial photosynthesis. Adv. Mater. 23: 1883–1888.

Sasaki, Y., A. Iwase, H. Kato and A. Kudo. 2008. The effect of co-catalyst for Z-scheme photocatalysis systems with an Fe^{3+}/Fe^{2+} electron mediator on overall water splitting under visible light irradiation. J. Catal. 259: 133–137.

Savolainen, J., R. Fanciulli, N. Dijkhuizen, A.L. Moore, J. Hauer, T. Buckup, M. Motzkus and J.L. Herek. 2008. Controlling the efficiency of an artificial light-harvesting complex. Proc. Natl. Acad. Sci. USA 105: 7641–7646.

Sayama, K. and H. Arakawa. 1994. Effect of Na_2CO_3 addition on photocatalytic decomposition of liquid water over various semiconductor catalysis. J. Photochem. Photobiol. A 77: 243–247.

Sayama, K., K. Mukasa, R. Abe, Y. Abe and H. Arakawa. 2002. A new photocatalytic water splitting system under visible light irradiation mimicking a Z-scheme mechanism in photosynthesis. J. Photochem. Photobiol. A 148: 71–77.

Schneider, S.H. 1989. The greenhouse effect: science and policy. Science 243: 771–781.

Scholes, G.D., G.R. Fleming, A. Olaya-Castro and R. van Grondelle. 2011. Lessons from nature about solar light harvesting. Nat. Chem. 3: 763–774.

Sekizawa, K., K. Maeda, K. Domen, K. Koike and O. Ishitani. 2013. Artificial Z-scheme constructed with a supramolecular metal complex and semiconductor for the photocatalytic reduction of CO_2. J. Am. Chem. Soc. 135: 4596–4599.

Shen, W., M. Li, B. Wang, J. Liu, Z. Li, L. Jiang and Y. Song. 2012. Hierarchical optical antenna: Gold nanoparticle-modified photonic crystal for highly-sensitive label-free DNA detection. J. Mater. Chem. 22: 8127–8133.

Shibata, M., A. Kudo, A. Tanaka, K. Domen, K.-I. Maruya and T. Onishi. 1987. Photocatalytic activities of layered titanium compounds and their derivatives for H_2 evolution from aqueous methanol solution. Chem. Lett. 16: 1017–1018.

Steele, B.C. and A. Heinzel. 2001. Materials for fuel-cell technologies. Nature 414: 345–352.

Steinberg-Yfrach, G., P.A. Liddell, S.-C. Hung, A.L. Moore, D. Gust and T.A. Moore. 1997. Conversion of light energy to proton potential in liposomes by artificial photosynthetic reaction centres. Nature 385: 239–241.

Su, J., X. Feng, J.D. Sloppy, L. Guo and C.A. Grimes. 2010. Vertically aligned WO_3 nanowire arrays grown directly on transparent conducting oxide coated glass: synthesis and photoelectrochemical properties. Nano Lett. 11: 203–208.

Subramanian, V., E. Wolf and P.V. Kamat. 2001. Semiconductor-metal composite nanostructures. To what extent do metal nanoparticles improve the photocatalytic activity of TiO_2 films? J. Phys. Chem. B 105: 11439–11446.

Subramanian, V., E.E. Wolf and P.V. Kamat. 2004. Catalysis with TiO_2/gold nanocomposites. Effect of metal particle size on the Fermi level equilibration. J. Am. Chem. Soc. 126: 4943–4950.

Suga, M., F. Akita, K. Hirata, G. Ueno, H. Murakami, Y. Nakajima, T. Shimizu, K. Yamashita, M. Yamamoto and H. Ago. 2014. Native structure of photosystem II at 1.95 Å resolution viewed by femtosecond X-ray pulses. Nature 517: 99–103.

Sun, W.T., Y. Yu, H.Y. Pan, X.F. Gao, Q. Chen and L.M. Peng. 2008. CdS quantum dots sensitized TiO_2 nanotube-array photoelectrodes. J. Am. Chem. Soc. 130: 1124–1125.

Tachibana, Y., L. Vayssieres and J.R. Durrant. 2012. Artificial photosynthesis for solar water-splitting. Nat. Photonics 6: 511–518.

Tada, H., T. Mitsui, T. Kiyonaga, T. Akita and K. Tanaka. 2006. All-solid-state Z-scheme in CdS-Au-TiO_2 three-component nanojunction system. Nat. Mater. 5: 782–786.

Thimsen, E., F. Le Formal, M. Grätzel and S.C. Warren. 2010. Influence of plasmonic Au nanoparticles on the photoactivity of Fe_2O_3 electrodes for water splitting. Nano Lett. 11: 35–43.

Thomas, A., A. Fischer, F. Goettmann, M. Antonietti, J.-O. Müller, R. Schlögl and J.M. Carlsson. 2008. Graphitic carbon nitride materials: variation of structure and morphology and their use as metal-free catalysts. J. Mater. Chem. 18: 4893–4908.

Varghese, O.K., M. Paulose, T.J. LaTempa and C.A. Grimes. 2009. High-rate solar photocatalytic conversion of CO_2 and water vapor to hydrocarbon fuels. Nano Lett. 9: 731–737.

Varley, J., A. Janotti and C. Van de Walle. 2011. Mechanism of visible-light photocatalysis in nitrogen-doped TiO_2. Adv. Mater. 23: 2343–2347.

Wang, C., Z. Xie, K.E. deKrafft and W. Lin. 2011. Doping metal-organic frameworks for water oxidation, carbon dioxide reduction, and organic photocatalysis. J. Am. Chem. Soc. 133: 13445–13454.

Wang, C., K.E. deKrafft and W. Lin. 2012. Pt nanoparticles@photoactive metal–organic frameworks: efficient hydrogen evolution via synergistic photoexcitation and electron injection. J. Am. Chem. Soc. 134: 7211–7214.

Wang, F., W.G. Wang, X.J. Wang, H.Y. Wang, C.H. Tung and L.Z. Wu. 2011. A highly efficient photocatalytic system for hydrogen production by a robust hydrogenase mimic in an aqueous solution. Angew. Chem. Int. Ed. 50: 3193–3197.

Wang, H. and J. Lewis. 2006. Second-generation photocatalytic materials: anion-doped TiO_2. J. Phys.: Condens. Matter 18: 421.

Wang, J.-L., C. Wang and W. Lin. 2012. Metal-organic frameworks for light harvesting and photocatalysis. ACS Catal. 2: 2630–2640.

Wang, P., B. Huang, X. Qin, X. Zhang, Y. Dai, J. Wei and M.H. Whangbo. 2008. Ag@AgCl: a highly efficient and stable photocatalyst active under visible light. Angew. Chem. Int. Ed. 47: 7931–7933.

Wang, X., J.C. Yu, C. Ho, Y. Hou and X. Fu. 2005. Photocatalytic activity of a hierarchically macro/mesoporous titania. Langmuir 21: 2552–2559.

Wang, X., K. Maeda, A. Thomas, K. Takanabe, G. Xin, J.M. Carlsson, K. Domen and M. Antonietti. 2009. A metal-free polymeric photocatalyst for hydrogen production from water under visible light. Nat. Mater. 8: 76–80.

Wang, Y., X. Wang and M. Antonietti. 2012. Polymeric graphitic carbon nitride as a heterogeneous organocatalyst: from photochemistry to multipurpose catalysis to sustainable chemistry. Angew. Chem. Int. Ed. 51: 68–89.

Wasielewski, M.R. 1992. Photoinduced electron transfer in supramolecular systems for artificial photosynthesis. Chem. Rev. 92: 435–461.

Wasielewski, M.R. 2009. Self-assembly strategies for integrating light harvesting and charge separation in artificial photosynthetic systems. Acc. Chem. Res. 42: 1910–1921.

Woolerton, T.W., S. Sheard, E. Reisner, E. Pierce, S.W. Ragsdale and F.A. Armstrong. 2010. Efficient and clean photoreduction of CO_2 to CO by enzyme-modified TiO_2 nanoparticles using visible light. J. Am. Chem. Soc. 132: 2132–2133.

Wrighton, M.S., D.S. Ginley, P.T. Wolczanski, A.B. Ellis, D.L. Morse and A. Linz. 1975. Photoassisted electrolysis of water by irradiation of a titanium dioxide electrode. Proc. Natl. Acad. Sci. USA 72: 1518–1522.

Wu, H.B., H.H. Hng and X.W.D. Lou. 2012. Direct synthesis of anatase TiO_2 nanowires with enhanced photocatalytic activity. Adv. Mater. 24: 2567–2571.

Wu, X., Z. Chen, G.Q.M. Lu and L. Wang. 2011. Nanosized anatase TiO_2 single crystals with tunable exposed (001) facets for enhanced energy conversion efficiency of dye-sensitized solar cells. Adv. Funct. Mater. 21: 4167–4172.

Yablonovitch, E. 1987. Inhibited spontaneous emission in solid-state physics and electronics. Phys. Rev. Lett. 58: 2059.

Yanagida, T., Y. Sakata and H. Imamura. 2004. Photocatalytic decomposition of H_2O into H_2 and O_2 over Ga_2O_3 loaded with NiO. Chem. Lett. 33: 726–727.

Yang, H.G., C.H. Sun, S.Z. Qiao, J. Zou, G. Liu, S.C. Smith, H.M. Cheng and G.Q. Lu. 2008. Anatase TiO_2 single crystals with a large percentage of reactive facets. Nature 453: 638–641.

Yang, H.G., G. Liu, S.Z. Qiao, C.H. Sun, Y.G. Jin, S.C. Smith, J. Zou, H.M. Cheng and G.Q. Lu. 2009. Solvothermal synthesis and photoreactivity of anatase TiO_2 nanosheets with dominant {001} facets. J. Am. Chem. Soc. 131: 4078–4083.

Yang, M.-Q., N. Zhang, M. Pagliaro and Y.J. Xu. 2014. Artificial photosynthesis over graphene-semiconductor composites. Are we getting better? Chem. Soc. Rev. 43: 8240–8254.

Yang, N., Y. Zhang, J.E. Halpert, J. Zhai, D. Wang and L. Jiang. 2012. Granum-like stacking structures with TiO_2-graphene nanosheets for improving photo-electric conversion. Small 8: 1762–1770.

Yang, Q., J. Liu, H. Li, Y. Li, J. Hou, M. Li and Y. Song. 2015. Bio-inspired double-layer structure artificial microreactor with highly efficient light harvesting for photocatalysts. RSC Adv. 5: 11096–11100.

Yang, S., Y. Gong, J. Zhang, L. Zhang, L. Ma, Z. Feng, R. Vajtai, X. Wang and P.M. Ajayan. 2013. Exfoliated graphitic carbon nitride nanosheets as efficient catalysts for hydrogen evolution under visible light. Adv. Mater. 25: 2452–2456.

Yang, S., B.X. Yang, L. Wu, Y.H. Li, P. Liu, H. Zhao, Y.Y. Yu, X.Q. Gong and H.G. Yang. 2014. Titania single crystals with a curved surface. Nat. Commun. 5: 5355.

Ye, M., Q. Zhang, Y. Hu, J. Ge, Z. Lu, L. He, Z. Chen and Y. Yin. 2010. Magnetically recoverable core-shell nanocomposites with enhanced photocatalytic activity. Chem. Eur. J. 16: 6243–6250.

Yehezkeli, O., R. Tel-Vered, J. Wasserman, A. Trifonov, D. Michaeli, R. Nechushtai and I. Willner. 2012. Integrated photosystem II-based photo-bioelectrochemical cells. Nat. Commun. 3: 742.

Yehezkeli, O., R. Tel-Vered, D. Michaeli, R. Nechushtai and I. Willner. 2013. Photosystem I (PSI)/Photosystem II (PSII)-based photo-bioelectrochemical cells revealing directional generation of photocurrents. Small 9: 2970–2978.

Youngblood, W.J., S.-H.A. Lee, Y. Kobayashi, E.A. Hernandez-Pagan, P.G. Hoertz, T.A. Moore, A.L. Moore, D. Gust and T.E. Mallouk. 2009. Photoassisted overall water splitting in a visible light-absorbing dye-sensitized photoelectrochemical cell. J. Am. Chem. Soc. 131: 926–927.

Yu, J., Y. Su and B. Cheng. 2007. Template-free fabrication and enhanced photocatalytic activity of hierarchical macro-/mesoporous titania. Adv. Funct. Mater. 17: 1984–1990.

Yu, J., G. Dai and B. Huang. 2009. Fabrication and characterization of visible-light-driven plasmonic photocatalyst Ag/AgCl/TiO_2 nanotube arrays. J. Phys. Chem. C 113: 16394–16401.

Yu, J., J. Low, W. Xiao, P. Zhou and M. Jaroniec. 2014. Enhanced photocatalytic CO_2-reduction activity of anatase TiO_2 by coexposed {001} and {101} facets. J. Am. Chem. Soc. 136: 8839–8842.

Yuhas, B.D., C. Prasittichai, J.T. Hupp and M.G. Kanatzidis. 2011. Enhanced Electrocatalytic Reduction of CO_2 with Ternary $Ni-Fe_4S_4$ and $Co-Fe_4S_4$-Based Biomimetic Chalcogels. J. Am. Chem. Soc. 133: 15854–15857.

Yuhas, B.D., A.L. Smeigh, A.P. Samuel, Y. Shim, S. Bag, A.P. Douvalis, M.R. Wasielewski and M.G. Kanatzidis. 2011. Biomimetic multifunctional porous chalcogels as solar fuel catalysts. J. Am. Chem. Soc. 133: 7252–7255.

Zhai, Q., S. Xie, W. Fan, Q. Zhang, Y. Wang, W. Deng and Y. Wang. 2013. Photocatalytic conversion of carbon dioxide with water into methane: platinum and copper (I) oxide co-catalysts with a core-shell structure. Angew. Chem. Int. Ed. 52: 5776–5779.

Zhang, C. and Y. Zhu. 2005. Synthesis of square Bi_2WO_6 nanoplates as high-activity visible-light-driven photocatalysts. Chem. Mater. 17: 3537–3545.

Zhang, L.W., Y.J. Wang, H.Y. Cheng, W.Q. Yao and Y.F. Zhu. 2009. Synthesis of porous Bi_2WO_6 thin films as efficient visible-light-active photocatalysts. Adv. Mater. 21: 1286–1290.

Zhang, T. and W. Lin. 2014. Metal-organic frameworks for artificial photosynthesis and photocatalysis. Chem. Soc. Rev. 43: 5982–5993.

Zhang, X., Y.L. Chen, R.-S. Liu and D.P. Tsai. 2013. Plasmonic photocatalysis. Rep. Prog. Phys. 76: 046401.

Zhang, Z., L. Zhang, M.N. Hedhili, H. Zhang and P. Wang. 2012. Plasmonic gold nanocrystals coupled with photonic crystal seamlessly on TiO_2 nanotube photoelectrodes for efficient visible light photoelectrochemical water splitting. Nano Lett. 13: 14–20.

Zhao, T., Z. Liu, K. Nakata, S. Nishimoto, T. Murakami, Y. Zhao, L. Jiang and A. Fujishima. 2010. Multichannel TiO_2 hollow fibers with enhanced photocatalytic activity. J. Mater. Chem. 20: 5095–5099.

Zhao, Z.G. and M. Miyauchi. 2008. Nanoporous-walled tungsten oxide nanotubes as highly active visible-light-driven photocatalysts. Angew. Chem. Int. Ed. 47: 7051–7055.

Zhou, H., X. Li, T. Fan, F.E. Osterloh, J. Ding, E.M. Sabio, D. Zhang and Q. Guo. 2010. Artificial inorganic leafs for efficient photochemical hydrogen production inspired by natural photosynthesis. Adv. Mater. 22: 951–956.

Zhou, H., T. Fan and D. Zhang. 2011. Biotemplated materials for sustainable energy and environment: current status and challenges. ChemSusChem 4: 1344–1387.

Zhou, W., W. Li, J.-Q. Wang, Y. Qu, Y. Yang, Y. Xie, K. Zhang, L. Wang, H. Fu and D. Zhao. 2014. Ordered mesoporous black TiO_2 as highly efficient hydrogen evolution photocatalyst. J. Am. Chem. Soc. 136: 9280–9283.

Zhu, M., P. Chen and M. Liu. 2011. Graphene oxide enwrapped Ag/AgX (X = Br, Cl) nanocomposite as a highly efficient visible-light plasmonic photocatalyst. ACS Nano 5: 4529–4536.

Züttel, A., A. Borgschulte and L. Schlapbach. 2011. Hydrogen as a Future Energy Carrier. Wiley-VCH Verlag GmbH & Co. KGaA, Weinheim, Germany.

14

Advanced Materials for Reflective Display Applications

Mingliang Jin, Guofu Zhou and *Lingling Shui**

ABSTRACT

Reflective display shows information on demand by reflecting visible light via changing colors of one material or different moving colored materials at the scale of pixels. The increase in the interaction between human beings and machines has made display devices indispensable for visual communication. Based on the materials properties, reflective displays can be driven electrically by electrophoresis, electrowetting, electrochemical reactions, or electromechanical forces. Displaying information by reflecting light is regarded as a green technology because of the advantages of low energy consumption, comfortable reading and outdoor usage. In this chapter, the materials and devices of several popular and potential reflective displays, including electrophoretic display, electrowetting display, electrochromic display, cholesterol liquid crystal display, photonic crystal display, interferometric modulator display and liquid powder display, will be discussed in details based on the requirements of light manipulation and device function. The chapter

Institute of Electronic Paper Displays, South China Academy of Advanced Optoelectronics, South China Normal University, High Educational Mega Center of Guangzhou, Guangzhou 510006, China.
Emails: jinml@scnu.edu.cn; zhougf@scnu.edu.cn
* Corresponding author: shuill@m.scnu.edu.cn

will be involved in the interdisciplinary studies on materials and devices. It will be of interest to many—from the graduate students getting started in their research to materials science and technology to the researchers looking for applications of functional materials and devices.

1. Introduction

Displays show information on demand by manipulating visible light via changing colors of one material or moving different colored materials at the scale of pixels. Depending on the interaction between light and materials, the light can be manifested by different materials via reflection, transmission or emission mechanism. The electrophoretic display (EPD), the liquid crystal display (LCD) and the organic light emission display (OLED) is the representative of reflective, transmissive and emissive displays, respectively. The focus of this chapter is the reflective display materials which can reflect light with wavelength in the range of about 400–700 nm by electrical actuation, showing different colors and grayscales in the display devices.

A common display device is basically bi-level in nature with individual display pixels/cells, all of the same materials and sizes, arranged in a regular array. Displays can present information like text, images or videos in color (visible light) by activating the materials in pixels and segments, as shown in Fig. 1. Each pixel or segment is aligned and controlled by a display driver integrated circuit (IC) which can transfer the electrical signal to the materials in the pixel to show the desired color or light intensity. Based on different materials and device structures, reflective displays can be driven electrically by electrophoresis, electrowetting, electrochemical reactions, or electromechanical forces. The materials and working principles of

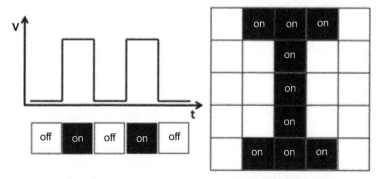

Figure 1. Schematic drawing of the electronic digital display principle. The left-top is the curve of driving signal and the left-bottom is the sketch of the front-plane material showing white and black colors driven by the signal to switch on and off, respectively. Right: a letter "I" is shown in black by switching on the corresponding pixels or segments.

several popular and potential reflective display technologies are described in details, including electrophoretic display, electrowetting display, electrochromic display, cholesterol liquid crystal display, photonic crystal display, interferometric modulator display and liquid powder display.

The reflective displays, also called "electronic paper", are different from LCDs and OLEDs on the working principles and the materials as well. The most used and one of the oldest reflective displays is printed ink on paper, such as books or newspaper. The electronic paper technology shares a word of "paper", holding the similar optical properties as the printed paper and at the same time, has the "electronic" speed and controllability. Unlike conventional backlit flat panel displays which emit light, the reflective displays reflect environmental light like paper. Such displays are superior with respect to readability (particularly in high light conditions), viewing angle and flexibility and are based on the high contrast between (mostly white) paper and dark colored ink (liquid containing pigments or dyes). On the other hand, many electronic paper technologies hold static text and images indefinitely without electricity, which holds the advantages of low energy consumption. Therefore, the reflective display is regarded as a green technology because of the advantages of low energy consumption, comfortable reading and outdoor usage.

This chapter focuses mainly on the property requirements, the design and development of the advanced materials for reflective display applications. The basic working principle and display device structures will be introduced to the readers. Based on working principles and the fabrication technologies, materials are designed and developed to satisfy device manufacturing and functioning requirements. From this, the readers can get to know how this technology has been started, and how materials have been developed driven by applications. The main technologies of reflective displays and related materials, including the display functional materials and the supporting materials for device integration, will be discussed from the molecular structure level to the device structure level. It covers the basic knowledge for reflective display applications, giving the hint about how to find out existing materials or designing new materials to fulfill a functional device application.

2. Advanced Materials for Reflective Display Applications

Reflective displays work by electrically driven advanced materials to reflect the ambient light through either locally changing position of different materials or turning the optical properties of one material. Summarizing the current technologies available, they vary in working principles and therefore, in molecular structures and device structures.

2.1 Electrophoretic Displays (EPD)

The electrophoretic display was first presented in 1973 (Ota et al. 1973). It was an idea about placing two bi-chromo-color pigments in the glass cavities and controlling the movement of particles to switch between two colors for display applications. In the following decades, people have tried to understand the mechanism and develop a model for commercializing the electrophoretic displays (Ploix et al. 1976, Dalisa 1977, Comiskey et al. 1998, Huitema et al. 2005, Huitema et al. 2006, Yu et al. 2007, Henzen 2009, Christophersen et al. 2010). The major breakthrough was made in 1998 by Jacobson et al. who reported a novel material and design overcoming the most critical shortcoming of the electrophoretic display at that moment (Comiskey et al. 1998). The original sample was made of the microcapsules containing white particles and black oil. Plenty of work has been done to improve the materials and the devices (Ploix et al. 1976, Dalisa 1977, Comiskey et al. 1998, Huitema et al. 2005, Huitema et al. 2006, Lenssen et al. 2008, Christophersen et al. 2010). The matured mode of the electrophoretic display is encapsulating the dispersion of the black and the white pigments in a microcapsule, and controlling the movement of the pigments in the microcapsule by the electrical field, which is called electrophoretic ink (E-Ink). The electrophoretic ink basically uses pigments similar to those in regular ink for paper such as books and newspapers. Therefore these displays have the same agreeable readability as printed paper.

Nowadays, the microcapsule-based electronic paper displays occupies almost 90 percentage of the electronic paper products market. The working principle of the microcapsule-based electrophoretic display is demonstrated in Fig. 2.

Electrophoresis describes the motion of a charged surface submerged in a liquid under the action of an applied electric field. Considering the case of a charged particle, the electrophoretic velocity (U) of the particle is described by the Helmholtz-Smoluchowski equation,

$$U = \frac{\varepsilon \xi_{EP} E_x}{\mu}, \tag{1}$$

where ε is the dielectric constant of the liquid, ξ_{EP} is the zeta potential of the particle, E_x is the applied electrical field and μ is the mobility of the particle. The electrophoretic zeta potential (ξ_{EP}) is a property of the charged particle. Electrophoresis causes movement of charged particles through a stationary solution. The key factors affecting electrophoretic display function are the charge density and size of the white and black particles, the viscosity and dielectricity of the transparent medium, the size of the microcapsules, and the thickness and dielectricity of the microcapsule shell. Considering the

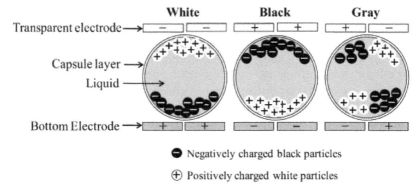

Figure 2. Schematic of electrophoretic displays structure and working principle.

stability of the particles in the liquid medium, sedimentation needs to be reduced as much as possible by compensating the gravity between particles and the dispersion solvent. According to the working principle and device structure, the colored particles/pigments, the microcapsule shell and the insulation oil are the key materials of this technology, and the charge control agents and stabilizers as well.

(1) Colored Particles/Pigments

As discussed before, the colored nano- to micro- meter particles are the key materials for realizing electrophoretic display functions. The basic requirements for the pigment are: specific density compatibility with the suspension solvent in order to reduce sedimentation, low solubility in the solvent for a lifetime, high brightness for efficient optical appearance, easily chargeable surface, good stability and easy purification to ensure manufacture and production volume. When encapsulated into pixels or microcapsules, the particles should not be absorbed in the pixel or on the capsule surface. Different types of materials have been investigated for EPD applications (Ota et al. 1973, Orsaev et al. 2000, Strohm et al. 2003, Yu et al. 2004, Song et al. 2005, Cho et al. 2007, Nagamine et al. 2007, Song et al. 2007, Song et al. 2007, Guo et al. 2009, Sgraja et al. 2010, Tan et al. 2010, Wang et al. 2012, Guang et al. 2013, Tseng et al. 2013). The most studied inorganic materials include TiO_2, black carbon, SiO_2, Al_2O_3, chromium yellow, chromium red, ironic red and magnesium purple. The organic particles of toluidine reds, phthalocyanine blue and phthalocyanine green have also been investigated. The black particles can also be fabricated by polymers with extra double bonds for which the black is dependent on the density of the double bonds, thickness of the particle shell, ratio of core to shell, and the size of the particles.

In general, the particles/pigments are dispersed in a solution in the original states with nanometer size and then coated with polymeric materials to form a core-shell structure. The shell materials are typically long chain organic materials with hydrogen bonds, such as materials with alkoxy group, acetyl group or halogens. For a long period of time, even now, the black carbon and titanium dioxide is used for the black and white particles in EPD devices because of their high brightness and easy accessibility in nature. Both black carbon and white titanium dioxide are conductive which have to be coated by polymers to reach the requirements (Werts et al. 2008).

(2) Capsule Shell Materials

The electrophoretic display device is composed by microcapsules or micropixels, for which the shell/pixel wall becomes one of the key materials in this technology. According to the mechanism of the electrophoretic display, the shell is for encapsulation of colored particles and the electrophoretic medium; therefore, this material needs to be transparent, mechanically stable but flexible, have low conductivity and be compatible with the inner materials. For these reasons, the most used materials are organic polymers like polyamine, polyurethane, polysulfones, polyethylene acid, cellulose, gelatin, arabic gum, etc. (Sun et al. 2002, Song et al. 2007, Wang et al. 2009, Li et al. 2011, Zandi et al. 2011, Mao et al. 2012, Xing et al. 2012, Campardelli et al. 2013, Fan et al. 2013, Guang et al. 2013, Song et al. 2013, Yuan et al. 2013, Bai et al. 2014). The microcapsule fabrication method is based on the chosen materials. The typical examples are *in situ* polymerization of formaldehyde and urea to form urea-formaldehyde resin (Li et al. 2012), and composite coagulation of gelatin and arabic gum to form composite film (Wang et al. 2009).

(3) Suspension Medium

Inside the microcapsules of the electrophoretic display devices, the colored particles are suspended in the liquid medium. To fulfill the requirements of electrophoretic displays, the medium needs to be environment friendly, have good insulation property (with dielectric constant larger than 2 preferably), low resistance to particle transportation, thermal and electrical stability and similar density and reflectivity with particles. These requirements could be satisfied by using single or formulated organic solvents such as alkylene, aromatic hydrocarbon, aliphatic hydrocarbon, oxosilane, and so on (Park et al. 2001, Nagamine et al. 2007, Suryanarayana et al. 2008, Werts et al. 2008, Wang et al. 2009, Li et al. 2011). The formulation of 2-phenylbutane/ tetrachloroethylene and n-haxane/tetrachloroethylene have been obtained and used widely. To tune the density, the high density fluorinated solvent and low density hydrocarbon is typically mixed together.

(4) Charge Control Agents

Another key step for manufacturing electrophoretic display materials is to charge the particles with high density. As seen from Equation 1, the electrophoretic velocity is proportional to the zeta potential of the particle. Hence, for a fast display response, the particles could have high charge density. In the non-aqueous medium, the charges could be done via the following technologies: particle surface dissociation in solvent, surfactant dissociation, absorption of ions or polarization by friction. Typical charge control agents include sulfate (vitriol), sulfonate, metallic soap, organic acid amide, organophosphate, phosphate ester, polymers, copolymers or graft polymers.

(5) Stabilizers

In a microcapsule, it is actually a suspension system with oppositely charged particles suspended in an insulation liquid. The oppositely charged particles are naturally attracted to each other because of the electrostatic force. Besides, two other degradation modes exist in the suspension: agglomeration which is caused by an insufficient repulsive barrier between particles and clustering which is caused by fluid motion within the microcapsule. All forms of instabilities are detrimental to the life of the display device. The stabilizer is therefore very necessary in this system (Murau et al. 1978, Yu et al. 2007). Polymer coating is one of the most used methods to bring steric repulsion among particles. In the non-aqueous suspension medium, the low molecular weight surfactants cannot stabilize the particles like in aqueous solution. The super dispersant, showing affinity to particles and the solvent has been synthesized. The particle affinity is part of the super dispersant links to the particles via ionic pairing, hydrogen bond, van der Waals force. The solvent affinity part can be solvated to form a polymer chain to become a steric stabilizer among particles. The most used steric stabilizers or polymer coating-stabilizers are poly(vinyl alcohol) (PVA), Chevron OLOA370 and OGA 472.

In conclusion, in the simplest implementation of an electrophoretic display, negatively charged titanium dioxide particles and positively charged carbon black particles are dispersed in a dielectric oil. The surfactants and charging agents are added to the oil, which charge the particles and hold them with a gap due to the steric resistance. This mixture is then placed between two parallel, conductive plates separated by a gap of ten to hundreds of micrometres. When a voltage is applied across the two plates, the particles migrate electrophoretically to the plate that bears the opposite charge from the one on the particles. When the white particles are located on the front (viewing) side of the display, it appears white, because light is scattered back to the viewer by the high-index titanium oxide particles. When the black particles are located on the front side of the display, it appears black, because the incident light is absorbed by the black particles. The gray scale can be realized by mixing the white and black particles in a ratio via electrical driving schemes. If the rear electrode is divided into a number of small picture elements (pixels), an image can be formed by applying the appropriate voltage to each region of the display to create a pattern of black and white regions.

2.2 Electrowetting Displays (EWD)

Electrowetting is a technology which can change the contact angle between a liquid droplet and a hydrophobic surface when a voltage is applied. The electrowetting concept for display application was first recognized by Beni and co-workers more than three decades ago (Beni et al. 1981, Beni et al. 1981, Jackel et al. 1982, Jackel et al. 1983). The reflective display technology based on electrowetting was first realized and published in 2003 by Hayes and Feenstra at Philips Research Labs (Hayes and Feenstra 2003).

The schematic drawing of the electrowetting display is demonstrated in Fig. 3. The hydrophobic insulator layer and micropixels are normally

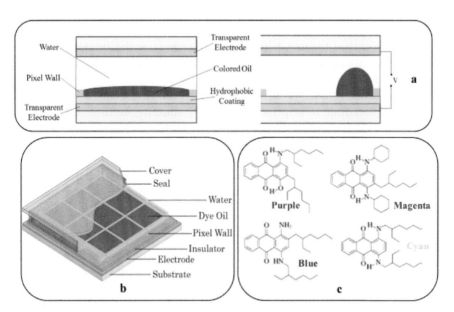

Figure 3. Schematic displays of electrowetting working principle and device structure. **(a)** In an EWD pixel, without applied voltage, a homogeneous oil film spreads over the pixel area showing the color of the dyed oil (left), with an applied voltage of V, the oil film contracts to one corner of the pixel, showing the color of the bottom substrate (right). **(b)** The 3D structure of an electrowetting display device. **(c)** Molecular structures of dyes for EWD applications.

made by using micro-fabrication technologies, then filled with colored oil immersed in water and sealed with a waterproof material. The colored oil sticks to the hydrophobic surface at the bottom of the pixel forming a thin film without electrical field. By applying a voltage through the water and the electrode underneath the dielectric layer, the oil film contracted to the corner of the pixel because of changing of contact angle. The display can therefore be operated by showing the color of the oil by spreading the colored oil over the pixel or showing the substrate color by driving the oil into the corner of the pixel. EWD has shown its potential for high quality information displays with reflective mode for using ambient light, quick response (< 2 ms switching speed has been reached) for video display (Smith et al. 2009), good optical performance (> 50% white state reflectance (Heikenfeld et al. 2009) and full color) (Heikenfeld et al. 2009) and fluidic and soft display materials for flexible displays in the future.

The electrowetting function can be described by Young-Lipmann's equation, which is

$$\cos \theta_v = \cos \theta_0 + \frac{\varepsilon V^2}{2\sigma\lambda}. \tag{2}$$

In this equation, θ_v is the contact angle between the liquid and the hydrophobic surface at applied voltage V, θ_0 is the static contact angle without applied voltage, ε is the dielectric constant of the insulator material, σ is the interfacial tension between two fluids, λ is the thickness of the insulator. To obtain the same value of contact angle the applied voltage is determined by the properties of materials (dielectric constant and fluidic interfacial tension) and the thickness of the dielectric material. Extensive work has been done to improve the display stability, lifetime, brightness, contrast and color gamut. Typically, in an electrowetting display device, the pixel size is in the range of tens to hands of microns, and the pixel walls are a few microns high. The key materials for electrowetting displays are the properties of colored non-conductive liquid (dyed oil), the transparent conductive liquid, the insulation layer, the hydrophobic surface, and the pixel walls.

(1) Dyed Oil

The selection of the dye materials is critical for the electrowetting display application. The dye screening needs to ensure good electrowetting performance, good solubility in non-conductive liquid and insoluble in the conducting liquid, high optical density and a specific color point, the lifetime including oil film stability at a certain temperature and resistance to sunlight exposure, and the integrity of the two liquid phases. The dodecane and silicone oil (Staicu and Mugele 2006) are typically chosen as the nonconductive liquid for its interfacial tension, viscosity and dielectricity.

The commercially available dyes include Oil Blue N, Sudan Red and Sudan Blue 673 from BASF, but these materials are weak to light bleaching. Other dyes like Blue 98, Red 164 have been created by Motorola cooperated with Cincinnati and the yellow liquid dye from Keystone, for which the black dye could be formulated (Sun et al. 2007, You and Steckl 2010). Liquavista has designed and synthesized the anti-UV dye materials including purple, magenta, blue and cyan as shown in Fig. 3c. Samsung has recently made the dye series of SK1-SK4, specifically for electrowetting display applications (Massard et al. 2013).

(2) Insulator and Hydrophobic Materials

Another key material in electrowetting display is the hydrophobic insulator. As seen from the Lipmann's equation, the hydrophobicity and the dielectric constant of the insulator determine the working voltage and the reversibility of the electrowetting. In the previous years, the single layer organic hydrophobic insulators have been investigated intensively, including Teflon AF (DuPont), Fluoropel 1601V (DuPont), Cytop (Asahi), Hyflon (Solvay). These materials have good solubility in low surface tension fluorocarbon solvent for which it is easy to be coated on substrates via spin-coating, screen-printing or dip-coating, forming the film with a thickness of 0.1–1 μm. This strategy could achieve good electrowetting performance. However, according to the low dielectric constant of these Teflon materials of about 2.0, higher voltage needs to be applied for the electrowetting performance, and the pin-hole of single layer is still the problem for the EWD devices. Therefore, the double or multiple layers of insulator plus hydrophobic coating have also been proposed and investigated. This typically include a layer of insulator like SiO_2 (Moon et al. 2002), Si_3N_4 (Malk et al. 2011), SiOC (silicon-oxide-carbon) (Malk et al. 2011) or ONO (oxide-nitride-oxide) (Papathanasiou et al. 2008); and the hydrophobic layer could be either a thin layer of Teflon or other types of vapor deposited hydrophobic organic layer. This strategy somehow solved the problem of the pin-hole, however, it needs to be carried at high temperature, making the device of fabrication more complicated.

(3) Pixel Wall Materials

Another important material in electrowetting display is the pixel wall which should have the wettability contrast to the hydrophobic bottom to hold the oil film inside the pixel, and at the same time, to avoid the oil sticking to or climb over it. The pixel wall material needs to resist to both conductive liquid phase and the non-conductive oil phase. The photo-patternable resin like SU-8 has been chosen as the pixel wall materials (Hayes and Feenstra 2003, Heikenfeld et al. 2009, Massard et al. 2013). Recently, other materials like polyimide have also been investigated for materials compatibility and easy fabrication purpose.

(4) Conductive Liquid

As seen from the Fig. 4, the conductive liquid fills the gap between the oil and the top plate (electrode), acting as the common electrode and the dye oil moving medium. The conductivity, viscosity and interfacial tension needs to be tuned to fit for micropixel based dual-liquid electrowetting display applications. Salts like $NaCl$, $NaSO_4$, Na_2CO_3, have normally been added to tune the conductivity of this phase, and surfactants like sodium dodecyl sulfate, Tween 20 or Tween 80 are added to the solution to reduce the interfacial energy. However, the small molecules like alkali metal ions carry a major charge for electrolysis and breakdown of the insulator layer. And small ions can easily penetrate into the pinholes or defects of the insulator, which is detrimental to the devices. Therefore, the polyelectrolytes and new designed surfactants have been investigated to improve the device stability (Choi et al. 2013). Moreover, ionic liquid has also been considered in the EWD to avoid liquid evaporation (Chevalliot et al. 2011).

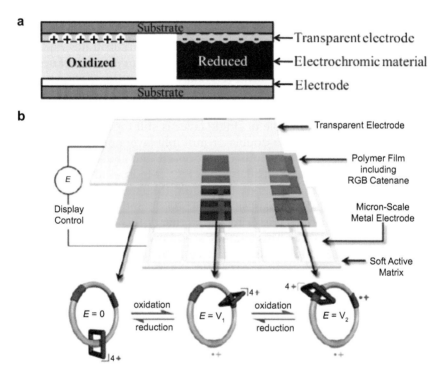

Figure 4. (a) Schematic figure of electrochromic display device structure and working principle. **(b)** The example of multicolor (RGB) electrochromic material ([2]catenane) and its usage in a display device, reprinted with permission from ref. (Ikeda and Stoddart 2008), copyright 2009 IOP.

2.3 Electrochrome Displays (ECD)

Electrochromic materials have the property of a change, evocation, or bleaching of color as affected either by an electro-transfer (redox) process or by a sufficient electrochemical potential. It can change color in a persistent but reversible manner by an electrochemical reaction. Electrochromism was discovered in 1969 by S.K. Deb (Deb 1969). The principle of electrochromic materials is explained simply as a redox reaction:

$$O + ne \longleftrightarrow R, \tag{3}$$

in which O means oxidized state, ne means the number of electrons and R means reduced state of an electrochromic material. Depending on the manipulation of light in the visible or invisible range, electrochromic materials can demonstrate colored or transparent states. Therefore, it can work either in transmissive or reflective mode. The transmissive mode is commonly considered to be used as smart window material, and the reflective mode is a potential candidate for reflective display application.

An electrochromic device is essentially a rechargeable battery in which the electrochromic electrode is separated by a suitable solid or liquid electrolyte from a charge balancing counter electrode, and the color changes occur by charging and discharging the electrochemical cell with applied potential of a few volts. The simplest way of making a working cell is to sandwich the electrochromic material between two electrodes. The technology for making a working electrochromic cell is very similar to the technology used in LCD displays. An electrochromic display device principle is demonstrated in Fig. 4a.

For display applications, the electrochromic materials are deposited onto the electrode via different methods depending on the materials themselves or the specific application reasons. In this device, the electrochromic material is able to change its color reversibly when it is placed in a different electronic state. So, by absorbing an electron (the materials is reduced) or by ejecting one (the material is oxidized), the material is able to change its color. The appearance of an electrochromic display would be that of a painted surface, it is easily viewed from any angle. The state of an electrochromic material is determined by the injected charges, which means the ECD device can be bi-stable (no power is needed until the state is changed). Moreover, the ECD device is simple and easy to fabricate, which bears the advantage of large area fabrication on flexible substrates.

It is clear that the electrochromic material is the key of this device. The materials are considered to be electrochromic when light is modulated by reflectance or absorbance in the visible region of the electromagnetic spectrum-color changes perceptible to the human eye. Reflective display applications require electrochromic materials with a high contrast ratio, coloration efficiency, write-erase efficiency, long cycle life, and short response time. From the application and device fabrication perspective, electrochromic materials can be classified into three types based on their solubility of each redox state. Type I is soluble both in the reduced state and the oxidized state, Type II is soluble in one redox state but that form a solid film on the surface of an electrode following electron transfer to oxidized state, and Type III is solid in both or all redox states. Various types of materials and structures can be used to construct electrochromic devices, depending on the specific applications (Chang et al. 1975, Arman 2001, Mortimer 2011). Here, we simply discuss the electrochromic materials for reflective display applications divided into three main categories: inorganic materials that change color with a double injection of ions and electrons; organic materials that change color by an electro-redox reaction and complex materials that change color via electron/charger transfer among different states. For those who are interested in electrochromic materials themselves, the review articles are recommended (Arman 2001, Ikeda and Stoddart 2008, Mortimer 2011).

(1) Inorganic Electrochromic Materials

Transition metal oxides represent a large family of materials possessing various interesting properties in the field of electrochromism. Among them, tungsten oxide (WO_3), has been the most extensively studied material, used in the production of electrochromic windows/smart glass and more recently electrochromic displays on paper substrate. The inorganic electrochromic materials are normally deposited on a conducting substrate in the form of a thin film. This film is held in contact with an electrolyte that is conductive to the relevant ions. The other side of the electrolyte is in contact with an ion reservoir containing the ions. A small voltage is applied between the conducting substrate and an electrode is in contact with the ion reservoir. This forces the ions into the electrochromic film from the ion reservoir through the electrolyte, at the same time the electrons are injected into the film from the substrate. The electrons are needed to preserve charge neutrality.

Prussian blue (PB) is a dark blue pigment with the idealized formula of $Fe_7(CN)_{18}$ which has a long history of being used in the formulation of paints, printing inks, typewriter ribbons and carbon paper. It is not soluble in water and employed as very fine colloidal dispersion—complex with variable amounts of other ions. The intense blue color of Prussian blue is associated with the energy of the transfer of electrons from Fe(II) to Fe(III). The mixed-valence compounds absorb orange-red light of ~ 680 nm wavelength according to the intervalence charge transfer, resulting in reflecting blue light. The appearance of it is also sensitive to the size of the colloidal particles. Prussian blue has been reviewed in books many times as an example of electrochromic materials.

Typical inorganic electrochromic materials include the anodically coloring materials of NiO, IrO_2, Prussian blue (PB); the cathodically coloring materials of WO_3, MoO_3, Nb_2O_5, TiO_2; and the materials colored at both states such as V_2O_5, CoOx, Rh_2O_3. Table 1 lists the coloring principle and electrochromic colors of these materials. These materials are metal oxide for which the film forming methods are vacuum deposition (thermal evaporation, electron beam evaporation and magnetron sputtering) and electrochemical deposition. The metal oxides are typically deposited on a conducting substrate as a thin film with typical thickness of 0.1–1.0 μm. Such thin films are either amorphous or polycrystalline, sometimes both admixed, and the morphology depends strongly on the mode of film preparation.

Table 1. The coloring principle, name and color of some inorganic electrochromic materials.

Coloring principle	Name	Color
Anodically coloring	IrO_2	Transparent/Black
	PB	Transparent/Dark-blue
	V_2O_5	Grey/Yellow
Cathodically coloring	WO_3	Transparent/Dark-blue
	MoO_3	Transparent/Dark-blue
	Nb_2O_5	Transparent/Pale-blue
	TiO_2	Transparent/Pale-blue
	NiO	Transparent/Dark-bronze
Colored at both states	CoO_x	Red/Blue
	Rh_2O_3	Yellow/Green

(2) Organic Electrochromic Materials

The example of the organic electrochromic material is viologen which is hoped to be the potential candidate material for display applications, which shows dark blue with a high contrast to the bright white, providing high visibility. Viologen radical cations are colored intensely, with high molar absorption coefficients, owing to optical charge transfer between the (formally) +1-valent and 0-valent nitrogen. Different colors can be achieved, depending on the nature of the substituent(s) on the viologen molecule. Suitable choice of nitrogen substitutes in viologens to attain the appropriate molecular orbital energy levels will give the color choice of the radical cation.

One of the attracting facts for organic material is the easy processability via solution-phase or polymeric phase. Viologen based electrochromic materials have enabled cost effective manufacturing of printed electronic displays on a variety of flexible substrate materials using industry standard printing techniques. For example, NTERA Limited has developed NanoChromics™ Ink Systems, and the monolith construction of multilayer printed structure on a single substrate has also been obtained in NanoChromics™ displays. To enhance the contrast ratio and the coloration efficiency, the viologen based electrochromism device has been developed by making use of nanostructured electrodes such as the TiO_2-nanoparticle coated electrodes. Via the nanostructured surface, the surface area to volume ratio has been increased dramatically, for which a higher volume/number of viologen molecules can be concentrated on a relatively small surface area, leading to a high coloration efficiency.

Polymers with intermediate bandgaps have distinct optical changes throughout the visible region and can exhibit several colors. Conjugated conducting polymers showing color change or contrast between doped and undoped forms depends on the bandgap magnitude. For the undoped polymer, the optical bandgap between the highest-occupied π-electron band (the valence band) and the lowest-unoccupied band (the conduction band) determines the electrochromic properties. In the conducting oxidized state, conjugated conducting polymers have positive charge carriers, are charge balanced with counter-anions (p doped) and have delocalized π-electron-band structures. Table 2 summarizes some conjugated polymers and their colors in different states.

Table 2. Some conjugated conductive polymers and their colors at different states.

Coloring principle	Name	Color
Electron/ charge doping	Poly (3-methylthiophene)	Purple/Pale-blue
	Poly[3,4-(ethylenedioxy)thiophene] (PEDOT)	Deep-blue/Light-blue
	[3,4-(propylenedioxy) thiophene (ProDOT)]-(benzothiadiazole) heterocyclic polymer	Green/Light-blue
	Poly[3,4 - (ethylenedioxy)pyrrole] (PEDOP)	Bright-red/Light-blue
	Poly[3,4 - (propylenedioxy)pyrrole] (PProDOP)	Orange/Brown/Gray-blue
	N-methyl-PProDOP	Purple/Dark-green/Blue
	N-[2-(2-ethoxy-ethoxy)-ethyl]PProDOP (N-GlyPProDOP) N-propanesulfonatePProDOP (N-PrSP-ProDOP)	Colorless/Multiple-colors
	Poly(thiophene)	Blue/Red

(3) Complexes of Electrochromic Materials

Metal coordination complexes show promise as electrochromic materials because of their intense coloration and redox reactivity. Chromophoric properties arise from low energy metal-to-ligand charge transfer, intraligand excitation and related visible-region electronic transitions. Because these transitions involve valence electrons, chromophoric characteristics are altered or eliminated upon oxidation or reduction of a complex. These spectroscopic and redox properties would be sufficient for direct use of metal coordination complexes in solution-phase ECDs and thin-film systems. Their polymeric phase has often been investigated for their possible applications in type III devices (all-solid-state). Table 3 shows some metal coordination complex electrochromic materials and their colors at different states.

Table 3. Some metal coordination complexes and their colors at different states.

Coloring principle	Name	Color
Electron/charge transfer	Poly(ZnPc-SNS) SNS=4-(2,5-di-2-thiophen-2-yl-pyrrol-1-yl) Pc=phthalocyanine	Green/Black
	Poly(H$_2$Pc-SNS)	Green/Turquoise-green
	[Ru(terpy)-(box)PVP$_{20}$]PF$_6$ terpy=2,2':6',2"-terpyridine box=2-(2-hydroxyphenyl)benzoxazole	Wine-red/Red-orange/ Light-green
	Poly[Lu(4,4',4",4'''-tetra- aminophthalocyanine)$_2$]	Green/Gray/Blue
	Poly(phthalocyaninatocobalt)	Blue-green/Yellow-brown/Red-brown
	Poly(phthalocyaninatonickel)	Green/Blue/Purple
	Poly[tetrakis (2-hydroxyphenoxy) phthalocyaninato cobalt(II)]	Light-green/Yellowish-green/Dark-yellow
	Lutetium bis(phthalocyanine) [Lu(Pc)$_2$]	Violet-blue/blue/Vivid-green/Blue/Violet-blue

For display application purpose, the multicolor electrochromic materials are promising, which can show various colors in one material, making a full color reflective display. Mechanically interlocked compounds which consist of a pair of mutually interlocked ring components, are referred to as [2] catenanes which can electrochemically control red–green–blue (RGB) colors. The design and synthesis of electrochromic materials based on switchable three-station [2]catenanes has been summarized and its application in reflective display is shown in Fig. 4b (Ikeda and Stoddart 2008).

2.4 Other Reflective Displays

As discussed in the previous sessions, the materials or technologies which can manipulate the visible light reflection can find their application in reflective displays. Besides EPDs, EWDs, ECDs, other materials like cholesteric liquid crystals, liquid powders and photonic crystals have also been investigated as reflective display materials and the technology like interferometric modulator has also pushed its way through reflective display technology. Figure 5 shows the materials, devices and working principles of the reflective displays based on the cholesteric liquid crystal (Fig. 5a), photonic crystal (Fig. 5b), liquid powder (Fig. 5c) and interferometric modulator (Fig. 5d).

(1) Cholesteric Liquid Crystal Displays (ChLCD)

Cholesteric liquid crystals are also known as chiral nematic liquid crystals which organize in layers with no positional ordering within layers, but a director axis varies with layers. The variation of the director axis tends to be periodic in nature. The period of this variation (the distance over which a full rotation of 360° is completed) is known as the pitch which determines the wavelength of light reflected (Bragg reflection) as shown in Fig. 5a.

The molecular orientation naturally tends to twist, so that the orientation of the long axes of the molecules traces out helical domains. When the helix of the twist aligns along the normal of the substrate, a planar texture can be achieved. This configuration can reflect a component of the incoming light. When the component of the circularly polarized light has the same sense of the helical twist, it will be reflected to show different colors to human eyes; and the other opposite component is transmitted (Yang et al. 1994, Hongbo et al. 2014, Oh and Yoon 2014, Suk-Hwan et al. 2014, Yu et al. 2014).

The ChLCD is a bi-stable display, no power is dissipated to maintain the contents displayed. The device manufacture can be compatible with existing LCD production lines. Kent displays and Fujitsu are the main forces commercializing the ChLCD technique. A flexible e-Paper called i2R based on ChLCD technology has been developed by the Industrial Technology Research Institute. The German company BMG MIS has developed ChLCDs in full color and monochrome in various sizes and resolutions, named "Geameleon". The major disadvantage of the ChLCD is the high driving voltage and relatively low optical performance compared with other reflective display techniques, which limits its applications in display areas, but could find applications requiring static or slow switching.

(2) Photonic Crystal Displays (PHD)

Photonic crystal display technology is based on the electrical actuation of photonic crystal materials. Photonic crystals are periodic optical

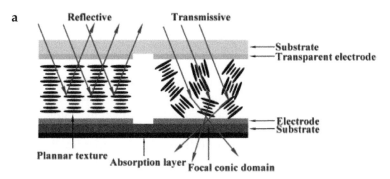

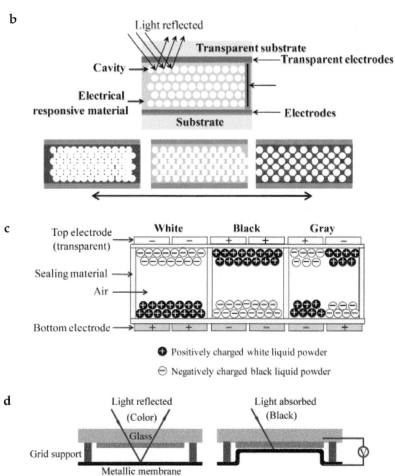

Figure 5. The working principles and device structures of the cholesteric liquid crystal display (**a**), the photonic crystal display (**b**), the liquid powder display (**c**) and the interferometric modulator display (**d**).

nanostructures that affect the motion of photons in much the same way that ionic lattices affect electrons in solids. Photonic crystals occur in nature in the form of structural coloration and promise to be useful in different forms in a range of applications. Photonic crystals can, in principle, find uses wherever light can be manipulated. Electrically tunable photonic crystals can provide electronic displays clearly with unique properties (Arsenault et al. 2003, Arsenault et al. 2003, Arsenault et al. 2007).

The working principle of a photonic crystal display is shown in Fig. 5b. Photonic crystals materials with a periodic modulation in refractive index, are of exceptionally bright and brilliant reflected colors arising from coherent Bragg optical diffraction. The photonic crystals can be fabricated into 1D, 2D, 3D via different methods: particle self-assembly, photolithography, drilling, direct laser writing, etc. The periodicity of the photonic crystal structure must be around half the wavelength of the electromagnetic waves that are to be diffracted. That is ~ 200 nm (blue) to 350 nm (red) for photonic crystals operating in the visible part of the spectrum. The contrast of the refractive index of the periodic structure must therefore be fabricated at this scale to obtain better color efficiency and sensitivity.

Electrical actuation of a photonic crystal ink film was enabled by its incorporation into a sealed thin-layer electrochemical cell. The device consists of the photonic crystal composite supported on ITO–glass as the working electrode, a hot-melt ionomer spacer, and an ITO–glass counter-electrode. The cell is filled with an organic solvent-based liquid electrolyte by vacuum filling and sealed with epoxy. It is known that the electrically responsive polymer in solution as well as in supported films display reversible electrochemical oxidation and reduction, with the partial electronic delocalization along the polymer backbone leading to a continuously tunable degree of oxidation, inducing changes in volume. By virtue of their continuously tunable state of oxidation, the photonic crystal films display voltage-dependent continuous shifts in reflected colors.

This technology was initiated by Arsenault et al. in University of Toronto. At the same time they founded the Opalux Company to commercialize this technology. For display application, the device fabrication and driving mode is simple. Each pixel of the photonic crystal display can show its own color by electrical actuation. Apart from devices, the photonic crystal material itself is sophisticated, and the stability and range of controllability is quite limited. Currently, the switching speed is still low and the color is not bright enough. According to its full color range, photonic crystals may find applications in other fields besides displays.

(3) Liquid Powder Displays (LPD)

Liquid powder display technology has been developed by Bridgestone. He called it quick response liquid powder display (QR-LPD) (Sakurai et al.

2006). The electronic liquid powder is created by manipulating black and white polymer nanoparticles as illustrated in Fig. 5c. The principle is similar to EPD, however, its medium is air rather than a liquid. The low viscosity medium makes the particle move fast, and the response time can then be shortened. The panel fabrication process is simple and low temperature, which means that flexible plastic materials can be used. Similar to EPD device, the positively charged black and negatively charged white powder is enclosed in each pixel between front and back electrodes. The electronic powder is driven by the applied voltage to display white or black colors. The gray level can be modulated by the electric field. Liquid powder display doesn't require a backlight and has a quick response time (< 1 ms). It can also be bi-stable. The front panels are thin and flexible (Sakurai has realized a roll-to-roll flexible liquid powder display in 2006). However, the liquid powder display has the disadvantage of high driving voltage and limited switching times of < 1 million which makes it difficult to make the display products, and Bridgestone has recently withdrawn it.

(4) Interferometric Modulator Displays (IMOD)

Interferometric modulator is not a specific material but a structure which can modulate the light transfer by nanostructures. It makes color in the same way as the wings of iridescent butterflies or peacock feathers—by being an imperfect mirror that tunes the color of incoming light before reflecting it back to the viewer. The principle is show it in Fig. 5d. The basic elements of an IMOD-based display are microscopic devices that act essentially as mirrors that can be switched on or off individually. Each of these elements reflects exactly one wavelength of light, such as a specific hue of red, green or blue when turned on, and absorbs light (appears black) when off. Elements are organized into a rectangular array in order to produce a display screen (Miles 1997).

Mirasol display is the example of IMOD. It is done by small cavities known as interferometric modulators, tens of microns across and a few hundred nanometers deep, beneath the glass surface of the display. Mirasol modulators have been made using techniques similar to those used to pattern metals and deposit materials in computer chip manufacturing. It's the air gap between the back of a glass and a mirror membrane at the bottom of the modulator that sets the color. The mirror membrane of each modulator can snap upwards against the glass when a small voltage is applied, closing the cavity and displaying a black color to the viewer. This display can show multi-color in one pixel, showing color reflective display with fast switching without using color filter. However, according to its sensitivity to the nanometer size cavity, the fabrication is sophisticated and difficult, resulting in low production yield and therefore high price. So, the Mirasol appeared with high expectation, but disappeared quickly due to this fatal disadvantage.

3. Other Materials and Technologies for Electronic Display Devices

3.1 Electrodes

All display devices require at least one optically transparent electrode. The electrode substrates comprise an optically transparent, electrically conducted film coated onto glass or flexible substrates. The optically transparent, electrically conducted film is usually a transparent conducting oxide such as tin-dopedindium oxide (ITO), F-doped tin oxide (FTO), or antimony-doped tin oxide. ITO is ubiquitously used worldwide in numerous optoelectronic applications, including flat-panel displays; however the indium used in ITO is scarce. The alternative materials with lower indium content, such as zinc-indium-tin oxide, are being investigated (Kim et al. 2011). Films of pure single-walled carbon nanotubes have also emerged with comparable transparency in the visible region and higher transparency in the 2–5 µm infrared region (Wu et al. 2004, Wang and Fugetsu 2015). Graphene on substrate has recently attracted a lot of attention, and a large transparent area has been fabricated (Bae et al. 2010, Woo et al. 2014). Another option for transparent area is the nanowire/nanofiber network based transparent electrode which has recently been developed quickly and is ready for commercialization for touch panel and smart window applications. At the same time, bio-mimic nanowire network has also been explored (Han et al. 2014). The transparent nano-network electrode is possibly being used as a whole plate-based touch panel or the common electrode of reflective displays like an electrowetting display. However the unpatternability limits its applications for patterned electrodes.

For display applications, the other very important "electrode" is the thin film transistor (TFT) for which the fast switch or video speed switch can be realized on a display device. TFT is a special kind of field effect transistor made by depositing thin films of an active semiconductor layer as well as the dielectric layer and metallic contacts over a supporting (not conducting) substrate. The most beneficial aspect of TFT technology is its use of a separate transistor for each pixel on a display. Since each transistor is small, the amount of charge needed to control it is also small, which allows it for very fast switching and controlling display pixels. TFT has been widely used in LCDs, OLDs; and the reflective display devices such as electrophoretic displays, electrowetting displays and electrochromic displays are all addressable by TFT technology. TFTs can be made using a variety of semiconducting materials including silicon-based TFT (amorphous silicon or microcrystalline silicon), oxide TFT including metal oxides like zinc oxide and cadmium selenide, or organic TFT made of organic materials. Driven by the huge market of displays, TFT technologies have developed very quickly and have become a big field. The solution-processed TFTs have been reported in 2003. Recently, the transparent TFTs and paper transistors have also been demonstrated and would be applied either in display technology or other electrical devices.

3.2 Sealants

To make up the device stay in place and not be degraded by contact with the environment, sealing and fixation have to be applied. From this perspective, the fabrication of a leak-proof device might be even more important than other parts since it determines the lifetime and portability of a device. We can imagine that, nowa days, millions of prototypes and products are produced everyday and go for testing or to the markets. All need to pass the environmental test. The key requirements for the sealant is chemical durability, the temperature durability in the range of –40 to 70°C, the anti-UV properties (surviving outdoor usage), hydrostatic pressure (especially in a large device containing organic solvent, inorganic solvent or both). From the materials aspect, although there are a lot of reviews or books focused on functional materials, the special focus of sealant for device applications is not enough. In general, the same or a similar sealant will be chosen based on an existing functional device. The normal sealants are thermo-curable, photo-curable or ultrasonic curing plastics. Silicon based thermo-curable sealants are typically used for electrical device sealing. This type of sealant is resistant to the effects of sunlight, rain, snow, ozone, UV and temperature extreme.

3.3 Electronic Contacts

Electronic contact is the interconnection from functional materials to electrical driving end. Although electronic contacts come to play the role at the end of a device assembly, it is actually started from the design of a device and the selection of materials for fabrication at the beginning. The electronic contacts are the communication medium between the front-plane and the back-plane of a display device. Although it seems the work of the back-plane of a display, the interconnection needs to be settled before the work of the front-plane starts. The size of the electronic contacts, their conductivity, optical properties and uniformity among contacts are all considered and need to be satisfied to realize a fully functional device.

3.4 Choice of Processing Technologies for Materials

Materials exist with a variety of appearances: solid or liquid or gas, crystalline or amorphous, conductive or nonconductive, transparent or opaque. When making a device, both the choice of materials and the fabrication technologies are very important. For applications of the previously discussed materials, the application in flat-panel displays are based on film technology with or without microstructures. Based on the process status, these technologies can be simply divided into the vacuum deposition technology and the wet process technology. Vacuum deposition is a family of processes used to deposit layers of material atom-by-atom or molecule-by-molecule on a solid surface. The deposited layers can range from a thickness of one atom up to meters. Multiple layers of different materials can be applied. When the source of vapor is a liquid or solid the process is called physical vapor deposition (PVD). When the source is a chemical vapor precursor the process is called chemical vapor deposition (CVD). Wet process is also called wet coating or printing, based on solution-phase materials. The deposited layers can also range from a thickness of one molecule to meters. Recently, the wet process has obtained extensive attention according to its flexibility in choice of substrate materials, larger area fabrication, and mild working conditions.

Some materials function differently when prepared via different technologies. For example, the tungsten oxide, depending on the deposition method, could be at a very different crystallinity which influences the electrochromic properties essentially. Vacuum evaporation onto unheated substrates will produce films with an amorphous structure and high degree of porosity. Sputtering on substrates heated to 400°C will produce dense films with a high degree of crystallinity. Amorphous and porous films have in general a faster switching speed and a higher optical efficiency. On the other hand, crystalline films are more stable and will last longer.

For the display applications, when considering the large area film deposition, the vacuum deposition technologies like electro-deposition and thermal deposition can be applied; and the wet process technologies of dip-coating, slit-coating and screen-printing are widely used. When considering the resolution of a display, the micropatterning technologies are critical. Currently, the most popular micropatterning technology is photolithography which can create structures precisely down to a few nanometer size, and up to micro-, mini- and centi-meters with high flexibility. This technology is widely used in the display fabrication process for almost all displays either for pixilation, electrodes patterning, or for both. Other micropatterning technologies like laser writing, screen-printing and flexography could also be applied, but with the limitations of slow process, low resolution and thin film thickness, respectively. Recently, with the development in both materials and fabrication technologies, wet processes like solution-based printing technologies have been paid a lot of attention and are developing quickly to minimize materials consumption and cost in order to fulfill the requirements of production volume and market.

4. Summary and Perspectives

For reflective displays, the materials which can reflect visible light at different wavelength (color) could be applied. EPDs, EWDs and LPDs show color switching by mechanically driving different colored materials to the front of the display. ECDs change colors by electrochemical reaction in the electrochromic materials. ChLCDs manipulate reflection light by changing its intrinsic molecular arrangement. PCDs and IMODs demonstrate structural coloring by Bragg diffraction. These materials and functional devices have been discussed as reflective displays in this chapter, for which the readers can quickly catch how a material could be applied to a functional device and how the functions of the device could be satisfied by the advanced materials.

Reflective displays have the advantages of being applicable in both indoor and outdoor environments, the possibility of bi-stable, low power consumption, and easy readability due to the principle of reflecting ambient light. Applications of visual electronic displays existing in the market include e-readers, watches, electronic pricing labels in retail shops, digital signages, time tables at bus stations, electronic billboards, newspaper, mobile phone displays. Among these reflective display technologies, EPDs have been successfully commercialized with a big market occupation; ECDs have been successful not only in display area but also for smart window or glasses applications; EWDs are also developing quickly to hit video-speed reflective color display market; ChLCDs and PHDs are struggling with some disadvantages like slow switch and weak color efficiency, but will settle down where there is no need for high brightness and fast switching speed; LPDs and IMODs have been highly expected and have already brought prototypes and products to markets, but with difficult situation due to some intrinsic properties.

The main requirements for display applications are: pixel level controllability including reversibility, speed, brightness, power consumption and flexibility. In order to satisfy those requirements, the material science and technology, together with optics, electronics and physics, is forming a multidisciplinary research community. When the corresponding large community of interest builds around a technology, it would drive the technology to irresistible development and maturity and the functional materials are the key. In the future, reflective displays would be expected to show like a paper with the speed of electronics. The desire for reflective displays is very strong. The future market will not only be in the normal electronic displays, but also expanding quickly for wearable electronic devices, and even the bigger market for smart window/building materials. The challenge for a display technology is not only to develop suitable front-plane materials and processing technologies, but also the back-plane materials like active matrix with fully functioning electronics.

References

Arman, S. 2001. Electrochromic materials for display applications: An introduction. J. New. Mat. Electrochem. Systems 4(3): 173–179.

Arsenault, A.C., H. Miguez, V. Kitaev and G.A. Ozin. 2003. Towards photonic ink (P-Ink): A polychrome, fast response metallopolymer gel photonic crystal device. Macromol. Symp. 196: 63–69.

Arsenault, A.C., H. Miguez, V. Kitaev, G.A. Ozin and I. Manners. 2003. A polychromic, fast response metallopolymer gel photonic crystal with solvent and redox tunability: A step towards photonic ink (P-Ink). Adv. Mater. 15(6): 503–507.

Arsenault, A.C., D.P. Puzzo, I. Manners and G.A. Ozin. 2007. Photonic-crystal full-colour displays. Nat. Photonics 1(8): 468–472.

Bae, S., H. Kim, Y. Lee, X. Xu, J.-S. Park, Y. Zheng, J. Balakrishnan, T. Lei, H.R. Kim, Y.I. Song, Y.-J. Kim, K.S. Kim, B. Ozyilmaz, J.H. Ahn, B.H. Hong and S. Iijima. 2010. Roll-to-roll production of 30-inch graphene films for transparent electrodes. Nat. Nanotechnol. 5(8): 574–578.

Bai, W.Y., Y. Wang, X.L. Song, X.Y. Jin and X.H. Guo. 2014. Modification of urea-formaldehyde microcapsules with lignosulfonate-Ca as Co-polymer for encapsulation of acetochlor. J. Macromol. Sci. A 51(9): 737–742.

Beni, G. and S. Hackwood. 1981. Electrowetting displays. Appl. Phys. Lett. 38(4): 207–209.

Beni, G. and M.A. Tenan. 1981. Dynamics of electrowetting displays. J. Appl. Phys. 52(10): 6011–6015.

Campardelli, R., G. Della Porta, V. Gomez, S. Irusta, E. Reverchon and J. Santamaria. 2013. Encapsulation of titanium dioxide nanoparticles in PLA microspheres using supercritical emulsion extraction to produce bactericidal nanocomposites. J. Nanopart. Res. 15(10): 1978.

Chang, I.F., B.L. Gilbert and T.I. Sun. 1975. Electrochemichromic systems for display applications. J. Electrochm. Sco. 122(7): 955–962.

Chevalliot, S., J. Heikenfeld, L. Clapp, A. Milarcik and S. Vilner. 2011. Analysis of nonaqueous electrowetting fluids for displays. J. Disp. Technol. 7(12): 649–656.

Cho, S.H., Y.R. Kwon, S.K. Kim, C.H. Noh and J.Y. Lee. 2007. Electrophoretic display of surface modified TiO_2 driven by Poly (3,4-ethylenedioxythiophene) Electrode. Polym. Bull. 59(3): 331–338.

Choi, S., Y. Kwon, Y.S. Choi, E.S. Kim, J. Bae and J. Lee. 2013. Improvement in the breakdown properties of electrowetting using polyelectrolyte ionic solution. Langmuir 29(1): 501–509.

Christophersen, M. and B.F. Phlips. 2010. Recent patents on electrophoretic displays and materials. Recent Pat. Nanotech. 4(3): 137–149.

Comiskey, B., J.D. Albert, H. Yoshizawa and J. Jacobson. 1998. An electrophoretic ink for all-printed reflective electronic displays. Nature 394(6690): 253–255.

Dalisa, A.L. 1977. Electrophoretic display technology. IEEE T. Electron. Dev. 24(7): 827–834.

Deb, S.K. 1969. A novel electrophotographic system. Appl. Optics 8(S1): 192–195.

Fan, C.J., J.T. Tang and X.D. Zhou. 2013. Effects of process parameters on the physical properties of poly (urea-formaldehyde) microcapsules prepared by a one-step method. Iran. Polym. J. 22(9): 665–675.

Guang, Y. and J.K. Lee. 2013. Microcapsules containing self-healing agent with red dye. Polym-Korea 37(3): 356–361.

Guo, X.F., Y.S. Kim and G.J. Kim. 2009. Fabrication of SiO2, Al2O3, and TiO2 microcapsules with hollow core and mesoporous shell structure. J. Phys. Chem. C 113(19): 8313–8319.

Han, B., Y. Huang, R. Li, Q. Peng, J. Luo, K. Pei, A. Herczynski, K. Kempa, Z. Ren and J. Gao. 2014. Bio-inspired networks for optoelectronic applications. Nat. Commun. 5.

Hayes, R.A. and B.J. Feenstra. 2003. Video-speed electronic paper based on electrowetting. Nature 425(6956): 383–385.

Heikenfeld, J., N. Smith, M. Dhindsa, K. Zhou, M. Kilaru, L.L. Hou, J.L. Zhang, E. Kreit and B. Raj. 2009. Recent Progress in Arrayed Electrowetting Optics. OPN 4: 20–26.

Heikenfeld, J., K. Zhou, E. Kreit, B. Raj, S. Yang, B. Sun, A. Milarcik, L. Clapp and R. Schwartz. 2009. Electrofluidic displays using Young–Laplace transposition of brilliant pigment dispersions. Nat. Photonics 3(5): 292–296.

Henzen, A. 2009. Development of E-paper color display technologies. Sid. Int. Symp. Dig. Tec.: 28–30.

Hongbo, L., Z. Jun, S. Zhigang, Z. Guobing, W. Xianghua, Q. Longzhen and L. Guoqiang. 2014. Submillisecond-response light shutter for solid-state volumetric 3D display based on polymer-stabilized cholesteric texture. J. Disp. Technol. 10(5): 396–400.

Huitema, E., G. Gelinck, P. van Lieshout, E. van Veenendaal and F. Touwslager. 2005. Flexible electronic-paper active-matrix displays. J. Soc. Inf. Display 13(3): 181–185.

Huitema, H.E.A., G.H. Gelinck, P.J.G. van Lieshout, E. van Veenendaal and F.J. Touwslager. 2006. Flexible electronic-paper active-matrix displays. J. Soc. Inf. Display 14(8): 729–733.

Ikeda, T. and J.F. Stoddart. 2008. Electrochromic materials using mechanically interlocked molecules. Sci. Technol. Adv. Mat. 9(1): 014104.

Jackel, J.L., S. Hackwood and G. Beni. 1982. Electrowetting optical switch. Appl. Phys. Lett. 40(1): 4–6.

Jackel, J.L., S. Hackwood, J.J. Veselka and G. Beni. 1983. Electrowetting switch for multimode optical fibers. Appl. Opt. 22(11): 1765–1770.

Kim, Y.H., D.W. Kim, R.I. Murakami, D. Zhang, S.W. Yoon, S.H. Park and K.M. Moon. 2011. The study of transmittance and conductivity by top ZnO thickness in ZnO/Ag/ZnO transparent conducting oxide films. Adv. Sci. Lett. 4(4-5): 1570–1573.

Lenssen, K.M.H., P.J. Baesjou, F.P.M. Budzelaar, M.H.W.M. van Delden, S.J. Roosendaal, L.W.G. Stofteel, A.R.M. Verschueren, J.J. van Glabbeek, J.T.M. Osenga and R.M. Schuurbiers. 2008. Invited paper: Novel design for full-color electronic paper. Sid. Int. Symp. Dig. Tec. 39: 685–688.

Li, H.Y., R.G. Wang and W.B. Liu. 2011. Toughening self-healing epoxy resin by addition of microcapsules. Polym. Polym. Compos. 19(2-3): 223–226.

Li, J., S.J. Wang, H.Y. Liu, N. Liu and L. You. 2011. Preparation and application of Poly(melamine-urea-formaldehyde) microcapsules filled with sulfur. Polym-Plast. Technol. 50(7): 689–697.

Li, J., S.J. Wang, H.Y. Liu, L. You and S.K. Wang. 2012. Preparation of Poly(urea-formaldehyde) microcapsules containing sulfur by *in situ* polymerization. Asian J. Chem. 24(1): 93–100.

Malk, R., Y. Fouillet and L. Davoust. 2011. Rotating flow within a droplet actuated with AC EWOD. Sens. Actuators, B 154(2): 191–198.

Mao, J., H. Yang and X. Zhou. 2012. *In situ* polymerization of uniform poly(urea-formaldehyde) microcapsules containing paraffins under the high-speed agitation without emulsifier. Polym. Bull. 69(6): 649–660.

Massard, R., J. Mans, A. Adityaputra, R. Leguijt, C. Staats and A. Giraldo. 2013. Colored oil for electrowetting displays. J. Inform. Disp. 14(1): 1–6.

Miles, M.W. 1997. A new reflective FPD technology using interferometric modulation. J. Soc. Info. Display 5(4): 379–382.

Moon, H., S.K. Cho, R.L. Garrell and C.J. Kim. 2002. Low voltage electrowetting-on-dielectric. J. Appl. Phys. 92(7): 4080–4087.

Mortimer, R.J. 2011. Electrochromic materials. Annu. Rev. Mater. Res. 41: 241–268.

Murau, P. and B. Singer. 1978. The understanding and elimination of some suspension instabilities in an electrophoretic display. J. Appl. Phys. 49(9): 4820–4829.

Nagamine, S., A. Sugioka and Y. Konishi. 2007. Preparation of TiO2 hollow microparticles by spraying water droplets into an organic solution of titanium tetraisopropoxide. Mater. Lett. 61(2): 444–447.

Oh, S.W. and T.H. Yoon. 2014. Fast bistable switching of a cholesteric liquid crystal device induced by application of an in-plane electric field. Appl. Opt. 53(31): 7321–7324.

Orsaev, A.M., T.M. Orsaev and D.S. Gaev. 2000. New reflective type electrophoretic display. SPIE 4511: 190–192.

Ota, I., J. Ohnishi and M. Yoshiyam. 1973. Electrophoretic image display (Epid) panel. P. IEEE 61(7): 832–836.

Papathanasiou, A.G., A.T. Papaioannou and A.G. Boudouvis. 2008. Illuminating the connection between contact angle saturation and dielectric breakdown in electrowetting through leakage current measurements. J. Appl. Phys. 103(3): 034901.

Park, S.J., Y.S. Shin and J.R. Lee. 2001. Preparation and characterization of microcapsules containing lemon oil. J. Colloid Interf. Sci. 241(2): 502–508.

Ploix, J.L. and M. Moulin. 1976. Electrophoretic display device. J. Electrochem. Soc. 123(8): C257–C257.

Sakurai, R., S. Ohno, S.I. Kita, Y. Masuda and R. Hattori. 2006. Color and flexible electronic paper display using QR-LPD technology. Sid. Int. Symp. Dig. Tec. 37(1): 1922–1925.

Sgraja, M., J. Blomer, J. Bertling and P.J. Jansens. 2010. Thermal and structural characterization of TiO2 and TiO2/Polymer micro hollow spheres. Chem. Eng. Technol. 33(12): 2029–2036.

Smith, N.R., L.L. Hou, J.L. Zhang and J. Heikenfeld. 2009. Fabrication and demonstration of electrowetting liquid lens arrays. J. Disp. Technol. 5(11): 411–413.

Song, J.K., H.J. Choi and I. Chin. 2007. Preparation and properties of electrophoretic microcapsules for electronic paper. J. Microencapsul. 24(1): 11–19.

Song, J.K., H.C. Kang, K.S. Kim and I.J. Chin. 2007. Microcapsules by complex coacervation for electronic ink. Mol. Cryst. Liq. Cryst. 464: 845–851.

Song, J.K., H.J. Myoung, K. Kim, J.H. Sung, H.J. Choi and I.J. Chin. 2005. Preparation of TiO2 based microcapsules for electrophoretic ink. Abstr. Pap. Am. Chem. S. 229: U1138–U1138.

Song, Y.K. and C.M. Chung. 2013. Repeatable self-healing of a microcapsule-type protective coating. Polym. Chem. 4(18): 4940–4947.

Staicu, A. and F. Mugele. 2006. Electrowetting-induced oil film entrapment and instability. Phys. Rev. Lett. 97(16): 167801.

Strohm, H., M. Sgraja, J. Bertling and P. Lobmann. 2003. Preparation of TiO2-polymer hybrid microcapsules. J. Mater. Sci. 38(8): 1605–1609.

Suk-Hwan, J., K. Jang-Kyum, K. Hoon, S. Ki-Chul, K. Heesub and S. Jang-Kun. 2014. Tri-Stable polarization switching of fluorescent light using photo-luminescent cholesteric liquid crystals. Mol. Cryst. Liquid Cryst. 601: 29–35.

Sun, B., K. Zhou, Y. Lao, J. Heikenfeld and W. Cheng. 2007. Scalable fabrication of electrowetting displays with self-assembled oil dosing. Appl. Phys. Lett. 91(1): 011106.

Sun, G. and Z. Zhang. 2002. Mechanical strength of microcapsules made of different wall materials. Int. J. Pharm. 242(1-2): 307–311.

Suryanarayana, C., K.C. Rao and D. Kumar. 2008. Preparation and characterization of microcapsules containing linseed oil and its use in self-healing coatings. Prog. Org. Coat. 63(1): 72–78.

Tan, T.F., S.R. Wang, S.G. Bian, X.G. Li, Y. An and Z.J. Liu. 2010. Novel synthesis and electrophoretic response of low density TiO-TiO2-carbon black composite. Appl. Surf. Sci. 256(22): 6932–6935.

Tseng, W.J. and P.S. Chao. 2013. Synthesis and photocatalysis of TiO2 hollow spheres by a facile template-implantation route. Ceram. Int. 39(4): 3779–3787.

Wang, D.W. and X.P. Zhao. 2009. Microencapsulated electric ink using gelatin/gum arabic. J. Microencapsul. 26(1): 37–45.

Wang, S.R., Y.M. Mei, X.G. Li and T.F. Tan. 2012. Preparation of high efficiency hollow TiO2 nanospheres for electrophoretic displays. Mater. Lett. 74: 1–4.

Wang, Y. and B. Fugetsu. 2015. Mono-dispersed ultra-long single-walled carbon nanotubes as enabling components in transparent and electrically conductive thin films. Carbon 82: 152–160.

Werts, M.P.L., M. Badila, C. Brochon, A. Hebraud and G. Hadziioannou. 2008. Titanium dioxide-polymer core-shell particles dispersions as electronic inks for electrophoretic displays. Chem. Mater. 20(4): 1292–1298.

Woo Sik, K., M. Sook Young, K. Hui Jin, P. Sungjin, J. Koyanagi and H. Hoon. 2014. Large-scale graphene-based composite films for flexible transparent electrodes fabricated by electrospray deposition. Mat. Res. Exp. 1(4): 046404.

Wu, Z.C., Z.H. Chen, X. Du, J.M. Logan, J. Sippel, M. Nikolou, K. Kamaras, J.R. Reynolds, D.B. Tanner, A.F. Hebard and A.G. Rinzler. 2004. Transparent, conductive carbon nanotube films. Science 305(5688): 1273–1276.

Xing, R.Y., Q.Y. Zhang and J.L. Sun. 2012. Preparation and properties of self-healing microcapsules containing an UV-curable oligomers of silicone. Polym. Polym. Compos. 20(1-2): 77–82.

Yang, D.K., J.L. West, L.C. Chien and J.W. Doane. 1994. Control of reflectivity and bistability in displays using cholesteric liquid crystals. J. Appl. Phys. 76(2): 1331–1333.

You, H. and A.J. Steckl. 2010. Three-color electrowetting display device for electronic paper. Appl. Phys. Lett. 97(2).

Yu, C.J., S.I. Jo, Y.J. Lee and J.H. Kim. 2014. Continuous pitch stabilization of cholesteric liquid crystals and display applications. Mol. Cryst. Liq. Cryst. 594(1): 63–69.

Yu, D.G. and J.H. An. 2004. Titanium dioxide core/polymer shell hybrid composite particles prepared by two-step dispersion polymerization. Colloid. Surface. A 237(1-3): 87–93.

Yu, D.G., S.H. Kim and J.H. An. 2007. Preparation and characterization of electronic inks encapsulation for microcapsule-type electrophoretic displays (EPDs). J. Industr. Eng. Chem. 13(3): 438–443.

Yuan, L., A.J. Gu, S. Nutt, J.Y. Wu, C. Lin, F. Chen and G.Z. Liang. 2013. Novel polyphenylene oxide microcapsules filled with epoxy resins. Polym. Advan. Technol. 24(1): 81–89.

Zandi, M., S.A. Hashemi, P. Aminayi and F. Hosseinali. 2011. Microencapsulation of disperse dye particles with nano film coating through layer by layer technique. J. Appl. Polym. Sci. 119(1): 586–594.

Index

Printed and bound by CPI Group (UK) Ltd, Croydon, CR0 4YY

01/11/2024

01782622-0014